Tom Schuppe

RELIABILITY:
MANAGEMENT, METHODS, AND MATHEMATICS

RELIABILITY:
MANAGEMENT, METHODS, AND MATHEMATICS

David K. Lloyd and Myron Lipow

TRW Systems and Energy

SECOND EDITION

Published by the Authors
Redondo Beach, California

COVER DESIGN BY T. RACHELS
COVER FIGURE FROM "CATASTROPHE THEORY" BY E. C. ZEEMAN
COPYRIGHT © APRIL 1976 BY SCIENTIFIC AMERICAN, INC.
ALL RIGHTS RESERVED.

© Prentice-Hall, Inc., 1962, Englewood Cliffs, NJ;
© David K. Lloyd and Myron Lipow, 1977, Redondo Beach, California. All rights reserved. Printed in the United States of America. No part of this book may be reproduced in any form, by mimeograph or any other means, without permission in writing from the authors.

Second Printing 1979

Library of Congress Catalog Card Number 77-80554
Printed in the United States of America
ISBN 0-9601504-1-2

TO MY PARENTS, HARRY AND DORIS
D.K.L.

TO MY WIFE, SUSAN-LEE
M.L.

ABOUT THE AUTHORS:

David K. Lloyd, M.A., Mathematics, Oxford University, England, M.A., Urban Planning, U.C.L.A., and Myron Lipow, B.S., Mathematics, California Institute of Technology. Both are presently with the Defense and Space Systems Group, TRW Systems and Energy, Redondo Beach, California.

PREFACE TO FIRST EDITION

This book is intended to acquaint the reader with the evolving methodology of reliability. Its writing presented many problems—notably how comprehensive it should be, and to how extensive an audience it should be addressed.

Reliability is a relatively new subject, continuously developing and expanding; it is not a limited, established subject like college algebra or elementary statistics. Moreover, much of the published work applies only to specific problems and is of limited use and interest. Consequently, its extent as a subject is not clearly defined. We have attempted to resolve this problem by presenting the fundamental concepts and considerations of the subject—including such topics as important statistical methods, communication systems, and reliability activities. These techniques are now evolving into the status of a reliability methodology; consequently, the reader will find the tools for the solution of his reliability problems rather than the solutions themselves. How the reader will choose among specific methods will depend on his awareness of the available methodology, the type of equipment with which he is concerned, and also the specific answers to the fundamental reliability questions he raises as they are applicable to his program.

The book's purpose having been thus defined, what of its audience? Since at the commencement of its writing, no organized, comprehensive presentation of the subject existed, it was felt that the main exposition should be fairly extensive and should be directed primarily to the reliability engineer. Consequently, there is an emphasis on probability and statistical theory as it has been adapted to or is appropriate for reliability purposes. Although this emphatically is not a book on probability and statistics, nevertheless these subjects are perhaps the most important tools the reliability engineer has at his disposal.

The authors' experience indicated that besides the reliability engineers, at least one further group must be addressed. This group is management—for it is the attitude of management which establishes the extent and effectiveness of reliability activities throughout the organization. Management's attitude, in turn, depends upon its understanding of the impact of reliability on organization, of the utility of reliability techniques in design and development work, and of statistical methodology, particularly of its implications with respect to contractual obligations. Consequently, each of these subjects is touched upon from a nontechnical management point of view.

A further audience directly concerned consists of design, development, test, manufacturing, and quality control engineers, who will become involved in some of the activities here described. Experience shows that if they can be made aware of the reliability techniques that are available and of how these techniques can aid them, then their acceptance, appreciation, and—more important—their participation in the use of the methodology can be enlisted.

The book is also directed to the engineering and mathematics students who have completed a course in elementary mathematical statistics. They will find in Section II a somewhat different orientation and extension of probability and statistical methods than have been presented in most statistical text books. Exercises of varying degrees of difficulty have been provided where they are appropriate; however, the instructor should determine which of the more advanced topics and exercises may be omitted until a second reading.

As a consequence of these considerations, the style, level, and depth of the writing vary. For this the authors make no apology. Reliability has become a broad subject concerning many people who have different backgrounds and different understandings of technical and nontechnical languages. Thus there are many different levels of reliability problems, ranging all the way from "very mathematical" to "common sense-practical." The reliability engineer in particular must appreciate this in order to help alleviate the technical communication barrier that so frequently exists in engineering today.

The book presents some statistical methods which are appearing between hard covers for the first time. Some of these methods are the authors' own work, others appear by the kind permission of many writers in the field. Where material has been quoted, we have made specific reference to the authors in the text. However, we would like to express our thanks in particular to several authors who have generously given us permission to abstract considerable portions from their works. These writers are B. Epstein, N. R. Garner, D. E. Hartvigsen, T. A. Budne, and G. E. P. Box for contributions to sections in Chapters 10 and 13. Also, we are indebted to O. L. Davies, Macclesfield, Cheshire, and to Messrs. Oliver and Boyd, Ltd., Edinburgh, for permission to reprint Table No. 11.21 from their book *Design and Analysis of Industrial Experiments*.

PREFACE

We must also acknowledge the indirect contributions made in the course of many lengthy and frequently controversial reliability discussions held with our colleagues at Space Technology Laboratories, Inc., as well as with representatives of those contractors working on the Air Force's Ballistic Missile and Space Systems Programs.

The technical contents of the manuscript were reviewed by E. M. Scheuer and R. L. Eidemiller, both Members of the Technical Staff at Space Technology Laboratories, Inc., to whom we are extremely grateful. However, the authors accept sole responsibility for any technical errors which may have been included.

Finally, we wish to express our most sincere thanks to Miss Ruthann Cruikshank, who typed the major portion of the manuscript, and also to Mrs. Carol Means and Mrs. Colleen Conway, who assisted her in this project.

1962

D.K.L.
M.L.

PREFACE TO SECOND EDITION

If reliability can be considered as the science of estimating, controlling, and managing the probability of failure, it follows that any complex, high technology system, whether or not it is a military or space system, can advantageously use the methods of reliability. It is particularly important when there are concerns over the consequences of system failure in terms of reliability, safety or cost. It is only necessary to cite the public apprehension over the reliability and safety of nuclear power plants, off-shore oil drilling, high speed transportation systems and medical equipment, to list but a few, to recognize that the need for controlling failures in these fields is vitally important. There is also a growing corporate concern over product liability cases and with the rising insurance rates to alleviate their impact. Department of Defense regulations requiring procurement based on total life cycle cost, together with reliability improvement warranties, are new and significant factors in producing military equipment. Each of these topics calls for an expanding role for the application of reliability methods.

In the years that have passed since the hard-cover edition was published in 1962, the field of reliability has experienced growth and widespread application in the aerospace industry; however its application in the civil and commercial sectors has so far been somewhat limited. The purpose of publishing the soft-cover edition is to make the methods of reliability more readily available to the new generation of graduate engineering students and practicing engineers who will be or are working in the civil and commercial sectors. The methodology and techniques presented in the book have had successful and continuing application since 1962, and it is with this experience in mind that we would like to see them adopted as standard engineering and technical management disciplines in all high technology systems.

In addition reliability has been consociated with the similar disciplines of maintainability, availability, and safety, which share to a great extent probability theory, mathematical and applied statistics, and management techniques with an emphasis not usually presented in the standard textbooks on statistics, probability theory or operations research. In this respect the original book still fulfills its intention of formalizing the subject of reliability and presenting a fundamental and comprehensive treatise suitable both as an intermediate textbook for graduate engineering students and as a reference book for practicing engineers.

Before issuing a new edition we were faced with the question of how much rewriting should be done, how much new material should be added, and whether we should include chapters on maintainability, availability and safety. Most recent developments in reliability have taken place in fault tolerant circuit analysis, "Bayesian reliability," computer simulation, and computer software reliability. To add all of these foregoing topics, particularly maintainability, availability and safety, to the existing contents of the book would have been a monumental task demanding experience and knowledge beyond the abilities of the authors and outside the scope of the book, and would, to some extent, be discounting the fine work of other authors who have addressed some of these subjects in varying degrees of detail. For example, advanced reliability analysis of electronic circuits and systems is discussed by B.V. Gnedenko et al., 1969 and M.L. Shooman 1968. It requires specialized knowledge of electronic engineering which is beyond the scope of this book. For those readers interested in the fields of maintainability, availability and safety the texts by B.S. Blanchard and E.E. Lowrey (1969), R.E. Barlow and F. Proschan (1965, 1975), W.P. Rodgers (1971) and G.A. Peters (1971) are recommended.

With regard to "Bayesian reliability," this is still a somewhat controversial subject and we prefer to refer the reader to the discussion by N.R. Mann et al., 1974. Whatever the merits of Bayesian methods we presume to make the suggestion that they should be treated as an advanced topic and not applied at least until the reader has mastered the "classical" technique of this book.

Computer simulation of more complex reliability models has become a powerful tool for the reliability analysts; however, discussion of these techniques is better suited to books on numerical analysis. Monte Carlo analysis is a frequently used computer approach to examine the probability behavior of the system as a function of the model relationships and the failure distribution of its component/parts. Occasionally, the cost of running such simulations can get out of hand; to overcome this situation, techniques utilizing fractional factorial statistical designs and analysis of variance have recently been developed by several authors, including ourselves. They represent a novel application of the methods of Chapters 12 and 13 (see Additional Reading Chapter 13).

We have, however, added a new chapter on computer software reliability reflecting its growing importance, as evidenced by the substantial number of articles in the professional journals and symposia on this subject.

Lastly, we would have preferred to illustrate the methods with more examples taken from a variety of civil and commercial fields but unfortunately such data are limited. We trust that readers who are working in those fields can take the methods described in this book and successfully apply them to their own particular set of circumstances.

In preparation of the new chapter on software reliability, the authors are indebted to their colleagues at TRW Defense and Space Systems Group for many sources of new material, and many helpful discussions. We particularly wish to express our thanks to Barry Boehm, John Brown, Bill Buck, Annette Frimtzis, Eldred Nelson, Heinie Shaw, Tom Thayer and Ben White. Thanks are also due to Gene Gerstner, Bernadean Schwab, and especially Christie Campbell, Pat Chaney, Sandra Cronan, Vera Dodder, Gene Melynis and Margaret Milton of TRW publications services for their assistance with this chapter.

D.K.L.
M.L.

1977

TABLE OF CONTENTS

0	INTRODUCTION: THE GROWING IMPORTANCE OF RELIABILITY	1
0.1	The reasons for unreliability	3
0.1.1	The limits of experience	4
0.1.2	The complexity of equipment	4
0.1.3	The complexity of organization	5
0.1.4	Human error	6
0.2	Preventing unreliability	7
0.2.1	Redundancy	7
0.2.2	Systems engineering	7
0.2.3	Management control and communication	8
0.2.4	Human engineering and reliability education	9

SECTION I. MANAGEMENT, ORGANIZATION, AND COMMUNICATION 11

1	MANAGEMENT AND ORGANIZATION	13
1.1	The responsibility of management	13
1.2	The place of a reliability group within an organization	16
1.3	An independent reliability group	18
1.4	Personnel capabilities	18
1.5	Reliability budgeting	19
2	PROBLEMS AND ACTIVITIES OF PLANNING AND OPERATING A RELIABILITY PROGRAM	20
2.1	Introduction	20
2.2	Reliability estimation	20
2.3	Reliability as a probability concept	20
2.4	Reliability as a function of success criteria	22
2.5	Nonrepresentative configurations and tests	23
2.6	Contractual implications of reliability evaluation	23
2.7	Reliability goals	25
2.8	Reliability apportionment	25

2.9	Design review for reliability	27
2.9.1	An example of a reliability design review checklist	28
2.10	Reliability knowledge	30
2.11	Test planning and analysis	30
2.12	Design control and specification, materials, and processing review	31
2.13	Vendor control	32
2.14	Reliability and failure reporting systems	32
2.15	Mathematical and statistical activities	33
2.16	Internal coordination	33
2.17	Reliability education and awareness	34
2.18	Conclusion and summary: The reliability program plan	34
3	RELIABILITY DATA SYSTEMS	37
3.1	Reliability data collection, processing, and reporting	37
3.2	Collection: Type and amount of data	37
3.2.1	Failure report forms	38
3.2.2	Failures due to human error	39
3.3	Distribution of data	39
3.4	Analysis, reporting, and follow-up	40
3.5	An example of a data reporting system	41
3.6	Distribution of failure forms	43
3.7	Failure analysis initiation	44
3.8	Follow-up, analysis, and corrective action	45
3.9	Operating time reporting	46
3.10	The time log	47
3.11	A method for collecting time information	50
3.12	The reliability central file	51
3.12.1	Examples of reliability information in the central file	51
3.13	Pretest declaration	51
3.14	Periodic reliability and failure summaries for project management and engineering	55
3.15	Periodic reliability status reports for top management	57
3.16	Special reliability reports	57

SECTION II. THE MATHEMATICS OF RELIABILITY 59

4	RELIABILITY MODELS AND ANALYSIS: A NONMATHEMATICAL INTRODUCTION	61
4.1	Types of data	61
4.2	Statistical distributions	63
4.3	Uncertainty of observation	64
4.4	Statistical estimation and reliability estimation	65
4.5	Reliability structure models, redundancy, and apportionment	65
4.6	Component interaction	67
4.7	Sampling and reliability demonstration	69
4.8	Reliability prediction and growth	70

5	BASIC PROBABILITY AND STATISTICS	72
5.1	Introduction	72
5.2	Sample spaces	72
5.2.1	Combination of events	73
5.3	Probabilities of events	77
5.3.1	Conditional probability	78
5.3.2	Independence of events	80
5.3.3	Example: Analysis of a space vehicle battery power supply	81
5.4	Discrete probability distributions	83
5.5	Continuous probability distributions	88
5.6	Distribution functions for discrete distributions	91
5.7	Expected values and moments	92
5.7.1	Tchebycheff's inequality	93
5.8	Moment generating functions	95
5.9	Probability distributions in two or more dimensions	98
5.9.1	Conditional distributions	102
5.9.2	Covariance	103
5.10	Sampling theory	105
5.10.1	Simple random sampling—random sample	106
5.10.2	Distribution of the sample	106
5.10.3	Moments of the sample distribution	107
6	DISCRETE AND CONTINUOUS DISTRIBUTION MODELS	112
6.1	Introduction	112
6.2	The binomial distribution	113
6.2.1	Moments of the binomial distribution	115
6.2.2	The law of large numbers	116
6.2.3	The normal approximation	116
6.2.4	Unbiased estimator of p	117
6.3	The multinomial distribution	118
6.4	The geometric and negative binomial distribution	121
6.4.1	The mean and variance	122
6.4.2	Application of the negative binomial distribution	123
6.4.3	Unbiased estimator of p	124
6.4.4	Limiting exponential and gamma distributions	125
6.5	The Poisson distribution	126
6.6	Generalized binomial success-failure models	129
6.6.1	Unbiased estimator of p	133
6.7	Continuous distribution models	134
6.8	Hazard functions	135
6.8.1	The exponential distribution	137
6.8.2	The Weibull distribution	137
6.8.3	The extreme value distribution	139
6.8.3.1	A particular form of the hazard function	140
6.8.3.2	A particular application of the extreme value distribution	140
6.9	Death process models	142

6.9.1	Derivation of the exponential distribution	142
6.9.2	Derivation of a more general distribution of time-to-failure	145
6.9.2.1	The gamma distribution	148
6.9.2.2	Limiting cases of the general time-to-failure distribution	150
6.9.2.3	Initial failure	151
6.9.2.4	Physical interpretation of the model parameters	152
6.10	The normal distribution	153
6.10.1	The central-limit theorem	155
7	**RELIABILITY ESTIMATION: PART I**	**159**
7.1	Introduction: The reliability function R	159
7.1.1	Example for the normal distribution	159
7.1.2	Example for the exponential distribution	160
7.1.3	Example for the binomial distribution	160
7.1.4	The problem of estimating R	160
7.2	Properties of estimators	161
7.2.1	Unbiasedness	161
7.2.2	Consistency	162
7.2.3	Minimum variance	162
7.3	Methods for finding estimators	163
7.4	The method of maximum likelihood (one unknown parameter)	163
7.4.1	Applications to continuous distributions	164
7.4.1.1	A special example	165
7.4.2	Application to discrete distributions	167
7.4.3	The variance of the maximum likelihood estimator	167
7.4.4	Confidence limits for parameters	168
7.5	The application of maximum likelihood to distributions with two or more parameters	170
7.5.1	Estimation of parameters of the normal distribution	170
7.5.2	Estimation of parameters of the gamma distribution	171
7.5.3	Method of matching moments	173
7.5.4	Variances and covariances of maximum likelihood estimators	174
7.5.5	Estimation of parameters of the Weibull distribution	177
Appendix 7A	*Solution of equations of the form* $x = f(x)$	183
Appendix 7B	*Solution of n equations in n unknowns*	186
8	**RELIABILITY ESTIMATION: PART II**	**190**
8.1	Reliability estimates and confidence limits from the reliability function	190
8.2	A method of reliability estimation for single-parameter distributions	191
8.3	A general method for obtaining approximate reliability estimates and confidence limits	191
8.3.1	The mean of the reliability function	192
8.3.2	The variance of the reliability function	192
8.3.3	Examples	193

CONTENTS   xvii

8.4	Extension of the approximate method to two-parameter distributions	195
8.4.1	Examples	195
8.5	Estimator and confidence limit for R in the case of the Weibull distribution	197
8.5.1	Exact confidence limit for R in the case of the Weibull distribution	197
8.6	Estimator and confidence limit for R in the case of the gamma distribution	199
8.6.1	Exact confidence limit for R in the case of the gamma distribution	201
8.7	Estimator and confidence interval for R in the case of the normal distribution	203
8.7.1	Examples	204
8.7.1.1	Confidence limits for R (two-sided specification)	205
8.8	Confidence limits for parameters of binomial, negative binomial, and Poisson distributions	206
Appendix 8A	*One-sided binomial confidence limits*	209
Appendix 8B	*Exact one-sided binomial confidence limits*	212
8B.1	Introduction	212
8B.2	Theory	213
Appendix 8C	*Upper confidence limits on p for negative binomial sampling*	217
Appendix 8D	*Upper confidence limits on λ for the Poisson distribution*	218
9	**RELIABILITY STRUCTURE MODELS**	220
9.1	Introduction	220
9.2	Serial systems	221
9.2.1	Independent serial systems	222
9.2.1.1	An example of a serial system	224
9.2.2	Lower confidence limits on reliability of independent serial systems	224
9.2.3	A preferred alternate method of computing serial system confidence limits	226
9.3	The "weakest link" model	229
9.3.1	The chain model	231
9.3.2	Chain strength	233
9.3.2.1	Limits of R_n	234
9.3.2.2	Upper and lower limits of R_n achieved	236
9.3.3	Stress versus strength analysis	237
9.4	Parallel systems and redundancy	239
9.4.1	Example of a partially parallel system	240
9.4.1.1	Example: Stand-by redundancy	242
9.4.2	Example of a mixed parallel-serial system	243
9.4.3	An example of reliability design analysis	249
9.4.3.1	Part application	249
9.4.3.2	Redundancy	251
9.4.3.3	Example of reliability design analysis of a space vehicle's temperature control subsystem	255

9.5	System performance variability as a function of subsystem performance variability	260
9.5.1	Method of calculating means and variances of a function of several variables	261
9.5.2	Evaluation of tolerances	263
Appendix 9A	*A technique for reliability apportionment*	267
Appendix 9B	*Reliability of a system with component replacement*	271
10	**RELIABILITY DEMONSTRATION AND DECISIONS**	279
10.1	Introduction	279
10.1.1	Nonstatistical factors in selection of sampling plans	280
10.2	Reliability decisions based on binomial sampling	281
10.2.1	The operating characteristic function	282
10.2.2	General properties of a binomial sampling plan	283
10.2.3	The average sample number function	285
10.2.3.1	Formula for the ASN function for a curtailed binomial sampling plan	287
10.2.4	Modifications to optimize the curtailed sampling plan	291
10.3	Criteria for specification of a binomial sampling plan (fixed sample size or curtailed)	295
10.4	Application of Wald sequential binomial plans to reliability demonstration	297
10.5	Reliability demonstration based on time-to-failure	302
10.6	Reliability demonstration when test time is unequal to required time	303
10.6.1	Generalization to partially known hazard functions	307
10.7	Reliability demonstration utilizing time-to-failure information—exponential distribution	308
10.7.1	Testing until all items fail	308
10.7.2	Testing until r items fail $(r \leq N)$	310
10.7.3	Cost model for reliability demonstration, and example	312
10.8	A truncated and censored life test procedure	314
10.9	Application of Wald sequential life test plans to reliability demonstration	319
10.9.1	Specifications and criteria of the plan	320
10.9.2	Properties of the plan	321
10.9.3	Expected number of failures and expected waiting time to reach a decision	323
10.9.4	Example of Wald sequential life test for reliability demonstration	324
10.9.5	Cost model	327
11	**RELIABILITY GROWTH MODELS**	330
11.1	Introduction	330
11.2	A simple model of reliability growth	331
11.2.1	Further properties of the model	333
11.3	Fitting a curve to a reliability growth model	338

11.3.1	Maximum likelihood estimators	338
11.3.1.1	Variances and covariances	342
11.3.2	Least-square estimators	344
11.3.3	Summary of estimation procedure	346
11.4	Concluding remarks	347
11.4.1	Problems of non-independence	347
11.4.2	Other possible types of models	347
12	EXPERIMENTATION AND TESTING	349
12.1	The reasons for testing	349
12.2	Philosophies of testing	349
12.3	System and component testing	350
12.4	Techniques of testing	351
12.4.1	Time testing	352
12.4.2	Event testing	352
12.4.3	Peripheral testing	353
12.4.4	Environmental testing	353
12.5	The scientific method	353
12.6	Statistical experimentation	354
12.6.1	Experimental considerations	354
12.6.2	The elements of valid comparison	355
12.6.3	A simple example	356
12.6.4	Components of variation	358
12.6.5	An example of a nonstatistical experiment	359
12.6.6	A statistically designed program for solid rocket engines	365
12.6.7	An example of statistical experimentation with electronic equipment	368
12.6.7.1	Establishing the relative importance of the effects of the environmental factors	369
12.6.7.2	The estimate of reliability	370
12.6.7.3	A practical consideration when applying the technique	371
12.6.8	The role of statistical experimentation in reliability	373
13	STATISTICAL DESIGNS FOR RELIABILITY	375
13.1	Introduction	375
13.2	Continuous experimental designs	376
13.2.1	Example and analysis	377
13.3	Sensitivity testing	383
13.3.1	The method of sensitivity testing	385
13.3.1.1	Standard deviations of the mean and variance	387
13.3.1.2	Confidence limits for μ	388
13.3.1.3	Percentage limits	388
13.3.1.4	Level of assurance of estimate	388
13.3.2	Considerations when using the technique	389
13.4	Random balance designs	389
13.4.1	Multiple balance designs	394
13.4.1.1	Example of multiple balance designs	395

13.5	Evolutionary operation	403
13.5.1	Example	404
13.6	Response surface experimentation	412
13.6.1	Geometrical representation with two input variables	413
13.6.2	Examples of application	416
13.6.3	The methodology of experimentation	417
13.6.3.1	The linear equation and method of steepest ascent	418
13.6.3.2	An example of the method of steepest ascent	420
13.6.3.3	The quadratic equations	424
13.6.3.4	Reduction to canonical form	425
13.6.3.5	An example of determining the optimum from a quadratic surface	427

SECTION III. EXAMPLES OF RELIABILITY EVALUATION AND DEMONSTRATION PROGRAMS 433

14	A RELIABILITY EVALUATION PROGRAM FOR A LARGE SOLID PROPELLANT ROCKET ENGINE DURING ENGINEERING DEVELOPMENT	435
14.1	Objectives and problems	435
14.2	Discussion of problems	436
14.3	Description and conduct of the program	438
14.3.1	System apportionment	438
14.3.2	Applicability of principal subsystems	439
14.3.3	Declaration policy	440
14.3.4	Test result classification	441
14.4	Reliability reporting and estimation	444
14.4.1	Representative and current data	444
14.4.2	Engine reliability estimation	446
14.4.2.1	Technique of estimation	446
14.4.2.2	An example of estimation	446
14.5	Conclusion	448
15	A RELIABILITY EVALUATION PROGRAM FOR A TURBO-GENERATOR DEVICE	449
15.1	Objectives and problems	449
15.2	Test conditions and ground rules	449
15.3	Statistical considerations	450
15.4	Statistical analysis of the data	453
15.4.1	The statistical test for a significant change	454
15.4.1.1	An example	454
15.4.2	Reliability demonstration	455
15.4.3	Reliability reporting	456
15.5	Conclusions and comments	457
Appendix 15A	*One-sided sampling acceptance and reporting lines for reliability estimation*	458
Appendix 15B	*Method of computation for change in failure rate*	460

16	A RELIABILITY EVALUATION PROGRAM FOR A LARGE LIQUID ROCKET ENGINE DURING ENGINEERING DEVELOPMENT	463
16.1	Objectives and problems	463
16.2	Discussion of problems	463
16.2.1	Reliability ground rules	463
16.2.2	The master list and excludable components	467
16.2.3	Scope of reliability evaluation	467
16.3	Statistical analysis of data	468
16.3.1	Weighting factors	469
16.3.2	An example of computing weighting factors	471
16.3.3	An example of computing reliability using weighting factors	472
16.3.4	Use of the weighting factors and reliability evaluation	473
16.3.5	Type of statistical estimate	474
16.4	Reliability reporting	474
16.5	Reliability demonstration	475
Appendix 16A	*Maximum likelihood equations for p_i, ϵ*	479
Appendix 16B	*Variances and covariances of the $\hat{p}_i$, $\hat{\epsilon}$*	479

17	SOFTWARE RELIABILITY	484
17.1	Introduction	484
17.2	The software reliability problem	485
17.2.1	The cost of unreliable software	485
17.2.2	Traditional views of software reliability	488
17.2.3	Differences between software and hardware reliability	489
17.2.4	Software quality characteristics	490
17.3	Software errors	495
17.3.1	Genesis of software errors	495
17.3.1.1	Design errors	496
17.3.1.2	Coding errors	498
17.3.2	Error categorization	499
17.3.3	Frequencies of error categories	502
17.3.4	Methods of error detection and prevention	502
17.3.4.1	Techniques	503
17.3.4.2	Tools	503
17.3.5	Error data collection	504
17.4	Software reliability evaluation methods	510
17.4.1	Introduction	510
17.4.2	Reliability prediction	511
17.4.2.1	Background to the phenomenological approach	511
17.4.2.2	Early prediction of errors	512
17.4.2.3	Cost model	512
17.4.2.4	Prediction of operational failures from pre-operational tests	514

17.4.3	Reliability estimation	514
17.4.3.1	The basic model and assumptions	514
17.4.3.2	Estimation of parameters	516
17.4.3.3	Model validation	520
17.4.4	Reliability measurement	521
17.4.4.1	Partitioning model	521
17.4.4.2	Functional partition of the input data space	522
17.4.4.3	Interval partition of the input data space	524
17.4.4.4	Sampling theory for partition models	528
17.4.4.5	Approximate confidence limits for R	531
17.4.4.6	Exact confidence limits for R	532

APPENDIX **543**

Table A.1	Number of tests without failure vs. reliability and confidence	545
Table A.2	Minimum size of sample to be tested for a time t to assure a mean life of at least $\hat{\theta}_L$ when confidence $\gamma = 75$ per cent when F is the allowable number of failures	546
Table A.3	Minimum size of sample to be tested for a time t to assure a mean life of at least $\hat{\theta}_L$ with confidence $\gamma = 80$ per cent when F is the allowable number of failures	547
Table A.4	Minimum size of sample to be tested for a time t to assure a mean life of at least $\hat{\theta}_L$ with confidence $\gamma = 85$ per cent when F is the allowable number of failures	548
Table A.5	Minimum size of sample to be tested for a time t to assure a mean life of at least $\hat{\theta}_L$ with confidence $\gamma = 90$ per cent when F is the allowable number of failures	549
Table A.6	Minimum size of sample to be tested for a time t to assure a mean life of at least $\hat{\theta}_L$ with confidence $\gamma = 95$ per cent when F is the allowable number of failures	550
Table A.7	Sample size and criteria to demonstrate reliability at a given confidence	551
Table A.8	Percentage points of the χ_n^2/n distribution	552
Table A.9	Tolerance factors for normal distributions	554
Figure A.1	Upper confidence limit on unreliability (one minus lower confidence limit on reliability) number of trials N, observed failures F, confidence coefficient $\gamma = 0.50$	556
Figure A.2	Upper confidence limit on unreliability (one minus lower confidence limit on reliability) number of trials N, observed failures F, confidence coefficient $\gamma = 0.80$	557
Figure A.3	Upper confidence limit on unreliability (one minus lower confidence limit on reliability) number of trials N, observed failures F, confidence coefficient $\gamma = 0.90$	558

CONTENTS

Figure A.4 Upper confidence limit on unreliability (one minus lower confidence limit on reliability) number of trials N, observed failures F, confidence coefficient $\gamma = 0.95$ 559

Figure A.5 Upper confidence limit on unreliability (one minus lower confidence limit on reliability) number of trials N, observed failures F, confidence coefficient $\gamma = 0.99$ 560

Figure A.6 50 per cent lower confidence limit on system reliability for observed failure combinations of a two-subsystem serial system, N trials per subsystem 561

Figure A.7 90 per cent lower confidence limit on system reliability for observed failure combinations of a two-subsystem serial system, N trials per subsystem 562

Figure A.8 95 per cent lower confidence limit on system reliability for observed failure combinations of a two-subsystem serial system, N trials per subsystem 563

Figure A.9 50 per cent lower confidence limit on system reliability for observed failure combinations of a three-subsystem serial system, N trials per subsystem 564

Figure A.10 90 per cent lower confidence limit on system reliability for observed failure combinations of a three-subsystem serial system, N trials per subsystem 565

Figure A.11 95 per cent lower confidence limit on system reliability for observed failure combinations of a three-subsystem serial system, N trials per subsystem 566

BIBLIOGRAPHY 567

INDEX OF AUTHORS 570

SUBJECT INDEX 573

INTRODUCTION
THE GROWING IMPORTANCE OF RELIABILITY

In the last few years reliability has suddenly become important. It has become of primary concern in the development of most large weapon systems, and this emphasis will soon be reflected in the civilian market. Industry has become reliability-conscious: "Reliability is everybody's business." Reliability conferences are organized, reliability departments formed, reliability programs written, and reliability requirements appear in specifications and contracts. We may ask, Why? What has happened to create this concern? What is so radically different now compared with a few years back?

The answer is to be found in the technological revolution which we have been experiencing in the last two or three decades. In turn this revolution was significantly accelerated by the Second World War, the Korean war, and the stress on military preparedness since that time. But besides accelerating technology, the wars only too vividly emphasized the importance of reliability. They did this by showing us the consequences of unreliability—we learned the inevitable in circumstances not to our advantage. We learned "the hard way."

Unreliability has consequences in cost, time wasted, the psychological effect of inconvenience, and in certain instances personal and national security. Generally, the cost of unreliability is not only the cost of the failing item but of the associated equipment which is damaged or destroyed as a result of the failure. The reason is the interdependency between components in complex systems. As an example, the failure of a transistor in a home radio would generally cost the amount needed to replace it; on the other hand, the failure of a similar transistor in a ballistic missile might prevent a staging from occurring, leading to the subsequent loss of the missile at tremendous cost. This, of course, is an obvious example. A little less obvious is the expense of maintenance. Although the immediate results of unreliability may not be so disastrous, the necessity of having trained personnel continuously check out the equipment is obviously expensive. In addition, any item which needs replacing has to be paid for, and this must include transportation as well as the cost of communication and logistical organization. If we need two or three times as many items to be available for use as would be theoretically necessary if the reliability were 100%, then the expenses of unreliability become unnecessarily high (unless, of course, *all* the cost functions have been optimized).

Time wasted is a frequent consequence of unreliability; in industry it is almost always synonymous with money wasted. For example, there is the cost of "down-time" in large computers during the search for and repair of failures. Or consider airliners which develop mechanical or electronic faults necessitating a delay in departure: this might entail the cost of alternative transportation or of overnight lodging and extra meals for the passengers.

The almost classical example of the psychological effect of unreliability was that of the ill-famed Vanguard satellites. The United States, smarting from Russia's success with Sputnik I, attempted to compete using a relatively untested vehicle, which was required to operate successfully almost at the limits of its ability. The failure and the subsequent dismay and loss in prestige were very serious.* In the commercial field, frequent or unpleasant inconveniences resulting from unreliability will soon bring customer dissatisfaction and loss of business. For example, a company producing a television set or car which is continuously in the repair shop will find that customers will turn elsewhere for their next purchase. It is not only the inconvenience of losing the use of an item upon which we begin to depend, but we are also at the mercy of the repair men who always seem to find more things wrong or charge more money than we expected.

Lastly, perhaps the most important consequence of unreliability is its effect on national security. National security depends on the nation's resources, industrial potential, level of military technology, preparedness, and military budget. Reliability is concerned mainly with the last two items on this list. Reliability itself does not directly advance the "state of the art" of military technology, although its methodologies can help increase the rate of advancement. However, preparedness is a direct concern. Generally, when we discuss the numbers of missiles on the launching pads, the number of airplanes in the Strategic Air Command, or the number of radar stations we are actually thinking of the physical number and not the *effective* number. Thus if the reliability of a weapon system is 50% the effective number available for use is only one-half the physical number. This fact is generally ignored, which is tantamount to assuming a 100% reliability for that weapon system when in fact we are only half prepared. Equally important, frequently we do not know specifically which weapon systems comprise this 50% effectivity; thus we can never

---

* The third attempt to launch a Vanguard Satellite, on March 17, 1958, 3 months after the initial attempt, was very successful.

deploy the weapons for an immediate 100% effective retaliation. This situation tends to emphasize the importance of reliability being as high or as close to 100% as is feasible—especially, for instance, if this weapon were to be used as a vehicle to launch a man into space. On the other hand, it is important to know the level of reliability so that we can plan in terms of effective numbers rather than actual numbers. For various reasons—manufacturability, total cost, performance, etc.—a 50% reliable weapon system might be considered preferable to a 90% reliable weapon system from the point of view of national economy, strategy, and security. Such a decision would require a good understanding of reliability and of the relative efforts and expense required in order to reach any desired level of reliability.

The budget (except in times of emergency when time is the overriding factor) is generally the limiting consideration in the development of any item or system. Reliability's share in the budget has become somewhat controversial. At the outset reliability is costly; it requires certain expensive activities such as organized and efficient planning, testing, and reporting without immediately being able to demonstrate its worth compared with the initial outlay. However, as soon as this initial period is passed, the higher reliability obtained will begin to save money because, for instance, a missile does not explode, or a test stand is not damaged, and so on. Since we cannot hypothesize about what did not happen, it is difficult therefore to demonstrate the net savings due to reliability; consequently, budgeting for reliability activities is frequently insufficient.

Reliability, then, is an important problem. We may therefore ask that since we apparently recognize the problem can we not apply the necessary solutions and thereby render it unimportant? It has been said that any problem becomes trivial once it has been stated properly.

Unfortunately, there is no panacea for unreliability. We cannot expect the reliability equation to equal unity when its factors are human beings, communications, understanding, and requirements beyond current experience. However, the first approximation to the solution of the problem can be obtained by use of the means now at our disposal. It is the purpose of this book to discuss some of the methods available; it is the responsibility of the reliability engineer and engineering management to be able to choose and apply the most efficient methods for their particular problems.

0.1 THE REASONS FOR UNRELIABILITY

The specific causes of unreliability are many. However, the situation at the root of the problems is due to *the dynamic complexity of system development concurrent with a background of urgency and budget restrictions.*

0.1.1 THE LIMITS OF EXPERIENCE

In the last few decades our engineering abilities have improved, spurred on by two wars, to such an extent that we are now within reach of being able to create engineering systems and devices which previously existed only within the realms of our scientific imagination. For instance, the basic laws of motion have been known for several centuries and so therefore have the mathematics of space travel; however, the feasibility of such travel has only just begun to be realized. Again, we have been capable of the manipulations of binary arithmetic, but it was not until the advent of the various electronic devices that the construction of a digital computer became possible, leading to the tremendous advances in the handling of complicated and tedious arithmetic computation.

However, what is feasible and possible is not necessarily reliable. Devices and systems are not perfect; they do not operate in the same manner in all circumstances. Our total knowledge may be insufficient about any item so that when it is placed into an environment, about which we also have an insufficiency of information, failure occurs. In other words, we are working at the limits of our technological knowledge. Of course, given time we acquire and organize our knowledge so that this situation changes. Our learning process does this for us. Notable examples are in the missile programs, where failures in any one specific program have become less and less frequent. Were we able to let this evolution take place at a natural pace, our reliability problem might be relatively minor. Unfortunately, there is an urgency which prevents us from giving sufficient time to all of the many considerations. The evolutionary process conflicts with the "revolutionary" atmosphere. Before we have time to experience, synthesize, and apply our knowledge we are developing another system or device. We eliminate some of the mistakes, but these are usually compensated for by the new ones introduced with the differences of the new system and its applications. Such is the dynamic framework within which we must operate.

0.1.2 THE COMPLEXITY OF EQUIPMENT

The implications of complexity are extensive. Complexity is not only an attribute of the physical system itself but also of the processes which are necessary for its creation. Let us discuss the physical system first. We are referring to a system which requires—with certain exceptions to be discussed later—that for a successful operation all its subsystems, components, and parts shall function successfully also. This interdependency introduces *probability theory*, which in its simplest form states that the probability of a successful operation of the system is the probability that all lesser devices within the system operate successfully. The mathematical formula expressing this situation can be expressed as follows: *The probability of a successful operation of the system, i.e., the reliability of the system, is equal to the product of the probabilities of successful operation of each device, i.e., the product of the reliabilities of the devices, providing that these devices are statistically independent.* This rule, usually referred to as the *product rule,** tends to imply that the greater complexity of the system, the lower will be its reliability.

Another aspect of complexity which produces unreliability is *subsystem interaction*. This interaction may be *environmental*, such as vibration of an engine causing electronic failure, or it may be *functional*, where the output surges of one component may be greater than the input specifications of its mate. This latter example is a problem of *systems engineering*, but nevertheless it creates unreliability. Complexity requires concurrent development. We are advancing the state of the art in many fields; therefore, frequently we have insufficient knowledge of the limits of the operation of parts and components, yet at the same time we know that the uncertain behavior of any one component may affect the behavior of its functional or environmental neighbor in a detrimental manner. Thus there may be so many contingencies and interactions that the designer cannot comprehend or accommodate all of them in his initial designs. He may be able to do this eventually, but he has to go through the learning process mentioned earlier, with its consequential mistakes and failures.

0.1.3 THE COMPLEXITY OF ORGANIZATION

In addition to the physical complexity of the system, there are associated complexities which are equally as important. A complex system demands well-organized management which in turn requires an efficient system of communication. If the total experience which any one type of component had accumulated in all its varying applications were gathered together and properly interpreted, we could reduce its failure rate significantly. We would do this by suitable application, by modifying the components, and by techniques of *redundancy*. However, the limitation of our communication system will not permit a complete data interchange. Further, there is the problem of transmitting proprietary and classified information. This situation results in our having to learn by our own experiences rather than from someone else's. No less serious than the lack of information is the problem of misinterpretation of facts by the recipient, who may not have been close to their generation. This has the consequence of sidetracking us, involving loss of time and money until our learning process brings us back into the main stream of advancement.

* See Sec. 9.2 for the mathematical derivation of the product rule.

0.1.4 HUMAN ERROR

This brings us to a very important topic in reliability: human error. Failures from this source occur not because the person involved did not understand what he was doing (unless, of course, he was a trainee without adequate supervision), but because he was simply careless or forgetful. Generally, procedures, manuals, and instructions will specify the correct operating procedure which must be followed in the course of maintenance and operation. Nevertheless, as the personnel become more familiar with the instructions, there is a tendency to ignore them, so that occasionally an operation is missed or performed out of sequence. Again, owing to pressure of time during a perhaps unscheduled adjustment, wiring may be reversed, tools left in the machinery, and other obviously foolish actions performed which result in failure.

But though man's presence may produce failures, his absence perhaps creates more. Man's ability to control an operation represents an almost irreplaceable factor in the decision-making function. His judgment and experience permit him to change or anticipate situations and to compensate for deteriorating conditions. However, in his quest for higher speeds, shorter times, and greater outputs, he has almost programmed himself out of the operation of today's modern complex systems. For example, the pilot of today's high-speed jet aircraft is much less the absolute master than was the pilot of yesterday's slower-flying piston engine planes. He cannot make decisions fast enough; there are too many factors with complicated interrelationships for him to analyze and coordinate in the time available. Thus many of his decisions must be made for him by high-speed automatic computing equipment.

Such, then, are the general reasons for unreliability. Most specific failures can be traced back to the general causes discussed above. One type of failure which may be conspicuous by its absence from the discussion is the so-called *random failure*. It is questioned whether such an event ever truly manifests itself; this depends on its definition. If we define random failure as a failure due to an *unassignable cause*, then the writers assert there is no such thing. We must differentiate between there being a cause and our ability to relate it to the particular effect of failure. If, on the other hand, random failure relates to an event which is random with respect to the *time of its occurrence*, then the definition can be held to be valid. Thus if we state the premise that all failures have a cause, we can, by intelligent anticipation, analysis, and application, endeavor to eliminate the causes and/or the belligerent environment and reduce the rate of failure to a tolerable incidence and the consequences to insignificance.

0.2 PREVENTING UNRELIABILITY

The dynamics of progress and the increasing complexity of our systems cannot be expected to become less dynamic or complex. That is, the factors which create the setting for unreliability will not become less severe. Therefore, we must expect the solution to take on some of the characteristics of the problem.

Thus, as the particular problems appear on a sequential basis, so should our decisions be made and be expected to be made on a sequential basis. Our plans should be laid to obtain the maximum information per observation while encompassing the fact that in a development program the inherent character of the system will be undergoing a continuous change. Sampling plans and sequential experimental and production techniques are being developed and are available which have the dynamic characteristics required. Some of these procedures are described in this book.

0.2.1 REDUNDANCY

The basic rule of reliability with respect to complexity is: Try to keep the system as simple as is compatible with the performance requirements. While this rule might seem obvious, it is not in fact always followed. We should make our aims compatible with our abilities; perhaps from a reliability viewpoint our aims should be a little less than our imagined abilities. In fact for "safety" we can introduce the concept of *redundancy*. Redundancy can take several forms. Overdesign is one technique. Alternate back-up components is another. As a most simple example we can design a system which, to operate successfully, requires that only one of its several components work successfully. This implies that the more alternate components there are, the greater the probability that one of them—and therefore the system—will operate satisfactorily. In reliability theory this relationship is the mathematical counter to the *product rule* mentioned in Sec. 0.1.2. Of course, by adding components to take advantage of the redundancy rule we introduce such disadvantages as increased complexity and extra weight, and so there is a limit to the utilization of this technique.

0.2.2 SYSTEMS ENGINEERING

The problem of failures produced by interaction comes under the realm of *systems engineering*. This requires careful test planning and design so that the components can be integrated into the system while they still have a flexibility of operation and configuration for system adjustment. If the components have too much separate development without system testing or consideration, their designs will become frozen and inadaptable for system requirements. On the other hand, if the components are assembled into a system at too early a stage in their development, so many will fail owing to questionable reactions to unknown causes that the system test will provide a negligible amount of information. Statistical experimentation will help provide estimates of functional and environmental effects and variations in components' behavior from both separate and within-system tests. The most suitable time for integration can be established, since by comprehending the limits of the behavior of the components we can estimate the extent of overlap of their weak areas and therefore the system's weaknesses.

0.2.3 MANAGEMENT CONTROL AND COMMUNICATION

Organizational complexity makes a control necessary so that the technological knowledge generated within the various subgroups is efficiently coordinated, analyzed, and disseminated. It is the function of systems engineering or management science to establish and exercise the controls necessary to optimize the efficiency of the organization. That part of systems management which is concerned with the manifestation of failures, their prevention, and the associated disciplines of reliability activities is by definition the reliability group. Anything which might contribute to a failure or potential failure must be surveyed. Poor communication represents a major hazard in this respect in a complex organization; the reliability group must therefore concern itself with this problem. Poor communication can be the result of several different types of deficiencies: (1) inadequate coverage of the information, i.e., not all the relevant data are being reported; (2) all the data are being reported but not all are being received by the responsible or cognizant personnel; (3) the data are being misinterpreted; (4) the data are not being generated properly at the source. Any single one and all combinations of these situations may create failure and thus must be controlled. Since the generation, transmittal, and interpretation of reliability data and consequential decisions and actions based thereupon cover the whole spectrum of system development activities, it is essential that the reliability group be effectively placed to be familiar with the data-associated functions. Once this is done, it becomes their responsibility to review, modify, and establish where needed the disciplines necessary to minimize the communication problem.

Let us take the specific communication deficiencies mentioned earlier. We can control the adequacy of the coverage only by clearly defining the type of data to be reported as well as emphasizing the importance of reporting at every occurrence the event being monitored. To insure that the responsible personnel are receiving the information, transmittal channels, frequencies, and distribution of reports should be determined.

In short, an efficient and well-organized data collecting, analysis, and reporting system should be established. The data interpretation depends to a significant extent on the analytical abilities of the engineering staff; however, the amount of information derived from observations as well as its validity can be greatly increased by efficient testing techniques. By education and consultation the best methods of scientific inference and statistical methodology can be applied to test planning and test results so that real knowledge will accumulate during product development.

0.2.4 HUMAN ENGINEERING AND RELIABILITY EDUCATION

We attempt to control human error in several ways. We can try to design so that the assembly, operation, and maintenance of the system are as straightforward as possible. This comes within the province of human engineering and is beyond the scope of this book. However, reliability education and the reporting of human errors are essential so that greater responsibility and understanding can be exercised by all those who contribute to the engineering and production programs. In this era of mass production and immensely large and complicated programs the average worker is apt to feel that his contribution is unimportant. The consequence is a lack of enthusiasm in his job and with it a carelessness which produces mistakes and sometimes failures. Reliability can present films, posters, and lectures to demonstrate how these can be prevented and help encourage the spirit of cooperation and eagerness.

To summarize, we have suggested how unreliability may be lessened, controlled, or prevented. The general techniques described are the basis for the evolving methodology of reliability. Although reliability is still not a well-defined subject, techniques exist which, with suitable application, can greatly improve—both with respect to time and inherently—the reliability of any system and its components.

ADDITIONAL READING

Lambert, J. S., "Air Force Electronic Reliability Program," *I.R.E. Trans. Reliability and Quality Control*, **14,** 17–21 (September 1958).

Och, H. G., and Tinus, W. C., "Systems Engineering for Usefulness and Reliability," *I.R.E. Trans. Military Electronics*, **3,** 8–12 (January 1959).

Schlager, Kenneth J., "Systems Engineering—Key to Modern Engineering," *I.R.E. Trans. Engineering Management*, **EM-3:** 3, 64–66 (July 1956).

Soucy, Chester I., "A Broad Survey of the Military Electronic Equipment Reliability Problem and Its Controlling Factors," *Proc. 1956 Electronic Components Symposium, Washington, D.C., May 1–3, 1956*, 8–23.

Sparling, Rebecca H., "Testing in the Guided Missile Industry," *ASTM Bull.*, No. 218, 52–56 (December 1956).

SECTION I
MANAGEMENT, ORGANIZATION, AND COMMUNICATION

The totality of specialized knowledge needed for each and all of the many interrelated components within a complex system tends to transcend the capabilities of small organizations. Consequently, just as the systems have grown in technical complexity, so also have the organizations grown. This situation has created new problems which in many ways are no less troublesome than many of the technical difficulties with which we are faced—problems of the management and efficiency of organizationally independent groups responsible for separate but functionally related equipment.

The reliability group is very much affected by this situation so that in addition to the scientific problems, nontechnical managerial considerations must also be taken into account. This section therefore includes a discussion of the position and effectivity in an organization of various reliability groups, based on experience in different companies.

The many activities of a large engineering program must be arranged so that no relevant considerations are omitted. Reliability is the subject which endeavors to cover any potential omissions. As a result, techniques have emerged which have become specialized reliability activities. These are presented and discussed in the second chapter of this section.

The third chapter deals with the problem of communication, a vital link in the chain of engineering activities. Without adequate information and without proper organization of information, our efforts are to some extent wasted—especially in the field of reliability where ineffective communication can be tremendously expensive in both time and money. Therefore, a reliability communication system which is in successful use for a highly complex system and a very large organization is described here in detail.

CHAPTER ONE
MANAGEMENT AND ORGANIZATION

1.1 THE RESPONSIBILITY OF MANAGEMENT

Management has the ultimate responsibility for any major decisions and policies established by a company or an organization. The soundness of management decisions is reflected in the quality of products produced and in customer satisfaction. Reliability is merely one quality of the product; others might be performance, style, convenience, economy, and so on. However, reliability differs from these qualities in a major respect: reliability is not an obvious attribute. On the contrary it is a most abstruse quality. It is open to many interpretations and is in need of many qualifying and defining provisions. Thus, although management can expect to make the correct decisions most of the time concerning the other qualities because of its experience and understanding of them, frequently reliability represents something of an enigma.

Reliability as a subject is not well defined. On one hand, it might be thought of as simply the statistical estimation of numerical reliability. On the other extreme, it might be thought of as encompassing the whole development program. As we shall see, it is neither of these extremes; rather, it is a set of techniques generated by an attitude of anticipation of unreliability and an appreciation of the necessity of preplanned elimination of the problems.

Reliability covers such a broad spectrum of activities that its effectivity is very much dependent not only on the capabilities of the personnel but also on their position within the organization. If the reliability group is staffed with capable men but placed at a level or in a section of the organization in which they have no authority or effectivity, they are an expensive luxury. It might seem foolish for management to do this, but frequently lip service is paid to reliability. Reliability groups and their talents should not be used only for window dressing in contracts or proposals. However, in practice sometimes certain elements of management will endeavor to restrict the efforts of the reliability personnel within the organization. This can be prevented by the correct placement of the reliability group. Since reliability is an integral or associated part of many of the functions of a development and operational program, one of the major problems of management is to establish the most effective delegation of the various reliability functions and the degree of responsibility and authority that the reliability personnel will possess. Thus the problem of the reliability group's position within a company is a major and pertinent one.

One of the consequences of the broadness of the activities is a potential irritation or conflict of responsibility. Reliability, as will be made clearer in later sections, has many areas of responsibility which are not readily distinct from the responsibilities of existing groups. Wherever there exists a double responsibility, even though these responsibilities are theoretically supplementary, as in the case of reliability, then a potential management problem exists. There is always a natural reluctance to accept advice when it is not asked for or to resent a change in an established system. Frequently, both of these situations occur with the introduction of a reliability group. For example, it is only too natural that Quality Control does not like having its data collecting system modified or extended any more than Test Planning likes being reminded that it is not fully utilizing the concepts of statistical experimentation. Reliability is not just one more separate activity to add to engineering, manufacturing, quality control, testing, etc., but is a subject which is an integral part of all of these functions. Thus reliability poses the threat of interference, conflict, and usurpation of authority.

The problem of funding produces management's major dilemma in reliability. Intelligent reliability approaches can evolve significant cost-cutting techniques; and it is here that management's appreciation and the reliability group's effort should find a basis for mutual respect and benefit. On the other hand, management faces the conflict between the apparent high cost of an extensive reliability program and the consequences of producing a product with an unsatisfactory reliability. The cost trade-off problem is certainly one of the most elusive. In many instances a product of 99% reliability is too unreliable, yet as the percentage reliability is increased, so is the cost of the reliability program. There are the obvious costs such as the salaries of the reliability engineers and the clerical workers, and with a limited understanding of reliability technology it is natural for management to ask why they should be paying these salaries when they have design engineers and quality control people who are supposed to "take care of reliability." However, the major expenses produced by reliability are not in the salaries but arise from the need to establish the *assurance* of reliability. Assurance demands greater numbers of tests, which in turn require equipment, facilities, and personnel. The results of these tests must then be analyzed and organized, and this demands expensive data collection and processing systems. Thus management sees reliability as an expensive overhead and its introduction into the organization as an added complexity. These, then, are the obvious objections which produce apathy or reluctance in management to wholeheartedly endorse reliability activities.

Management has every right to ask, "If we are to pay the above-mentioned price for organized reliability, what are we buying, and is it worth the price?" The term *organized reliability* is used because every management will claim that it is producing reliable equipment. Management will ask, "Why do we need a reliability group? Are not our design engineers sufficiently analytical, our test engineers sufficiently competent, our manufacturing techniques adequately controlled?" The simple and honest answer is: "No." The techniques and procedures are not adequate to assure success at all times. If they were, reliability would not have acquired its current prominence.

Reliability is a new methodology—one of many appearing in various fields. As we learn to appreciate and use these methodologies and reap their benefits, we are more anxious to accept them. Reliability is not an obvious quality nor is proof of its worth immediately obvious; however, its worth can be demonstrated. For example, we can point to the statistical design (Sec. 12.6.6) by which we reduce the sample size in a certain test program and prevent wasteful testing and also possible misinterpretation of information. As another example, we can take the case of a failure picked up by an efficient failure reporting system which prevented later similar expensive failures and also thereby reduced damage costs.

The burden of the proof is on the reliability group. They must convince management of their effectivity and the long-range savings in money. Management thinks in these terms, and it is in terms of money, time, and effectivity that reliability will be evaluated. Military management considers these factors and the additional one of security. Security and reliability are close to being synonyms, and it is perhaps for this reason that the importance of reliability has been recognized by the military. The manner of its recognition is by contractual reliability specification. This is a sure way of stimulating "interest" in the subject.

If we are able to succeed in explaining to and convincing management of the uses of reliability methodology, how does management then react? Possibly the most important reaction is one of attitude. If management does not offer encouragement, then all reliability activities are severely compromised. Management also should establish the reliability group at its most effective position within the company in regard to the over-all program. This aspect is discussed in subsequent sections. However, possibly the most telling means of recognition is by separate funding for adequate reliability activities. By these methods management will fulfill its responsibilities and assure that the concepts and disciplines associated with reliability will be introduced into the development program.

Management must accept the evolution of reliability as a major subject as one of the consequences of increased technology.

1.2 THE PLACE OF A RELIABILITY GROUP WITHIN AN ORGANIZATION

What is the most efficient position of a reliability group within a company? The answer depends on the program in question. We shall discuss here in general terms the advantages and disadvantages of various positions of a reliability group as they have been experienced in a variety of organizations.

Let us first list those activities for which a reliability group either has direct responsibility or else participates in an advisory or secondary capacity. These functions are:

1. Reliability evaluation
2. Reliability apportionment
3. Design review
4. Design control
5. Specification, material, and processing review
6. Vendor control
7. Test planning, operation, and analysis
8. Reliability knowledge
9. Reliability and failure reporting systems
10. Mathematical and statistical services for reliability problems
11. Reliability education
12. Internal coordination of reliability activities

If we consider as a simple model a company composed of the following subdivisions

1. Management
2. Engineering
3. Testing
4. Manufacturing
5. Quality Control
6. Purchasing and Contracts

then it can be seen that the activities of reliability overlap and enter into all the major departments within a company. Consequently, it might be asked why the responsibility for reliability cannot be divided among the appropriate departments. This approach is completely unsatisfactory for several reasons. First, reliability is now a subject in its own right and has methodologies with which most engineers are unfamiliar. As such, it demands full-time attention and adequate recognition which it will not receive under this arrangement. Second, it is a systems function of management coordination and engineering details which will be weakened by such subdivision. To assign reliability to design engineers or quality control inspectors brings us back to the original situation which was one of the reasons for the reliability problem.

It is generally agreed that best results are obtained when a single group has reliability as its responsibility. It must be an independent and forceful group. We must decide where this group should be placed, bearing in mind that it must coordinate or cover all of the above activities, be effective, and possess sufficient authority. Let us then consider some of the advantages and disadvantages of placing the reliability group within one of the subdivisions of a company as listed above.

When reliability is organized as a management function, it has the authority for demanding action for reliability approvement but does not have the capability for evaluating the "reliability cost and worth" of the action taken. Nor is management itself equipped to handle the technical and sometimes minor details with which the reliability group must be concerned. It is the aggregation of these details which result in unreliability; therefore, though it is imperative that management be closely cognizant of reliability activities, achievements, and estimates, it is not sufficiently close to the work-level problems.

When reliability is placed in the engineering division with an in-line responsibility, it has the advantage of being close to the people who early in the program have the major responsibility of providing a system of high reliability; i.e., the design and development engineers. However, this very closeness sometimes has the disadvantage that the reliability group cannot be independent in its criticism. Another disadvantage might be that it is ineffectual because of the policy of engineering management who are not familiar nor in agreement with reliability methodology; therefore, there is little chance for independent arbitration. Also, as the development program phases into the production program, there will sometimes occur a lack of continuity that cannot be controlled adequately by reliability when organized as part of engineering.

Quite often the reliability group is established with quality control since it is frequently considered that the capabilities of quality control meet the requirements of a reliability group. This is only partially the case. It is true that the data reporting system, vendor control, statistical services, and some of the other functions can be or are operated by quality control; but design review, test planning, etc. are not within the realm of quality control responsibility. Quality control plays too passive a role. A reliability group's most important contributions are in improving the designs and operations to increase the level of reliability. Quality control, on the other hand, is mostly concerned with retaining the level which exists; it is essentially a subsection of reliability, although it will be some time before this is acknowledged.

Thus whenever the reliability group is a subgroup of one of the existing company divisions, either it is inadequate in some way by not covering all the activities, or else in attempting to do so it will create jealousy or conflict between the various divisions and bring forth accusations of "empire building."

1.3 AN INDEPENDENT RELIABILITY GROUP

It appears, therefore, that reliability ideally should be organizationally independent of all of the other major divisions but guaranteed by management of close cooperation with each of those divisions so that it can work effectively and in sufficient detail to accomplish the technical aspects of its job. Its head should also report to top management so that his authority cannot be ignored.

The group's effectivity will also be a function of the abilities of the personnel involved. It should include both systems engineers and technical specialists. Since it will participate as an independent group in the whole spectrum of company activities, from design review to in-the-field failure reporting, there will be a great deal of interaction necessitating much coordination. The amount of coordination and integration of the many activities will depend on the extent of the activities delegated to the reliability group, the size of the group, and the abilities of its people. Thus, for example, a member of the reliability group, upon reviewing a test plan or surveillance program, might see the advisability of statistical test planning. If he is not able to design the test himself, he should have the authority to call upon the statistical services group or quality control or whoever has the capability so that an acceptable plan is achieved.

Therefore, whether the reliability group is large and self-sufficient to perform all the required activities, or small and "management-like"—delegating, directing, and coordinating its specific activities to outside technical groups—it must have certain abilities in the form of experience and education. Having these abilities, it must then have authority, effectivity, and over-all responsibility for the reliability program activities of the organization.

1.4 PERSONNEL CAPABILITIES

The reliability group should consist of people thoroughly experienced in the following backgrounds:

1. Systems engineering and operational analysis

2. Component design
3. Specification and materials review
4. Manufacturing operations and quality control
5. Test planning and environmental testing
6. Design and analysis of statistical experiments
7. Data collection, evaluation, analysis, and follow-up
8. Project management and coordination
9. Mathematical and statistical probability theory

Detailed discussion of how these experiences are utilized is deferred to subsequent chapters.

1.5 RELIABILITY BUDGETING

Reliability is an intrinsic quality which evolves with the development of the system. Consequently it can be argued that the reliability funds are "integrated" into the development costs of the over-all program. However, the most effective manner of supporting the reliability group is to give it a separate budget, because then reliability itself is a separate entity which stands on its own merits. In addition to the salaries and support activities, provision should be made for a certain amount of funding for hardware. This allows the reliability group to conduct or have conducted independent tests specifically for reliability investigations. An independent and productive reliability group performing the reliability activities listed represents the most effective tool that management can provide to aid in reducing unreliability.

ADDITIONAL READING

Beaton, G. N., "Putting the R and D Reliability Dollar To Work," *Proc. 5th National Symposium, I.R.E. Reliability and Control in Electronics, January 12–14, 1959*, 65–72.

Cohen, J., and Okun, A. M., "Organizing for Reliability," *I.R.E. Trans. Reliability and Quality Control*, **PGRQC–9,** 1–8 (January 1957).

Goode, Harry H., "The Analogy Between the Problems of Systems Engineering and Management," *Chem. Eng. Progress*, **55,** 48–50 (January 1959).

Kuehn, R. E., "Organizing for Reliability," *Proc. 3rd National Symposium on Reliability and Quality Control in Electronics (I.R.E.), Washington, D. C., January 14–16, 1957*, 123–25.

CHAPTER TWO
PROBLEMS AND ACTIVITIES OF PLANNING AND OPERATING A RELIABILITY PROGRAM

2.1 INTRODUCTION

In Chap. 1 we listed the reliability activities and discussed the attitude an organization must take in order to run a reliability program. In this chapter we shall discuss these activities in greater detail and illustrate the problems with which we are faced in the establishment of such a program.

The tasks associated with the reliability group can be organized into two complementary sets of activities: (1) those necessary for improving the reliability of the product, (2) those used in defining and assessing reliability as a latent characteristic of the product. The latter activities are no less essential than the former; without a quantitative measure with which to evaluate the effects of productive activities, there is no objective way of knowing whether they are being used effectively.

2.2 RELIABILITY ESTIMATION

The reliability of a device is a quality of that device; however, it is not a quality which can be measured directly. In fact, except for some trivial cases mentioned later, it cannot be "measured" at all. We are limited to its estimation. Reliability is defined as *"the probability of a successful operation of the device in the manner and under the conditions of intended customer use."* Let us examine this definition more closely. We see that it is a probability statement and contains qualifying conditions which are open to interpretation.

2.3 RELIABILITY AS A PROBABILITY CONCEPT

First, it is seen that reliability is a probability concept. To evaluate probabilities it is generally necessary to utilize statistical estimation theory in all except trivial cases, such as when the whole population has been used; e.g., "In 1957 the reliability of rockets used for aircraft assisted take-off was 99.9%." This figure is simply the ratio of all successes to all attempted operations. There is no attempt being made in this statement to predict what might have happened in 1959, or if in 1957 only half the rockets had been used what would have been the expected reliability of the remainder. Probability is introduced when we try to make statements about a population based on observations obtained from only a sample or portion of that population or when we are trying to predict in advance the outcome of events.

When we are making predictions, statistical theory allows us to associate levels of assurance with any reliability estimate we might make. Thus, we can be 50% sure that the true reliability will be above or between certain specified values. On the other hand, we might wish to be more confident about our predictions. We might want to be 90% sure or 95% sure or perhaps 99% sure or even higher. We can never be 100% sure except where (as in the trivial case above) we have complete experience of our item, and of course, here we are just measuring reliability and no longer predicting it. However, the greater the assurance we wish to have in any predictions, the more conservative our predictions must be if the observed sample size is not increased. Thus, the "reliability limits" of our predictions must be lower or wider. Alternatively, if we wish to remain above or within prespecified limits, such as in contractual reliability demonstration, we must be prepared to take a larger sample. To illustrate these ideas with numbers consider the following example.*

Suppose we had tested 50 items and experienced two failures. What could we say about its reliability? One estimate could be 96%, another estimate 90%, and still another 84%. It will seem strange to any reader not acquainted with statistics that there could be three different values of reliability (in fact, there are an infinity of values) estimated from our observations. However, there is no conflict here, just a necessity of understanding the ground rules for what these estimates represent. The first value, 96%, is simply the ratio of successes to the number of tests. The second and third reliability numbers are both confidence limits; i.e., lower bounds for the true reliability. In the second case, we are 90% confident that the true reliability is greater than 90%. In the third case, we are 99% confident that the true reliability is greater than 84%. As we pointed out earlier, with higher confidence, our statements become more conservative. The first value, 96%, is a point estimate of reliability and, as such, can not have any specific confidence level associated with it. It would also represent the true reliability if, for instance, 50 items consisted of the total number of items of that kind manufactured and only two failed in all the

* The numbers quoted in this discussion are obtained from Figs. A.3 and A.5 in the Appendix. (The Appendix, pp. 545-566 contains Tables A.1–A.9 and Figs. A.1–A.11.)

experience. However, if we suppose this observation is only a sample of perhaps a thousand such items, we might ask, What is wrong with our point estimate? Why should we not always use this estimate? The answer is that nothing is wrong. However, it is not very discriminating. The point estimate is the same whether we have one failure in 25 tests or twenty failures in 500 tests; i.e., both sets of numbers give estimates of 96%. It is obvious, though, that we should have more confidence in any statement made from 500 observations than from 25 or 50. Statistical theory provides us with this confidence in a quantitative form. Thus, with twenty failures in 500 tests, we can state with 90% confidence that the true reliability is greater than 95% whereas, with one failure in 25 tests and still with 90% confidence, we can state that the true reliability is greater than only 85%.

It is evident that reliability estimates are open to a great amount of statistical manipulation. None of the above figures is inconsistent, but it is obvious from these very simple exercises that when we are specifying, estimating, or otherwise quoting reliability numbers, we must be quite clear as to their statistical meaning as well as their practical implications in terms of numbers and cost of tests.

2.4 RELIABILITY AS A FUNCTION OF SUCCESS CRITERIA

In our definition of reliability there is the phrase "successful operation." Let us look at the definition of success and at some of the questions it raises. If we have a piece of equipment which operates without failing for a given time, and if this time is not equal to the customer's expected use time, is this test a success? It is not a failure. If it is a success, is it a complete success? If not, should it have some *weighting factor* associated with it before being placed into a reliability equation? For example, what degree of success is an hour's operation of a computer which is required to operate for only two minutes? Again, what is the degree of success of a rocket engine which is test fired without failure for 20 seconds, when it has to fire for two minutes in operational use? Further, what is the degree of success when under test conditions we cannot simulate all the environmental conditions which would be experienced in operational use?

To see how the success criteria should relate to the end usage we can consider a dramatic example. A rocket engine giving only 99% of the total impulse necessary to get its satellite into orbit is a complete failure; on the other hand, a rocket engine giving 99% of the total impulse necessary to get its warhead to its destination might destroy 50% of the target. We see that it is necessary to qualify or define the phrase "successful operation." We must therefore be careful in choosing our ground rules for success and failure of any test and relate them to our various modes of operation. There is however no reason why several sets of criteria cannot be established to evaluate the data against different end requirements. As an example, we could estimate the probability of a missile's being able to get off the ground without exploding. A second estimate might be that of its hitting the target. Each of these figures is important, but it is also equally as important to know which one is being quoted. This is not known unless the basic criteria are clear on this point, and this fact is not always known or appreciated—especially by people who are not close to, or working with, the data.

2.5 NONREPRESENTATIVE CONFIGURATIONS AND TESTS

A further feature of reliability evaluation with which we must contend is that of interpreting test results of nonrepresentative devices. During any development program we have a continuing change of items or configurations and yet we still must know how good our product is and how rapidly it is improving. Therefore, we must consider how to utilize the results of systems or subassemblies which have components that are nonrepresentative of the final configuration. Similarly, we must determine the use we can make of separate component tests and how subscale test results are related to the full-scale unit.

Again, we must consider how to use data from tests which have specifically been made to fail or have failed due to human error or test facility malfunction.

These, then, are some general problems of reliability estimation. It is obvious that great care must be taken to use the pertinent data, make the correct interpretation of the results, and apply the most appropriate statistical methods for reliability estimation.

Many of the problems can be answered by statistical techniques, others by the introduction of careful controls of the reliability information, by common sense, or by arbitrary decision. However, any answer will relate to the needs of particular situations; consequently Section III of this book devotes considerable discussion to these problems in terms of specific examples.

2.6 CONTRACTUAL IMPLICATIONS OF RELIABILITY EVALUATION

It can be seen that the difficulties of reliability evaluation arise not with its definition but rather with the interpretation of criteria and the statistical problems of estimation. Therefore, when reliability is specified as a contractual requirement, the importance of the various criteria increases considerably.

We can think of reliability estimation as having two uses. First, it is used as a *relative* measure; that is, by applying it at various times along a development program, we can measure the growth of reliability; i.e., as a measure of increase by repeatedly applying the same criteria. Thus, we would be interested in knowing that reliability increased 5% last month compared with 3% the previous month. A second use is as an *absolute* measure. We can use the same criteria to estimate the absolute value of reliability, but this value is absolute only within the frame of references of those criteria. Consequently, if the criteria are not correct, then the absolute value is not valid either. As a relative measure of increase, the inaccuracies of the criteria are subtracted out to a first approximation. However, as a contractual reliability requirement it is the *absolute* estimate which must be demonstrated; therefore, the ground rules and statistical method of estimation must be carefully and completely specified.

Statistical techniques add precision to reliability assessment, but their limitations must also be understood. First, estimates can be no better than the data upon which they are based. Second, if the product is truly devoid of reliability, then no *valid* statistical manipulation can indicate otherwise. For instance, if the true reliability of a device is, say, 90%, then no amount of testing will *prove* it is, say, 95%. As with any statistical sampling plan there are certain probabilities of *observing* an apparent reliability of 95%, or greater, or any other number which might be specified.

It might at first sight appear as an anomaly that while we can never be 100% sure of being able to demonstrate a certain reliability, we can sometimes be 100% sure of *not* being able to demonstrate it. This fact can be quite important. For instance, it is important to realize that with a sample size of 20 items at a 90% confidence level it is impossible to demonstrate a minimum reliability of 95%.* The "best" that can be estimated from this sample, even with all successful tests at a 90% confidence level, is a minimum reliability of just over 89%. If we are too statistically unsophisticated we might be contractually agreeing to do the impossible! This situation sometimes results in confusion for those unfamiliar with statistics. This can be prevented if we think of statistics as the mechanism for estimating the probabilities of statistically demonstrating, by observation, such an event as s successes in n tests. We could compute the probabilities of observing this event for various hypothetical values of the true reliability. *Engineering ability creates the true value of reliability of the device; our statistical techniques can only estimate this intrinsic characteristic.* The higher our potential value of reliability, the greater, of course, is the probability of meeting the requirements, or conversely the smaller the risk of not meeting them. These probabilities can be computed by statistical methods and in quality control terminology are known as "operating characteristics." We can only hypothesize,

---

* See Fig. A.3.

compute, and predict the probabilities of meeting the contractual requirements. Our real probability of meeting them lies with our engineering abilities and the manner in which they are harnessed and organized.

2.7 RELIABILITY GOALS

A reliability goal is the most effective way of directing attention towards reliability whether or not it is a contractual requirement. The most important goal will be that for the highest level of assembly or system for which the organization concerned has responsibility. Thus, if the organization were one of the armed services, the reliability of prime concern could be that of the complete weapon system. If it were a small company it might be the reliability of, say, a miniaturized computer.

Our next step is to see what the system reliability requirement demands from its subsystems and components. Again, we must then relate these demands to what can be achieved. In order to do this, we must simultaneously review the reliability interrelationship between the components of the system, their actual and potential or expected reliabilities, and the amount and adequacy of the planned development program. The techniques for carrying out this study are *reliability apportionment, reliability design review, reliability knowledge review*, and *test* and *experimentation review*. As the development program is followed, these reliability objectives are continuously reappraised and, in addition, those activities necessary for the efficient handling of the program are introduced. These are *design control, vendor control, reliability data reporting and follow-up, reliability activity and information coordination, reliability analyses and education*. We shall now discuss these topics in greater detail.

2.8 RELIABILITY APPORTIONMENT

Reliability apportionment is the name given to the process of subdividing the system reliability goal into subsystem and component goals. The mathematics of this subject are discussed in Chap. 9 and Appendix 9A; for the moment we will assume that we are able to carry on the mechanics of this procedure so that we may discuss its utility. First, apportionment represents an exercise such that the reliability objectives established for each of the subsystems must be compatible with their current state of development, expected improvement, and the amount of testing and money budgeted for their development. When these components goals are recombined, their reliability interrelationship should be such that they satisfy the system goal. Apportionment has its greatest value at the first level of breakdown of a system into its major subsystems. It is also often necessary at this level, for frequently each of the major subsystems is produced by a separate contractor. And, it is also desirable to put a numerical reliability requirement into each contract. This would be the situation, for example, in the case of an ICBM weapon system. The entire system should have a certain reliability which would be apportioned into requirements for the propulsion contractor, requirements for the guidance contractor, etc. A still further breakdown of these subsystems into major components might be made and component reliability requirements established. Again, for example, if the propulsion system consisted of several rocket engines, we could establish requirements for each engine and similarly for other systems.

Reliability apportionment takes on great importance if we require these numbers (or some equivalent numbers) to be demonstrated. It is one thing to go through the exercise of estimating subsystem goals for an indication of how important that subsystem is or how much work has to be done; it is an entirely different state of affairs when the fulfillment of a contract is at stake. The reason is that apportionment has its limitations. As it is frequently applied, its mathematics are based on the assumptions of statistical independence and the form of the mathematical model which describes the relationship between the components and the system. At the higher levels of assembly, independency and the model may be obvious; at lesser levels the model may not be discernible nor will statistical independency necessarily apply. It is not necessary in apportionment mathematics for independence to exist, but the mathematics become increasingly difficult if independence does not hold. It can be seen that these arguments apply only when we are referring to reliability numbers estimated from separate component or subsystem tests. When we are referring to subsystem tests as part of the system tests, then correct mathematical and statistical interrelationships are automatically present and the objection does not hold. As an example, it is perfectly valid (though it may not be feasible) to require an igniter to demonstrate 99.9% reliability on rocket engine tests, where 99.9% is the apportionment for the igniter of the complete engine's required reliability. On the other hand, it is not valid (in this example) to state that we have demonstrated our apportioned reliability by igniter tests which are conducted independently of the engine. The reason is that we do not know the extent of the "interaction" between the igniter and the propellant. Generally, we attempt to establish the limits of interaction by means of our specifications, but these cannot always sufficiently define the system as in the above case. This comment would always apply to those systems in which the components exhibit strong mutual interdependency, such as solid propellant rocket engines. Generally, electronic equipment would have weak mutual interdependency, and liquid rocket engines would be somewhere in-between.

The physical makeup of the system also plays an important role in determining how far down into the assembly we should apportion our system requirement. Electronic equipment, for example, frequently consists of many repeated components such as tubes and printed circuits which are identical. In addition, the functional and environmental effects are generally "mild." Under these circumstances, then, the evaluation or demonstration of component reliability is worthwhile since in the process of evaluating one component we realize that this information frequently represents information relating to all the other components of that type. On the other end of the scale is the example of a liquid propellant rocket engine. Here we have a large number of completely different components, which in addition, experience severe system-generated environments of vibration and temperature. Even if we were able to establish the components' reliability goals, the difficulty of separately demonstrating the necessarily high number for every component and simulating the environments makes this task impractical In the light of these comments, how then should we use the techniques of apportionment?

We have seen that under certain circumstances we can establish a subsystem or component reliability goal which can be demonstrated by valid test results. When reliability goals cannot be validly demonstrated, either because of the unrealistic test circumstances or lack of numbers of units available for testing, the apportioned figures are still useful in indicating the relative stage of development of the various components. This may be done by comparing the relative magnitude of the per cent defective or failure rate with the apportioned per cent defective or failure rate (where we have statistically translated the reliability goals into the appropriate failure measure). In addition, we should compare the failure rates for any single component in its different environments. Reliability apportionment forces us to look closely at our test schedules for all levels of testing to see how much information and assurance we can derive from the test program. As a consequence, it will be evident whether we have organized or budgeted our test plan as wisely as possible; and, if not, how we should change the test plan in the light of the above considerations and those of Sec. 12.6.1.

2.9 DESIGN REVIEW FOR RELIABILITY

In the initial stages of conception of a system, the reliability group should review the proposed configuration with respect to the reliability of its major assemblies and components. In order to be able to do this it is necessary to make surveys of different types of assemblies to determine their relative merits with regard to performance, weight, reliability, cost, etc. For instance, we might compare pump-fed systems with pressure-fed systems, turbo-generator power supplies with chemical-electrical power supplies, liquid rocket engines with solid rocket engines, one type black box with a second type, one computer manufacturer with another. With this comparison we conduct a paper study to obtain an approximate estimate of the expected reliability of the system.

The system configuration which is decided upon is then reviewed in detail. Analytical studies are made to determine whether redundancy concepts have been used. These concepts include the use of items at derated outputs, the establishment of safety factors in marginal cases, and the use of parallel systems where possible. We also insure that the design is sufficiently flexible to accommodate later modifications and that modified designs or alternate items are available as back-up in the event of difficulties in the program. When new designs are used, we ascertain that they are significantly better than standard or off-the-shelf items. To do this, we must compare the new design with approximately similar designs about which we have failure experience. We review the new design to see which modes of failure have been eliminated by modification or elimination of components or parts. We must not forget to estimate by design analysis and judgment the potential failure modes introduced by any new design. The reliability group also ascertains that the design takes into account human factors and ease of maintenance by requiring that the design be coordinated with human engineering consultants.

Reliability design review is an essential activity; it is not a duplication of the effort of the design engineer. The designer's mind tends to be oriented towards new designs and configurations and he is not "probability conscious." As we have mentioned in the introduction, a design which is feasible is not necessarily reliable.

The designer's drive for maximum performance is a natural one; however, if there is too great a tendency to change and an insufficiency of experience or opportunity to learn, we must expect initial expensive failures which we cannot always afford. Optimum performance is frequently associated with low reliability. We must, therefore, evaluate our performance objectives with techniques of design review performed independently of the original designer. In this way we will arrive at a more *effective* system.

2.9.1 AN EXAMPLE OF A RELIABILITY DESIGN REVIEW CHECKLIST

1. Is the item an off-the-shelf device or especially developed for a particular function?
2. Does the item perform more than one function?
3. How many critical and noncritical parts and/or characteristics?
4. Could an existing off-the-shelf item be used? If not, why?
5. Could an existing off-the-shelf item be suitably modified? How might the modification affect its performance and behavior?

6. Is the item to be used at the limits of its strength or performance capabilities?
7. If an off-the-shelf item, is it being used as in previously experienced environments and at normal operating levels? What are the limits of satisfactory performance and environmental usage?
8. What have been problems on similar or subscale designs? Are these likely to reoccur on the present design? What can be done to eliminate them?
9. What are previous hypothesized modes of failure? What are new expected modes of failure? Why do these differ?
10. What is the failure history of this item? Is this a critical item; i.e., would its failure create system failure?
11. What steps have been taken in the item's application or system design to eliminate these types of failure?
12. What steps have been taken to prevent new types of failures from being introduced?
13. What is the expected or estimated numerical value of reliability of the item?
14. Is this sufficient for the item's apportioned or allocated reliability requirements?
15. Is it possible to introduce the concepts of redundancy and/or use the item at derated performance levels?
16. Is the part completely interchangeable with that of another manufacturer? If not, how does it differ in its failure experience?
17. What is known about its storage life, operating time or cycles; i.e., how much time or cycles operating and nonoperating may be accumulated without significantly, if at all, degrading the reliability?
18. State methods of inspecting and testing for modes of failure in 9.
19. If the item is a newly developed design, what are its critical weaknesses? What provision has been made in the design so that modifications can be made at the earliest possible time if these or other weaknesses show up in testing?
20. Has the item been designed as simply as possible? Have human factors been considered to prevent error, such as reversed wiring, mis-assembly, etc.?
21. Is it physically and functionally compatible with its neighboring components; i.e., is the performance likely to drift outside the range of its neighbors, are tolerances and clearances adequate? Will the physical location affect the performance or reliability?
22. Have physical features which might adversely affect performance or reliability been taken into account; i.e., are there any sharp corners which might damage parts or short-circuit wiring? Is the design sufficiently stressed against vibration, temperature, humidity, dust, etc.?

23. How might the design be modified to improve reliability? Would this compromise such factors as performance, cost, weight, availability, schedules, maintainability, etc.?
24. Has the item or system been designed for ease of production and assembly, maintenance, inspection?
25. Are unusual quality control or vendor problems expected?
26. Have factors of handling, transportation, packaging, and environments, other than the specified operational environments, been taken into account?

2.10 RELIABILITY KNOWLEDGE

As part of the design review activities, the reliability group will collect reliability information on various types of designs and configurations. As the development program progresses, further information will become available from the test program. This additional information is organized by the reliability group and integrated with the existing data to provide material in the form of reliability knowledge for analysis by the reliability and engineering personnel.

Such knowledge would list modes of failure, causes of failure, and consequences of failure. The conditions, environmental and otherwise, experienced when failure occurs should also be stated. Primary and secondary* failures should be distinguished. It is desirable that the various methods of testing for the different weaknesses be given. Those parameters which are indicative of reliability measurement should be listed, as should the ground rules which define success and failure and their relationship to reliability. If possible, estimates of failure rates and the type of failure distributions associated with the failures should be obtained. Examples of such reliability information are shown in Sec. 3.12.1.

2.11 TEST PLANNING AND ANALYSIS

At the beginning of the development program, all test plans are reviewed and schedules checked to insure that as balanced and integrated a program exists as it is possible to establish. A general test philosophy will be evident from the test plan, and it should be determined whether it is truly the most effective way of achieving reliability within the restrictions of time and cost. Thus, we can see whether an emphasis is being placed on component testing or subscale testing or system testing and whether the

---

* A secondary failure is the term given to a failure which occurs in a component as the result of failure occurring elsewhere; i.e., a secondary failure does not occur of its own accord but as the result of a primary failure.

forthcoming information is sufficiently relevant or whether there is too much "dead-end" testing. Further, the reliability group determines whether environmental effects and production-process considerations are incorporated sufficiently early in the program.

Where an insufficiency of realistic tests are to be performed owing to expense, destructive nature of the tests, inadequacy of test facilities, or for other reasons, then other methods of assurance should be sought. Inspection methods should be reviewed to see whether they can help compensate these inadequacies. Nondestructive test methods are investigated, and afterward failure analyses can be made to see whether they are pertinent and sufficient and adequate inspection standards can be established for them. The test plans should be evaluated to determine that the most efficient statistical methods are used, that the test criteria are valid, and that sample sizes are suitable for the requirements of the program. The activities in this area are discussed at length in the chapters on experimentation and testing (Chap. 12 *et seq.*).

2.12 DESIGN CONTROL AND SPECIFICATION, MATERIALS, AND PROCESSING REVIEW

These functions are the controls by which the physical hardware is made to conform to the design. Component and system specifications are reviewed to determine their compatibility, i.e., so that output limits of one component are not greater than the input limits of its functional neighbors. Specifications are reviewed for tightness to insure that the statistical build-up of component tolerance limits will not result in system inadequacy. Dimensions, weights, pressures, currents, times, and tolerances which are critical with respect to modes of failures are determined.

Material inspection and manufacturing procedures are examined to assure that the quality of material being used and the sampling plans are compatible with reliability requirements. Processing input levels are investigated to determine whether they are operating at an optimum combination. Processing outputs are reviewed for reproducibility and adequacy.

Design drawings are checked and copies retained in a central file. A master list of system components can be drawn up at the time when the prototype configuration is established. This list contains those components which in the judgment of the component engineer and reliability group represent the most reliable at any given point in time. Only one component design should be listed for each component of the system, and all components must appear on the list. The purpose of the list is to completely define the most reliable system at any given time. It also allows us to keep check on any test configuration and to ascertain whether any experimental or obsolete components are part of the configuration and whether failure may be attributed to them. A record of all engineering change orders is maintained and the consequences on the reliability noted. The master list is reviewed periodically and brought up to date. Its utility is described in Sec. 16.2.2. With these controls we can establish whether it is the basic design which is deficient rather than the hardware or procedures.

2.13 VENDOR CONTROL

Vendors are assessed for their engineering capability of meeting performance requirements. Under certain circumstances when cost and feasibility permit, they may be required to demonstrate apportioned or specified reliability goals or provide reliability input data on their products. The contractor's reliability group should be represented on the vendor survey team, whose other members are usually from the engineering, quality control, and purchasing groups. The contractor's survey team evaluates the vendor's program to determine whether sufficient emphasis is placed on reliability. If necessary, as it might be in the case of small companies, the prime contractor should help organize and monitor the vendor's reliability program. The reliability group should review all purchase orders to insure that reliability is not being unnecessarily compromised on account of cost.

For design and process control of both contractor and vendor components reliability tests and quality tests should be performed at specified standards to assure that all component designs and manufacturing techniques will satisfy the system performance requirements.

2.14 RELIABILITY AND FAILURE REPORTING SYSTEMS

The mechanism for obtaining reliability knowledge is the reliability and failure data reporting system. This subject is discussed in detail in Chap. 3. The reliability group which will analyze and be responsible for the data must establish the system at the beginning of the program so that valuable experience is not wasted or lost. Such data would include knowledge on all conditions—i.e., times and/or cycles, environments, type of configuration, test, etc.—which are relevant to the reliability of the item. The reliability group has the responsibility of making the reporting system as simple and inexpensive but as comprehensive as the amount of required reliability knowledge demands. This means that the degree of integration of equipment to which the data systems apply should be clearly stated, as well as the stage of development of that equipment. The establishment of the data flow and the determination of the adequacy of the follow-up is also the responsibility of the reliability group. Procedures are set up to control the integrity of the data. Great care should be taken so that the data are interpreted properly. Reliability reports and summaries suitable for all levels of management and engineering are issued on a periodic basis. Reliability program status reports should also be distributed periodically. All reliability and failure information should be located in the reliability knowledge file and be easily accessible.

2.15 MATHEMATICAL AND STATISTICAL ACTIVITIES

Optimum use must be made of all data. The data are analyzed to see whether they follow any mathematical pattern. From these analyses estimates of reliability are obtained, reliability growth curves established, and correlations of information from various locations and different configurations and stages of development are measured.

Reliability apportionment and design reliability analyses, as well as statistical experimentation are among the statistical and mathematical activities undertaken.

These techniques are discussed in Section II.

2.16 INTERNAL COORDINATION

As reliability covers such a large number of interrelated functions, close surveillance must be maintained over the reliability-oriented activities of all the separate departments. Coordination, therefore, represents an important role in keeping management, engineering, testing, and manufacturing abreast of the reliability status of the various components and the over-all system. More important, it assures that all possible means of increasing the reliability are being used. Work statements, contract proposals, schedules, and program plans should all be reviewed to ascertain that reliability is adequately covered. The methods for insuring coordination will vary depending on the activity; however, positive controls *must* be established which guarantee that coordination is obtained.

A signature is the most effective method of insuring that the reliability group has reviewed the document or activity. Committee membership is another formal method of providing the reliability group's representative with an opportunity of coordinating all the reliability activities. Other methods, by personal contact or other informal closeness, are generally not satisfactory since personnel change and lack of time frequently prevent informal contacts from being made. It is therefore mandatory that a formal and cooperative working relationship between reliability and the remainder of the departments be established.

2.17 RELIABILITY EDUCATION AND AWARENESS

There are two major considerations in this area. One involves creating an awareness of the importance of reliability and an understanding of the consequences of unreliability. This may be done by films, lectures, and posters on a general, nontechnical level and presented to all personnel. The second aspect of reliability education involves the description, on a more technical basis, of the methodology of reliability. For instance, the concepts of probability can be described to engineers and the reasoning behind statistical design and analysis of experiments to test planners. Similarly, the value of obtaining all relevant information from an efficient failure reporting system can be shown to design engineers. It is not necessary nor even desirable to become involved in the very technical details of these subjects with the nonreliability personnel, but it is important to describe how the methodology can aid them, prevent them from making wrong decisions, and provide them with a better understanding of the equipment with which they are working. This activity is important, for it should be noted that with the frequent attitude towards reliability the burden of proof of the desirability of its activities falls upon the reliability group.

2.18 CONCLUSION AND SUMMARY: THE RELIABILITY PROGRAM PLAN

We have discussed some of the problems and activities which are associated with the preparation and operation of a reliability program. These activities will change their emphasis as the development program progresses and also will vary according to the type and state of the art of the equipment in question.

Experience has shown that it is valuable at the beginning of the development program to formalize the reliability activities in the form of a reliability program plan. This plan serves at least four purposes. First, a detailed document represents a valuable exercise in organizing the reliability group's approach to the problem. Second, during its coordination, it provides the reliability personnel with an indication of the amount of cooperation or enthusiasm they may expect from the other departments, the amount of backing they will receive from management, as well as an opportunity to "sell" their contribution to the development program. Third, a detailed reliability document can be regarded as a "charter" which delineates the areas of responsibilities and describes the activities of the reliability group as well as the associated reliability activities of other departments. Fourth, it is an important contractual document which should explain to the procuring agency or customer the specific details of the contractor's internal reliability program as it relates to his specific equipment or system and permit the recipient to determine its adequacy.

The reliability program plan should:

1. state the system reliability requirements
2. specify the tests, ground rules, and statistical methods of reliability demonstration
3. describe the size, technical capabilities, organization, and responsibilities of the reliability group
4. define the reliability group's position, effectivity, and formal relationships within the company or organization
5. list the test plans and schedules for all levels of equipment
6. describe objectives, sample sizes, and state of development of equipment in these various tests
7. define which apportionment goals will be demonstrated and the circumstances of demonstration
8. describe in detail the reliability activities for evaluation, analysis, and product improvement as discussed in general under the subheadings of this chapter.

ADDITIONAL READING

Brown, R. W., "Reliability and the Component Engineer," *Elec. Mfg.*, **60:**5, 126–35 (November 1957).

Connor, J. A., "A Systematic Plan for Predicting Equipment Reliability," *Proc. 1956 Electronic Component Symposium, May 1956*, 233–35.

Davison, W. R., "Latent Reliability—An Approximation Method," *National Convention Trans., ASQC, 1959*, 99–103.

Dertinger, E. F., and Pertschuk, D. W., "Current Military Reliability Specifications," *I.R.E. Trans. Reliability and Quality Control*, **14,** 6–8 (September 1958).

Dreste, F. E., "A Reliability Handbook for Design Engineers," *Electronic Engineers*, **77,** 508–12 (June 1958).

Garbarino, H. L., "Selection of Reliability Levels in Equipment Design," *I.R.E. Trans. Industrial Electronics*, **IE–5,** 76–81 (April 1958).

Gottfried, P., Schneider, L. L., and Xavier, M. A., "Utilization of Component Part Reliability Information in Circuit Design," *I.R.E. Trans. Reliability and Quality Control*, **14,** 60–68 (September 1958).

Jacquemard, F. C., "Start Quality Control at the Source," *Mill and Factory*, **60,** 104–106 (February 1957).

Lamb, J. J., "RACER—A Proposed Rating System for Electronic Components and Devices," *I.R.E. Trans. Reliability and Quality Control*, **PGRQC–6,** 1–10 (February 1956).

Leubbert, W. F., "A Systems Approach to Electronic Reliability," *Proc. I.R.E.*, **44,** 523–28 (April 1956).

Lusser, R., "Components—Key to Reliability," *Military Electronics*, **3**:3, 38–42 (September 1957).

Ordemann, F. A., "How to Design Reliable Computers," *Electronic Equipment*, **4**, 22–25 (June 1956).

Ruther, F. J., and Smith, L. D., "Proof of Reliability," *Proc. I.R.E. National Conference on Aeronautical Electronics, Dayton, Ohio, May 1958*, 449–53.

CHAPTER THREE
RELIABILITY DATA SYSTEMS

3.1 RELIABILITY DATA COLLECTION, PROCESSING, AND REPORTING

This activity represents a most important part of a reliability program. A well-organized system for collecting, analyzing, disseminating, and following-up reliability data is one of the basic mechanisms which the reliability group utilizes. Much thought must be given to the system when it is being established. On one hand, too little or inadequate information is almost useless; on the other hand, we can create a monster of paperwork which requires all one's time and energy merely to control it. The data system is not an end in itself but the means to the end of providing all levels of engineering activity and management with pertinent and valid information for the benefit of the development and production programs.

The handling of data can be conveniently classified into three categories:

1. Collection
2. Distribution
3. Analysis, Reporting, and Follow-up

3.2 COLLECTION: TYPE AND AMOUNT OF DATA

In setting up any form of data reporting system we are faced with such considerations as the following: What kind of data are to be collected and why do we want them? Or perhaps a better way of directing the questions is: What type of data are meaningful and how do we intend to analyze them? Well-considered answers to both of these questions will do much to focus the data requirements.

How broad and detailed a coverage do we desire without being overwhelmed with masses of data? What is the minimum amount of data required to satisfy the minimum needs? If we are dealing with many different types of items, complexities of devices, and a variety of producers, what degree of uniformity of procedures, forms, etc., becomes important? On one hand, a form might be suitable for information on an electronic failure; it might not be appropriate for other devices which are hydraulic or mechanical, etc. If, however, a comprehensive form is used, it will probably contain a great deal of redundant space when utilized for particular types of devices.

Another question of similar implications which needs clarification is: How deep into the development program of a component or system do we wish to go using the proposed data system? For example, do we want or can we obtain the same kind of information from components involved in separate component tests as we might want from components in system tests? Do we wish to apply a particular data system to prototype configurations or only to production items destined for delivery? Do we want data on 100% of all production components, or do we want sampled data on production items, or do we want data on each new lot of a production item (such as each shipment of a particular tube)? Shall each component be identified with its shipping lot and its end production unit? How much data and what type of data should be recorded so that histories of contingencies which later turn out to be important are not lost through insufficient coverage? What methods can we use to insure both the accuracy and validity of the data—since conclusions can be no more accurate or complete than the data from which they were drawn? Should the reliability man periodically check the quality control man to see how effectively the assembly process instructions are being followed?

The answers to many of these questions should be governed by the cost and difficulty of gathering the information versus the danger of not. With automatic sorting methods available, it is relatively simple to select all sorts of "interesting" tab runs. On the other hand, where data must be reduced or sorted by hand, it is important to limit the coverage considerably, and even greater care must be applied in judging what are pertinent data to collect.

3.2.1 FAILURE REPORT FORMS

These forms should not be complicated or be open to misinterpretation by the originator. One method of assuring this is by making most of the entries check points such as in the sample failure form Fig. 3.1 in Sec. 3.5. However, there should be adequate spacing for remarks since not all information can be categorized.

It should be made quite clear to all possible users of the forms when they should and when they should not use them. The reliability education program plays an important role here. If many different people are initiating and completing the forms, there is bound to be some inconsistency. This can be overcome by having one member of the reliability group review all the forms to make sure they are properly completed. Also, by comparing the numbers of forms used in the different areas and by different people it can be determined whether all the required information is being reported conscientiously.

3.2.2 FAILURES DUE TO HUMAN ERROR

Failures due to human error begin to take on an increasing importance as the design evolves, until this source of trouble frequently becomes the most significant of all. It is mandatory, therefore, that all such failures be reported in addition to those due to design inadequacy. The term *human error* as used here covers not only errors such as reversed wiring, wrong part installation, etc.—that is, errors due to gross carelessness—but also discrepancies produced by failures of the control procedures associated with manufacturing, inspection, and operation. Numerous examples can be cited where Standard Operating Procedures (S.O.P.) or Standard Inspection Procedures (S.I.P.) are inadequate or incomplete so that critical measurements are not taken or—because of lack of definition—operations are performed which result in failure at a later time. Thus, wiring may be twisted or "kinked" if the assembler is left to his own devices; two accepted components may be forced together in assembly and later crack due to thermal expansion; two "thin" O-rings may be installed in place of one "thick" O-ring; certified material may get all the way through inspection simply because it is certified although comparison with design drawings would show it to be the wrong material.

Further failures can be produced by not keeping the S.I.P.'s and S.O.P.'s, Engineering Change Orders, Drawings, etc. up to date and also distributing them to all cognizant personnel to prevent the use of obsolete parts and inspection methods.

Since it is impossible to anticipate all contingencies which might cause failure, the best that can be done is to review all the procedural controls initially and modify them as their inadequacies become known. The only way to do the latter is to insist that all human errors be reported and prevent the natural tendency to relax once the design part of the problem appears close to solution. This subject is an excellent topic for a reliability education lecture.

3.3 DISTRIBUTION OF DATA

Our next concern is to make sure that the data are sent to the right people. Consequently, we must determine a distribution list and ascertain how many copies are necessary. This latter requirement is accommodated in the design of the form by having sufficient carbon copies or a reproducible form. Generally, the following personnel will require at least one copy: the originator of the report, the reliability department, the component engineer responsible for the component from which the information is derived, the department to which the information is relevant (quality control, maintenance, etc.), the central data file; also, one copy should remain with the equipment until its disposition is finally determined. The distribution is, of course, dependent on the type of information being reported and is discussed in greater detail in Sec. 3.6 with specific examples.

3.4 ANALYSIS, REPORTING, AND FOLLOW-UP

The next step is to analyze the data or make sure that the analysis is performed. Some of the analyses will be initiated by single reports. In other instances the aggregation of reports or summary reports will indicate the need for investigations as information trends begin to appear. In some instances, the reliability department may perform the actual analysis or participate as a member of a committee. However, generally, a development engineer will perform the analysis. In this case, reliability must insure by some method of follow-up that the analysis is being performed and that they will receive copies of the completed report.

Reporting represents one of the reliability group's greatest responsibilities. It furnishes all levels of engineering activities and management with information relevant to their needs. Consequently, the possibility of misinterpretation must be minimized. This is an important consideration—especially when numbers are being reported. If, for instance, reliability numbers are being quoted, a description of how they were generated and the circumstances of the original data upon which they were based should accompany them or at least be referenced. (Again, the methods of evaluation, criteria, etc. would be a good topic for a reliability education presentation.) Otherwise, such questions as the following might be asked: Were the data taken from development tests, field tests, separate component tests, system tests? Were environments present? Were the data homogeneous and representative? How large was the sample size? What assumptions were made concerning the shape of the failure distribution? This is not an exhaustive list but includes some of the more important considerations to be accounted for. If this information is not known the data might be compromised and invalid conclusions drawn. A further complication is that reliability numbers have the habit of being quoted out of context; therefore it is advisable to standardize the method of estimation and report to perhaps a single number (perhaps the lower 90% confidence limit) and so prevent confusion.

3.5 AN EXAMPLE OF A DATA REPORTING SYSTEM

Let us consider a data reporting system. The example presented approximates a failure and reliability reporting system which has been established to meet the requirements of a complex organization. This organization is in the process of developing and producing a large and complex weapon system, consisting of a variety of types of equipment including electrical, electronic, mechanical, hydraulic, and hot gas systems. The systems and subsystems, which are in various stages of development and assembly, are produced by a number of separate companies. While the method presented is fairly specific, the ideas, forms, and procedures described can be modified to apply to most if not all similar situations.

The basic form of the reporting system is the failure report form (Fig. 3.1). This form is initiated whenever a failure occurs on equipment. For this purpose a failure is defined as follows: *A failure is the inability of the equipment to satisfy performance or design specifications once the equipment has experienced successful operation or acceptance or has the expectation of successful performance without adjustment or rework.** Special tests such as tests-to-failure would not utilize this form. On the other hand, unscheduled maintenance would require a form to be initiated. The level of assembly would have to be specified for each subsystem, where *subsystem* refers to equipment—the computer subsystem, the engine subsystem, the guidance subsystem, etc.

The form is used on those subsystems which are ready for delivery or are in an equivalent state. It is also used on components which have been accepted, but could not be integrated into the higher assembly. The level of disassembly at which the failure form is used is agreed to by each producer so that coverage is consistent. This generally presents no problem for failures in the field since they occur on the integrated system, but in-plant failure reporting will be inconsistent unless its coverage is clearly specified. For instance, failure reports would be initiated on failures discovered on a turbopump assembly, which had been checked out separately from the complete rocket engine; but on the other hand, failure reports would not be initiated on, say, a printed circuit which had not yet passed through quality control inspection. Again, failures occurring on operating equipment tests made at receiving inspection or at the vendor's plant in final assembly checkout and acceptance testing are reported on this form. However, induced failures are not reported on this form but would appear in a more detailed engineering report. Other failures not covered by the foregoing situations might be reported on quality control forms, which could of course be identical or similar to the basic failure report form.

---

* This definition permits the reporting of *failures*, which have been noted during operation, as well as the reporting of *conditions*, which would result in a failure if operation were permitted.

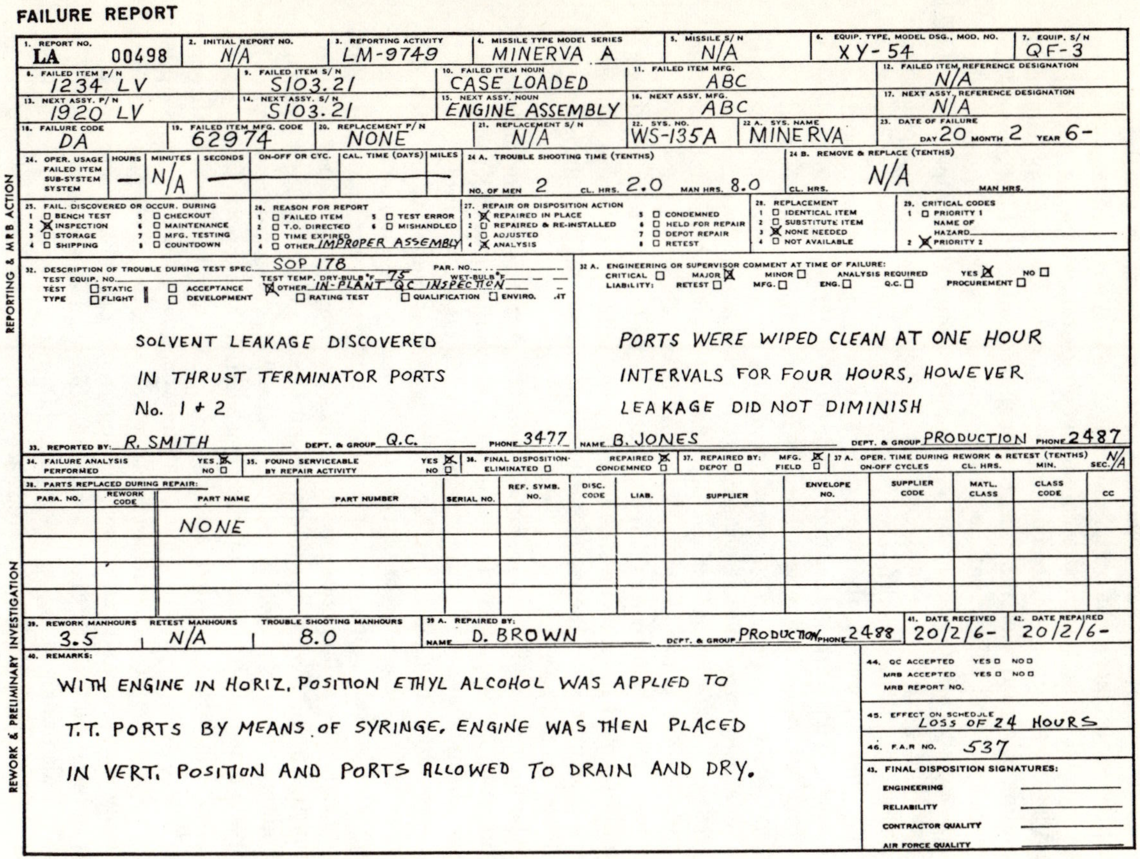

FAILURE REPORT

REPORTING & MRB ACTION

1. REPORT NO. LA 00498 | 2. INITIAL REPORT NO. N/A | 3. REPORTING ACTIVITY LM-9749 | 4. MISSILE TYPE MODEL SERIES MINERVA A | 5. MISSILE S/N N/A | 6. EQUIP. TYPE, MODEL DSG., MOD. NO. XY-54 | 7. EQUIP. S/N QF-3

8. FAILED ITEM P/N 1234 LV | 9. FAILED ITEM S/N S103.21 | 10. FAILED ITEM NOUN CASE LOADED | 11. FAILED ITEM MFG. ABC | 12. FAILED ITEM REFERENCE DESIGNATION N/A

13. NEXT ASSY. P/N 1920 LV | 14. NEXT ASSY. S/N S103.21 | 15. NEXT ASSY. NOUN ENGINE ASSEMBLY | 16. NEXT ASSY. MFG. ABC | 17. NEXT ASSY. REFERENCE DESIGNATION N/A

18. FAILURE CODE DA | 19. FAILED ITEM MFG. CODE 62974 | 20. REPLACEMENT P/N NONE | 21. REPLACEMENT S/N N/A | 22. SYS. NO. WS-135A | 22 A. SYS. NAME MINERVA | 23. DATE OF FAILURE DAY 20 MONTH 2 YEAR 6-

24. OPER. USAGE FAILED ITEM SUB-SYSTEM SYSTEM: HOURS — MINUTES N/A SECONDS | ON-OFF OR CYC. | CAL. TIME (DAYS) | MILES | 24 A. TROUBLE SHOOTING TIME (TENTHS) NO. OF MEN 2 CL. HRS. 2.0 MAN HRS. 8.0 | 24 B. REMOVE & REPLACE (TENTHS) CL. HRS. N/A MAN HRS.

25. FAIL. DISCOVERED OR OCCUR. DURING: 1 ☐ BENCH TEST 2 ☒ INSPECTION 3 ☐ STORAGE 4 ☐ SHIPPING 5 ☐ CHECKOUT 6 ☐ MAINTENANCE 7 ☐ MFG. TESTING 8 ☐ COUNTDOWN

26. REASON FOR REPORT: 1 ☐ FAILED ITEM 2 ☐ T.O. DIRECTED 3 ☐ TIME EXPIRED 4 ☐ OTHER IMPROPER ASSEMBLY 5 ☐ TEST ERROR 6 ☐ MISHANDLED

27. REPAIR OR DISPOSITION ACTION: 1 ☒ REPAIRED IN PLACE 2 ☐ REPAIRED & RE-INSTALLED 3 ☐ ADJUSTED 4 ☒ ANALYSIS 5 ☐ CONDEMNED 6 ☐ HELD FOR REPAIR 7 ☐ DEPOT REPAIR 8 ☐ RETEST

28. REPLACEMENT: 1 ☐ IDENTICAL ITEM 2 ☐ SUBSTITUTE ITEM 3 ☒ NONE NEEDED 4 ☐ NOT AVAILABLE

29. CRITICAL CODES: 1 ☐ PRIORITY 1 NAME OF HAZARD 2 ☒ PRIORITY 2

32. DESCRIPTION OF TROUBLE DURING TEST SPEC. SOP 178 PAR. NO. TEST EQUIP. NO. TEST TEMP. DRY-BULB °F 75 WET-BULB °F TEST TYPE ☐ STATIC ☐ FLIGHT ☐ ACCEPTANCE ☐ DEVELOPMENT ☒ OTHER IN-PLANT QC INSPECTION ☐ RATING TEST ☐ QUALIFICATION ☐ ENVIRO. .IT

SOLVENT LEAKAGE DISCOVERED IN THRUST TERMINATOR PORTS No. 1 + 2

33. REPORTED BY: R. SMITH DEPT. & GROUP Q.C. PHONE 3477

32 A. ENGINEERING OR SUPERVISOR COMMENT AT TIME OF FAILURE: CRITICAL ☐ MAJOR ☒ MINOR ☐ ANALYSIS REQUIRED YES ☒ NO ☐ LIABILITY: RETEST ☐ MFG. ☐ ENG. ☐ Q.C. ☐ PROCUREMENT ☐

PORTS WERE WIPED CLEAN AT ONE HOUR INTERVALS FOR FOUR HOURS, HOWEVER LEAKAGE DID NOT DIMINISH

NAME B. JONES DEPT. & GROUP PRODUCTION PHONE 2487

REWORK & PRELIMINARY INVESTIGATION

34. FAILURE ANALYSIS PERFORMED YES ☒ NO ☐ | 35. FOUND SERVICEABLE BY REPAIR ACTIVITY YES ☒ NO ☐ | 36. FINAL DISPOSITION: ELIMINATED ☐ REPAIRED ☒ CONDEMNED ☐ | 37. REPAIRED BY: DEPOT ☐ MFG. ☒ FIELD ☐ | 37 A. OPER. TIME DURING REWORK & RETEST (TENTHS) ON-OFF CYCLES CL. HRS. MIN. SEC. N/A

38. PARTS REPLACED DURING REPAIR:

PARA. NO.	REWORK CODE	PART NAME	PART NUMBER	SERIAL NO.	REF. SYMB. NO.	DISC. CODE	LIAB.	SUPPLIER	ENVELOPE NO.	SUPPLIER CODE	MATL. CLASS	CLASS CODE	CC
		NONE											

39. REWORK MANHOURS 3.5 | RETEST MANHOURS N/A | TROUBLE SHOOTING MANHOURS 8.0 | 39 A. REPAIRED BY: NAME D. BROWN DEPT. & GROUP PRODUCTION PHONE 2488 | 41. DATE RECEIVED 20/2/6- | 42. DATE REPAIRED 20/2/6-

40. REMARKS:

WITH ENGINE IN HORIZ. POSITION ETHYL ALCOHOL WAS APPLIED TO T.T. PORTS BY MEANS OF SYRINGE, ENGINE WAS THEN PLACED IN VERT. POSITION AND PORTS ALLOWED TO DRAIN AND DRY.

44. QC ACCEPTED YES ☐ NO ☐ MRB ACCEPTED YES ☐ NO ☐ MRB REPORT NO.

45. EFFECT ON SCHEDULE LOSS OF 24 HOURS

46. F.A.R. NO. 537

43. FINAL DISPOSITION SIGNATURES: ENGINEERING RELIABILITY CONTRACTOR QUALITY AIR FORCE QUALITY

Fig. 3.1 An example of a completed failure report.

The form itself consists of one hard copy with several attached carbons or reproducibles. Information is reported on the form by means of both check-type entries and supplementary narrative. Most of the cell numbers are self-explanatory; those which are not need not concern the reader. However, since the form is quite detailed, and the amount of information which can be entered varies according to where it is used, the reliability department should prepare, as a guide for the initiator, a list of cell numbers and types of information that should be completed and entered on the form at his particular inspection or test station. The remainder of the information can then be filled in by a designated member of the reliability department. This procedure also has the advantage of tending to improve the consistency of reporting.

3.6 DISTRIBUTION OF FAILURE FORMS

As it is the intent to use the form from the R&D phase through the operational phase of the program, its distribution will vary according to the initiator of the form. There will generally be three organizations initiating the form:

1. The equipment contractor observing failures on his own equipment

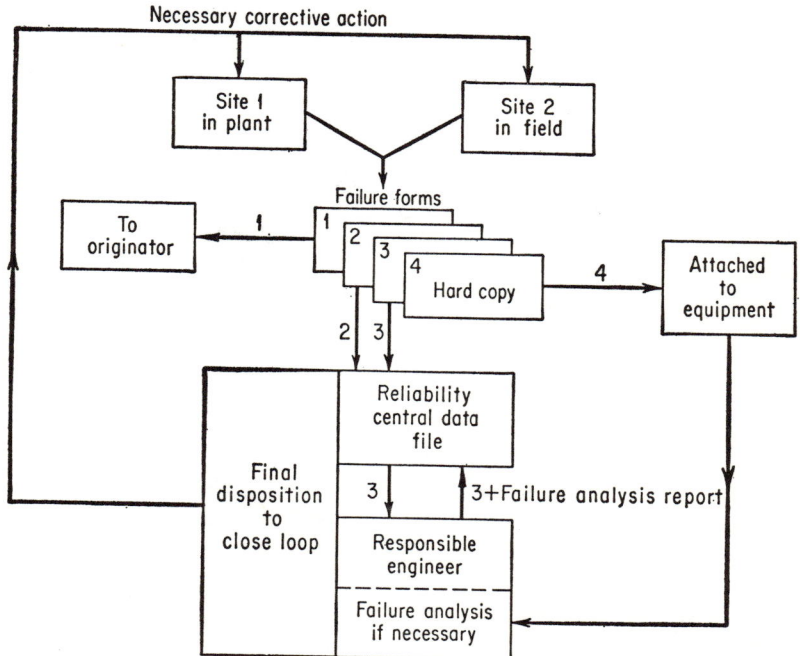

Fig. 3.2 Distribution of copies of failure forms initiated at major contractor's manufacturing and test sites.

at his own facilities both in-plant and in-field. The failure data flow is shown in Fig. 3.2.

2. The test or integrating contractor initiating the form on other contractors' equipment in-plant and in the field. The failure data flow is shown in Fig. 3.3.

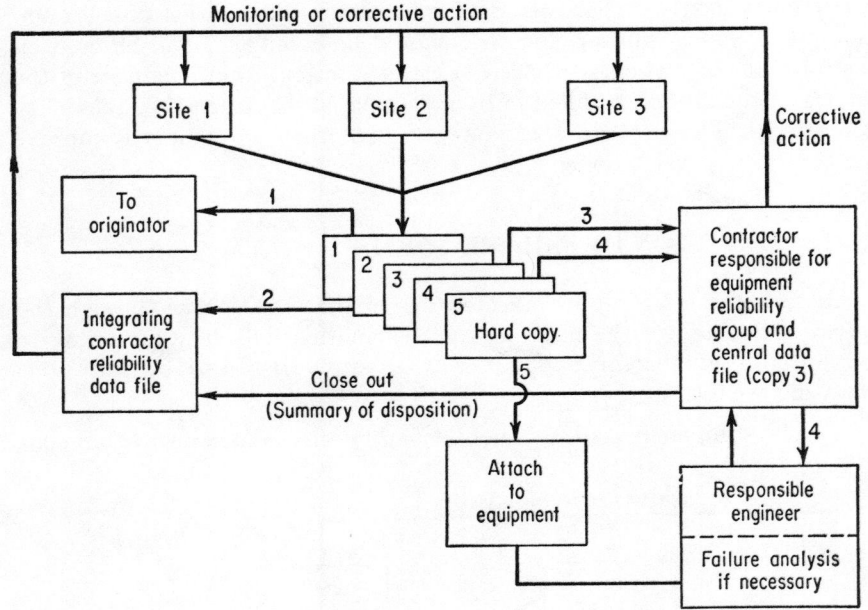

Fig. 3.3 Distribution of copies of failure forms initiated by integrating contractor for other contractor's equipment.

3. Service or operating personnel initiating failure forms at operational sites. The failure data flow is not shown for this situation; however, it will be very similar to that of the integrating contractor with an additional emphasis on the analysis of the data for logistics purposes.

3.7 FAILURE ANALYSIS INITIATION

As the failure forms are received, they are reviewed and/or completed by a member of the reliability group. This person has a good understanding of the complete system so that he is able to judge the consequences of a failure with respect to the system and establish its criticality. He also initiates the necessary corrective action. One method is to stamp the word *ACTION* over the failure form and submit it together with any other pertinent information to the responsible engineer. An alternative method

RELIABILITY DATA SYSTEMS 45

```
                    RELIABILITY ACTION REQUEST No. _____
  Date   1/20/6—                       Date of Required Response  1/25/6—

  To:       Name           Department      Location    Action   Anal.   Info.
       R. L. HIGGINS         DESIGN                     [X]     [X]     [ ]
       L. M. COX             PRODUCTION                 [ ]     [ ]     [X]
                                                        [ ]     [ ]     [ ]
  From                                                  [ ]     [ ]     [ ]
       C. B. RUSSEL          RELIABILITY  EXT 3478      [ ]     [ ]     [ ]
```

If "Action" is checked, return one copy of this form with suggested or
completed corrective action outlined on reverse to Reliability-Engineering
within ___ days. If "Action" and "Failure Analysis" are checked, make any
necessary laboratory examinations of the failed item, including any special
objectives as documented by the requesting reliability engineer on this form.

·PROBLEM DESCRIPTION
(Reference all pertinent history, failure reports, related cases, etc.)

See enclosed failure report LA 00498 Fig 3.1 .
There are no other previously known failures
of this type.

This Reliability Action Request conveys information regarding a problem
affecting weapon system reliability. It is intended to acquaint concerned
persons with the need for definite corrective action and to provide a means
for coordinating and recording such action. Further information may be
obtained from the reliability engineer originating this request.

Fig. 3.4 Reliability action request.

is to initiate a Reliability Action Request (R.A.R.) Form (Fig. 3.4). In each case a date for reply should be indicated.

The Reliability Action Request form can be used in other circumstances. The accumulation or increased frequency of a failure which is significant collectively, but not singly, would be one such circumstance. Other occasions occur after reliability design reviews when marginal areas and potential weaknesses are discovered; and again, after specifications are reviewed and interpreted against reliability requirements. In other words, a Reliability Action Request should be issued concerning any problem related to reliability including management procedures and scheduling.

3.8 FOLLOW-UP, ANALYSIS, AND CORRECTIVE ACTION

A positive form of control is *mandatory* to insure that action is being taken. The urgency and priority will vary from problem to problem, but the reliability coordinator should establish the status of the R.A.R. within three days of its issuance, or sooner, if it is significant. In addition to this, the reliability group issues a weekly report which lists the number of open problems, their relative urgency, and the length of time they

```
┌─────────────────────────────────────────────────────────────────┐
│ 1. General:          Failure Report No. LA 00498 [Fig.3.1] Attach. │
│    Facility  ABC     Reliability Act.Req.No. 187 [Fig.3.3]       │
│    Location  L.A.    Date of Analysis  10 Feb.196-               │
│    Page    1   of    1                                           │
├─────────────────────────────────────────────────────────────────┤
│ 2. Item Identification    Manufacturer    ABC                    │
│    Name  Case-Loaded      Part Number     1234 LV                │
│    Serial No. S103.21     Type or Model   Minerva A              │
├─────────────────────────────────────────────────────────────────┤
│ 3. History                                                       │
│        No previous analysis                                      │
├─────────────────────────────────────────────────────────────────┤
│ 4. Analysis Methods and Techniques                               │
│        Visual examination                                        │
├─────────────────────────────────────────────────────────────────┤
│ 5. Results and Conclusions and Actions                           │
│        The problem of solvent leakage from Thrust Termination    │
│        ports was investigated. The cause of the leakage was      │
│        determined to be improper fit and improper potting of     │
│        the Thrust Termination port flaps. The T.T. port flaps    │
│        have been redesigned and all leaks discovered during      │
│        the air test will be potted with a mixture of 60%         │
│        activator "W" and 40 % "C.7". The redesigned T.T.         │
│        port flaps will be available March 5,196-, and S.O.P.     │
│        179 has been revised to incorporate the new procedure.    │
├─────────────────────────────────────────────────────────────────┤
│ 6. Signature of Analyst  R. L. Higgins    Date 16 February 196-  │
└─────────────────────────────────────────────────────────────────┘
```

Fig. 3.5 Example of completed F.A.R.

have been open. This report is submitted to all at the supervisory level and is similar to the sample form in Fig. 3.11.

The closing of the loop takes place when the failure analysis is completed and a *failure analysis report* is returned to the reliability group failure form coordinator. An example of a completed form corresponding to the failure report form (Fig. 3.1) and the R.A.R. form (Fig. 3.4) is shown in Fig. 3.5. Not every failure report will require a detailed analysis but all failure forms must be closed out. Table 3.1 (pp. 48–49) illustrates types of failures, causes of failures, and necessary corrective action.

3.9 OPERATING TIME REPORTING

While failure and discrepancy reporting will initiate corrective action and thereby progressively reduce the failure rate, it does not cover the

RELIABILITY DATA SYSTEMS 47

complete story; i.e., it takes no note of success. The type of success information in which we are interested is how much time and/or how many cycles a functioning device might experience before failure. The circumstances under which it has been operating should also be known. The data system for collecting this information is called *operating time logging*. Operating time logging may be thought of as a tool valuable in anticipating or estimating reliability rather than as a tool for curing or preventing failures, as is the failure reporting system.

Whereas failure reporting is used for all components whenever a failure occurs, the operating time reporting system has application to only those systems deemed time- or cycle-critical. It is a continuous monitoring system; consequently, although it is a vital activity, it is expensive and therefore should be applied to a carefully chosen number of components.

3.10 THE TIME LOG

The basic document is the *operational time log*. There is a separate log for each major assembly. Each assembly usually contains a number of time-critical components, some of which may be replaced during the accumulation of operating time on the assembly. We therefore require a method which will keep track of the operating time of assemblies as well as histories of those components which have or will experience environments different from and in addition to the environment of the assembly. For instance, such information as the amount of bench testing, component qualification testing, etc., that the component may have received before assembly is important. A form for keeping this record is shown in Fig. 3.7.

The first page or portion of the log contains the reference information such as assembly number, type, model location, and designated system number, if known. The next section contains a failure history summary of the assembly. A sample page is shown in Fig. 3.6.

Failure	Component Type	Component Serial No.	Time on Assembly	No. of Cycles on Assembly	Failure Report No.	Environment Remarks
1st						
2nd						
3rd						

Fig. 3.6   History of assembly.

The entries under "Time on assembly" and "No. of cycles on assembly" are cumulative time and number of cycles.

The remainder of the log contains the histories of each of the time-critical components which are or have appeared on the assembly. Figure 3.7 illustrates a typical entry for a particular component type.

Table 3.1. Examples of Reported Failure, Analysis, and Corrective Action

Component, Part, or Assembly, Name and Number	Description of Failure	Cause of Failure	Corrective Action
Helium Start Valve 2-456792 YLR42-RE-4	Low pressure decay rate.	Pressure trapped in spring cavity prevents start valve opening fully.	Hole drilled in pintle shaft to vent pressure. Poppet face chamber lengthened to permit opening of valve sooner. Addition of .008 inches extension to stroke of solenoid for greater opening force.
TPA Fuel Pump Seal, 3-439491A XLR45-RE-4	Fuel pump seal bellows cracking in convolute welds.	Poor welding techniques.	Improved control of fabrication procedures has been instituted. Redesign of the seal to assure bellows concentricity and a positive deflection on the seal test rig.
Valve Assy., Motor Oper., Helium M/B 34-01114-27	S/N's 109, 907-0060 failed to open. S/N 398 failed to close.	Valves S/N 109 and 398 failed because of a corroded motor lead. Valve S/N 907-0060 failed because of a failure of micro switch No. 2.	A protective cover has been placed over the valve to prevent moisture from entering the valve, causing corrosion. Missile instrumentation has been installed to determine electrical loads on the valve switches.
Rate Integ. Gyro 9-12345-0	Spin motor draws excessive current.	Spin motor stator shorted to cover.	Vendor is performing over-voltage tests on stators, has decreased maximum stator turn height and changed gages used in measuring stators, and re-educated factory personnel in soldering techniques.

RELIABILITY DATA SYSTEMS

Solenoid Valve 31-07398	Solenoid valve failed to operate.	Improper plating and lubrication on the solenoid plungers.	Process control specification has been requested to insure proper plating and lubrication of the solenoid plungers.
MM Battery	Battery failed to activate on signal.	Inadequate squib power, caused by an insufficient explosive charge.	Vendor now fluoroscopes all squibs to determine the presence of adequate charge. This unit built prior to above procedure.
Marman Clamp 32403-130	The clamp "T" bolt failed.	The "T" bolt was overtorqued to failure.	The amount of torque has been reduced from 220 to 150 inch-pounds for present clamp installations.

Note: This information is abstracted and summarized from the *failure reports* and their corresponding *failure analysis reports*.

Component Name	History Prior to Installation in Assembly	Experience dur. First Operation of Assembly	Experience dur. Second Operation of Assembly	Experience dur. Third Operation of Assembly	Total Experience
Component Part No. I	Serial No.1234 Total time: 180 sec. Total Cycles:1 Total Failures:0 Environment: Production Environmental Bench Test	Time:160 sec Cycles:1 Environment: Ambient	Time:56 sec Cycles:0 Failure:1 Failure Rpt. No.7683	Removed for failure analysis. Replaced by component S/N 2345	Number of Components:2 Total Time:556 Total Cycles:3 Total Failures:1
	Serial No.2345 Zero	Time:160 Cycles:1 Environment: Humidity Test			
Component Part No. II	(Similar to Above)				

Fig. 3.7 Example of an operational time log.

3.11 A METHOD FOR COLLECTING TIME INFORMATION

The above mentioned information, which is required on each individual component as it experiences the various environments of bench checkout, environmental testing, testing in the assembly, rework and retesting, etc., is collected on a card which is attached to the component and is brought up to date each time a failure or change of location (environment) occurs. When the component is installed into an assembly, the cards are then taken off the components and placed inside a stiff envelope which is attached to the assembly for this purpose. The front of the envelope might contain the first two sections of the previously described time log; the back of the envelope might include the details of the component experience, as shown in Fig. 3.7. Thus, when a component is replaced or extracted from the assembly, its card is replaced or extracted from the envelope, the information being updated on both envelope and card.

Since the information contained on the time logs is for a continuing reliability evaluation (in general it would become an action item only by virtue of the failure reporting system), it is not necessary that a specific data flow system be discussed. The information is reviewed periodically by the groups responsible for the reliability of the components and assemblies for appropriate design improvement. An example of the type of summaries obtained from this information is given in Fig. 3.10. Methods described in other sections of the book indicate how reliability estimates can be computed from these data.

3.12 THE RELIABILITY CENTRAL FILE

As the information accumulates from such various sources as routine failure reports, operational time logs, environmental tests, engineering development tests, it is organized to provide a background of experience and knowledge for design engineers and reliability engineers. For each component type that has experienced failure under normal or other than normal stress conditions a card is established in the file which lists, where possible, the following information:

1. Critical design or manufacturing features
2. Modes of failure
3. Causes of failure
4. Stresses at failure
5. Methods of detection or test
6. Recommended necessary preventive or corrective action
7. Types of failure distributions
8. Estimates of reliability for various applications
9. Prime manufacturer and alternate sources

Although much of this information may be well known to the personnel closely associated with the equipment, it represents a valuable source of knowledge of reliability problems, design features, etc., for present/future systems using identical or similar components under similar or quite different conditions.

3.12.1 EXAMPLES OF RELIABILITY INFORMATION IN THE CENTRAL FILE

Sample reliability information sheets are shown on pages 52–54.

3.13 PRETEST DECLARATION

For the sake of completeness, although this subject is covered by examples in more detail in Section III, it is appropriate to discuss methods of insuring the validity of data. This is particularly necessary in the early phases of development when much experimentation is being performed. We do not wish to confuse failures occurring on production components under operational conditions with failures on components which are experimental or are being tested under non-normal conditions—e.g., surveying input and output limits, environmental extremes, experimental procedures, etc. Thus, it is imperative that the circumstances which differentiate any test from a normal routine test be known. This can be

Sample Reliability Information Sheet

Assembly __Rocket Nozzle__ Vendor _____ Vendor P/N _____

Drawing No. _____ Specification No. _____

Next Higher Assembly __Thrust Vector Control System__ Subsystem __Propulsion__

Design Feature:
Lightweight design to operate 100 sec without ballistic performance failure

Critical Characteristics:
Movable nozzle, split-line gap area

Mode of Failure:
1. Jamming or high torque due to differential expansion in split-line gap area
2. Jamming or high torque due to propellant particle fusing in split-line gap area
3. Cracking of throat inserts due to differential thermal expansion and material restraint
4. Throat insert slippage or ejection
5. Erosion of throat
6. Burn-through of exit cone

Failure Experience:
1. High torque values stalled actuator
2. Heavy erosion of graphite throat insert
3. Cracking of throat insert
4. High-temperature gas flow melted throat insert
5. High-temperature gas flow melted wiper ring producing high torque values

Control Methods:
1. Proper selection of design and material with correct thermal expansion properties
2. Protect gap by wiper rings and "redundant" seals, and modify design of propellant for different flow characteristics
3. Proper selection of materials
4. Choice of design for firmer "anchorage" of throat insert
5. Selection of material to provide erosion barrier
6. Higher quality control standards for graphite cloth on surface of exit cone

Alternate Concepts:
Nonmovable nozzles using jetavators
Nonmovable nozzles using fluid injection

Note: This example is for illustration only and does not attempt to be comprehensive in its information.

RELIABILITY DATA SYSTEMS 53

Sample Reliability Information Sheet

Assembly ___Automatically Activated Primary Battery___ Vendor _____ Vendor P/N _____

Drawing No. _____ Specification No. _____

Next Higher Assembly ___Airborne Electrical System___ Subsystem _____

Design Feature:
High-output, silver-zinc primary battery automatically activated by pressure system, low weight

Critical Characteristics:
Stand-by time, activation time and successful energization of battery

Mode of Failure:
1. Failure to energize
2. Slowness in energizing
3. Insufficient voltage outputs due to temperature effects
4. Leakage
5. Too short stand-by time

Failure Experience:
1. Low voltage output during cold tests
2. Failure to meet energizing time or to energize
3. Burst cases
4. Caustic leak at the seams

Control Methods:
1. Thermostatically controlled heater to maintain battery temperature between specified limits
2. Addition of insulation to battery case
3. Method of pressurization is compressed gas, piston and gas generator, and design to reduce internally generated pressure
4. Use of inert materials to prevent caustic leak

Alternate Concepts:
Energization by gravity flow
Energization by evacuation of battery cell area

Note: This example is for illustration only and does not attempt to be comprehensive in its information.

Sample Reliability Information Sheet

Assembly Vent and Relief Valve   Vendor _____   Vendor P/N _____

Drawing No. _____ Specification No. _____

Next Higher Assembly _____ Subsystem Pressurization

Design Feature:
  Butterfly valve with pneumatic control

Critical Characteristics:
  1. Leakage
  2. Crack and reseat band
  3. Ice contamination at low temperature
  4. Quality of diaphragm material

Mode of Failure:
  1. Leakage
  2. Improper crack and reseat
  3. Inoperative due to freezing or galling

Failure Experience:
  1. Leakage
  2. Will not crack and reseat within specification
  3. Inoperative due to icing, galling, and foreign matter

Control Methods:
  1. Certification of material
  2. Control of diaphragm manufacturing
  3. Cleanliness control
  4. Dimension and surface-finish control

Safety Margin:
  Proof pressure is 1.3 times operating pressure

Estimate Failure Rate:
  MTBF—25 cycles

Required Action for Higher Reliability Requirements:
  1. Investigate use of different diaphragm material
  2. Investigate a new seal design
  3. Investigate an integral filter
  4. Tighten control over manufacturing and assembly
  5. Reorient valve to eliminate condensation problem

Note: This example is for illustration only and does not attempt to be comprehensive in its information.

RELIABILITY DATA SYSTEMS 55

done by describing the intent of the test and listing the deviations on a *Declaration Form* (Figs. 14.1, 16.1) prior to the initiation of the test. It should be emphasized that the purpose of this procedure is not to question the integrity nor limit the activities of the development or test engineer but rather to help him in discerning the effects of the experimentation. Conclusions about the data will consequently be more valid and reliability estimates more realistic.

3.14 PERIODIC RELIABILITY AND FAILURE SUMMARIES FOR PROJECT MANAGEMENT AND ENGINEERING

These summaries are issued on a periodic basis, usually monthly, by the reliability group. They inform engineering and management of the reliability status of a system or component with regard to meeting its end requirements or some intermediate milestone. For instance, one report might indicate the probability of a completely successful operation—i.e., of performing within specification limits; whereas another estimate might give the probability of noncatastrophic-failure type success such as the ability to fly without exploding—i.e., without requiring the performance specification limits to be met. Each of these estimates will be obtained by applying a different set of ground rules. Care must be taken to see that this is clearly understood by the recipients of the reports. The reports should be sufficiently comprehensive without presenting too much detail, and they must provide sufficient information to permit valid estimates of reliability to be computed. They also should provide sufficient identification so that the test classifications can be questioned and if necessary reclassified upon further investigation.

Examples of various types of reliability summaries are shown in Figs. 3.8, 3.9, and 3.10. Figure 3.8 reports a summary of the behavior of the major subsystems tested within the system. Figure 3.9 reports the behavior of a typical major subsystem tested separately and as part of the complete system. Figure 3.10 reports a typical component failure summary.

1	2	3	4	5	6			7	8	
Subsystem Name	Missile or Serial Number	Location of System	Failure Discovered During	Operating Time or Cycles	Total Number of Failures			Starts This Month	No. of Start Failures	
					Crit.	Major	Minor		Crit.	Major
Control	B-5	AR	C'Down Other	10Hr 20Hr	1 0	2 3	4 12	16 18	2 1	1 0
	B-6	A6P/AR	C'Down Other	40Hr 36Hr	2 3	3 3	5 16	22 13	1 0	0 0
Airframe	B-5	AR	C'Down Other	X X	0 0	1 0	2 3	X X	0 0	0 0

Fig. 3.8 System failure summary report (sample).

1	2	3	4	5	6	7	8	9	10	11	12	13
					\multicolumn{2}{c	}{Run Duration}		\multicolumn{4}{c	}{Failures (including those discovered after run)}	Failure Description, Remarks and Exclusions
Engine Serial No.	Test Run No.	Test Stand No.	Test Run Date	Results	Scheduled	Actual	Environmental Temp., °F	Time	Part Name	Report No.	Failure Type	
02014	546	E-1 B-9	3	S	8/12/12	9/12/13	60					Acceptance Test-Retrofit Engine
02014	549	E-1 AR	5	S	8/10/10	8/10/12	65					
02014	552	E-1 AR	9	F	7/156/66	7/156/4	85	167	G.Line Bkt.	0189	Eng.	Loose B Nut on Line to GG Fuel Purge OK. Valve Caused Fire & Loss of Instrumentation.
02019	513	G-4	2	E	7/40/20	0/0/0	75	0		0124	Human	-77FS1 Inadvertently Pressed before Toj GG Reached Spec. Temp.(Human Error)
02019	516	G-4	20	E								Declared Inapplicable;Limits Survey

Fig. 3.9 System failure summary report, propulsion systems (sample).

1	2	3	4	5	6				7	8	9		
					\multicolumn{4}{c	}{CURRENT MONTH}			\multicolumn{3}{c	}{LAST SIX MONTHS}
Subsystem	Component	Part Number	Failure Discovered During	Operating time or cycles	\multicolumn{4}{c	}{Number of Failures}	No. Units Operating	Operating time or cycles	\multicolumn{3}{c	}{No. of Failures}
					Crit.	Major	Min.	Total			Crit.	Major	Min.
Control	3-Axis Ref.	P583060 0010-1	PLT-ENV PLT-ROOM FLD-OTHER FLD-C'DOWN	142H 168H 126H 120H	1 0 2 0	3 0 4 2	10 0 4 14	14 0 10 6	4 16 4 2	372H 2261H 521H 214H	15 12 12 5	26 26 18 10	35 62 34 16
Rate Gyro	Rate Gyro	P583060 001D-5,-6	PLT-ENV PLT-ROOM FLD-OTHER FLD-C'DOWN										
Rate Gyro	Rate Gyro	P583060 001D-7, -8,-9	PLT-ENV										

Fig. 3.10 Component failure summary report (sample).

Reliability Problem Status Report

Open Problems at Start of Month _____						Date _____			
New Problems Added _____						System _____			
Problems Closed This Period _____									
Net Problems Still Open _____									
Average Age of Open Problems _____									

Problem Number	Comp., Part, or Assembly No.	Failure Report No.	Date	Description of Problem	No. Critical Failures		Corrective Action	Scheduled Effectivity	Problem Closed
					This Month	Six Months			

Fig. 3.11 Reliability problem status report.

A fourth report which is of major importance is the Reliability Problem Status Report (Fig. 3.11). This is a report similar to the one that is sent to supervisory personnel monthly (Sec. 3.8) but lists only critical problems.

3.15 PERIODIC RELIABILITY STATUS REPORTS FOR TOP MANAGEMENT

These reports are much more limited than those discussed in Sec. 3.14, upon which they will generally be based. In addition to the current estimates of reliability and numbers of critical failures, the previous periodic estimates should be either tabulated or, better, plotted to illustrate the relative growth of reliability towards the program reliability milestones.

3.16 SPECIAL RELIABILITY REPORTS

The reports so far discussed are issued on a routine basis; however, the reliability group has further responsibilities. There are many special investigations which must be made and reported upon. Such items as the study of the failure patterns, correlation between field and test site experiences, reliability design reviews, recommendations for design changes, paper studies on proposed new systems, and so on—all these analyses are issued in addition to the routine reports. In other words, reliability reports are not restricted merely to a passive recording and organization of events and information but should include more positive activities and constructive analyses.

In conclusion, the reporting methods and procedures described in this chapter have been developed to satisfy the needs of a specific program. The particular formats, procedures, and situations quoted may not apply to any one given organization; nevertheless, similar ideas and the same basic principles will generate analogous requirements for most reliability and failure data reporting systems.

ADDITIONAL READING

Culbertson, J. E., and Vorhees, H. A., "Control Charts and Automation Applied to Analysis of Field Failure Data," *Proc. 2d National Symposium on Quality Control and Reliability in Electronics, Washington, D.C., January 9–10, 1956*, 18–46 (available from Institute of Radio Engineers, 1 E. 79th St., New York 21, N.Y.).

Dertinger, E. F., "Quality Control and Reliability in Guided Missile Production," *Rutgers Quality Control Conference Proceedings, ASQC, 1956*, 85–99.

Eaton, W. R., "Achieving Reliability Through Integrated Quality Control," *American Society Quality Control, 4th All-Day Conference, Dayton Sect.*, March 5, 1960, 13 pp.

Matosoff, H. I., "Corrective Action in a Quality Control Program," *Industrial Quality Control*, **12,** 8–12 (January 1956).

Warner, W. K., "Benefits of Time Recording for Producer and Consumer," *National Convention Trans., ASQC, 1959*, 597–602.

SECTION II
THE MATHEMATICS OF RELIABILITY

It has been stated that our knowledge about an item can become significant only if we can make quantitative measurements of that item. Unfortunately, the reliability of a device is not a characteristic which can be measured in the same way that performance or physical dimensions can be measured. However, reliability can be put in a quantitative form. This means that we can improve on statements such as "the item is very reliable" by, for example, such statements as "we are 95% sure that its reliability is greater than 90%." We are able to associate a number with the quality of reliability by processes of estimation. Mathematical statistics and probability theory allow us to do this by relating observed, past events to future, similar events by means of probability statements. We can then use these reliability estimates as though they were exact measurements, and therefore we can make decisions as to their adequacy just as we would use any other parameters for this purpose. Thus the theory of reliability estimation forms an exceedingly important part of reliability methodology.

However, the role of mathematics in reliability is not only as a measuring device. It is also an important analytical tool which can be used with great effectivity both in the evaluation of the basic design and for improving the efficiency of the development and qualifying test programs. We shall see, by the use of examples, how the reliability of a system can be increased simply by arranging its components according to certain configurations based on mathematical relationships, i.e., by using the techniques of redundancy. We shall also discuss statistical experimentation and sampling, and we shall note how sample sizes can be reduced and how at the same time we can derive more valid and informative data. In a research and development program one of the major products is information. If we can reduce the cost and time involved in its generation, improve its validity and thereby lessen the probability of wrong and costly decisions, then an important contribution will have been made.

CHAPTER FOUR
RELIABILITY MODELS AND ANALYSIS: A NONMATHEMATICAL INTRODUCTION

This chapter serves as an introduction to the subsequent chapters in which we shall discuss particular topics in statistical methodology which have applications in the field of reliability.

The importance of obtaining complete and valid data has been stated in Chaps. 2 and 3. In the chapters which follow we shall see how to interpret the data statistically and discuss how reliability techniques may be used to benefit a development program.

4.1 TYPES OF DATA

The first concern is the type of data upon which reliability predictions and analyses can be based. We see from the definition of reliability that we are trying to estimate the probability of a certain event occurring. We can assume that the event itself can be well defined even though its definition might be somewhat arbitrary. For example, a certain required event might be that the operating time of a given device be greater than T seconds, when the device operates in an environment with temperature exceeding $\theta°F$. Or the event might be that of impacting a missile within a specified radius of the target, and so on. If we are able to observe a sample of devices experiencing exactly the event in which we are interested, then it is relatively simple to make a probability statement concerning the whole population—providing that the sample is representative of the total population.

Consider the type of data which might be recorded for these two events. For the second example, our data might simply be numbers of successes and failures—the number of times the missile impacted within or missed the target area. On the other hand, the data from the first example might be in a more quantitative form; e.g., the equipment failed after 500 seconds after it had operated successfully while exposed to a −60°F. temperature.

How we use this information will depend on the *mathematical model* that has been chosen to describe the behavior of the device and upon how the model is related to the required behavior for a future successful event. It should be noted that although the above two types of information are similar, there is a distinction which can be very important.

Data in the form of "success" or "failure," giving no information other than the fact that the device passed or did not pass the success (or failure) criteria, are called *attribute data*.

Data giving additional specific information—e.g., such as successful operation for a measured length of time—are referred to as *variables data*. Occasionally, we may wish to ignore the additional variables information or to use it solely for classifying the test as a success or failure (i.e., as attribute data). Though this step may be undesirable, as wasting information, it is sometimes necessary. For instance, we may not know the form of the mathematical model which enables us to relate the variables information to reliability information. Another reason may be that there are several variables being measured, not necessarily independent, whose functional interrelationship is known only approximately. If more than one of these variables is indicative of reliability, then it might be impossible or imprudent to utilize the detailed information simply because of the complexity of the mathematics. In this case we might require all of the variables measurements to fall within prescribed limits before classifying the test as an attribute success.

However, if we are reasonably sure, by experience or from the data itself, that the variables data follow a known, preferably simple, mathematical form, we can advantageously make use of this knowledge in estimating reliability because such a model permits more precise estimation. This is intuitive, but it also can be shown mathematically that the use of variables data leads to a more efficient utilization of the data. This may be very important if experimentation is costly and sample sizes are limited; however, it must be emphasized that if evidence is lacking to show that in fact the data do follow the chosen model, then any apparently precise estimates may be vitiated. Reliability estimates obtained from attribute data do not have this drawback, but it is difficult to make general rules whether to use attribute or variables data.

Chapter 5 contains a review of some of the elements of statistics and probability theory as they will be used in later chapters. The important concepts of sample space, random variables, and distribution functions are presented as are also the probabilistic relationships and notations which provide the mechanics for the algebraic manipulations of the various statistical models.

4.2 STATISTICAL DISTRIBUTIONS

The form of the data will generally determine which of two classes of statistical probability distributions will be used for reliability estimation. These two classes are *discrete distributions* and *continuous distributions*. If reliability is considered as a parameter, generally it will not appear *explicitly* in the distribution. One exception would be in the *binomial distribution* where $q = (1 - p)$ would represent the probability of success in a single trial. This would be the direct measure of reliability if a *single* successful trial represented the event of interest. In general, therefore, reliability can be considered as *some function of the parameters of the mathematical model*, and we are faced with several steps before any reliability estimate can be obtained. The first step is to establish the type of statistical distribution which describes the failure phenomenon. The second step is to estimate the parameters which completely define the distribution. The last step is to utilize the knowledge of the statistical distribution together with the estimates of its parameters to obtain reliability estimates.

Let us consider these steps in greater detail. The first problem is to determine the form of the model, that is, to establish the general failure pattern. For instance, we can imagine that if we were able to plot some function of the whole population of observations, or of a very large sample, we would obtain the failure curves which describe the behavior of the device. There are a large variety of functions for which the failure curves could be drawn—for example, number of failures versus length of operating time, or number of failures versus number of cycles, or number of failures versus level of stress. Also there are many ways in which the data might be sampled. Thus observed data might be times-to-failures, number of failures in number of trials, number of successes before the first failure, total time before the first failure; etc.

It is important to note, therefore, *that in addition to the actual failure characteristics of the device, the method of sampling for the observations also determines the failure model*. This statement has two implications. As an example of the first implication, regarding the failure characteristics, the reader is acquainted with the fact that so-called "chance failures" frequently follow an *exponential probability distribution* whereas "wearout failures" can often be represented by the *normal distribution*. As an example of the second implication, concerning the method of sampling, a binomial distribution model might be used to estimate reliability from data such as "one failure in n tests," but it would not be used if it were known that the above data were obtained by testing until the first failure occurred. The reader will appreciate that these observations, although giving similar results, are from two completely distinct experiments. In this latter case the *geometric distribution* (Sec. 6.4) would be used as the failure model.

The determination of such models may be from *first principles*, that is, by mathematical formulation of the probability of the event being sampled for, or by *empiricism*, in which the observed failure phenomena for that particular type of device had been noted as having occurred in similar patterns previously. However, whether the model is theoretical or empirical, the observations of current data must be examined to determine that they approximate the chosen model. Inspection of the data might reveal that it had been obtained carelessly or that the previously designed sampling procedure had been changed. Under these latter circumstances the model, too, might have to be changed, either to fit the data or to more validly describe the modified event being sampled.

It is not within the scope of this book to discuss curve fitting to statistical distributions, but familiarity with the failure models presented in Chap. 6 will allow, by inspection, a narrowing down of the potential candidates to describe the observed or expected failure phenomena. In those situations where the model cannot be formulated from first principles or there is no particular reason to know that the data follow a specific pattern, we can arrive at a model describing the data only by a process of trial and error. It might be possible to reduce this task by short cut methods such as plotting the data on special graph paper or transforming the data first, but it must be recognized that only a substantial amount of data will verify that the indicated model is indeed truly representative. Of course, this is a problem which is general to all data-curve fitting situations and is not peculiar in any way to the field of reliability.

The most used probability distributions contain one or two and occasionally three parameters which completely define the distribution's shape, location, and other characteristics. Since the properties of these probability distributions are generally well known and frequently tabulated, we can make inferences from the observed data based on our knowledge of the distributions. Consequently, the most useful statistical probability distributions are examined, and their properties and applications to reliability estimation discussed in Chap. 6.

Once the general form of the model is established, i.e., whether it is a *Weibull distribution, gamma distribution, Poisson distribution*, or some other distribution, we reach the second of the problems mentioned earlier—that of completely determining the distribution by means of its parameters. This brings us to the subject of estimation and the statistical techniques associated with it.

4.3 UNCERTAINTY OF OBSERVATION

When we attempt to estimate any parameter by observation or experiment, we accept the fact that there is some error or uncertainty associated with that estimate. For example, with attribute data, if N independent trials

RELIABILITY MODELS AND ANALYSIS 65

are made with the result 9 successes and 1 failure, an *estimate* of the true probability of success, R, is 0.9. However, the *true* probability of success could conceivably be any number between zero and unity; e.g., if R were really 0.1, the result would still have had a small, but finite, probability of occurring. However, it is *likely* that R is near 0.9; we would very likely be correct if we stated that R is greater than, say, 0.75 or that R lies between the limits of, say, 0.75 and 0.95. Similarly, consider some variables data. If N independent tests to failure are made and an average time to failure of, say, 600 seconds is estimated, then it is *likely* that the *true* mean time to failure is close to 600 seconds. In addition, the larger the value of N the greater would be the confidence in the estimated mean-time-to-failure as being representative of the true value.

We are also concerned, therefore, with *precision* of the estimate. Knowledge of the statistical distribution involved and the general theory of sampling statistics allows us to associate *confidence limits* with the estimates of the distribution parameters.

4.4 STATISTICAL ESTIMATION AND RELIABILITY ESTIMATION

In Chap. 7 we discuss types of estimators for the distribution parameters and desirable properties of these estimators. In addition, it is important to discuss the various statistical techniques of obtaining estimators for the parameters in terms of the sample observations.

The *reliability function*, being defined as the probability of the event *success*, is obtained by summation or integration over that portion of the *sample space* which defines the occurrence of success. The reliability function depends on the distribution parameters. Since the parameters will be generally available only as sample estimates of measurable variation, it becomes necessary to examine how these estimates and their variations are used to obtain reliability estimates and confidence limits. Thus parameter estimation and its relationship to reliability estimation forms the main topic of Chaps. 7 and 8, where it is discussed with applications to the distributions described in Chap. 6.

4.5 RELIABILITY STRUCTURE MODELS, REDUNDANCY, AND APPORTIONMENT

We have so far been referring to the reliability estimation for a single item. Whether this item was a single component or a complex of components, we have been concerned with it only as an entity and in terms of its behavior as a unit. However, we are also interested in the interrelationship of component reliability and its effect on the system's reliability. If we have a system consisting of a number of components of known, or essentially known, reliability, it is important to consider how the components interact to affect system reliability. For instance, if we have a system consisting of two components with true reliabilities R_1 and R_2, what is the reliability of the system? Is it equal to the product of R_1 and R_2, the average of R_1, R_2, the minimum of R_1, R_2, or some other function of R_1 and R_2? The answer will depend on the functional relationship of the components and the probabilistic effect of their behavior on the system. This relationship is the system's *reliability structure*. Chapter 9 presents various types of structures. Knowledge of the structure of the system enables the reliability group to make important analytical contributions to the design and development program. One immediate example is the concept of *redundancy*—the idea of having several components available as part of the system design to act as "back-up" in the case of failure of one or more of the original components. An example might be the use of two squibs for the ignition of a rocket engine when only one is known to be necessary for ignition. Thus the structure model will enable us to compute for the design engineer the probability of a successful ignition or alternatively establish the number of redundant squibs necessary to ensure ignition with a required probability of occurrence.

An analogous problem exists with variables data. Consider the situation in which the output of a device must operate above a certain fixed limit for success. Although the design output may be greater than this limit, generally there will exist an inherent variability of the output which could on occasion result in the output falling below the required level and therefore produce failure. The probability of this occurrence can be computed and therefore so can the probability of success. In addition the reliability group can establish the output level versus the degree of variation which would be necessary to assure successful performance with a required probability of occurrence.

In this latter situation we have a fixed level beyond which failure occurs if the output parameter exceeds this critical stress. However, the level need not always be fixed. Alternately, the critical level itself also may be a random variable with a statistical distribution. An example would be the variation in the chamber pressures produced by solid propellant rocket engines compared with the variability of strengths of cases as measured by burst pressure. If the stress caused by the pressure exceeded the strength of the case, then failure would occur. The use of reliability *stress versus strength analysis* provides the reliability engineer with a method of estimating the probability of failure and of furnishing the design engineers with desirable levels of operation and variability for design criteria, in order to prevent unsatisfactory failure rates.

In addition to the aforementioned type of design information the knowledge of the reliability structure is inherent in *reliability apportionment analysis*. The reliability group must be able to interpret the system's reliability requirements in terms of component reliability requirements for the design engineer. The first set of reliability numbers generated this way would be reviewed against the state of the art, development costs, time, and feasibility (using some of the design review techniques just mentioned) and perhaps subsequently modified to obtain an optimum apportionment. In later phases of the program as difficulties manifested themselves the apportionment would be reviewed in the light of the changed circumstances. We shall see by an example how by applying greater proportions of the development effort to certain components we can "buy" a more rapid system reliability growth.

One problem with which we are frequently faced is that of *combining results of component tests to obtain estimates of system reliability*. An answer to this question is provided for certain types of structures in Chap. 9. There it is shown how component test results can be artificially synthesized to produce system reliability estimates when the component results are obtained under conditions which would validly represent their environment in the system. This would not always be the case. For example, if the results were obtained from separate component tests (see Sec. 4.6) under conditions which did not sufficiently simulate the system environment, then the component test results could not be combined by any relatively simple method to give valid system estimates. However, in the development of a large complex system, testing will be frequently carried out on a system which at different times contains many representative major subassemblies even though the total system cannot be considered as being representative in any one test. In this case the information generated by the subassemblies within the system tests can be artificially combined to give system reliability estimates. This situation is described in Chap. 14 where the results of Chap. 9 are applied to a development program of a large, solid-propellant rocket engine.

4.6 COMPONENT INTERACTION

On the other hand, when the results are obtained from separate component tests, there is not only the statistical problem of combining the test results but there is also the problem of validly interpreting the data. For instance, a whole assembly or groups of components can have properties not attributable or expressible as a function of similar properties of their separate components. For example, two systems, though similar or even identical in number and kinds of components, may be "packaged" differently. In one system, the influence of external vibration will result in considerable amplification of internal accelerations and cause failure in components not constructed to withstand this environment. In the other system, the same environment does not have the same effects because carefully designed supporting structures for the components damp out accelerations at frequencies met in the normal vibration environment. A similar type of interaction would be the situation in which two components may, when put together in the system, create an environment (heat vibration, etc.) which catastrophically fails either one or both and therefore the system, even though each component can separately perform its required function.

A good example of this last situation is shown by a hot gas driven turbine power supply. The hot gas generator performs its function by generating hot gases in a combustion chamber and supplying them through a nozzle to drive a turbine wheel. The turbine wheel is on a shift with an alternator delivering a-c electrical power. Separately, the gas generator shows that it can perform its function by producing an output of sufficient gas mass velocity. Also, separately, the turbine can be driven by *cold* gas from a pressurized source. However, when the two subsystems are put together, hot gases from the gas generator can penetrate shaft seals and destroy lubrication on bearings, with the expected consequences. Obviously, the turbine-power drive subsystem must be designed to be insensitive to this type of interaction.

Other types of interaction can be "enhancing" rather than "degrading" as the above examples indicate. Slight off-performance of one component can increase the chance that another will operate satisfactorily, or may result in a less severe environment. An interesting discussion of enhancing interactions in electronic equipment is given in Ref. 1.

If the attempt is made to predict or evaluate reliability of a given complex equipment, one should, if possible, evaluate the interactions present since these may have a preponderant effect on over-all system reliability. To begin with, it is recommended that the system be designed with the idea of "desensitizing" components to degrading interactions as well as of utilizing the enhancing ones. However, the problem of reliability apportionment and prediction can be very complicated mathematically when one wishes to allow for interactions in the model to be used. A discussion of the complications that might be involved is given in Ref. 2. A simplified model, which has been successful for many purposes, is the so-called *independent serial system model*, in which the over-all reliability is calculated as the product of the reliability of the components. This can be regarded as a first approximation to a more refined model which takes into account the above interactions. Furthermore, in view of the above recommended "desensitized" design, which implies a sort of independence or lack of interaction, the independence assumption becomes more reasonable with well-designed equipment.

4.7 SAMPLING AND RELIABILITY DEMONSTRATION

A frequent cause of misgivings in the minds of development engineers and management is an implication that "reliability" demands an excessive number of tests for reliability demonstration.

It is true in many instances, when high reliability is required, that a large number of tests or cycles or considerable time might have to be experienced by the device before we have the necessary statistical assurance. On the other hand, sole reliance on engineering intuition can also be very expensive. For example, the cost of demonstrating the reliability of an explosive bolt is small compared with the cost of the consequences of its failure in a missile staging operation. It is therefore of considerable importance that ample thought be given to test planning and objectives, not only on a test-to-test basis but as part of the over-all development program.

The first function of a test is to derive valid information for subsequent feedback and, if necessary, design improvement. Techniques for optimizing this activity are discussed in Chaps. 12 and 13. However, at various phases in a development program it is necessary to run demonstration or evaluation programs under rigidly controlled conditions to obtain a true picture of both performance and reliability unclouded by speculative experimentation with test procedures, conditions, and nonrepresentative configurations. In some cases these monitoring phases will justify a formal reliability demonstration program. Table A.1* illustrates that the minimum number of tests may be quite large before the higher levels of reliability can be demonstrated with a reasonable confidence. However, this table represents the basis for only one type of sampling plan, that of making N tests without failure. Chapter 10 examines various types of sampling procedures and discusses practical considerations which might lead to the choice of one plan over another. Detailed discussion is limited to plans based on binomial trials and the exponential distribution assumption for life test sampling. Apart from the normal distribution, which is extensively discussed elsewhere (see references, Chap. 10), and the previously mentioned exponential distribution, there are a limited number of variables-type sampling plans presently available in the literature. This is unfortunate, for although the exponential distribution plays a role in reliability statistics it has perhaps been overused. The possibility that large errors will result from assuming an exponential distribution when in fact the failure pattern followed a Weibull distribution has been pointed out in Refs. 3 and 4.

* Tables A.1–A.9 are contained in the Appendix.

4.8 RELIABILITY PREDICTION AND GROWTH

A consequence of reliability estimation is a study of reliability growth. Reliability, by virtue of its definition, necessarily implies prediction, but in this book we reserve the term *prediction* for use with the idea of future, projected reliability numbers not based directly on current observations. Reliability numbers obtained directly from current and previous observations are referred to as reliability *estimates*. The relationship between reliability prediction and reliability estimation introduces us to the subject of *reliability growth*.

We are familiar with the fact that reliability, during a development program, will, over a reasonable length of time, generally increase. However, we are also concerned with the rate of growth and whether it is adequate and will result in attainment of the end requirements. It is probable that the true reliability of the device increases in a series of steps as failure-producing components are redesigned and the development program progresses. The size and frequency of the steps will vary according to the type of equipment. For instance, large solid-propellant rocket engines with relatively few parts but with major engineering problems to be overcome will show fewer but steeper and higher steps in their growth curve. Liquid-propellant rocket engines with many more parts will generally indicate a higher initial development of reliability which will grow relatively more slowly than its solid-engine counterpart. Test results from which the reliability numbers are obtained will fluctuate randomly about the true reliability and hence, unless the data are very plentiful, the reliability points to which a reliability growth curve is to be fitted might show a large amount of variation. The points will also be "sensitive" depending on the method used in their estimation. For example, a *moving average** based on a hundred data points would be less "stable" than a moving average based on two hundred data points. The problem, therefore, is to fit curves to data of this type and to attempt to determine which factors significantly contribute to reliability growth. This is the topic of Chap. 11.

REFERENCES

1. M. A. Acheson, "The Whole Is Not the Sum of Its Parts," *Proc. 4th National Symposium on Reliability and Quality Control in Electronics (I.R.E.), Washington, D.C., January 6–8, 1958.*
2. J. R. Rosenblatt, "On Prediction of System Performance From Information on Component Performance," *Proc. Western Joint Computer Conference, February 1957*, 85–94.
3. M. Zelen, "Factorial Experiments in Life Testing," *Technometrics*, **1**, 269–288 (1959).

* See Chap. 16 for an example of an application of a "moving average."

4. M. Zelen and M. C. Dannemiller, "Are Life Testing Procedures Robust?," *Proc. 6th National Symposium on Reliability and Quality Control in Electronics*, (*I.R.E.*), Washington, D.C., January 11-13, 1960, 185-189.

ADDITIONAL READING

Lipow, M., "Recent Advances in Rocket Reliability Concepts," *Jet Propulsion*, **28,** 373-77 (June 1958).

Lusser, R., *Predicting Reliability*, Publication of Research and Development Division, Ordnance Missile Laboratories, Redstone Arsenal, October 1957.

Wilson, B. J., "Analyzing Missile Electric System Reliability," *Applications and Industry*, No. 26, 206-13 (September 1956).

Yueh, John H., "A Developmental Approach To Reliability in Missile System Equipment," *I.R.E. Trans. Reliability and Quality Control*, **PGRQC-8,** 44-54 (September 1956).

CHAPTER FIVE
BASIC PROBABILITY AND STATISTICS

5.1 INTRODUCTION

This chapter reviews the elements of probability and statistics, in order to provide the mathematical tools needed to deal with failure models and their application to reliability estimation problems. Included are only the most basic concepts and applications of probability and statistics which have direct bearing on the discussion in subsequent chapters. To supplement the material presented, the reader is urged to consult other references such as Refs. 1–7.

The concept of *sample space* is introduced in Sec. 5.2. The sample space represents the set of all possible outcomes of an experiment, and can be *discrete* (e.g., success, failure) or *continuous* (e.g., the set of real numbers in an interval representing all possible measurements of a performance parameter). Relations between subsets of the sample space (collections of outcomes, or *events*) are summarized briefly in Sec. 5.3. The notions of *random variable* and *probability distribution* for discrete and continuous sample spaces are introduced in Secs. 5.4 and 5.5. The concept of random variable is actually very simple; its use performs a service to clarify the language of probability. Sections 5.5 and 5.6 present distribution functions for continuous and discrete probability distributions. Sections 5.7 and 5.8 give some of the analytic aspects of probability distributions in relation to moments and moment generating functions. Section 5.9 presents probability distributions in two or more dimensions. Section 5.10 presents, for completeness, some concepts and applications of sampling theory.

The examples and exercises are an essential part of the text; the reader should be sure he understands them before proceeding to related topics.

5.2 SAMPLE SPACES

We review here the basic notions of probability, starting with those described by the terms *sample point*, a collection of which is an *event*, and

BASIC PROBABILITY AND STATISTICS 73

sample space which is the collection of all possible sample points or outcomes of an experiment. Actually the notion of *event* includes all three terms in the sense that an event is a well-defined set of sample points. It can consist of one sample point, several or all sample points. In fact it is convenient to define an ideal set which consists of no sample points, which is sometimes called the *impossible event*. Conversely, the event consisting of all sample points is called the *certain event*.

Sample spaces can consist of a finite or denumerable infinity of sample points, or of a nondenumerable infinity of sample points. The former are called *discrete* sample spaces and the latter are called *continuous* sample spaces. An example of a finite sample space is one which consists of two outcomes: Success, Failure. An example of a discrete sample space with a denumerable infinity of points is as follows: A coin is flipped until a heads (H) appears. The sample points can be written down as

$$O_1 = H, \quad O_2 = TH, \quad O_3 = TTH, \quad \cdots$$

since any number of successive tails (T) is conceivable before a head appears. In fact, we can even imagine a sequence in which H never appears. This sample point can be labeled O_0, and the sample space is still discrete.

An example of a continuous sample space is one which contains all the points (real numbers) in an interval. For example, the time-to-failure of a device can be thought of as any positive real number. The collection of all positive real numbers constitutes the sample space for an experiment in which we "turn on" the device at time $t = 0$, and measure time until the device fails. It is, of course, well known that this sample space is *nondenumerable*.

In the following sections it will be helpful to the reader to keep in mind one or more examples of sample spaces, events, and sample points. One such example is used in Table 5.1. Section 5.3.3 presents a detailed example which illustrates all of the principles to be given in Secs. 5.2 and 5.3. However, at first reading (of Secs. 5.2 and 5.3) it is suggested that only a very simple example be used to follow the presentation. For further details and examples, Refs. 6 and 7 are highly recommended.

5.2.1 COMBINATION OF EVENTS

Let $A, B, C, \cdots$ denote sets or collections of sample points in the sample space Ω. These are called *events* in the previous section. Associated with the event "A occurs" is the event "A does not occur," which will be denoted by $\bar{A}$, called *complement* of A, i.e., the sample points of Ω not contained in A. To symbolize "x (any sample point) is contained in A," we use $x \in A$. However, when we say "the event A is contained in the event B," we use $A \subset B$, and we mean "the occurrence of the event A

Table 5.1

Example for Verification

Let Ω be the set of all outcomes of a throw of a die; i.e., $\Omega = \{1, 2, 3, 4, 5, 6\}$. Let $A = \{1, 2\}$, $B = \{2, 3, 4\}$, and $C = \{2, 4, 6\}$.

	Property or Relation	Statement of Property or Relation	Example for Verification
(i)	$(\bar{\bar{A}}) = A$	The complement of the complement of A is A itself (the *involution* law).	$\bar{A} = \{3, 4, 5, 6\}$; hence $(\bar{\bar{A}}) = \{1, 2\} = A$
(ii)	$A \cap A = A$ $A \cup A = A$	The (intersection/union) of A with itself is A (the *idempotent* laws).	
(iii)	$A \cap B = B \cap A$ $A \cup B = B \cup A$	The *commutative* law for (intersection/union) of events. Can be extended to (intersection/union) of any number of events.	
(iv)	$A \cap (B \cap C) =$ $(A \cap B) \cap C$ $A \cup (B \cup C) =$ $(A \cup B) \cup C$	The *associative* law for (intersection/union) of events. Parentheses can be shifted at will when expression is (intersection/union) of any number of events.	$(B \cap C) = \{2, 4\}$; $A \cap (B \cap C) = \{2\}$ $(A \cap B) = \{2\}$; $(A \cap B) \cap C = \{2\}$ $(B \cup C) = \{2, 3, 4, 6\}$; $A \cup (B \cup C) =$ $\{1, 2, 3, 4, 6\}$ $(A \cup B) = \{1, 2, 3, 4\}$; $(A \cup B) \cup C =$ $\{1, 2, 3, 4, 6\}$
(v)	$A \cap (B \cup C) =$ $(A \cap B) \cup (A \cap C)$	The *distributive* law of *intersection with respect to union*. Note the analogy when A, B, C are considered as real numbers and $\cap$ and $\cup$ mean ordinary multiplication and addition, respectively.	$A \cap (B \cup C) = \{2\}$; $(A \cap B) \cup (A \cap C) =$ $\{2\} \cap \{2\} = \{2\}$
(vi)	$A \cup (B \cap C) =$ $(A \cup B) \cap (A \cup C)$	The *distributive* law of *union with respect to intersection*. Note that real numbers do *not* satisfy the analogous relationship.	$A \cup (B \cap C) = \{1, 2, 4\}$; $(A \cup B) \cap (A \cup C)$ $= \{1, 2, 3, 4\} \cap \{1, 2, 4, 6\} = \{1, 2, 4\}$

(vii) $A \cup (A \cap B) = A$
$A \cap (A \cup B) = A$

The *absorption* laws.

$\{1, 2\} \cup \{2\} = \{1, 2\}$
$\{1, 2\} \cap \{1, 2, 3, 4\} = \{1, 2\}$

(viii) $\overline{A \cap B} = \bar{A} \cup \bar{B}$
$\overline{A \cup B} = \bar{A} \cap \bar{B}$

The *dualization* laws.

$\overline{A \cap B} = \{1, 3, 4, 5, 6\}; \bar{A} \cup \bar{B} = \{3, 4, 5, 6\} \cup \{1, 5, 6\} = \{1, 3, 4, 5, 6\}$
$\overline{A \cup B} = \{5, 6\}; \bar{A} \cap \bar{B} = \{3, 4, 5, 6\} \cap \{1, 5, 6\} = \{5, 6\}$

(ix) $\bar{\phi} = \Omega$
$\bar{\Omega} = \phi$

The complement of the impossible event is the certain event, *or* the complement of the event consisting of no sample points is the event consisting of all sample points; and vice versa.

(x) $A \cap \Omega = A$
$A \cap \phi = \phi$
$A \cup \phi = A$
$A \cup \Omega = \Omega$

Properties of the events ϕ and Ω. Note that if ϕ and Ω are considered as the real numbers zero and one, respectively; and $\cap$ and $\cup$ mean ordinary multiplication and addition, respectively; the first three relationships have analogies, but the fourth does not.

(xi) $A \cup \bar{A} = \Omega$
$A \cap \bar{A} = \phi$

The *complementarity* relations. The first states that an event *or* its complement must occur. The second states that an event *and* its complement cannot occur.

$\{1, 2\} \cap \{3, 4, 5, 6\} = \{\ \} = \phi$

implies the occurrence of the event B." The relation $A \subset B$ is also equivalent to "$x \in A$ implies $x \in B$." Also, if $A \subset B$ and $B \subset A$, then the events A and B must be the same; i.e., $A = B$.

There are two fundamental operations that can be performed with events, *union* and *intersection*. The event $A \cup B$ is called the union of A and B, or the event "either A or B (or both) occur." This is equivalent to saying that if $x \in A$ or $x \in B$ (or both), then $x \in A \cup B$; and conversely, if $x \in A \cup B$, then $x \in A$ or $x \in B$ (or both). The event $A \cap B$ is called the intersection of A and B, or the event "both A and B occur." Thus $x \in A$ and $x \in B$ imply $x \in A \cap B$; and conversely, if $x \in A \cap B$, then $x \in A$ and $x \in B$.

We say events A and B are *mutually exclusive* if there are no sample points common to A and B; i.e., "both A and B cannot occur." This can be written symbolically as

$$A \cap B = \phi \qquad (5.1)$$

where ϕ denotes the event consisting of no sample points. Note that individual sample points (outcomes) are always mutually exclusive events. Table 5.1 presents some of the properties of events and relations between them which the reader can easily verify either "algebraically" or by example.

For example, to show that the relation (vi) in Table 5.1 holds, i.e.,

$$A \cup (B \cap C) = (A \cup B) \cap (A \cup C) \qquad (5.2)$$

let x be a sample point contained in $A \cup (B \cap C)$. Then $x \in A$ or $x \in B \cap C$. If the former, then $x \in A \cup B$ *and* $x \in A \cup C$, hence $x \in$ the intersection of $(A \cup B)$, $(A \cup C)$. If $x \in B \cap C$, then it is contained in *both* B and C. Thus x is contained in $A \cup B$ *and* $A \cup C$; i.e., in $(A \cup B) \cap (A \cup C)$, as before. Thus we have shown that

$$A \cup (B \cap C) \subset (A \cup B) \cap (A \cup C) \qquad (5.3)$$

The converse

$$(A \cup B) \cap (A \cup C) \subset A \cup (B \cap C) \qquad (5.4)$$

is shown by a similar argument. Thus the two events are equal.

The other relations may be demonstrated in a similar manner.

EXERCISE: Show that if $A \subset B$, then $A \cap B = A$, and conversely.

EXERCISE: Show that if $A \subset B$, then $A \cup B = B$, and conversely.

5.3 PROBABILITIES OF EVENTS

The probability of an event A is a non-negative real number attached to each event A, written as $P(A)$, with the following properties or defining axioms.

$$0 \leq P(A) \leq 1, \qquad P(\Omega) = 1$$

If
$$A = A_1 \cup A_2 \cup \cdots$$

and the A_i are mutually exclusive events, i.e.,

$$A_i \cap A_j = \phi, \qquad i \neq j$$

then
$$P(A) = P(A_1) + P(A_2) + \cdots$$

These properties are fundamental to the notion of probability measurement. In particular the latter relation states that probability is to be *completely additive*; that is, the probability of an event which consists of a denumerably infinite union of mutually exclusive subevents is simply the infinite series sum of the probabilities of the subevents. [A simple example is provided by the Poisson distribution, Eq. (5.22).] When the subevents are not necessarily mutually exclusive the latter relation can be modified to an inequality, with $P(A) \leq P(A_1) + P(A_2) + \cdots$. The following example shows for two subevents that equality may be restored by taking into account the probability of intersection of the subevents. For generalization to any finite number of events see Ref. 1, pp. 88–90.

EXAMPLE: Show that

$$P(A \cup B) = P(A) + P(B) - P(A \cap B)$$

This relation may be visualized by "counting sample points"; i.e., the sample points contained in $A \cup B$ consist of those contained in event A (which includes sample points common to or in the intersection of A and B) plus those contained in B (again which includes the sample points common to both A and B). Since the sample points in $A \cap B$ have been counted twice, we must then subtract them out once. This, incidentally shows a certain equivalence between "number of" and probability; if it happens that each of N sample points in $A \cup B$ is assigned probability $1/N$, then the number of sample points contained in any event when divided by N is numerically equivalent to the probability of the event. For comparison we present an analytical demonstration of the subject relation.

We have

$$A = (A \cap \bar{B}) \cup (A \cap B)$$

$$B = (\bar{A} \cap B) \cup (A \cap B)$$

$$A \cup B = (A \cap \bar{B}) \cup (A \cap B) \cup (\bar{A} \cap B)$$

The first of the above identities is obtained by first using the complementarity law [(xi) of Table 5.1] and the commutative law [(iii) of Table 5.1]: $\bar{B} \cup B = \Omega$. Then, since $A = A \cap \Omega$ [(x) of Table 5.1], we have $A = A \cap (\bar{B} \cup B)$. Then by the distributive law [(v) of Table 5.1] we get the first identity. The second and third identities are obtained similarly.

The parenthesized events on any one line are mutually exclusive; therefore, we can add probabilities obtaining

$$P(A) = P(A \cap \bar{B}) + P(A \cap B)$$

$$P(B) = P(\bar{A} \cap B) + P(A \cap B)$$

$$P(A \cup B) = P(A \cap \bar{B}) + P(A \cap B) + P(\bar{A} \cap B)$$

The desired result follows by algebraic manipulation.

The result in the next example finds continual application in reliability analysis. It states that if we know that an event B must occur when another event A occurs, then the probability attached to A cannot be greater than the probability attached to B. For example, if an electronic circuit must fail when one of its tubes fails, the probability of the latter event would be equal to or less than that of the former event, since the circuit could possibly fail for other reasons.

EXAMPLE: Let A and B be events in the sample space Ω. If $A \subset B$, show that $P(A) \leq P(B)$.

First, if $A = B$, the $P(A) = P(B)$. Otherwise, A is *strictly* contained in B; i.e., there exist sample points in B that are not in A. Let S denote the event consisting of all sample points in B that are not in A. Then $B = A \cup S$, and by definition A and S are mutually exclusive. Therefore $P(B) = P(A) + P(S)$. Since $P(S) \geq 0$, the desired inequality follows.

To prove this another way, define S as $S \equiv \bar{A} \cap B$. Then S and A are mutually exclusive and $B = A \cup S$. To show the former,

$$S \cap A = (\bar{A} \cap B) \cap A = A \cap (\bar{A} \cap B)$$

$$= (A \cap \bar{A}) \cap B = \phi \cap B = \phi$$

which states that S and A are mutually exclusive. To show that $B = A \cup S$, we have

$$A \cup S = A \cup (\bar{A} \cap B) = (A \cup \bar{A}) \cap (A \cup B) = A \cup B = B$$

The rest of the proof is the same as before. The reader should write down each step of the proof and carefully note the laws that are being used (*cf*. Table 5.1).

5.3.1 CONDITIONAL PROBABILITY

The notion of conditional probability can be introduced as follows: Let H, A be events of the sample space Ω. Suppose we know that H occurs;

BASIC PROBABILITY AND STATISTICS

what is the probability that A occurs? We denote this by the symbol $P(A \mid H)$ and define

$$P(A \mid H) = \frac{P(A \cap H)}{P(H)} \qquad (5.5)$$

except when $P(H) = 0$. Equation (5.5) states that to find the conditional probability that A occurs given that H occurs, one must sum up the probabilities for all sample points common to A and H and divide by the probability that H occurs. What this amounts to is choosing as a new sample space the event H. Thus, probabilities assigned to all events contained in H, based on the original sample space, are scaled up by the factor $1/P(H)$.

Generally we would know the probabilities of an event A *conditioned* on the occurrence of perhaps several mutually exclusive events. The problem would be to find the unconditional probability of the event A. Thus A might stand for "success" and $H_1, H_2, \cdots, H_n$ possible mutually exclusive events occurring that could lead to success. Let

$$H_1 \cup H_2 \cup \cdots \cup H_n = \Omega;$$

i.e., at least one of the events H must occur (and by the mutually exclusive property only one of the events H can occur). Then

$$A = A \cap \Omega = A \cap (H_1 \cup \cdots \cup H_n)$$
$$= (A \cap H_1) \cup \cdots \cup (A \cap H_n)$$

Since the latter events $(A \cap H_1) \cdots$ are mutually exclusive, we have

$$P(A) = \sum_i P(A \cap H_i) = \sum_i P(A \mid H_i) P(H_i) \qquad (5.5a)$$

EXAMPLE: A spacecraft is intended to land on the moon (H_1) and may then transmit lunar radiation data (A_1). It may instead (owing to unreliability) go into a long-life highly eccentric orbit around the Earth (H_2) and may then provide a very complete mapping of the Van Allen radiation belts (A_2). It may also, of course, not get off the Earth (owing to unreliability) H_3. The mission will be considered successful if either A_1 or A_2 occurs. What is the probability of a successful mission?

First, from Eq. (5.5),

$$P(A_1 \cup A_2 \mid H_1) = \frac{P[(A_1 \cup A_2) \cap H_1]}{P(H_1)} = \frac{P[(A_1 \cap H_1) \cup (A_2 \cap H_1)]}{P(H_1)}.$$

But $A_2 \cap H_1 = \phi$; i.e., landing on the moon and transmitting a good mapping of the Van Allen radiation belts are mutually exclusive events. Therefore $P(A_1 \cup A_2 \mid H_1) = P(A_1 \mid H_1)$. Similarly $P(A_1 \cup A_2 \mid H_2) = P(A_2 \mid H_2)$. Finally $P(A_1 \cup A_2 \mid H_3) = 0$. Thus, from Eq. (5.5a),

$$P(A_1 \cup A_2) = P(A_1 \mid H_1) P(H_1) + P(A_2 \mid H_2) P(H_2)$$

Reasonable values (1962–1965) of the probabilities might be $P(A_1 | H_1) = 0.80$,* $P(A_2 | H_2) = 0.95$,* $P(H_1) = 0.70$ and $P(H_2) = 0.25$. With these data $P(A_1 \cup A_2) = 0.7975$.

EXERCISE: An electronic assembly consists of two subsystems A and B. Each assembly is given one preliminary checkout test. Records on 100 preliminary checkout tests show that subsystem A failed 10 times. Subsystem B *alone* failed 15 times. Both subsystems A and B failed together 5 times.
(a) What is an estimate of the conditional probability of A failing knowing that B has failed?

Answer: 1/4
(b) What is an estimate of the probability that A alone fails?

Answer: 5/100

5.3.2 INDEPENDENCE OF EVENTS

Two events A, H are said to be independent if $P(A | H) = P(A)$; i.e., knowledge that H occurs does not change the probability that A occurs. From Eq. (5.5) we see that independence can also be expressed by

$$P(A \cap H) = P(A)P(H) \tag{5.6}$$

If $A_1, \cdots, A_n$ are events, they are said to be *pairwise independent* if for every $i \neq j$

$$P(A_i \cap A_j) = P(A_i)P(A_j) \tag{5.7}$$

This is not sufficient, however, to guarantee that the events—e.g., $A_i \cap A_j$ and A_k, $i \neq j \neq k$—are independent. We must introduce additional criteria for the latter result. To do this requires *mutual independence*, and we say that the events $A_1, \cdots, A_n$ are mutually independent if for all combinations $1 \leq i < j < k < \cdots \leq n$,

$$P(A_i \cap A_j) = P(A_i)P(A_j)$$
$$P(A_i \cap A_j \cap A_k) = P(A_i)P(A_j)P(A_k) \tag{5.8}$$
$$\cdot \cdot \cdot \cdot \cdot \cdot \cdot \cdot \cdot \cdot \cdot$$
$$P(A_1 \cap A_2 \cap \cdots \cap A_n) = P(A_1)P(A_2) \cdots P(A_n)$$

A theoretical example will be given in Sec. 5.4 which shows that of three events A_1, A_2, and A_3, each pair are independent; i.e., the first of Eqs. (5.8) is satisfied; but the last of Eqs. (5.8) is not; i.e.,

$$P(A_1 \cap A_2 \cap A_3) \neq P(A_1)P(A_2)P(A_3).$$

* $P(A_1 | H_1)$ would very likely be less than $P(A_2 | H_2)$ since the former event might involve damage to instrumentation due to the lunar landing.

5.3.3 EXAMPLE: ANALYSIS OF A SPACE VEHICLE BATTERY POWER SUPPLY

A battery in a space vehicle is to be activated by first firing a squib (a small explosive detonator) which ignites a chemical solid propellant charge. The gases generated by the burning propellant force a piston to move in a cylinder, thereby expelling a fluid electrolyte into the battery manifold and wetting the plates. When the latter event occurs, power is thereby generated to operate various controls of the space vehicle. Two squibs are used in a parallel* circuit—either squib detonation will ignite the propellant. The piston may jam immediately, or may jam (say) halfway down the cylinder, or will travel the full length. In the case of the latter two events, the electrolyte must correspondingly be either (1) incompletely expelled, thereby partially wetting the plates and producing limited power, or (2) completely expelled, thereby producing full power. There is a chance that only the limited power output will be necessary for successful control of the vehicle in a particular mission.

The event that sufficient power output occurs, S, is a function of the occurrence of the following events:

$A_1 \cup A_2$ the event that either one or both squibs detonate
B the propellant charge ignites and burns as designed
C_1 the piston travels the full length of the cylinder
C_2 the piston jams halfway down the cylinder
D_1 the electrolyte is fully expelled and produces full power
D_2 the electrolyte is partially expelled and produces limited power
E_1 full power is required
E_2 only limited power is necessary.

Thus we may write the event S as

$$S = [(A_1 \cup A_2) \cap B \cap C_1 \cap D_1 \cap (E_1 \cup E_2)]$$
$$\cup [(A_1 \cup A_2) \cap B \cap C_2 \cap D_2 \cap E_2] \quad (5.9)$$

Since the event C_1 implies that D_1 occurs; and, correspondingly, C_2 implies D_2, we have

$$C_1 \cap D_1 = C_1 \quad \text{and} \quad C_2 \cap D_2 = C_2 \quad (5.10)$$

Therefore we may delete D_1, D_2 from Eq. (5.9) which then becomes, after "factoring" $[(A_1 \cup A_2) \cap B]$,

$$S = [(A_1 \cup A_2) \cap B] \cap \{[C_1 \cap (E_1 \cup E_2)] \cup [C_2 \cap E_2]\} \quad (5.11)$$

* See Sec. 9.4.

The last expression can be reduced by use of the relations in Table 5.1 to

$$S = [(A_1 \cup A_2) \cap B \cap C_1 \cap E_1] \cup [(A_1 \cup A_2) \cap B \cap C_1 \cap E_2]$$
$$\cup [(A_1 \cup A_2) \cap B \cap C_2 \cap E_2] \quad (5.12)$$

The bracketed terms connected by the union operator in Eq. (5.12) are seen to be mutually exclusive, since $C_1 \cap C_2 = \phi$, $E_1 \cap E_2 = \phi$; and therefore their probabilities will add to give $P(S)$.

To determine the probability of success, $P(S)$, it is first assumed that A_1, A_2 are statistically independent and with equal probabilities $\equiv P(A)$. Thus

$$P(A_1 \cup A_2) = 1 - P(\overline{A_1 \cup A_2}) = 1 - P(\bar{A}_1 \cap \bar{A}_2)$$
$$= 1 - P(\bar{A}_1)P(\bar{A}_2) = 1 - (1 - P(A_1))(1 - P(A_2))$$
$$= P(A)(2 - P(A))$$

Secondly, we would expect for physical reasons that the event B (propellant ignition and burning) would be more probable if both squibs rather than just one detonated. This is already taken into account since all we need to know is $P(B \mid A_1 \cup A_2)$. However, data may be available in the form of separate estimates of $P(B \mid \bar{A}_1 \cap A_2)$, $P(B \mid A_1 \cap \bar{A}_2)$, and $P(B \mid A_1 \cap A_2)$ which may then be added to give $P(B \mid A_1 \cup A_2)$ since the three conditioning events are mutually exclusive.

The probabilities for the events C_1 or C_2 are each conditioned by the events $(A_1 \cup A_2) \cap B$. Consequently we would have to know

$$P(C_1 \mid (A_1 \cup A_2) \cap B) \quad \text{and} \quad P(C_2 \mid (A_1 \cup A_2) \cap B).$$

[*Remark*: The events C_1, C_2 would be considered as mutually exclusive and *therefore not statistically independent* (unless either or both $P(C_1)$, $P(C_2)$ were zero). (See the following example.)]

The events E_1 or E_2 would be statistically independent of all other events. The previous remark also holds for E_1 and E_2, since these events would be considered mutually exclusive.

Thus, the probability of each of the bracketed events in Eq. (5.12), e.g., the first one, would be written as

$$[2P(A) - \overline{P(A)^2}] \cdot [P(B \mid A_1 \cup A_2)]$$
$$\cdot [P\{C_1 \mid (A_1 \cup A_2) \cap B\}] \cdot [P(E_1)] \quad (5.13)$$

and similarly for the remaining events. In Chap. 9 we will discuss more generalized structure models and present further examples.

EXAMPLE: Mutually exclusive events C_1, C_2 are independent *if and only if* $([P(C_1) = 0] \cup [P(C_2) = 0])$ is true (occurs). First, since $C_1 \cap C_2 = \phi$, then

BASIC PROBABILITY AND STATISTICS

$P(C_1 \cap C_2) = 0$. Now we show the "only if" part. Assume C_1, C_2 are independent. Then

$$P(C_1 \cap C_2) = P(C_1)P(C_2) = 0$$

Hence, one or both of the numbers $P(C_1)$, $P(C_2)$ must equal zero. Now we show the "if" part. Assume that either $P(C_1) = 0$ or $P(C_2) = 0$ or both are equal to zero. Let $P(C_1) = 0$, but $P(C_2) > 0$. By Eq. (5.5)

$$P(C_1 \mid C_2) = \frac{P(C_1 \cap C_2)}{P(C_2)} \tag{5.14}$$

But the righthand side of Eq. (5.14) is zero. Therefore

$$P(C_1 \mid C_2) = P(C_1) = 0$$

and by definition C_1, C_2 are therefore independent. The proof is the same if we interchange C_1 and C_2. The last case is to assume that both $P(C_1)$ and $P(C_2)$ are equal to zero. Then

$$0 = P(C_1 \cap C_2) = P(C_1)P(C_2) = 0 \cdot 0 = 0 \tag{5.15}$$

and therefore by Eq. (5.6), C_1, C_2 are independent.

5.4 DISCRETE PROBABILITY DISTRIBUTIONS

Probability distributions are used to describe the structure of the sample space of observed outcomes. Almost the simplest type of probability distribution arises when we consider the sample space consisting of just two outcomes: S: Success and F: Failure. For obvious reasons we would like to talk about numbers rather than letters; therefore suppose we let the number 0 correspond to the letter S, and the number 1 correspond to the letter F. Remembering that the probability contained in the entire sample space is unity, we attach to the event S the probability $q \geq 0$ and to the event F the probability $p \equiv 1 - q$. It is therefore natural to attach the probabilities q and p to the corresponding numbers 0 and 1. What we have done, in essence, is to map the sample space onto the real numbers, carrying the attached probabilities from the points or events of the sample space onto the corresponding real-numbered points. The function that does this type of mapping is, in general, called a *random variable*. Let us call this particular random variable X. Thus

$$X(S) = 0 \equiv \text{the occurrence of } S \text{ implies } X = 0$$
$$X(F) = 1 \equiv \text{the occurrence of } F \text{ implies } X = 1 \tag{5.16}$$

and since

$$P(S) = q$$
$$P(F) = p \equiv 1 - q \tag{5.17}$$

then we say
$$P(X = 0) = q$$
$$P(X = 1) = p \equiv 1 - q \qquad (5.18)$$

Note that we do not necessarily make the correspondence one-one. For example, suppose the sample space consisted of three outcomes S, F_1, and F_2, where F_1 and F_2 denote two kinds of failure. Suppose also that $P(F_1) = p_1$, $P(F_2) = p_2$, and $p_1 + p_2 = p$; we could consider the correspondence:
$$X(S) = 0$$
$$X(F_1 \cup F_2) = 1 \qquad (5.19)$$

and we would also have Eqs. (5.18) hold. However, starting with the new sample space (0, 1), we could not map it back onto the old one. All we could do is map the number 1 back into $F_1 \cup F_2$. Thus if we were told "$X = 1$," we would only have the information that either F_1 or F_2 or both occurred; i.e., "a failure of one kind or the other occurred." However, a different correspondence could have been made. Depending upon the importance of distinguishing failure types, a random variable Y could have been defined as
$$Y(S) = 0, \qquad Y(F_1) = 1, \qquad Y(F_2) = 2 \qquad (5.20)$$
with respective probabilities $1 - p_1 - p_2$, p_1 and p_2. In this case, the correspondence is exactly one-one.

The above examples are of discrete probability distributions over a finite sample space. An example of a discrete probability distribution where the sample space is infinite is afforded by the "coin-flipping" example of Sec. 5.2. Define the random variable X by $X(O_j) = j$, and assume
$$P(O_j) = (\tfrac{1}{2})^j, \quad \text{hence} \quad P(X = j) = (\tfrac{1}{2})^j, \qquad j = 1, 2, \cdots$$
and we see that all of the probabilities add up to one.

One of the most important discrete distributions is the *binomial distribution* (Sec. 6.2). It is a two-parameter distribution with parameters n and p, where n is any positive integer and p any real number between 0 and 1. It can be symbolized by:
$$P(X = f) = \binom{n}{f} p^f q^{n-f}, \qquad f = 0, 1, 2, \cdots, n \qquad (5.21)$$
$$q = 1 - p$$

As generally used in this book, p denotes the probability of failure on a single trial; and n trials are made. The probability of any set of outcomes (failure or success) in any group of $k \leq n$ trials is the same whatever the results in the rest of the trials. This is one way of defining independence

BASIC PROBABILITY AND STATISTICS   85

of the outcomes of the trials; we say in this case that trials are independent. Equation (5.21) states that the probability that the number of failures, X, in n trials, is f, is given by the expression on the righthand side. The range of the random variable X is any integer from 0 to n. To obtain Eq. (5.21) from the definition of X, we consider the sample space of all possible outcomes of the experiment of making n trials as being represented by a set of parentheses (sample points) each containing a string of n letters S and F: $(S, S, \cdots, S), (F, S, \cdots, S), (S, F, \cdots, S)$, and so on to $(F, F, \cdots, F)$. The random variable X maps all of the sample points containing exactly f F's into the positive integer f. The probability carried by each sample point with f F's and $(n - f)$ S's is $p^f q^{n-f}$. Since there are

$$\binom{n}{f} \equiv \frac{n!}{(n-f)!f!}$$

such sample points, the total probability carried into the positive integer f is

$$\binom{n}{f} p^f q^{n-f}$$

It is a consequence of the independence assumption that the probability of any particular one of the outcomes $(S, \cdots, F, \cdots, F, S)$ with f F's and $n - f$ S's is

$$q \cdots p \cdots p \cdot q = p^f q^{n-f}$$

Recalling that individual sample points are mutually exclusive events and therefore that we can add probabilities, gives the result expressed by Eq. (5.21).

EXAMPLE: When $n = 1$, we have the distribution discussed previously: $P(X = 0) = q; P(X = 1) = p$.

EXERCISE: What is the distribution of the random variable $Y = X/n$, where X has the binomial distribution defined by Eq. (5.21)?

EXAMPLE: If X_1 and X_2 each have identical binomial distributions with parameters p and $n = 1$; and *also*

$$P[(X_1 = j) \cap (X_2 = k)] = P(X_1 = j)P(X_2 = k)$$

where $j = 0, 1; k = 0, 1$; show that the sum $X_1 + X_2$ has a binomial distribution with parameters $p, n = 2$.

We have: The event $X_1 + X_2 = 0$ is equivalent to the event $X_1 = 0$ *and* $X_2 = 0$. Hence

$$P(X_1 + X_2 = 0) = P[(X_1 = 0) \cap (X_2 = 0)]$$
$$= P(X_1 = 0) \cdot P(X_2 = 0) = q^2$$

The event $X_1 + X_2 = 1$ is equivalent to the event

$$[(X_1 = 0) \cap (X_2 = 1)] \cup [(X_1 = 1) \cap (X_2 = 0)]$$

Since the events indicated by square brackets are mutually exclusive (they cannot both happen) their probabilities add. Therefore

$$P(X_1 + X_2 = 1) = q \cdot p + p \cdot q = 2pq$$

A similar argument shows that

$$P(X_1 + X_2 = 2) = p^2$$

Thus, in general,

$$P(X_1 + X_2 = f) = \binom{2}{f} p^f q^{2-f}, \quad f = 0, 1, 2$$

Note how the assumption

$$P[(X_1 = j) \cap (X_2 = k)] = P(X_1 = j) P(X_2 = k)$$

is used in the example. This is precisely the definition of *independence* for two discrete random variables (j and k range over all possible values of X_1 and X_2, respectively).

EXERCISE: Using the definition of conditional probability in Sec. 5.3.1, show that

$$P[(X_1 = j) \mid (X_2 = k)] = P(X_1 = j)$$

if, and only if, X_1 and X_2 are independent random variables.

Hint: Recall that

$$P[(X_1 = j) \mid (X_2 = k)] P(X_2 = k) = P[(X_1 = j) \cap (X_2 = k)]$$

EXERCISE: Write down the conditions for mutual independence of n discrete random variables. Use the statement of mutual independence for events given in Sec. 5.3.2.

EXAMPLE: Find an example which illustrates that even if X_1 and X_2 are independent, X_1 and X_3 are independent, and X_2 and X_3 are independent, X_1, X_2, X_3 are not necessarily independent.

Such an illustration is given by the following (See Ref. 5, p. 162).* Assume that each of the points $(1, 0, 0)$, $(0, 1, 0)$, $(0, 0, 1)$, $(1, 1, 1)$ carries the probability $1/4$. First, we have in general:

$$P(X_1 = j, X_2 = k) = \sum_n P(X_1 = j, X_2 = k, X_3 = n)$$

also

$$P(X_1 = j, X_3 = n) = \sum_k P(X_1 = j, X_2 = k, X_3 = n)$$

and

$$P(X_2 = k, X_3 = n) = \sum_j P(X_1 = j, X_2 = k, X_3 = n)$$

* We will from here on use commas interchangeably with $\cap$. Also refer to Sec. 5.9.

Similarly
$$P(X_1 = j) = \sum_k \sum_n P(X_1 = j, X_2 = k, X_3 = n)$$
$$= \sum_k P(X_1 = j, X_2 = k)$$
and so on. We have, e.g.:
$$P(X_1 = 0, X_2 = 0) = \tfrac{1}{4}$$
$$P(X_1 = 1, X_2 = 0) = \tfrac{1}{4}$$
$$P(X_1 = 0, X_2 = 1) = \tfrac{1}{4}$$
$$P(X_1 = 1, X_2 = 1) = \tfrac{1}{4}$$
and the same relationship holds for the other combinations (X_1, X_3) and (X_2, X_3). By summing again, one obtains
$$P(X_1 = j) = P(X_2 = k) = P(X_3 = n) = \tfrac{1}{2}, \quad j, k, n = 0, 1$$
Hence
$$P(X_1 = j, X_2 = k) = P(X_1 = j)P(X_2 = k) = (\tfrac{1}{2})^2 = \tfrac{1}{4}$$
and similarly for the other combinations (X_1, X_3) and (X_2, X_3). However, we have, for example:
$$\tfrac{1}{4} = P(X_1 = 1, X_2 = 1, X_3 = 1) \neq P(X_1 = 1)P(X_2 = 1)P(X_3 = 1) = \tfrac{1}{8}$$

We will discuss independence again when we consider continuous random variables and probability distributions. Another important discrete probability distribution is the *Poisson* distribution.

$$P(X = j) = \frac{\lambda^j e^{-\lambda}}{j!}, \quad \lambda > 0; j = 0, 1, 2, \cdots \qquad (5.22)$$

EXERCISE: Verify that $\sum_{j=0}^{\infty} P(X = j) = 1$. (See Sec. 6.5.)

EXERCISE: Verify that $P(Y = j) = q^{j-1}p, j = 1, 2, \cdots$, where $q, p > 0$ and $p + q = 1$, is a probability distribution. This is the *geometric* distribution (see Sec. 6.4).

EXERCISE: Suppose that X denotes the number of independent trials, each with constant probability p of failure, that one must make to obtain the *first* failure. Show that X has the *geometric* distribution.

EXERCISE: Under the conditions given above, find the distribution of the number of trials to achieve the kth failure.

Answer:
$$P(X = n) = \binom{n-1}{k-1} p^k q^{n-k}$$
where $n = k, k + 1, \cdots$. (See Sec. 6.4.)

5.5 CONTINUOUS PROBABILITY DISTRIBUTIONS

We have discussed some of the more important discrete probability distributions that will be met with in this book. Now, we will consider the notion, and some examples, of continuous probability distributions.

When one makes "quantitative" measurements of something, such as the length of a rod, the weight of a rocket engine, the output voltage of a transformer, and so on, the result is some set of real numbers, e.g., 2.53, 2.58, 2.50, etc. The measuring apparatus we use may be so coarse, that the population of all such measurements may have the appearance of a discrete set of numbers. However, it is easy to imagine that our method of measurement can be infinitely fine, so that one could measure, for example, rods with length $2.530663\cdots$, $2.579149\cdots$, $2.504116\cdots$, and so on. In other words, we can imagine that a length can be any real number in some interval of real numbers $a \leq x \leq b$. This is an example of a continuous or nondenumerable sample space, as mentioned in Sec. 5.2.

Taking the viewpoint that every sample space is in reality discrete*— i.e., no matter how we refine our methods of measurement, at any given stage of refinement there will always be gaps or intervals which contain numbers that will not appear in the sample space—the concept of continuous sample spaces is merely a convenience; but it is a great convenience. It enables one to perform a considerable simplification in describing the structure of the sample space when "quantitative" measurements of the type discussed above are being used.

Now, if we look upon continuous sample spaces as a limiting case of discrete sample spaces, we can see that although a positive probability could be attached to any point of a discrete sample space, the probability for any one point in an interval of real numbers $a \leq x \leq b$ would in general have to be zero.† Nevertheless, if the sample space consists of all of the real numbers x contained in $a \leq x \leq b$, we still wish the total probability contained in the sample space to be unity. This brings up the notion of *probability density functions* and the use of integration to measure probabilities in continuous sample spaces. We define a probability density function in one dimension (on the x-axis) as follows:

$f(x)$ is a probability density function‡ if

$$f(x) \geq 0, \quad -\infty < x < \infty$$

and

$$\int_{-\infty}^{\infty} f(x)\ dx = 1$$

assuming that $f(x)$ is integrable.

* And also not dense in the set of real numbers.
† See the discussion on discontinuous distribution functions (Sec. 5.6).
‡ Abbreviated p.d.f., or sometimes called *frequency function*.

BASIC PROBABILITY AND STATISTICS   89

The quantity $f(x)\,dx$ is called the *probability density element* (or *frequency element*) at the point x; it measures the probability in the *interval* $(x, x + dx)$. The probability contained in the interval $a < x \leq b$ is defined as

$$\int_a^b f(x)\,dx$$

For intervals

$$a \leq x \leq b, \quad a \leq x < b, \quad a < x < b$$

the probability is still given by

$$\int_a^b f(x)\,dx$$

provided that the distribution is continuous on the interval. When we deal with discontinuous distribution functions, a distinction must be made between the various kinds of intervals above, since their probabilities may not be equal.

It is useful at this point to introduce random variables, as was done before for discrete probability distributions. However, we will not consider the notion of a random variable as a mapping from certain non-numerical continuous sample spaces onto intervals. Instead, the whole set of real numbers will be regarded as the original sample space. We say that the random variable ξ is contained in the interval $(a, b]$ by simply stating $a < \xi \leq b$.* We also speak of the probability density function of ξ, say $f(x)$, so that

$$P(x < \xi < x + dx) = f(x)\,dx$$

Furthermore

$$P(a < \xi \leq b) = \int_a^b f(x)\,dx \qquad (5.23)\dagger$$

EXERCISE: Let

$$f(x) = 0 \qquad x < 0$$
$$= \lambda e^{-\lambda x} \qquad x \geq 0$$

and

* A conventional notation for types of intervals, is:

$$(a, b) \equiv a < x < b$$
$$(a, b] \equiv a < x \leq b$$
$$[a, b) \equiv a \leq x < b$$
$$[a, b] \equiv a \leq x \leq b.$$

† See the previous footnote on types of intervals.

Show that $f(x)$ is a p.d.f. by integrating it from $-\infty$ to $+\infty$. This is the well-known exponential probability density function.

EXERCISE: Using the definition of $f(x)$ in the previous exercise, calculate $F(x) \equiv P(\xi \leq x)$, where the random variable ξ has $f(x)$ as its p.d.f.

The function $F(x) = P(\xi \leq x)$ is called a (cumulative) *distribution function*. A distribution function in one dimension, $F(x)$, must have the properties:

(1) $0 \leq F(x) \leq 1$

(2) $F(-\infty) = 0, \quad F(\infty) = 1$

(3) $F(x)$ is nondecreasing in x

(4) $F(x)$ is continuous at each point x on the right; i.e.,

$$\lim_{h \to 0} F(x + h) = F(x)$$

where $h > 0$.*

In addition, when $F'(x) = f(x)$ exists and is continuous for all x,† then $f(x)$ is a probability density function, with the property that

$$F(x) = \int_{-\infty}^{x} f(t) \, dt$$

The importance of property (4) will be illustrated when we consider distribution functions for discrete probability distributions in Sec. 5.6. A well-known example of a probability density function is the *normal* p.d.f. defined by

$$f(x) = (2\pi\sigma^2)^{-1/2} \exp\left[-(x-\mu)^2/2\sigma^2\right]$$
$$-\infty < x < \infty\,; \sigma > 0, -\infty < \mu < \infty \quad (5.24)$$

It is a two-parameter distribution with parameters μ and σ, where μ turns out to be the *mean*, and σ turns out to be the *standard deviation* of the random variable ξ which has $f(x)$ for its p.d.f.‡ If we define a *standardized* random variable η by

$$\eta = \frac{\xi - \mu}{\sigma} \quad (5.25)$$

then η has the standard normal p.d.f. given by

$$\phi(x) = (2\pi)^{-1/2} e^{-x^2/2} \quad (5.26)$$

* This is convention; $F(x)$ could just as well be defined to be continuous on the left.
† Continuity for *all* x is not necessary for this property; but we will not consider this question here (see Ref. 5, p. 169).
‡ Commonly, we also say that the *distribution* has mean μ, standard deviation σ. (See Sec. 5.7 for the definitions of mean and standard deviation.)

BASIC PROBABILITY AND STATISTICS 91

The standard normal distribution function is denoted by $\Phi(x)$ and defined by

$$P(\eta \leq x) = \Phi(x) = (2\pi)^{-1/2} \int_{-\infty}^{x} e^{-u^2/2}\, du \qquad (5.27)$$

Both $\phi(x)$ and $\Phi(x)$ are extensively tabulated (Ref. 10). Also see Sec. 6.10.

5.6 DISTRIBUTION FUNCTIONS FOR DISCRETE DISTRIBUTIONS

For any random variable ξ with a discrete probability distribution

$$P(\xi = x_i) = p_i, \qquad i = 1, 2, \cdots$$

where

$$\sum_i p_i = 1,$$

the distribution function $F(x)$ is defined as

$$F(x) = P(\xi \leq x) = \sum_{x_i \leq x} P(\xi = x_i) \qquad (5.28)$$

For example, if ξ has the binomial distribution with $n = 1$, then

$$P(\xi \leq x) = F(x) = \begin{cases} 0, & -\infty < x < 0 \\ q, & 0 \leq x < 1 \\ 1, & 1 \leq x < \infty \end{cases} \qquad (5.29)$$

The distribution function $F(x)$ is shown in Fig. 5.1. Note that

$$P(a < \xi \leq b) = F(b) - F(a)$$

for all a and b; but, e.g.,

$$P(a < \xi < b) \neq F(b) - F(a)$$

for all a and b. For, let $a = \tfrac{1}{2}$, $b = 1$, then

$$P(\tfrac{1}{2} < \xi \leq 1) = p = F(1) - F(\tfrac{1}{2})$$

but

$$P(\tfrac{1}{2} < \xi < 1) = 0$$

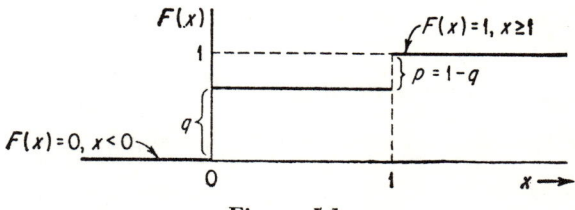

Figure 5.1

In general, if a point x_0 carries positive probability $p_0 = P(\xi = x_0)$, then $F(x)$ "jumps" by amount p_0 as $x \to x_0$ from the left. Conversely, if $F(x)$ is discontinuous at the point x_0 (it must by definition still be continuous on the right at x_0), and "jumps" by amount p_0, then the point x_0 carries positive probability $P(\xi = x_0) = p_0$. Thus, a discrete probability distribution has a distribution function $F(x)$ which is a step function.

EXAMPLE: How would one write an infinite series formula for the distribution function $F(x)$ of a general discrete distribution, defined by Eq. (5.28)?

Answer: Let $\epsilon(x - x_i)$ be the unit step function defined by

$$\epsilon(x - x_i) = \begin{cases} 0, & x < x_i \\ 1, & x \geq x_i \end{cases} \tag{5.30}$$

then

$$F(x) = \sum_{x_i \leq x} P(\xi = x_i) = \sum_{i=1}^{\infty} p_i \epsilon(x - x_i) \tag{5.31}$$

where $\qquad p_i \equiv P(\xi = x_i) \qquad$ and $\qquad \sum_{i=1}^{\infty} p_i = 1$

Note also that $\epsilon(x)$ is a distribution function of a probability distribution with all of the probability carried at one point, $x = 0$.

5.7 EXPECTED VALUES AND MOMENTS

Let ξ be a random variable with p.d.f. $f(x)$. Then if $g(\xi)$ is some function of ξ, the *expected value* of $g(\xi)$ is defined as

$$E(g) \equiv E[g(\xi)] = \int_{-\infty}^{\infty} g(x) f(x) \, dx \tag{5.32}$$

For a discrete distribution, the integral is replaced by a sum, and we have correspondingly

$$E[g(\xi)] = \sum_i g(x_i) P(\xi = x_i) \tag{5.32'}$$

When $g(\xi) = \xi^n$, then $E(g)$ is called the *n*th ordinary moment of the random variable ξ (or of the distribution whose p.d.f. is $f(x)$). When $n = 1$,

$$E(g) \equiv E(\xi) \equiv \mu, \text{ the } mean \tag{5.33}$$

EXERCISE: Show that μ is the mean of the normal p.d.f. defined by Eq. (5.24).

The value of $\alpha_n \equiv E(\xi^n)$ is also called the *n*th moment about the origin. It is very useful to consider the *n*th moment about the mean,

BASIC PROBABILITY AND STATISTICS

which is analogously defined as

$$\mu_n \equiv E[(\xi - \mu)^n] = \int_{-\infty}^{\infty} (x - \mu)^n f(x) \, dx \qquad (5.34)$$

When $n = 2$, we call $E[(\xi - \mu)^2] \equiv$ Variance of ξ or, abbreviated, Var ξ, μ_2, or, commonly, σ^2. The quantity

$$+\sqrt{\text{Var } \xi} = \sigma$$

is called the *standard deviation* of the variable ξ.

> EXERCISE: Show that the parameter σ is the standard deviation of the random variable ξ whose p.d.f. is given by Eq. (5.24).
>
> EXERCISE: Show that $\alpha_2 = \mu^2 + \sigma^2$.
>
> EXERCISE: Show that if ξ has mean μ, standard deviation σ, then $\eta = (\xi - \mu)/\sigma$ has mean zero, standard deviation unity.
>
> EXERCISE: Let the distribution function $F(x)$ be defined for $x > 0$; $\alpha > 0$, $\lambda > 0$, by
>
> $$F(x) = \frac{\lambda^\alpha}{\Gamma(\alpha)} \int_0^x u^{\alpha-1} e^{-\lambda u} \, du \qquad (5.35)$$
>
> Find the mean and variance of the random variable with $F(x)$ as its distribution function. Use the fact that the Laplace transform of t^β is
>
> $$\int_0^\infty t^\beta e^{-st} \, dt = \Gamma(\beta + 1) s^{-\beta-1}, \qquad \beta > -1$$
>
> and also the relation $E[(\xi - \mu)^2] = E(\xi^2) - \mu^2$.

5.7.1 TCHEBYCHEFF'S INEQUALITY

When all the moments of a distribution are known, one can show under fairly general conditions that a unique distribution exists with the given moments (see Ref. 5, p. 176). Knowledge of the first two moments $E(\xi) = \mu$ and Var $\xi = \sigma^2$ gives a great deal of information about the distribution, however, as the following inequality, due to Tchebycheff, shows. Tchebycheff's inequality is:

$$P(|\xi - \mu| \geq k\sigma) \leq \frac{1}{k^2}, \qquad k > 0 \qquad (5.36)$$

for any random variable ξ which has a mean μ and standard deviation σ. It will be proved here for a random variable ξ with p.d.f. $f(x)$; but it is true whether the probability distribution is continuous or not (see Ref. 5, p. 182).

Proof: Let K be greater than zero. Then $(x - \mu)^2 \geq K$ for either

$$x \geq \mu + \sqrt{K}, \quad x \leq \mu - \sqrt{K}$$

We have

$$\sigma^2 = E[(\xi - \mu)^2] = \int_{-\infty}^{\infty} (x - \mu)^2 f(x) \, dx \quad (5.37)$$

$$= \int_{-\infty}^{\mu - \sqrt{K}} + \int_{\mu - \sqrt{K}}^{\mu + \sqrt{K}} + \int_{\mu + \sqrt{K}}^{\infty}$$

Since the second integral is ≥ 0, we have

$$\sigma^2 \geq \int_{-\infty}^{\mu - \sqrt{K}} (x - \mu)^2 f(x) \, dx + \int_{\mu + \sqrt{K}}^{\infty} (x - \mu)^2 f(x) \, dx \quad (5.38)$$

Since $(x - \mu)^2 \geq K$ in the intervals $(-\infty, \mu - \sqrt{K})$ and $(\mu + \sqrt{K}, \infty)$, then the righthand side of Eq. (5.38) is not increased if we replace $(x - \mu)^2$ by K in both integrals. Hence

$$\sigma^2 \geq KP(\xi \leq \mu - \sqrt{K}) + KP(\xi \geq \mu + \sqrt{K})$$
$$= KP((\xi - \mu)^2 \geq K)$$

Since the event $(\xi - \mu)^2 \geq K$ is equivalent to the event $|\xi - \mu| \geq +\sqrt{K}$, we have

$$P(|\xi - \mu| \geq +\sqrt{K}) \leq \frac{\sigma^2}{K}$$

and letting $K \equiv k^2 \sigma^2$, we get the result expressed by Eq. (5.36).

A sharper inequality (due to Gauss) is that if $f(x)$ is continuous and has one *mode* (a mode is a value of $\xi = x_0$, such that $f(x)$ has a maximum at x_0), then

$$P(|\xi - x_0| \geq k\tau) \leq \frac{4}{9k^2} \quad (5.39)$$

where $\tau^2 = \sigma^2 + (x_0 - \mu)^2$. If $x_0 = \mu$, then

$$P(|\xi - \mu| \geq k\sigma) \leq \frac{4}{9k^2}$$

which is an improvement over the Tchebycheff inequality, Eq. (5.36). However, it cannot be as universally applied.

EXAMPLE: Compare the upper bounds on the probability $P(|\xi - \mu| \geq 2\sigma)$, when ξ has a normal p.d.f., obtained by Eq. (5.39) and obtained by finding the probability from tables of the normal distribution function (see Sec. 6.10).

Answer: From Eq. (5.39) the upper bound to the above probability is $\frac{1}{9} = 0.1111$. From the tables (Ref. 10) the upper bound is 0.0455. Why can inequality (5.39) be used in this case?

BASIC PROBABILITY AND STATISTICS 95

5.8 MOMENT GENERATING FUNCTIONS

Moment generating functions can be used to find general formulas for moments of a random variable (moments of a distribution). The moment generating function (m.g.f.) of the random variable ξ with p.d.f. $f(x)$ is defined by

$$\psi(t) \equiv E(e^{t\xi}) = \int_{-\infty}^{\infty} e^{tx} f(x) \, dx \tag{5.40}$$

We have

$$\psi(0) = \int_{-\infty}^{\infty} f(x) \, dx = 1 \tag{5.41}$$

$$\psi'(0) = \int_{-\infty}^{\infty} x f(x) \, dx = \alpha_1 \tag{5.42}$$

and generally

$$\psi^{(n)}(0) = \int_{-\infty}^{\infty} x^n f(x) \, dx = \alpha_n \tag{5.43}$$

When all the moments $\alpha_0 = 1, \alpha_1, \alpha_2, \cdots, \alpha_n, \cdots$ exist, then $\psi(t)$ can be written as an infinite power series:

$$\psi(t) = \sum_{n=0}^{\infty} \frac{\alpha_n}{n!} t^n \tag{5.44}$$

EXAMPLE: Let

$$f(x) = 1, \quad 0 \leq x \leq 1$$
$$= 0, \quad \text{all other } x$$

$$\psi(t) = \int_0^1 e^{tx} \, dx = \frac{e^t - 1}{t} \tag{5.45}$$

We have

$$\frac{e^t - 1}{t} = \sum_{n=0}^{\infty} \frac{t^n}{(n+1)!} = \sum_{n=0}^{\infty} \frac{1}{n+1} \frac{t^n}{n!}$$

Therefore

$$\alpha_n = \frac{1}{n+1} \tag{5.46}$$

EXAMPLE: Let $f(x)$ be the normal p.d.f. defined by Eq. (5.24); then

$$\psi(t) = \frac{1}{\sigma \sqrt{2\pi}} \int_{-\infty}^{\infty} e^{tx} e^{-(1/2)[(x-\mu)/\sigma]^2} \, dx \tag{5.47}$$

Let $(x - \mu)/\sigma = v$, $dx = \sigma\, dv$; then

$$\psi(t) = e^{\mu t + t^2\sigma^2/2} \frac{1}{\sqrt{2\pi}} \int_{-\infty}^{\infty} e^{-(v-t\sigma)^2/2}\, dv$$

$$= e^{\mu t + t^2\sigma^2/2} \tag{5.48}$$

Now let $\mu = 0$, then $\alpha_n = \mu_n$ and we have

$$\psi(t) = e^{t^2\sigma^2/2} = \sum_{n=0}^{\infty} \left(\frac{\sigma}{\sqrt{2}}\right)^{2n} \frac{(2n)!}{\left(\dfrac{2n}{2}\right)!} \frac{t^{2n}}{(2n)!}$$

$$= \sum_{n \text{ even}}^{\infty} \left(\frac{\sigma}{\sqrt{2}}\right)^{n} \frac{n!}{\left(\dfrac{n}{2}\right)!} \frac{t^n}{n!}$$

Therefore

$$\begin{aligned} \mu_n &= 0, & n &= 1, 3, 5, \cdots \\ \mu_n &= \sigma^n \frac{n!}{2^{n/2}\left(\dfrac{n}{2}\right)!} & n &= 2, 4, 6, \cdots \end{aligned} \tag{5.49}$$

EXERCISE: Show that if $\eta = a\xi + b$ and $\psi(t) = E(e^{\xi t})$, then

$$E(e^{\eta t}) = e^{bt}\psi(at) \tag{5.50}$$

Therefore the m.g.f. for the standard normal p.d.f., defined by Eq. (5.26) is $e^{t^2/2}$.

For a discrete probability distribution, the m.g.f. (cf. Eq. (5.32′)) is

$$\psi(t) \equiv E(e^{t\xi}) = \sum_i e^{tx_i} P(\xi = x_i) \tag{5.51}$$

EXAMPLE: The binomial probability distribution defined by Eq. (5.21) has the m.g.f.

$$\psi(t) = \sum_{i=0}^{n} \binom{n}{i}(pe^t)^i q^{n-i}$$

$$= (q + pe^t)^n \tag{5.52}$$

EXERCISE: What is the m.g.f. for the Poisson distribution (Eq. (5.22))?

Answer: $\psi(t) = e^{-\lambda + \lambda e^t}$. Calculate the mean and variance.

BASIC PROBABILITY AND STATISTICS

EXERCISE: Find the m.g.f. for the *exponential distribution*, defined by the p.d.f.

$$f(x;\theta) = \frac{1}{\theta} e^{-x/\theta}, \quad x > 0$$
$$= 0, \quad x \leq 0 \tag{5.53}$$

Answer: $\psi(t) = \dfrac{1}{(1-\theta t)}$

From Eq. (5.50) we also have the m.g.f. for the distribution of the random variable ξ/n, where n is some constant:

$$\psi_1(t) = \psi\left(\frac{t}{n}\right) = \frac{1}{1 - (\theta t/n)}$$

In general, the moment generating function $\psi_2(t)$ for the random variable

$$\xi = \sum_{i=1}^{n} x_i \quad \text{is} \quad \prod_{i=1}^{n} \psi_i(t)$$

where the x_i are independently distributed random variables with m.g.f.'s $\psi_i(t)$. If the distributions of the x_i are identical, then $\psi_2(t) = (\psi(t))^n$, where $\psi(t) = \psi_i(t)$, all i. For example, we shall have occasion to make use of the distribution of the quantity

$$\hat{\theta} = \sum_{i=1}^{n} \frac{x_i}{n}$$

where each x_i has the same exponential distribution (Eq. (5.53)). In this case the random variable $\hat{\theta}$ has the p.d.f.

$$g(x;\theta) = \frac{\left(\dfrac{n}{\theta}\right)^n x^{n-1} e^{-xn/\theta}}{\Gamma(n)}, \quad x > 0$$
$$= 0, \quad x \leq 0 \tag{5.54}$$

which can be shown directly by *inverting* the m.g.f.

$$\psi_2(t) \equiv \left[\psi\left(\frac{t}{n}\right)\right]^n = \left(1 - \frac{\theta t}{n}\right)^{-n}$$

by inverse Laplace transform methods. However, these methods are beyond the scope of the presentation in this book. Nevertheless, we can verify that $\psi_2(t)$ is actually the m.g.f. of the distribution defined by Eq. (5.54) by the direct Laplace transform, which is easier to apply.

This implies a certain uniqueness or one-to-one correspondence between m.g.f.'s and p.d.f.'s. Thus if we have obtained in any manner the m.g.f. for a random variable we are interested in, and the m.g.f. be identified with that for a known distribution, then the subject random variable must have the known distribution. Actually, from a theoretical viewpoint of the uniqueness problem, m.g.f.'s are not satisfactory; rather, *characteristic functions* (c.f.'s) are used, defined by $E(e^{i\xi t})$, where $i = +\sqrt{-1}$, since the m.g.f.'s do not always exist, whereas the c.f.'s always exist. (See Ref. 5, Chap. 10.)

EXERCISE: Show that the moment generating function of the distribution defined by Eq. (5.54) is $\psi_2(t) = (1 - \theta t/n)^{-n}$ by using the Laplace transform relation

$$\int_0^\infty e^{-su} u^m \, du = \frac{\Gamma(m+1)}{s^{m+1}}$$

EXERCISE: The probability distribution of the sum of k independent Poisson random variables each with parameter λ is also a Poisson distribution with parameter $k\lambda$. Generalize this result to the case where the parameters λ_i; $i = 1, 2, \cdots, k$ are different.

EXERCISE: Apply moment generating function methods to the binomial distribution. If $\nu_1, \nu_2, \cdots, \nu_k$ are independent and have binomial distributions with parameters (n_i, p), show that $\sum_{i=1}^{k} \nu_i$ has a binomial distribution with parameters $\left(\sum_{i=1}^{k} n_i, p\right)$.

5.9 PROBABILITY DISTRIBUTIONS IN TWO OR MORE DIMENSIONS

Probability distributions in two dimensions arise when one wishes to express the probability of two events simultaneously, i.e., the probability that both events occur. For example, two measurements of a rocket engine's performance are thrust, F, and mixture-ratio, r.* The frequency with which the *joint* event $F_1 < F \leq F_2$ and $r_1 < r \leq r_2$ occurs is measured by

$$P(F_1 < F \leq F_2, r_1 < r \leq r_2)\dagger$$

If the two random variables F and r are independent, the latter expression can be written as $P(F_1 < F \leq F_2) \cdot P(r_1 < r \leq r_2)$, as will be seen shortly.

---

* r = ratio of oxidizer-to-fuel weight flow rates.
† We are using commas instead of $\cap$.

BASIC PROBABILITY AND STATISTICS 99

The joint distribution function of random variables $\xi_1, \xi_2, \cdots, \xi_n$ is defined as

$$F(x_1, x_2, \cdots, x_n) \equiv P(\xi_1 \leq x_1, \xi_2 \leq x_2, \cdots, \xi_n \leq x_n) \quad (5.55)$$

with the following properties:

1. $F(x_1, x_2, \cdots, x_n)$ is nondecreasing, continuous on the right in each variable and $0 \leq F \leq 1$.
2. If any set of variables, say $x_k, \cdots, x_n$, approach $+\infty$, then the limiting value of both sides of Eq. (5.55) is the joint distribution function (called the *marginal* distribution function) of the remaining variables; e.g.,

$$\lim_{x_k, \cdots, x_n \to \infty} F(x_1, x_2, \cdots, x_n) \equiv G(x_1, \cdots, x_{k-1})$$

and

$$G(x_1, \cdots, x_{k-1}) = P(\xi_1 \leq x_1, \cdots, \xi_{k-1} \leq x_{k-1}) \quad (5.56)$$

Furthermore if *all* of the variables $x_1, \cdots, x_n \to +\infty$ then $F \to 1$.
3. If *any* of the variables $x_1, \cdots, x_n \to -\infty$, then $F \to 0$.

The above properties are obvious generalizations of the properties of a one-dimensional distribution function.

To calculate probabilities of events of the form

$$(a_1 < \xi \leq b_1) \cap (a_2 < \eta \leq b_2)$$

where $F(x, y)$ is the joint distribution function of ξ, η, we proceed as follows. The event to be measured is (geometrically) the shaded region exclusive of the dashed boundaries shown in Fig. 5.2. We have

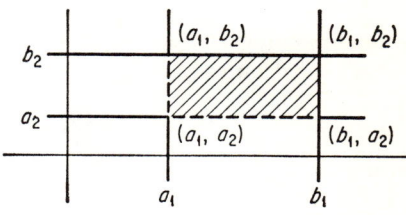

Figure 5.2

$$P(a_1 < \xi \leq b_1, a_2 < \eta \leq b_2) = P(\xi \leq b_1, \eta \leq b_2)$$
$$- P(\xi \leq b_1, \eta \leq a_2) - P(\xi \leq a_1, \eta \leq b_2) + P(\xi \leq a_1, \eta \leq a_2) \quad (5.57)$$
$$P(a_1 < \xi \leq b_1, a_2 < \eta \leq b_2)$$
$$= F(b_1, b_2) - F(b_1, a_2) - F(a_1, b_2) + F(a_1, a_2) \quad (5.58)$$

To see this, note that $P(\xi \leq b_1, \eta \leq b_2)$ measures the probability in the "southwest" quadrant, including the probability on the boundaries, whose corner is (b_1, b_2). $P(\xi \leq b_1, \eta \leq a_2)$ gives the probability in the southwest quadrant (plus probability on boundaries) whose corner is (b_1, a_2). Similarly for the next term $P(\xi \leq a_1, \eta \leq b_2)$. Finally the last term "adds back" the probability contained in the southwest quadrant, plus the probability on the boundaries, whose corner is (a_1, a_2)—since this probability was subtracted off twice by the second and third terms of Eq. (5.57).

If one writes $a_1 = x$, $b_1 = x + h_1$, $a_2 = y$, $b_2 = y + h_2$, where h_1 and h_2 are positive, then Eq. (5.58) becomes

$$P(x < \xi \leq x + h_1, y < \eta \leq y + h_2)$$
$$= F(x + h_1, y + h_2) - F(x + h_1, y)$$
$$- [F(x, y + h_2) - F(x, y)] \quad (5.59)$$

Dividing Eq. (5.59) by hk, if the partial derivative $\partial^2 F/\partial x\, \partial y$ exists at the point (x, y), then as $h, k \to 0$, the lefthand side tends to the probability density function at the point (x, y). Thus we define the p.d.f. for two variables ξ, η in terms of their joint distribution function as

$$f(x, y) = \frac{\partial^2 F}{\partial x\, \partial y} \quad (5.60)$$

Similarly, in n dimensions:

$$f(x_1, \cdots, x_n) = \frac{\partial^n F}{\partial x_1\, \partial x_2 \cdots \partial x_n} \quad (5.61)$$

Two random variables are said to be *independent* when their joint distribution function can be expressed as

$$F(x, y) = F_1(x) F_2(y) \quad (5.62)$$

where
$$F_1(x) = \lim_{y \to +\infty} F(x, y)$$

and
$$F_2(y) = \lim_{x \to +\infty} F(x, y)$$

i.e., $F_1(x)$ and $F_2(y)$ are the (marginal) distribution functions of the variables ξ and η respectively. Equation (5.62) can be taken as the definition of independence for two random variables ξ, η. The expression for n random variables $\xi_1, \xi_2, \cdots, \xi_n$ in the independence case is, similarly,

$$F(x_1, \cdots, x_n) = \prod_{j=1}^{n} F_j(x_j) \quad (5.63)$$

where $F_j(x_j)$ is the marginal distribution function of ξ_j. Applying the independence condition to Eq. (5.58),

$$P(a_1 < \xi \leq b_1, a_2 < \eta \leq b_2)$$
$$= [F_1(b_1) - F_1(a_1)] \cdot [F_2(b_2) - F_2(a_2)] \quad (5.64)$$

BASIC PROBABILITY AND STATISTICS

Equation (5.64) shows that in the *independence* case

$$P(I_1, I_2) = P(I_1)P(I_2) \tag{5.65}$$

where I_1, I_2 are half open intervals of the type $a < \xi \leq b$. Actually the multiplication rule, Eq. (5.65), holds for general sets of points, other than intervals, for which probability can be assigned. In fact, we have already used Eq. (5.65) where I_1, I_2 are considered as single points. From the previous remark, it also holds when I_1, I_2 denote a finite or denumerable set of points, which are events of a discrete sample space.

EXAMPLE: Let X and Y be independent and each have binomial probability distributions with the parameters $n = 2$ in each case and p_1, p_2 respectively. What is the probability that both X and Y are not equal to 0?

Answer:

$$\begin{aligned} P(X \neq 0, Y \neq 0) &= P_1(X \neq 0)P_2(Y \neq 0) \\ &= [1 - P_1(X = 0)][1 - P_2(Y = 0)] \\ &= (1 - q_1^2)(1 - q_2^2) \end{aligned}$$

When the joint p.d.f. exists, then in general

$$F(x, y) = \int_{-\infty}^{x} \int_{-\infty}^{y} f(t, u) \, du \, dt \tag{5.66}$$

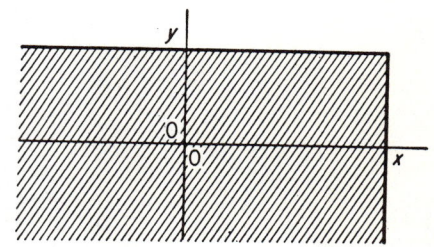

Figure 5.3

Equation (5.66) states that the double integration of the joint p.d.f. over the shaded region shown in Fig. 5.3 gives the joint distribution function of ξ and η. The (marginal) distribution function of ξ is obtained by letting $y \to \infty$ in Eq. (5.66), obtaining

$$F_1(x) = \int_{-\infty}^{x} dt \left[\int_{-\infty}^{\infty} f(t, u) \, du \right] = \int_{-\infty}^{x} f_1(t) \, dt \tag{5.67}$$

and similarly for $F_2(y)$. In Eq. (5.67) $f_1(x)$ is the (marginal) p.d.f. for ξ, where

$$f_1(x) = \int_{-\infty}^{\infty} f(x, u) \, du \tag{5.68}$$

Similarly

$$f_2(y) = \int_{-\infty}^{\infty} f(t, y) \, dt \tag{5.69}$$

is the (marginal) p.d.f. for η.

When the variables ξ and η are independent, application of Eqs. (5.60) and (5.62) yields

$$f(x, y) = f_1(x)f_2(y) \tag{5.70}$$

5.9.1 CONDITIONAL DISTRIBUTIONS

Suppose we are given that the random variable η lies in the strip $y < \eta < y + k$. The conditional probability that $\xi \leq x$ is then

$$P(\xi \leq x \mid y < \eta < y + k) = \frac{P(\xi \leq x, y < \eta < y + k)}{P(y < \eta < y + k)} \tag{5.71}$$

$$= \frac{\int_{-\infty}^{x} \int_{y}^{y+k} f(t, u) \, du \, dt}{\int_{y}^{y+k} f_2(u) \, du} \tag{5.72}$$

or

$$P(\xi \leq x \mid y < \eta < y + k) = \frac{\frac{1}{k} \int_{y}^{y+k} du \left[\int_{-\infty}^{x} f(t, u) \, dt \right]}{\frac{1}{k} \int_{y}^{y+k} f_2(u) \, du} \tag{5.73}$$

If, now, $k \to 0$, the righthand side of Eq. (5.73) tends to the limit

$$\frac{\int_{-\infty}^{x} f(t, y) \, dt}{f_2(y)}$$

and we have

$$P(\xi \leq x \mid \eta = y) = \frac{\int_{-\infty}^{x} f(t, y) \, dt}{f_2(y)} \tag{5.74}$$

which is called the conditional distribution function of ξ given that $\eta = y$. If ξ and η were independent then

$$\int_{-\infty}^{x} f(t, y) \, dt = f_2(y) \int_{-\infty}^{x} f_1(t) \, dt$$
$$= f_2(y) P(\xi \leq x)$$

so Eq. (5.74) would become

$$P(\xi \leq x \mid \eta = y) = P(\xi \leq x) \qquad (5.75)$$

which shows that the probability of the event $\xi \leq x$ is the same irrespective of any hypothesis on η, as it should be for independence. Going back to Eq. (5.74) again, we see that by differentiating with respect to x, we obtain the *conditional p.d.f.* of ξ, given $\eta = y$ (with obvious notation)

$$f(x \mid y) = \frac{f(x, y)}{f_2(y)} \qquad (5.76)$$

EXERCISE: Show that Eq. (5.74) defines a distribution function.

EXERCISE: Find $f(y \mid x)$, knowing $f_1(x)$ and $f(x, y)$.

EXERCISE:

$$f(x, y) = 2(x + y - 2xy), \qquad 0 \leq x \leq 1, 0 \leq y \leq 1$$
$$= 0, \qquad \text{all other } (x, y)$$

Find the marginal p.d.f.'s, the conditional p.d.f.'s, the joint and conditional distribution functions. What would the joint p.d.f. be if the variables were independent, with the same marginal p.d.f.'s?

EXERCISE: Show that the analogue of Eq. (5.68) for discrete distributions is (for two random variables):

$$P_1(\xi = x_i) = \sum_j P(\xi = x_i, \eta = y_j) \qquad (5.77)$$

5.9.2 COVARIANCE

In Chaps. 7 and 8 we shall be making use of *estimators* of parameters of one-dimensional probability distributions in estimating the reliability functions associated with the probability distribution (e.g., distribution of time-to-failure). If there are two (or more) parameters in the basic probability distribution, their estimators, being functions of the sample of data, would be regarded as random variables (cf. Sec. 5.10) which are generally *not independent*. The estimator for the reliability function will itself be a function of the parameter estimators; and in order to evaluate its variance, it will be necessary to evaluate the covariance (a measure of the degree of dependence) of the parameter estimators. The following paragraphs present the basic formulas for calculating covariance and correlation coefficient (dimensionless covariance).

Consider two random variables ξ, η with a joint probability density function $f(x, y)$. The covariance of the random variables ξ, η is defined as

$$\text{Cov}(\xi, \eta) = \int_{-\infty}^{\infty} \int_{-\infty}^{\infty} (x - \mu_x)(y - \mu_y) f(x, y) \, dx \, dy \quad (5.78)$$

where $E(\xi) = \mu_x$, $E(\eta) = \mu_y$ are the respective *means* of the variables ξ and η. For discrete distributions, the analogous formula is

$$\text{Cov}(\xi, \eta) = \sum_i \sum_j (x_i - \mu_x)(y_j - \mu_y) P(\xi = x_i, \eta = y_j) \quad (5.79)$$

In either case, we can write the covariance in terms of expected values (cf. Sec. 5.7) with respect to the joint distribution of ξ, η:

$$\text{Cov}(\xi, \eta) = E[(\xi - \mu_x)(\eta - \mu_y)] \quad (5.80)$$

Equation (5.80) may be simplified by expanding the terms in the brackets to obtain

$$\text{Cov}(\xi, \eta) = E[\xi\eta - \mu_x\eta - \mu_y\xi + \mu_x\mu_y]$$
$$= E(\xi\eta) - \mu_x\mu_y \quad (5.81)$$

EXERCISE: Derive Eq. (5.81) from (5.80) using (5.78) or (5.79).

When the random variables ξ, η are independent, then $E(\xi\eta) = \mu_x\mu_y$, thus making $\text{Cov}(\xi, \eta) = 0$. In general, however, the covariance may be positive or negative.

EXERCISE: Use Eqs. (5.78) and (5.70) to show that $E(\xi\eta) = E(\xi)E(\eta) = \mu_x\mu_y$ when ξ and η are independent random variables (the converse is not necessarily true; see Ref. 1, p. 222).*

The correlation coefficient $\rho(\xi, \eta)$ is defined as

$$\rho(\xi, \eta) = \frac{\text{Cov}(\xi, \eta)}{\sqrt{\text{Var } \xi \cdot \text{Var } \eta}} \quad (5.82)$$

It can be shown that $-1 \leq \rho \leq 1$ (Ref. 5, pp. 263–264). When $\rho = \pm 1$, we say that the variables ξ, η show perfect correlation, or to be more precise: ξ can be expressed as a linear function of η (and vice versa) with probability one if and only if $\rho = \pm 1$.

* The results of the two exercises can be summed up as follows:

$$\text{Independence} \Rightarrow [E(\xi\eta) = \mu_x\mu_y \Leftrightarrow \text{Cov}(\xi, \eta) = 0]$$

BASIC PROBABILITY AND STATISTICS 105

EXERCISE: Find the correlation coefficient for the random variables ξ, η whose p.d.f. is

$$f(x, y) = 2(x + y - 2xy), \quad 0 \leq x \leq 1$$
$$0 \leq y \leq 1$$
$$= 0, \quad \text{all other } (x, y)$$

(Refer to the exercise preceding Eq. (5.77)).

EXERCISE: Find the variance of the function $a\xi \pm b\eta$, where a and b are constants.

Answer: $\text{Var}(a\xi \pm b\eta) = a^2 \text{Var } \xi + b^2 \text{Var } \eta \pm 2ab \text{ Cov}(\xi, \eta)$.

EXERCISE: Find the variance of the function $\xi\eta$. (Assume that ξ^2 and η^2 are uncorrelated.)

Answer:

$$\text{Var}(\xi\eta) = \mu_x^2 \text{Var } \eta + \mu_y^2 \text{Var } \xi + \text{Var } \xi \cdot \text{Var } \eta$$
$$- \overline{\text{Cov}(\xi, \eta)}^2 - 2\mu_x\mu_y \text{Cov}(\xi, \eta)$$

5.10 SAMPLING THEORY

In this section we will cover some of the basic aspects of sampling theory. A *sample* is obtained when a series of observations or measurements is made in a series of *random experiments*. Fundamental to the notion of random experiment is the *sample space* which was pointed out in Sec. 5.1 as the set of all possible outcomes of the experiment.

Let ξ denote the outcome of a random experiment. For example, when the experiment is to flip a coin once, we agree beforehand that the sample space consists only of two outcomes: heads (H) or tails (T). The meaning of the term "random experiment" lies in the fact that we do not necessarily know beforehand which outcome will take place when the experiment is performed. However, there are definite, though possibly unknown, probabilities that can be assigned to each of the possible outcomes of the experiment (*cf.* Sec. 5.3). Thus the quantity ξ is a *function* of the experiment; we say it is a random function or more commonly *random variable*, since we cannot say with certainty before the experiment is performed what its value will be. As stated in Sec. 5.4, we deliberately make the random variable ξ a real number by mapping the outcome (H, T) into (say) (0, 1), respectively, and attach the probabilities assigned to H and T to the numbers 0 and 1, respectively. Thus, e.g., instead of saying "the flip of the coin resulted in heads," we equivalently say "the observed value of ξ is a '0'."

5.10.1 SIMPLE RANDOM SAMPLING—RANDOM SAMPLE

The process of *simple random sampling* has a particular definition which we shall state shortly. First consider a series of n random experiments connected with a one-dimensional random variable (n flips of a coin; n measurements of the diameter of a shaft; etc.).* The outcomes can be denoted by the series of real numbers $x_1, x_2, \cdots, x_n$. Thus on the first experiment we say that we would observe $\xi = x_1$, on the second $\xi = x_2$, and so on. Actually, then, the quantities $x_1, x_2, \cdots, x_n$ *considered before the experiment is undertaken* form a set of n random variables, evidently associated in some way with the "parent" random variable ξ.† We say that we have a process of simple random sampling when two conditions apply: (1) a particular outcome (say, the particular outcome: heads) has the same probability of being observed in each experiment; and (2) the random variables $x_1, \cdots, x_n$ are *mutually independent* (*cf.* Sec. 5.9). The set of *observed* values of $x_1, \cdots, x_n$ are also then said to form a *simple random sample*. (We will, as is common, abbreviate the latter term to *random sample*.) The referred-to association of the random variables x_i with ξ is that in a process of simple random sampling, each x_i has the same probability distribution as the random variable ξ. Thus the underlying distribution function $F(x) = P(\xi \leq x) = P(x_i \leq x); i = 1, \cdots, n$ (see Sec. 5.4).

There are many other types of sampling procedures discussed in textbooks on applied statistics; e.g., *sequential sampling*, *censored sampling*, *stratified sampling*, *purposive sampling*, *quota sampling*, etc. In particular the procedures used to achieve randomness and also the aforementioned types of sampling are discussed in Refs. 8 and 9. We will encounter the first two types in Chap. 10.

5.10.2 DISTRIBUTION OF THE SAMPLE

Throughout this section we will be speaking only of simple random sampling. We define the *distribution of the sample* (or *sample distribution*) as that probability distribution obtained by assigning probability $1/n$ to each of the n "points" $x_1, \cdots, x_n$. This distribution is evidently discrete and therefore its distribution function is a step function with steps of height $1/n$ at each of the points $x_1, \cdots, x_n$ (*cf.* Sec. 5.6). Figure 5.4 shows a typical *sample distribution function* for a random sample, denoted by

---

* A two-dimensional random variable would be associated with two observed outcomes for each random experiment; e.g., diameter *and* length of the shaft. Also note that we are talking about a *single* random variable; i.e., all n random experiments consist of doing the same thing and not, for example, flipping a coin on the first experiment, rolling a die on the second, and so on.

† We shall also use the terms *underlying random variable*, *underlying population*, *underlying distribution*, etc.

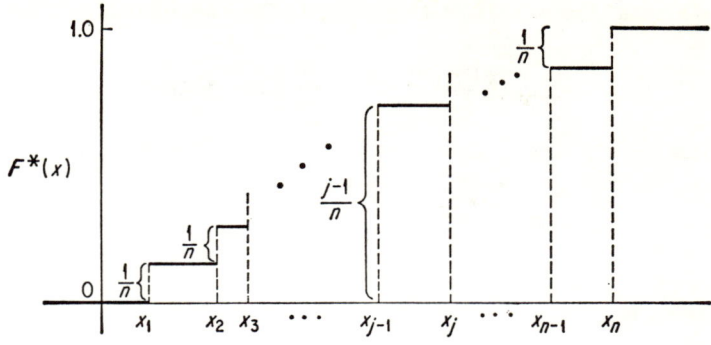

Figure 5.4

$F^*(x)$. Note that we have relabeled the points $x_1, \cdots, x_n$ so that $x_1 \leq x_2 \leq \cdots \leq x_n$.† Analytically, $F^*(x)$ may be expressed by

$$F^*(x) = \frac{j-1}{n}, \quad x_{j-1} \leq x < x_j \quad (j = 1, 2, \cdots, n+1) \quad (5.83)$$

where $\quad x_0 \equiv -\infty, \, x_{n+1} \equiv +\infty.$

Alternatively, if ν denotes the number of sample values $\leq x$, then

$$F^*(x) = \frac{\nu}{n} \quad (5.84)$$

We see that $F^*(x)$ is then the observed frequency in n experiments, of the event $\xi \leq x$, whose true probability is $F(x)$, the distribution function of the parent random variable ξ. It is also apparent that $F^*(x)$ is a random variable since it is a function of the outcomes of the experiments. The importance of the quantity $F^*(x)$ is that for large samples $(n \to \infty)$, $F^*(x)$ converges in probability to $F(x)$ for any fixed x. The meaning of this property is that by taking a large enough sample, we can, with probability approaching one, determine the underlying distribution function to as close an approximation as we wish. Section 6.2.2 contains a proof of the latter statement. Unfortunately, a large number of observations is generally required to do this (see Secs. 6.2.2, 6.2.3).

5.10.3 MOMENTS OF THE SAMPLE DISTRIBUTION

We now consider the sample distribution function $F^*(x)$ as *fixed* for the moment, i.e., ignoring the nature of the x_j as random variables. From

† $x_1, x_2, \cdots, x_n$ is not necessarily the order in which the values of ξ are observed.

Eq. (5.32') and the remark at the beginning of Sec. 5.10.2, it follows that

$$E[g(\eta^*)] = \frac{1}{n} \sum_i g(x_i) \qquad (5.85)†$$

Hence when $g(\eta^*) = \eta^{*k}$, the kth *ordinary moment of the sample distribution* is

$$a_k \equiv E(\eta^{*k}) = \frac{1}{n} \sum_i x_i^k \qquad (5.86)$$

Correspondingly the kth *moment about* the *mean* or kth *central moment* is

$$m_k \equiv E[(\eta^* - a_1)^k] = \frac{1}{n} \sum_i (x_i - a_1)^k \qquad (5.87)$$

Note that we distinguish between sample moments and underlying distribution moments by using Latin letters for the former (a_k, m_k, $\cdots$) and Greek letters for the latter (α_k, μ_k, $\cdots$). The *sample mean* is given by a_1 in Eq. (5.86) and is more commonly denoted by m or $\bar{x}$. The *sample variance* is given by m_2 in Eq. (5.87) and is commonly denoted by s^2. Thus, the sample mean and variance are

$$\bar{x} = \frac{1}{n} \sum_i x_i \qquad (5.88)$$

and

$$s^2 = \frac{1}{n} \sum_i (x_i - \bar{x})^2 \qquad (5.89)‡$$

We now reconsider the sample values as random variables, each having the same distribution function $F(x)$ as the parent random variable ξ, whose mean and variance are given by μ and σ^2, respectively (cf. Sec. 5.7). Then the sample moments (Eqs. (5.86) through (5.89)) are themselves regarded as random variables, since they are functions of the random variables $x_1, \cdots, x_n$. Evidently the probability distributions of the sample moments are uniquely determined by the parent distribution function $F(x)$. It follows, also, that the sample moments themselves have moments which are uniquely determined by $F(x)$. We shall only consider here the first two moments (denoted by $E(\cdot)$ and Var $(\cdot)$) of the sample moments $\bar{x}$ and s^2, For calculations of any moment of any sample moment, the reader is referred to Ref. 5, Chap. 27.

† Here we define η^* as the random variable with distribution function $F^*(x)$. Also note that summations are always from 1 to n.

‡ There is sometimes confusion between the sample variance given by Eq. (5.89) and an *unbiased* estimator for the parent distribution variance σ^2. An unbiased estimator for σ^2 is denoted by s'^2 in Chap. 5–9 of this book, where $s'^2 = [n/(n-1)]s^2$. See also Sec. 7.4.4.

BASIC PROBABILITY AND STATISTICS   109

First we note three properties of *expectation*. The first property is that the expected value of the sum of any finite number of random variables, whether independent or not, is the sum of the expected values of the random variables. This follows for our purposes from Eq. (5.32) or (5.32′), by the elementary properties of the integral (or summation sign) (i.e., the integral of a sum is the sum of the integrals). The second property is that the expected value of a product of any finite number of *mutually independent* random variables is equal to the product of the expected values of the random variables. This follows from the property of a multiple integral to be expressible as a repeated integral (see Ref. 5, p. 173). The third property is simply that the expected value of a constant times the random variable is equal to the constant multiplied by the expected value of the random variable. This also follows from the properties of integrals (or summations). We will label these three properties $P1$, $P2$, and $P3$, respectively.

EXERCISE: (a) Show that the expected value of a constant is the constant itself, (b) show that the variance of a constant is zero.

From Eq. (5.88), using $P1$ and $P3$,

$$E(\bar{x}) = \frac{1}{n} \sum E(x_i) = \frac{n\mu}{n} = \mu \qquad (5.90)$$

This shows that the expected value of the sample mean is the underlying distribution mean.

By the definition of second central moment (Sec. 5.7)

$$\text{Var}(\bar{x}) \equiv E[(\bar{x} - \mu)^2] \qquad (5.91)$$

By $P1$, $P3$, part (a) of the previous exercise and Eq. (5.90)

$$E[(\bar{x} - \mu)^2] = E[\bar{x}^2 - 2\mu\bar{x} + \mu^2]$$
$$= E(\bar{x}^2) - 2\mu E(\bar{x}) + \mu^2$$
$$= E(\bar{x}^2) - 2\mu^2 + \mu^2$$
$$= E(\bar{x}^2) - \mu^2 \qquad (5.92)$$

Now, it remains to find $E(\bar{x}^2)$. We have

$$\bar{x}^2 = \frac{1}{n^2} \left(\sum_{i=1}^{n} x_i \right)^2 \qquad (5.93)$$

$$= \frac{1}{n^2} \left[\sum_{i=1}^{n} x_i^2 + 2 \sum_{i<j} x_i x_j \right] \qquad (5.94)$$

For the first sum in the brackets, by $P3$

$$E\left(\sum x_i^2\right) = \sum E(x_i^2)$$

$$= n\alpha_2$$

$$= n(\mu^2 + \sigma^2) \qquad (cf.\ \text{Sec. 5.7}) \qquad (5.95)$$

The second sum consists of $n(n-1)/2$ products $x_i x_j$; each x_i is different from the x_j (all terms x_i^2 have already been put in the first sum) and the x_i, x_j are independent. By $P2$, $E(x_i x_j) = E(x_i)E(x_j) = \mu \cdot \mu = \mu^2$. Thus the summation yields

$$\frac{2n(n-1)}{2}\mu^2 = n(n-1)\mu^2$$

Taking into account the coefficient $1/n^2$ outside the bracket and property $P3$, we have

$$E(\bar{x}^2) = \frac{1}{n^2}\left[n(\mu^2 + \sigma^2) + n(n-1)\mu^2\right]$$

$$= \mu^2 + \frac{\sigma^2}{n} \qquad (5.96)$$

From Eq. (5.92), using (5.96), we finally have

$$\text{Var}(\bar{x}) = \frac{\sigma^2}{n} \qquad (5.97)$$

Thus the variance of the mean of a random sample is the parent distribution variance reduced by a factor $1/n$. Evidently when n is sufficiently large the uncertainty in $\bar{x}$ can be made arbitrarily small. A consequence of this property of the sample mean is given in Sec. 7.2.2.

By the same type of calculations it may be shown (Ref. 5, pp. 347–348) that

$$E(s^2) = \frac{n-1}{n}\sigma^2 \qquad (5.98)$$

$$\text{Var}(s^2) = \frac{\mu_4 - \sigma^4}{n} - \frac{2(\mu_4 - 2\sigma^4)}{n^2} + \frac{\mu_4 - 3\sigma^4}{n^3} \qquad (5.99)$$

When $F(x)$ is a normal distribution function with mean μ, variance σ^2, we found that $\mu_4 = 3\sigma^4$ (see Eq. (5.49)). In this case $\text{Var}(s^2)$ reduces to $2\sigma^4(n-1)/n^2$.

REFERENCES

1. W. Feller, *An Introduction to Probability Theory and Its Applications*, 2d ed., John Wiley & Sons, Inc., New York, 1957, vol. I.
2. E. Parzen, *Modern Probability Theory and Its Applications*, John Wiley & Sons, Inc., New York, 1960.
3. A. M. Mood, *Introduction to the Theory of Statistics*, McGraw-Hill Book Company, Inc., New York, 1950.
4. A. H. Bowker and G. J. Lieberman, *Engineering Statistics*, Prentice-Hall, Inc., Englewood Cliffs, N.J., 1959.
5. H. Cramér, *Mathematical Methods of Statistics*, Princeton University Press, 1946.
6. G. Birkhoff and S. MacLane, *A Survey of Modern Algebra*, The Macmillan Company, New York, 1947, chap. XI.
7. J. G. Kemeny, H. Mirkil, J. L. Snell, and G. L. Thompson, *Finite Mathematical Structures*, Prentice-Hall, Inc., Englewood Cliffs, N.J., 1959, chaps. 1–3.
8. F. E. Croxton and D. J. Cowden, *Applied General Statistics*, 2d ed., Prentice-Hall, Inc., Englewood Cliffs, N.J., 1955.
9. W. E. Deming, *Some Theory of Sampling*, John Wiley & Sons, Inc., New York, 1950.
10. *Tables of Normal Probability Functions*, National Bureau of Standards, Applied Mathematics Series 23, 1953.

ADDITIONAL READING

Hoel, P. G., *Introduction to Mathematical Statistics*, Second Edition, John Wiley & Sons, Inc., New York, 1954.

Siegel, S., *Nonparametric Statistics: For the Behavioral Sciences*, McGraw-Hill Book Company, Inc., New York, 1956.

Uspensky, J. V., *Introduction to Mathematical Probability*, McGraw-Hill Book Company, Inc., New York, 1937.

CHAPTER SIX
DISCRETE AND CONTINUOUS DISTRIBUTION MODELS

6.1 INTRODUCTION

As we have seen in Chap. 5, measures (probabilities) of events of the type "a device performs as intended," "the time-to-failure of a device exceeds T," "performance parameter(s) of a device are within certain limits," etc., combinations of such events, or other related events are defined by probability distributions. All three of the quoted events can be expressed as the simple event "Success" and their probability called "Reliability." However, because of the particular failure characteristics of the device, one may, for various reasons, choose to measure an event different from, but related to, the event "Success." An example which is almost trivial but directly to the point is that in many instances we would measure the event "Failure" as the ratio of the observed number of failures to trials. What we really want is a measure of "Success," however, but this is simply defined as one minus the measure of the event "Failure."

Another example which is not quite so simple nor trivial is to measure the event "time-to-failure exceeds T." We may choose to observe merely one of the two events: (1) the device fails on or before time T (Failure), (2) the device fails after time T (Success). This situation is exactly equivalent to that of the previous example. However, we may wish to extend the experiment to observations of the actual times at which failures occur. To make use of this additional information to measure the event "time-to-failure exceeds T" requires the specification of a probability distribution of time-to-failure for the sample space $t > 0$. To specify a relevant probability distribution may be a difficult task, as the mathematical form of the distribution would depend upon the "mechanism" of failure of the particular device. Fortunately, as a great deal of experience has shown, there are a few probability distributions whose applicability is almost universal. One of especial importance to reliability is the *exponential distribution* of time-to-failure. The aim of the latter half of this chapter will be to derive the exponential as well as related distributions of time-to-failure under plausible mathematical-physical assumptions. One method of deriving such distributions is by the use of *hazard functions*. The derivations are primarily based on the principle of choosing the simplest forms of hazard function, with the underlying philosophy that simplicity of assumptions should yield results of general application. Another model used is related to the so-called "pure death-process" (Ref. 1, p. 434). It will be seen that a surprising amount of generality is obtained by a simple modification of the *pure death-process* model, leading to a different derivation of the exponential distribution, the *gamma* distribution, as well as a combined *exponential-gamma* distribution of time-to-failure.

For almost all such time-to-failure distributions, there are certain limiting approximations to the important *normal* distribution, which will also be discussed briefly in Sec. 6.10 of this chapter.

The normal distribution is very generally applicable in describing performance characteristics or *outputs* of devices. In fact, when non-normality is detected, it is often indicated that "something is wrong"; and usually the nature of the discrepancy can be found and eliminated. Thus we can in a large variety of instances measure such events as "transconductance is within certain limits" by use of the normal probability distribution.

In the cases where a specific mathematical form of a distribution of time-to-failure or performance output is *a priori*, unknown, or difficult or inconvenient to use, we can always fall back to direct observation of the events "Success" or "Failure." This is a procedure of great generality which leads to the experimental concept of binomial *trials*. In the apparently simple case of binomial trials, it can be shown that the distribution model used clearly depends on the specified set of outcomes of the experiment, i.e., the *sample space*. Thus, e.g., the *binomial distribution* applies when the number of trials is fixed in advance; however, when the experiment is to be terminated at a fixed number of failures, the *negative binomial* distribution applies. Even in the case of such discrete distribution models, we find under certain limiting conditions that the normal distribution can be used to calculate "binomial" probabilities—another example of its universality.

Since the discrete distribution models are of greater generality, their discussion is given in the first part of this chapter, followed by the aforementioned sections on continuous distribution models.

6.2 THE BINOMIAL DISTRIBUTION

We have previously introduced the binomial distribution in Sec. 5.4, where it was shown that if there were two mutually exclusive outcomes of an experiment—e.g., Failure (F) and Success (S)—with respective probabilities p and $q \equiv 1 - p$, and N independent experiments (trials) were performed, then the number ν of failures in the N trials would be measured by the binomial distribution with parameters N, p:

$$P(\nu = f) = \binom{N}{f} p^f q^{N-f} \qquad (6.1)$$

Evidently $N - \nu$ is the number of successes observed in the N trials. Since the event $N - \nu = k$ is equivalent to the event $\nu = N - k$, the two events must have the same probability. Thus, from Eq. (6.1) the distribution of the number of successes in N trials is given by

$$P(N - \nu = k) = P(\nu = N - k) = \binom{N}{N-k} p^{N-k} q^k \qquad (6.2)$$

that is, we replace f in Eq. (6.1) by $N - k$. If we denote $\xi = N - \nu$, then using the well-known identity

$$\binom{N}{N-k} = \binom{N}{k} \qquad (6.3)$$

Equation (6.2) becomes

$$P(\xi = k) = \binom{N}{k} q^k p^{N-k} \qquad (6.4)$$

Thus the number of successes ξ has a binomial distribution with parameters N, q.

For a given value of p between zero and one, Eq. (6.1) or (6.4) enables us to calculate the probability of any specified number of failures or successes in a given number of trials or experiments conducted under the conditions stated in the beginning of this section. It is important to note that the number N either be fixed in advance of conducting the experiments or be independent of the results of any of the trials. Equations (6.1) or (6.4) would not give correct probabilities if, for example, we decided in the course of making trials to "wait for just one more success before stopping the experiment."

As an example, suppose we knew the reliability $R \equiv q$ of a guided missile to be 0.90. What is the probability of at least nine successful flights in ten attempts? The event E we wish to measure then is the *union* of the two events: "exactly nine successes," "exactly ten successes"; thus

$$P(E) = P[(\xi = 9) \cup (\xi = 10)] \qquad (6.5)$$

Each of these events is a sample point of the sample space of all possible outcomes of ten trials. Thus they are mutually exclusive, and the proba-

DISCRETE AND CONTINUOUS DISTRIBUTION MODELS 115

bility of their union is the sum of the individual probabilities. From Eq. (6.4)

$$P(E) = \binom{10}{9}(0.90)^9(0.10) + \binom{10}{10}(0.90)^{10}$$
$$= 10(0.387)(0.10) + 1(0.349) \tag{6.6}$$
or
$$P(E) = 0.736 \tag{6.7}$$

When the number N is large, a calculation of the above type becomes exceedingly tedious even if the number of successes is only moderately less than N. There are several very extensive tables of sums of binomial probabilities, e.g., Ref. 3, which can be used for this purpose.

6.2.1 MOMENTS OF THE BINOMIAL DISTRIBUTION

The moment generating function for the distribution of the number of failures, Eq. (6.1), was shown (Eq. (5.52)) to be

$$\psi(t) = (q + pe^t)^N \tag{6.8}$$

By differentiating Eq. (6.8) with respect to t, then setting $t = 0$, we know from Sec. 5.8 that the result is the mean of the random variable ν. Thus

$$\psi'(t) = N(q + pe^t)^{N-1}pe^t \tag{6.9}$$

and therefore

$$\psi'(0) = E(\nu) = pN(q + p)^{N-1} = Np \tag{6.10}$$

since $q + p = 1$.

Differentiating again, and setting $t = 0$, we obtain the second ordinary moment

$$\psi''(t) = N(q + pe^t)^{N-1}pe^t + N(N-1)(q + pe^t)^{N-2}p^2e^{2t} \tag{6.11}$$

and
$$\psi''(0) = Np + N(N-1)p^2$$

or
$$E(\nu^2) = Npq + N^2p^2 \tag{6.12}$$

Thus the variance or second moment of ν about the mean is

$$\text{Var } \nu = E(\nu^2) - E^2(\nu) = Npq \tag{6.13}$$

Higher moments may be calculated in the same manner, although the calculation rapidly becomes more lengthy.

The mean Np is the *expected* or average number of failures in N trials. In the previous example where $q = 0.9$ and therefore $p = 0.1$, we would expect to observe one failure in ten missile flights. This also means more generally that as the number of attempted launches, N, increases, the total number of missile failures, divided by N, should tend in some manner to the number $p = 0.1$. This is actually a theoretical result called the *law of large numbers*, which we shall derive next.

6.2.2 THE LAW OF LARGE NUMBERS

To prove the law of large numbers we shall use Tchebycheff's inequality (Sec. 5.7.1), which was proved for continuous random variables, but which was stated to be true for any random variable for which at least the first two moments exist. We actually wish to apply Tchebycheff's inequality to the random variable ν/N, for which $E(\nu/N) = p$ and Var $(\nu/N) = pq/N$. Thus, Tchebycheff's inequality gives

$$P\left(\left|\frac{\nu}{N} - p\right| \geq k\sqrt{\frac{pq}{N}}\right) \leq \frac{1}{k^2} \qquad (6.14)$$

where k is any positive number. If we let $k\sqrt{pq/N} \equiv \epsilon$, then Eq. (6.14) becomes

$$P\left(\left|\frac{\nu}{N} - p\right| \geq \epsilon\right) \leq \frac{pq}{N\epsilon^2} \qquad (6.15)$$

The latter inequality states that no matter how small we pick $\epsilon > 0$, and $\delta > 0$, then for any N (equal to or greater than the next larger integer in $pq/\epsilon^2\delta$), the observed failure ratio ν/N will differ from p by a quantity equal to or greater than ϵ, with probability less than δ. We say in this case that the random variable ν/N *converges in probability* to the constant p. For example, if $\delta = 0.01$ and $\epsilon = 0.01$ are chosen in advance, and $p = 0.1$, then for any

$$N \geq \frac{(0.1)(0.9)}{10^{-6}} = 90{,}000$$

the probability that the failure ratio differs from 0.1 by a quantity equal to or greater than 0.01 is less than 0.01. Incidentally, since $pq \leq \frac{1}{4}$ for any p between zero and one, then even if p were not known, the same inequality holds, provided that $N \geq 1/4\epsilon^2\delta$.

6.2.3 THE NORMAL APPROXIMATION

The result (Eq. (6.15)) can be replaced by a much more precise expression, involving the limiting approximation of the binomial distribution by the normal distribution (Sec. 6.10). We merely state one form of the result here:

For large N

$$P\left[\left|\frac{\nu}{N} - p\right| \geq \epsilon\right] \simeq 2\left[1 - \Phi\left(\epsilon\sqrt{\frac{N}{pq}}\right)\right] \qquad (6.16)*$$

* Much more precise forms of this limit law are available [see, e.g., W. Feller, "On the Normal Approximation to the Binomial Distribution," *Ann. Math. Stat.*, **16** (1945), 319–329]

where Npq should be (say) ≥ 25 and Φ denotes the standard normal distribution function

$$\Phi(z) = \frac{1}{\sqrt{2\pi}} \int_{-\infty}^{z} e^{-t^2/2}\, dt \qquad (6.17)$$

We can make a direct numerical comparison using Eq. (6.16) and the previous example. We set $\delta = 0.01$ equal to the right side of Eq. (6.16) and solve for N using tables of the normal distribution function (Ref. 2; see also Sec. 6.10)

$$0.01 = 2\left[1 - \Phi\!\left(\frac{1}{30}\sqrt{N}\right)\right] \qquad (6.18)$$

and N turns out to be approximately 5970. While $Npq = 5.4$ is less than 25, it is very likely that at most (say) 7000 trials would be needed to observe a failure ratio "close" to the true value of $p = 0.1$ in the sense of Eq. (6.15).

6.2.4 UNBIASED ESTIMATOR OF p

The preceding discussion shows that what we wish in a practical sense to hold true, namely the ratio of failures (successes) to total trials to be close to the true unreliability (reliability), holds true theoretically. Thus if the number of trials is large enough, the observed ratio of failures to trials will give as close an estimate as we desire for the true value of p. The estimator ν/N can be shown to have certain desirable properties (Sec. 7.2) that an estimator should have. In particular it is *unbiased*; i.e., its expected value is just the parameter p it estimates, for any finite value of N:

$$E\!\left(\frac{\nu}{N}\right) = \sum_{j=0}^{N} \frac{j}{N}\binom{N}{j} p^j q^{N-j} \qquad (6.19)$$

$$= \frac{1}{N} \sum_{j=1}^{N} \frac{N!}{(N-j)!(j-1)!} p^j q^{N-j} \qquad (6.20)$$

$$= \frac{Np}{N} \sum_{j=0}^{N-1} \binom{N-1}{j} p^j q^{N-1-j} \qquad (6.21)$$

Thus

$$E\!\left(\frac{\nu}{N}\right) = p \qquad (6.22)$$

since the sum in Eq. (6.21) is the sum of probabilities for all possible outcomes in $N - 1$ trials and therefore equal to one. We will consider the problem of estimation in more detail in Chap. 7.

6.3 THE MULTINOMIAL DISTRIBUTION

When more than two outcomes of an experiment of the same type as considered in Sec. 6.2 are possible, we are led to a simple generalization of the binomial distribution. The experiment consists of independent trials for which each trial must result in one of k mutually exclusive events. The probabilities $p_1, \cdots, p_k$ of the respective events are constant from trial to trial and we must have

$$\sum_{j=1}^{k} p_j = 1$$

Let $\nu_1, \nu_2, \cdots, \nu_k$ denote the numbers of times event 1, event 2, $\cdots$, event k occur in N trials; then

$$\tilde{P}(\nu_1 = n_1, \nu_2 = n_2, \cdots, \nu_k = n_k) = \frac{N!}{n_1! n_2! \cdots n_k!} p_1^{n_1} p_k^{n_2} \cdots p_k^{n_k} \quad (6.23)$$

called the *multinomial distribution* gives the joint distribution of the random variables $\nu_1, \cdots, \nu_k$, where

$$\sum_{j=1}^{k} \nu_j = N, \quad \sum_{j=1}^{k} n_j = N, \quad \text{and} \quad \sum_{j=1}^{k} p_j = 1$$

For the general multinomial distribution given by Eq. (6.23), one can deduce the various marginal and conditional distributions, as well as the correlation coefficients for any pair of variables.

EXERCISE: Show that the mean and variance of ν_j are given by $E(\nu_j) = Np_j$ and $\text{Var}(\nu_j) = Np_j q_j$, where $q_j \equiv 1 - p_j$.

As an example of calculations involving the multinomial distribution we now show that

$$\text{Cov}(\nu_1, \nu_2) = -Np_1 p_2$$

We proceed first to find the marginal distribution of the combined variable (ν_1, ν_2). To do so we rewrite Eq. (6.23) as the joint distribution of $\nu_1, \cdots, \nu_{k-1}$ only, since $\nu_k = N - \nu_1 - \cdots - \nu_{k-1}$, obtaining

$$P(\nu_1 = n_1, \cdots, \nu_{k-1} = n_{k-1})$$

$$= \frac{N! p_1^{n_1} \cdots p_{k-1}^{n_{k-1}} (1 - p_1 - \cdots - p_{k-1})^{N-n_1-\cdots-n_{k-1}}}{n_1! \cdots n_{k-1}!(N - n_1 - \cdots - n_{k-1})!} \quad (6.24)$$

DISCRETE AND CONTINUOUS DISTRIBUTION MODELS 119

where
$$0 \le \sum_{j=1}^{k-1} \nu_j \le N, \quad 0 \le \sum_{j=1}^{k-1} n_j \le N, \quad \sum_{j=1}^{k-1} p_j \le 1$$

First we sum over all possible values of ν_{k-1}; thus

$$P(\nu_1 = n_1, \cdots, \nu_{k-2} = n_{k-2}) = \frac{N! p_1^{n_1} \cdots p_{k-2}^{n_{k-2}}}{n_1! \cdots n_{k-2}!(N - n_1 - \cdots - n_{k-2})!}$$

$$\sum_{n_{k-1}=0}^{N-n_1-\cdots-n_{k-2}} \left\{ \frac{(N-n_1-\cdots-n_{k-2})! p_{k-1}^{n_{k-1}}(1 - p_1 - \cdots - p_{k-1})^{N-n_1-\cdots-n_{k-1}}}{(N - n_1 - \cdots - n_{k-2} - n_{k-1})! n_{k-1}!} \right\}$$

(6.25)

$$= \frac{N! p_1^{n_1} \cdots p_{k-2}^{n_{k-2}}(1 - p_1 - \cdots - p_{k-2})^{N-n_1-\cdots-n_{k-2}}}{n_1! \cdots n_{k-2}!(N - n_1 - \cdots - n_{k-2})!} \qquad (6.26)$$

since

$$\sum_{j=0}^{N-a} \frac{(N-a)!}{(N-a-j)! j!} p^j (b-p)^{N-a-j} = (b - p + p)^{N-a} = b^{N-a} \qquad (6.27)$$

by the well-known binomial theorem.

If we compare Eq. (6.26) with (6.24) we see they are the same except that all references to the variable ν_{k-1} are eliminated in (6.26). By induction, we obtain

$$P(\nu_1 = n_1, \nu_2 = n_2) = \frac{N! p_1^{n_1} p_2^{n_2}(1 - p_1 - p_2)^{N-n_1-n_2}}{n_1! n_2!(N - n_1 - n_2)!} \qquad (6.28)$$

where $0 \le \nu_1 + \nu_2 \le N$, $0 \le n_1 + n_2 \le N$, and $p_1 + p_2 \le 1$. Since

$$\text{Cov}(\nu_1, \nu_2) = E(\nu_1 \nu_2) - E(\nu_1) E(\nu_2)$$

we merely have to find $E(\nu_1 \nu_2)$:

$$E(\nu_1 \nu_2) = \sum_{n_1=0}^{N} \sum_{n_2=0}^{N-n_1} n_1 n_2 P(\nu_1 = n_1, \nu_2 = n_2) \qquad (6.29)$$

Consider first the summation with respect to n_2:

$$\frac{N! p_1^{n_1} n_1}{n_1!} \sum_{n_2=0}^{N-n_1} \frac{n_2 p_2^{n_2}(1 - p_1 - p_2)^{N-n_1-n_2}}{n_2!(N - n_1 - n_2)!} \qquad (6.30)$$

We can rewrite Eq. (6.30) as

$$\frac{N! p_1^{n_1} n_1 p_2}{n_1!(N-n_1-1)!} \sum_{n_2=0}^{N-n_1-1} \frac{(N-n_1-1)! p_2^{n_2}(1-p_1-p_2)^{N-n_1-1-n_2}}{n_2!(N-n_1-n_2-1)!} \quad (6.31)*$$

$$= \frac{N! p_1^{n_1} n_1 p_2 (1-p_1)^{N-n_1-1}}{n_1!(N-n_1-1)!} \quad (6.32)*$$

Now summing Eq. (6.32) with respect to n_1

$$E(\nu_1 \nu_2) = \frac{N!}{(N-2)!} p_2 \sum_{n_1=0}^{N-2} \frac{p_1^{n_1+1}(1-p_1)^{N-n_1-2}(N-2)!}{n_1!(N-n_1-2)!} \quad (6.33)*$$

$$= p_2 p_1 N(N-1) \quad (6.34)$$

Thus

$$\text{Cov}(\nu_1, \nu_2) = N(N-1) p_1 p_2 - N^2 p_1 p_2$$

$$= -N p_1 p_2 \quad (6.35)$$

which is the desired result.

The correlation coefficient (*cf*. Eq. (5.82)) is

$$\rho(\nu_1, \nu_2) = \frac{-N p_1 p_2}{[N p_1(1-p_1) N p_2(1-p_2)]^{1/2}}$$

or

$$\rho(\nu_1, \nu_2) = -\sqrt{\frac{p_1 p_2}{(1-p_1)(1-p_2)}} \quad (6.36)$$

To assure ourselves that $|\rho(\nu_1, \nu_2)| \leq 1$ we note that since $p_1 + p_2 \leq 1$, then $p_1 \leq 1 - p_2$ and $p_2 \leq 1 - p_1$; then $p_1/(1 - p_2) \leq 1$ and $p_2/(1 - p_1) \leq 1$. It follows by symmetry that to obtain Cov (ν_i, ν_j) and $\rho(\nu_i, \nu_j)$, one merely replaces 1 and 2 by i and j in Eqs. (6.35) and (6.36) respectively.

One application of the multinomial distribution is in deriving the probability distribution of the smallest (or largest) sample value in a sample drawn from a continuous probability distribution (see Sec. 9.3.2.). Another application to reliability evaluation would be as follows: Instead

---

* In going from Eq. (6.30) to (6.31), first the lower limit of the summation is changed from $n_2 = 0$ to $n_2 = 1$, since when $n_2 = 0$ the summand is zero owing to the factor n_2. Then $n_2/n_2!$ is written as $1/(n_2-1)!$. The variable of summation is then transformed by setting $n_2 = j + 1$, then relabeling $j \equiv n_2$. Finally the whole expression is multiplied and divided by $(N - n_1 - 1)!$. Equation (6.32) is obtained from (6.31) by using the binomial theorem. Similar manipulations are used to obtain Eq. (6.33); in this case, however, the upper limit of summation is also changed owing to the fact that $1/(N - n_1 - 2)! \equiv 0$ for $n_1 = N - 1$.

of considering as possible outcomes only success (S) and failure (F), one could consider F as a union of mutually exclusive events $F_1, F_2, \cdots, F_n$. Each of the events F_j could denote a different mode of failure. However we must be careful in defining the F_j that no more than one of the F_j can occur in any trial.

For any model in which the multinomial distribution is used, one can obtain estimates of the parameters $p_1, \cdots, p_{k-1}$ as $\hat{p}_j = \nu_j/N$ which are unbiased and converge in probability to the true values of p_j, for $j = 1, 2, \cdots, k - 1$. Also the variables ν_j/N obey a limit law analogous to Eq. (6.16); in this case the limiting probability distribution is a *multivariate normal distribution*. Further details are left to references.

6.4 THE GEOMETRIC AND NEGATIVE BINOMIAL DISTRIBUTION

In this section we consider a model in which binomial trials are made, but the sample space is different from that specified in Sec. 6.2. Previously the number of trials was specified in advance; the random variable observed was number of failures. In this section we do just the opposite—fix the number c of failures and make independent trials until c failures occur. Thus the number of trials ν until c failures occur is a random variable with probability distribution

$$P(\nu = n) = \binom{n-1}{c-1} p^c q^{n-c} \qquad (6.37)$$

where $n = c, c + 1, \cdots$; and p, q denote the probabilities of failure and success on a single trial. To derive Eq. (6.37), we note that the event that the cth failure occurs at trial n is equivalent to the joint event: exactly $c - 1$ failures occur in $n - 1$ trials *and* a failure occurs on the nth trial. Since the events are independent the respective probabilities

$$\binom{n-1}{c-1} p^{c-1} q^{n-c} \quad \text{and} \quad p$$

multiply, giving Eq. (6.37).

It is sometimes more convenient to define a new random variable $\xi \equiv \nu - c$, so that ξ is the number of trials in addition to the minimum number of trials c necessary to end the experiment. Also, ξ is the number of successes preceding the cth failure. Thus,

$$P(\xi = k) = P(\nu = k + c)$$

$$= \binom{k+c-1}{c-1} p^c q^k \qquad (6.38)$$

or, since $\binom{k+c-1}{c-1} \equiv \binom{k+c-1}{k}$,

$$P(\xi = k) = \binom{k+c-1}{k} p^c q^k \qquad (6.39)$$

where $k = 0, 1, 2, \cdots$.

It can be shown that

$$\binom{k+c-1}{k} = (-1)^k \binom{-c}{k}$$

(Ref. 1, p. 61), where

$$\binom{-c}{k} \equiv \frac{(-c)(-c-1) \cdots (-c-k+1)}{k!}$$

Thus Eq. (6.39) becomes

$$P(\xi = k) = \binom{-c}{k} p^c (-q)^k \qquad (6.40)$$

where $k = 0, 1, 2, \cdots$.

The distribution given by Eq. (6.40) is called the *negative binomial* distribution, for the reason that the righthand side of (6.40) is a term of the expansion, in powers of q, of the expression $p^c(1-q)^{-c}$.

We note that ξ (or ν) can be considered as a "waiting time" to the cth failure, if trials are made at a given time-rate; e.g., one trial per second.

When $c = 1$, we have from Eq. (6.39)

$$P(\xi = k) = q^k p \qquad (6.41)$$

called the *geometric* distribution.

6.4.1 THE MEAN AND VARIANCE

The mean and variance of the variable ξ can be obtained most easily using Eq. (6.40). We have

$$E(\xi) = \sum_{k=0}^{\infty} k \binom{-c}{k} p^c (-1)^k q^k \qquad (6.42)$$

$$= q p^c \frac{d}{dq} \sum_{k=0}^{\infty} \binom{-c}{k} (-q)^k \qquad (6.43)$$

$$= q p^c \frac{d}{dq} [(1-q)^{-c}] \qquad (6.44)$$

$$= q p^c c (1-q)^{-c-1} \qquad (6.45)$$

Finally
$$E(\xi) = \frac{qc}{p} \qquad (6.46)$$
since $p \equiv 1 - q$. For example, if $p = 0.10$ and $c = 1$, we should expect $0.9/0.1 = 9$ successes before the first failure, or a total of 10 trials to the first failure.

The variance of ξ is obtained by the relation
$$\text{Var } \xi = E(\xi^2) - E^2(\xi) \qquad (6.47)$$
Therefore we have to find
$$E(\xi^2) = \sum_{k=0}^{\infty} k^2 \binom{-c}{k} p^c(-1)^k q^k \qquad (6.48)$$
$$= q^2 \sum_{k=0}^{\infty} k(k-1) \binom{-c}{k} p^c(-1)^k q^{k-2} + \sum_{k=0}^{\infty} k \binom{-c}{k} p^c(-1)^k q^k \qquad (6.49)$$

The second term on the right side of Eq. (6.49) is by (6.42) and (6.46) equal to qc/p. The first term on the right side of (6.49) is
$$q^2 p^c \frac{d^2}{dq^2}(1-q)^{-c} = q^2 p^c c \frac{d}{dq}(1-q)^{-c-1} \qquad (6.50)$$
$$= q^2 p^c c(c+1)(1-q)^{-c-2} \qquad (6.51)$$
$$= \frac{c(c+1)q^2}{p^2} \qquad (6.52)$$

Thus, from (6.47)
$$\text{Var } \xi = \frac{c(c+1)q^2}{p^2} + \frac{qc}{p} - \frac{q^2 c^2}{p^2} = \frac{qc}{p^2} \qquad (6.53)$$

To summarize, the random variable ξ, with probability distribution given by Eq. (6.39) or (6.40) has expected value qc/p and standard deviation $\sqrt{qc}/p$. Note that the random variable $\nu = \xi + c$ has mean c/p and standard deviation $\sqrt{qc}/p$.

6.4.2 APPLICATION OF THE NEGATIVE BINOMIAL DISTRIBUTION

Suppose now we wish to calculate the probability of waiting at most n trials for the cth failure. This is obtained from Eq. (6.37) by summing over the probabilities that $\nu = c, c+1, \cdots, n$.
$$P(\nu \le n) = \sum_{j=c}^{n} \binom{j-1}{c-1} p^c q^{j-c} \qquad (6.54)$$

In order to calculate such sums consider the events $\nu \leq n$ and $X_n \geq c$, where X_n is defined as the number of failures occurring in n trials. With a little reflection it becomes evident that the two events are equivalent and hence have the same probability. The probability of the latter event is

$$P(X_n \geq c) = \sum_{j=c}^{n} \binom{n}{j} p^j q^{n-j} \qquad (6.55)$$

since X_n has the ordinary binomial distribution. Thus we have the interesting identity from Eqs. (6.54) and (6.55):

$$\sum_{j=c}^{n} \binom{j-1}{c-1} p^c q^{j-c} = \sum_{j=c}^{n} \binom{n}{j} p^j q^{n-j} \qquad (6.56)$$

Since we have tables of cumulative sums of binomial probabilities (the right side of Eq. (6.56)), the left side of (6.56) can also be evaluated.

EXAMPLE: What is the probability that we must make at most fifty trials until the third failure occurs, where $p = 0.1$? This probability P is given by the sum

$$P = \sum_{j=3}^{50} \binom{j-1}{2} (0.1)^3 (0.9)^{j-3} \qquad (6.57)$$

$$= \sum_{j=3}^{50} \binom{50}{j} (0.1)^j (0.9)^{50-j} \qquad (6.58)$$

by Eq. (6.56). The last sum can be computed directly or by using tables (Ref. 3). We obtain from p. 69, $P = 0.88827$.

6.4.3 UNBIASED ESTIMATOR OF p

When p is unknown it can be estimated from an experiment in which binomial trials are made until the cth failure occurs. In this case the unbiased estimator of p is

$$\hat{p} = \frac{c-1}{n-1}, \qquad c > 1 \qquad (6.59)$$

where n is the number of trials made. When $c = 1$ (geometric distribution of trials) the unbiased estimator is

$$\hat{p} = \begin{cases} 1, & n = 1 \\ 0, & n > 1 \end{cases} \qquad (6.60)$$

To show that Eq. (6.60) is correct, let us define $\hat{p}$ equal to some function of n, say f_n, such that $0 \leq f_n \leq 1$. Then for $\hat{p}$ to be unbiased we must have $E(\hat{p}) = p$; thus we set

$$p = \sum_{n=1}^{\infty} f_n q^{n-1} p \qquad (6.61)$$

or

$$1 = \sum_{n=1}^{\infty} f_n q^{n-1} \qquad (6.62)$$

Equation (6.62) must hold identically for all values of q, and f_n cannot be a function of q; thus, differentiating both sides with respect to q

$$0 = \sum_{n=0}^{\infty} (n + 1) f_{n+2} q^n \qquad (6.63)$$

The righthand side is a power series in q, convergent in the interval $0 \leq |q| < 1$ and therefore a regular function of q. Thus its power series coefficients are unique; hence they must all be zero. Therefore $f_n = 0$ for $n \geq 2$. Finally by Eq. (6.61) or (6.62) f_1 must be equal to unity, which gives the result expressed by Eq. (6.60).

To verify Eq. (6.59) we have

$$E(\hat{p}) = \sum_{n=c}^{\infty} \frac{c-1}{n-1} \binom{n-1}{c-1} p^c q^{n-c} \qquad (6.64)$$

$$= p \sum_{n=c-1}^{\infty} \binom{n-1}{c-1-1} p^{c-1} q^{n-(c-1)} \qquad (6.65)$$

The summand is just the probability of waiting n trials for the $(c - 1)$th failure; thus the summation, being over all possible values of n, must be one. Hence

$$E(\hat{p}) = p \qquad (6.66)$$

which is the desired result. It can also be verified that $\hat{p}$ defined by Eq. (6.59) is the only unbiased estimator of p (Ref. 4).

6.4.4 LIMITING EXPONENTIAL AND GAMMA DISTRIBUTIONS

The geometric distribution can be converted by a limiting process to the exponential distribution (Sec. 6.8.1). In the former case the random variable is number of trials to failure; in the latter it is time-to-failure. To do this we consider that trials are made at a faster and faster rate. This would result in time-to-failure approaching zero, unless the probability of failure p also tended to zero. To specify this limiting process, we first write

$$P(\nu > n) = \sum_{j=n+1}^{\infty} q^{j-1}p = (1-p)^n \qquad (6.67)$$

Now, define $\tau = \nu \, \Delta t$ and assume that $p = \lambda \, \Delta t$, where λ is a constant. Then, the probability density element of τ is

$$f(t) \, dt \equiv P(t < \tau < t + dt)$$

$$= P(\nu \, \Delta t > t) - P(\nu \, \Delta t > t + dt) \qquad (6.68)$$

$$= P\left(\nu > \frac{t}{\Delta t}\right) - P\left(\nu > \frac{t}{\Delta t} + \frac{dt}{\Delta t}\right) \qquad (6.69)$$

$$= (1 - \lambda \, \Delta t)^{t/\Delta t} - (1 - \lambda \, \Delta t)^{t/\Delta t + dt/\Delta t} \qquad (6.70)$$

$$= (1 - \lambda \, \Delta t)^{t/\Delta t}[1 - (1 - \lambda \, \Delta t)^{dt/\Delta t}] \qquad (6.71)$$

When $\Delta t \to 0$, the right side of Eq. (6.71) approaches

$$e^{-\lambda t}[1 - 1 + \lambda \, dt + o(dt)] \qquad (6.72)$$

Thus

$$f(t) \, dt = \lambda \, e^{-\lambda t} \, dt + o(dt) \qquad (6.73)$$

Finally dividing through by dt and letting $dt \to 0$, we obtain

$$f(t) = \lambda e^{-\lambda t} \qquad (6.74)$$

the exponential probability density function.

Similarly, for the distribution of trials to the cth failure, one can show as a limiting time-to-failure probability density function

$$f_c(t) = \frac{\lambda^c t^{c-1} e^{-\lambda t}}{\Gamma(c)} \qquad (6.75)$$

which is recognized as the gamma probability density function (Sec. 6.9.2.1).

6.5 THE POISSON DISTRIBUTION

The Poisson distribution is a discrete probability distribution over a denumerably infinite sample space. If the points of the sample space are

DISCRETE AND CONTINUOUS DISTRIBUTION MODELS

$0, 1, 2, \cdots$, then we define

$$P(\nu = n) = \frac{\lambda^n e^{-\lambda}}{n!}, \qquad \begin{array}{l} n = 0, 1, 2, \cdots \\ \lambda > 0 \end{array} \qquad (6.76)$$

as the Poisson distribution. We easily see that Eq. (6.76) defines a proper probability distribution for any value of the parameter $\lambda > 0$, since

$$\sum_{n=0}^{\infty} \frac{\lambda^n e^{-\lambda}}{n!} = e^{-\lambda} \cdot e^{\lambda} = 1 \qquad (6.77)$$

In applications, ν generally denotes "number of events" when there are a large number of opportunities for an event to occur, but small probability that any one of the opportunities yields the occurrence of the defined event. For example, many of the hundreds of components of a missile guidance subsystem may be sensitive to vibration environment. Suppose that "critical" acceleration peaks occur at a certain time-rate δ, and for each component there is a small probability p that an acceleration magnitude equal to or greater than "critical" causes component failure. Within a fixed time interval t, we should expect approximately $\delta p t$ component failures. The number of component failures in the fixed interval t would tend to have the Poisson distribution with parameter $\lambda = \delta p t$. Further, if failure of one or more components in time T, where T is the required period of operation, caused failure of the guidance subsystem as a whole, then the reliability of the guidance subsystem would be defined as the probability of no failure in time T, or

$$R = e^{-\delta p T} \qquad (6.78)$$

The expression in Eq. (6.78) is simply the first term of the Poisson distribution (Eq. 6.76)).

While the above example is more complicated than it need be for illustration, we note that both quantities p and δ are capable of being determined in separate experiments; either or both could conceivably be decreased by different means, thus increasing R. In a general sense Eq. (6.78) expresses the interaction of the inherent properties of a component characterized by p, with its environment characterized by δ.

In Sec. 6.8.1 we reinterpret Eq. (6.78) by defining a random variable τ as the time-to-failure. Thus Eq. (6.78) is the probability that τ exceeds T when τ has the *exponential distribution*. The quantity δp is called the failure rate, or reciprocal of the mean time-to-failure.

EXERCISE: Show that ν defined by Eq. (6.76) has mean λ, standard deviation $\sqrt{\lambda}$.

From the remark following Eq. (6.77) we would expect that the binomial distribution defined by Eq. (6.1) approaches the Poisson distribution as a limit when the number of trials N becomes large and the probability p of failure on any trial becomes small. To see that this is so, let $N \to \infty$ and $p \to 0$, but in a manner such that $Np \to \lambda$.* Then Eq. (6.1) can be written:

$$P(\nu = f) = \binom{N}{f}\left(\frac{\lambda}{N} + o\left(\frac{1}{N}\right)\right)^f \left(1 - \frac{\lambda}{N} + o\left(\frac{1}{N}\right)\right)^{N-f} \quad (6.79)$$

Now

$$f!\binom{N}{f} = \frac{N!}{(N-f)!} = \frac{\Gamma(N+1)}{\Gamma(N+1-f)} \quad (6.80)$$

Since

$$\frac{\Gamma(N+a)}{\Gamma(N+b)} \sim N^{a-b} \quad \text{as } N \to \infty \quad (6.81)$$

(Ref. 5) then

$$\frac{\Gamma(N+1)}{\Gamma(N+1-f)} \sim N^f \quad \text{as } N \to \infty \quad (6.82)$$

This is the same as saying that

$$\frac{\Gamma(N+1)}{\Gamma(N+1-f)} = N^f(1 + o(1)) \quad (6.83)*$$

Thus

$$f!P(\nu = f)$$

$$= (1 + o(1)) \frac{\left(\lambda + No\left(\frac{1}{N}\right)\right)^f \left(1 - \frac{\lambda}{N} + o\left(\frac{1}{N}\right)\right)^N}{\left(1 - \frac{\lambda}{N} + o\left(\frac{1}{N}\right)\right)^f} \quad (6.84)$$

Therefore as $N \to \infty$

$$P(\nu = f) \to \frac{\lambda^f e^{-\lambda}}{f!} \quad (6.85)$$

---

* As usual $o(f(N))$ denotes a function of N that tends to zero faster than $f(N)$; thus, $o(f(N))/f(N) \to 0$ as $N \to \infty$. The function $o(1) \to 0$ as $N \to \infty$; also we note that "$Np \to \lambda$" means $p = \lambda/N + o(1/N)$.

DISCRETE AND CONTINUOUS DISTRIBUTION MODELS   129

When observed values of ν from k independent experiments are given as $\nu_1 = n_1, \nu_2 = n_2, \cdots, \nu_k = n_k$, then the *unbiased estimate* of the parameter λ is given by

$$\hat{\lambda} = \frac{1}{k} \sum_{j=1}^{k} n_j \qquad (6.86)$$

EXERCISE: A reliability group in a certain company received failure reports covering valve failures occurring in a period of 10 weekdays. The numbers of reports received for the given days were 3, 3, 2, 5, 4, 3, 2, 1, 3, 4. What is an estimate of the daily failure rate?

Answer: $\hat{\lambda} = 3$.

EXERCISE: What is an estimate of the probability, P_0, that no failures will occur on a given day?

Answer: $\hat{P}_0 \simeq e^{-\hat{\lambda}} = 0.05$.

EXERCISE: Suppose that additional data giving an average number N of attempts at valve operation per day were obtained, with $N = 500$. What is an estimate of the probability that a valve operates successfully on a given attempt?

Answer: $\hat{R} = 1 - \hat{p} \simeq 1 - \dfrac{\hat{\lambda}}{N} = 0.994$.

EXAMPLE: Suppose that previous test data gave the estimate $\hat{R} = 0.97$. One conclusion might be that 80% of all valve failures are not being reported. A further investigation by the reliability group might reveal that only "critical-type" failures are being reported, whereas the previous test data evaluation considered both "critical-type" as well as "minor" failures to be included in the definition of failures.

EXERCISE: A guided missile is counted as a complete success when none of a large number of possible independent modes of failure materializes. Data from ten flight tests show 1, 1, 0, 2, 0, 0, 2, 0, 1, 0 as the respective numbers of such failures occurring in the tests. What is an estimate of the probability of complete success? Explain how this model is different from a binomial success-failure model.

6.6 GENERALIZED BINOMIAL SUCCESS-FAILURE MODELS

In Secs. 6.2 and 6.4 the importance of the sample space in binomial trials was brought out by showing that the distribution model used depends on the choice of sample space or specification of all possible outcomes of the experiment. When the number of trials is fixed in advance it was shown that the number of failures (or successes) has a *binomial* distribution. On the other hand, when the number of failures was fixed in advance of the experiment, the number of trials was shown to have a *negative binomial distribution*.

We now consider an important generalization of this problem by pictorializing a sequence of binomial trials in the following manner: Consider the usual rectangular coordinates with ordinate denoting *failures* and abscissa denoting *trials*. Starting from the origin we plot a point one unit horizontally to the right when a success occurs on the first of a sequence of binomial trials; or one unit horizontally to the right and one unit vertically up if a failure occurs on the first of a sequence of binomial trials. Thus the point (1, 0) represents "first trial success," and the point (1, 1) represents "first trial failure."

The second trial results in a point either one unit horizontally to the right or one unit to the right and up one unit (diagonally up and to the right) depending on whether a success or a failure occurs, using the result

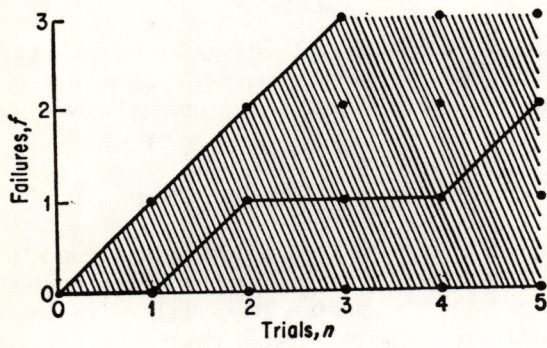

Figure 6.1

of the first trial as origin. Thus there are three possible points reached on the second trial: (2, 2), (2, 1), (2, 0). Continuing in this way, if an indefinite series of trials is permitted, any point with integral coordinates can be reached within the shaded region as well as on its boundaries as shown in Fig. 6.1. The zigzag line or *path* shown within the shaded region of Fig. 6.1 denotes the result $SFSSF \cdots$, where S denotes success and F, failure.

It is evident that there are

$$\binom{n}{f}$$

different paths, each corresponding to f F's and $(n - f)$ S's, to the point (n, f). Before we can define probabilities attached to a point (n, f) we must specify the sample space of an experiment involving binomial trials.

The sample space for an experiment is specified by defining a *boundary* of points on the plot of failures versus trials such that when one of the

DISCRETE AND CONTINUOUS DISTRIBUTION MODELS 131

points on this boundary is reached the experiment is ended. This can also be called a *stopping rule*. For example, if the total number of trials is fixed in advance equal to N, the boundary points lie on the vertical line $n = N$, for $0 \leq f \leq N$ (see Fig. 6.2).*

The circled points of Fig. 6.2 are the boundary points. Since the point N, f corresponds to f F's and $(N - f)$ S's, the probability attached to this boundary point is

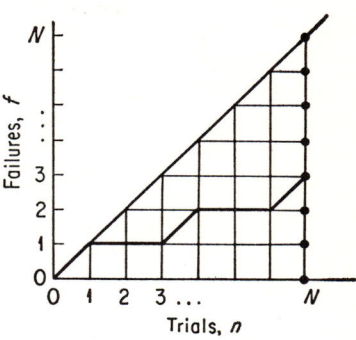

Figure 6.2

$$\binom{N}{f} p^f q^{N-f},$$

where p is the probability of failure and $q = 1 - p$.

If we specify that the experiment ends when the number of failures equal c, then the boundary points are shown in Fig. 6.3, on the horizontal line $f = c$, for $n \geq c$.

The boundary points (n, c) have the attached probabilities

$$\binom{n-1}{c-1} p^c q^{n-c}, \quad \text{not} \quad \binom{n}{c} p^c q^{n-c}$$

since the point (n, c) is not accessible from the point $(n - 1, c)$ horizontally one unit to its left (the experiment would have already ended). Thus the number of paths to a particular *boundary* point (n, c) is just equal to the number of paths to the point $(n - 1, c - 1)$ indicated for each n by the points marked x in Fig. 6.3.

A sample space of noteworthy importance is shown in Fig. 6.4; this is an example of Wald's sequential sampling scheme (Ref. 6, Chap. 5). In the figure, the lower set of the boundary points can be specified by the equation

$$n = 2f + 5, \quad f \geq 0 \tag{6.87}$$

---

* It is evident that testing does not always have to be on a sequential basis; i.e., testing in groups is allowed, provided that the result (e.g., total number of failures in the group) cannot lead to a point (f, n) which is "outside" the specified boundary. The only way to *guarantee* that the probabilities attached to the boundary points are not changed is to make the group size at most equal to the minimum number of trials necessary to reach any boundary point from the point at which group testing is initiated. For example, in the plan of Fig. 6.2, one could test all N items at once, without changing the probabilities (see paragraph after Fig. 6.2). However, if, e.g., $N = 10$, and the point $(n, f) = (5, 2)$ had already been reached in some manner, then at most 5 items could then be tested as a group, so as not to "overshoot" the boundary. In the plan of Fig. 6.3, at most c items in one group could be tested initially. In the plan of Fig. 6.4, at most 3 items could be tested as a group when the testing is first commenced.

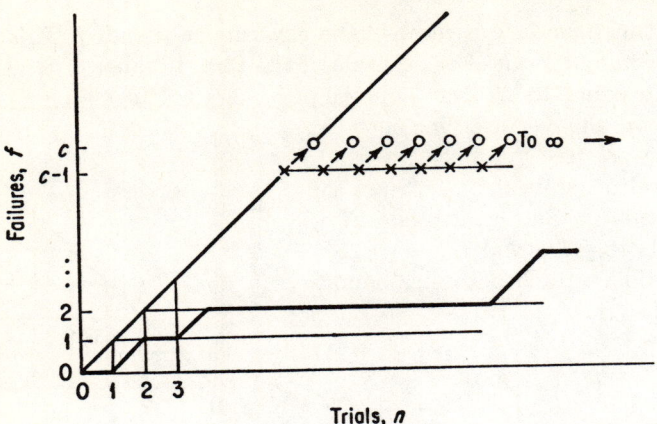

Figure 6.3

and the upper set by the equation

$$n = 2f - 3, \quad f \geq 3 \tag{6.88}$$

Although it is extremely difficult to find a general formula for the probabilities attached to the boundary points in an experiment such as shown in Fig. 6.4, *we can nevertheless determine these probabilities by actually counting paths to any particular point*. We first note that the number of paths to a particular point (n, f) is equal to the number of paths to the point $(n - 1, f)$ plus the number of paths to the point $(n - 1, f - 1)$. We must be careful that the path to any particular boundary point we are interested in does not go through another boundary point, otherwise the experiment would have already ended.

To illustrate the procedure, Fig. 6.5 shows the set of all possible paths to the point (11, 3) for the sample space of Fig. 6.4 with numbers attached to each point denoting the number of paths to that point. The point (11, 3) is the result of 3 failures and 8 successes in all possible orders *except* those

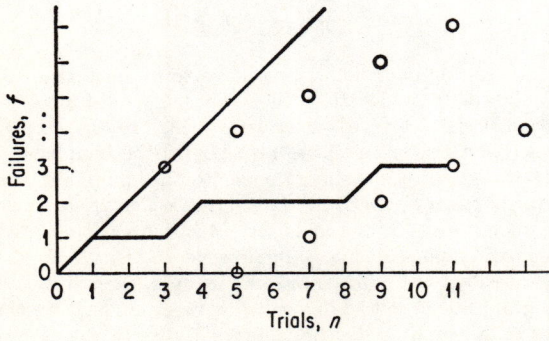

Figure 6.4

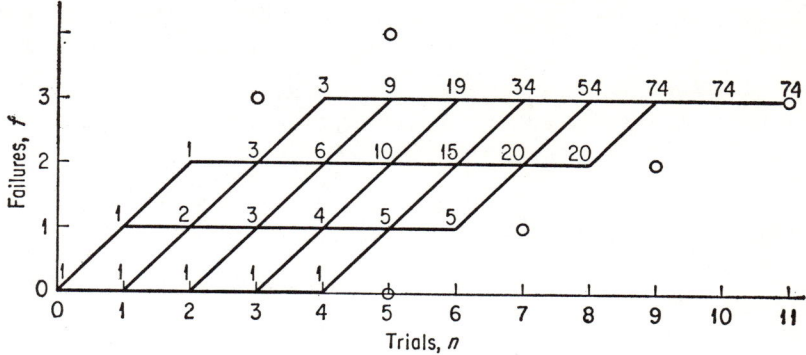

Figure 6.5

represented by paths going through other boundary points. The number of *possible orders* or, as we say, *possible paths* to the point (11, 3) is precisely 74, as shown in Fig. 6.5. Thus the probability attached to the point (11, 3) is 74 p^3q^8.

EXERCISE: Compute in the same manner the probabilities of all boundary points such that $f \leq 5$ using the sample space of Fig. 6.4.

We note that in Fig. 6.4, as in Fig. 6.3, the sample space contains a denumerable infinity of points. The question may arise in either case, could sampling go on indefinitely without coming to an end? The answer is that we can discount this event, since the probability that a path does not reach a boundary point in a finite number of trials is zero. Wald (Ref. 6) has shown this answer to be true for sample spaces more general but similar to that of Fig. 6.4. For the sample space of Fig. 6.3 this result is a direct consequence of the fact that $P(\nu > n) \to 0$ as $n \to \infty$, which can be verified using Eq. (6.18). However, a rigorous proof will not be given here.*

6.6.1 UNBIASED ESTIMATOR OF p

The method of counting paths can be used to obtain an unbiased estimator of p (Ref. 4). To do this one merely counts the number of paths to the given boundary point starting from the point (1, 1) and divides this by the number of paths to the given boundary point starting from (0, 0).

For the example of Fig. 6.5, the number of paths starting from the point (1, 1) to the point (11, 3) is easily found to be 26. Thus if the experiment ends with three failures in eleven trials, the unbiased estimate of

---

* A special theorem is needed, namely, if A_n denote events, and $A_1 \supset A_2 \supset \cdots$; and if $A \equiv \bigcap_{n=1}^{\infty} A_n$, then $P(A) = \lim_{n \to \infty} P(A_n)$. See Ref. 12, p. 33. In this case $A_n \equiv (\nu > n)$, and A is the event "$\nu > 1$ and $\nu > 2$ and $\cdots$"; i.e., "$\nu >$ every finite integer."

probability of failure, p, is $\hat{p} = \frac{26}{74} = 0.351$. Note that this value is different from the value one would obtain using the sample space of Fig. 6.2.

> EXERCISE: Compute the number of paths from the point $(1, 1)$ to all boundary points such that $f \leq 5$ using the sample space of Fig. 6.4. Compute the unbiased estimate of p for these boundary points.

> EXERCISE: Verify the general formula for the unbiased estimate $\hat{p} = f/N$ for the sample space of Fig. 6.2. Also verify $\hat{p} = (c - 1)/(n - 1)$ for the sample space of Fig. 6.3 when $c > 1$. Note that when $c = 1$ and $n > 1$, there are no paths from $(1, 1)$ to a boundary point, which confirms Eq. (6.60).

The general methods of this section should make clear to the reader the importance of specifying all possible outcomes of an experiment, i.e., the sample space, before the experiment is undertaken. If not, it is not possible to make valid inferences about the true value of p, which is generally the purpose of making binomial trials. We will apply the methods of this section to the problem of reliability demonstration in Chap. 10.

6.7 CONTINUOUS DISTRIBUTION MODELS

Models of this type are used to measure events such as E_1: "time-to-failure exceeds T," where T represents the required or intended operating time of a device, or E_2: "performance parameter is within certain limits." Thus, the probability of either of the above events could be called the reliability of the device. (In fact we might define "reliability" as $P(E_1 \cap E_2)$; that is, as the probability that *both* events occur.

Three types of failures have been generally recognized as having a time-characteristic. If a device is "turned on" at time $t = 0$, it may already be in a failed state. Infant mortality (at and after birth) provides a readily understood example of this type of failure. We choose to call it *initial failure*. The probability of initial failure of a device can be estimated by the fraction defective of a number of devices at the time they are put into operation. It is possible, however, that an initial-type failure will manifest itself shortly after time $t = 0$; e.g., for certain high-quality electron tubes, a "burn-in" period is used to screen-out tubes which could fail during the first few hours of operation. After the burn-in period, the tubes will fail at a much lower rate than during the initial period of operation. Another example in which initial failure is recognized is that in many medical and health insurance policies covering all dependents of the "breadwinner," a newborn dependent is not completely covered by the insurance until two weeks after birth. Furthermore, the standard human mortality tables recognize that up to the age of 10 years a child can die of congenital or hereditary defects; but having lived past this age the child is almost assuredly free of such defects.

Following this initial period, many devices exhibit a constant failure rate, generally lower than during the initial period. For example, human beings from age 10 years to 30 years die generally because of accidents; after this period wearout processes begin to take over and an increasing proportion of deaths can be considered as due to "old age." The former period, whether in the life of a young child and adult, an electron tube, or whatever, is called the *chance failure* period, in that failure occurs by reason of unusually severe, unpredictable, and/or unavoidable environmental conditions occurring during the operating time of the "device." The latter period is termed the *wearout* failure period and is perhaps associated with gradual "depletion of a material," or "accumulated shocks," "fatigue," and so on.

Thus we term the three types of failure *initial*, *chance*, and *wearout* failure. Each has associated with it a period of operation of a device, the initial type of failure occurring at time zero or shortly thereafter, the chance type of failure occurring in the "middle" period, and finally the wearout type of failure which eventually occurs for most, if not all, devices whether they are human beings, electromechanical devices, or whatever.

First we consider *hazard functions* as a general approach to derivation of time-to-failure distributions, in particular the *exponential*, *Weibull*, and *extreme-value* distributions. In addition a different derivation of the *extreme-value* distribution, as well as an application, will be given. Following this, a model will be given based on a modified *pure death process*, which again leads to the exponential distribution, as well as to a more general time-to-failure distribution containing both the exponential and the *gamma* distribution as special cases. Finally, the important and well-known *normal* distribution is discussed.

6.8 HAZARD FUNCTIONS

The *hazard function* $h(t)$ of a probability distribution of time-to-failure is defined as the (conditional) probability density function of time-to-failure, given that the device has not failed prior to time t. If $f(t)$ is the (absolute) probability density function, then $f(t)\,dt$ will represent the proportion of a population of devices starting at time $t = 0$, which fail in the time interval $(t, t + dt)$. On the other hand $h(t)\,dt$ represents the proportion of a population of devices which have not failed prior to time t, but which do fail in the interval $(t, t + dt)$. The relationship between $h(t)$ and $f(t)$ is easily shown to be

$$h(t) = \frac{f(t)}{1 - F(t)} \qquad (6.89)$$

where

$$F(t) = \int_0^t f(x)\,dx$$

To show Eq. (6.89), we have, from the definition of $h(t)$,

$$P[(t < \tau < t + dt) \mid (\tau > t)] \equiv h(t) \, dt \tag{6.90}$$

But

$$P[(t < \tau < t + dt) \mid (\tau > t)] = \frac{P[(t < \tau < t + dt) \cap (\tau > t)]}{P(\tau > t)} \tag{6.91}$$

(This is the equation defining conditional probabilities.) The numerator of the righthand side is, however

$$P(t < \tau < t + dt) \equiv f(t) \, dt$$

since the event $(t < \tau < t + dt)$ is contained in the event $(\tau > t)$. Since $P(\tau > t) = 1 - F(t)$, we see that Eq. (6.89) is correct (cancelling out the dt).

Before we consider possible forms that could be assigned to $h(t)$, let us solve Eq. (6.89) for $f(t)$ or $F(t)$, assuming $h(t)$ known. We have

$$\frac{dF}{1 - F} = h(t) \, dt \tag{6.92}$$

or

$$\left[-\log(1 - F)\right]_{F(0)}^{F(t)} = \int_0^t h(x) \, dx$$

Then

$$\log \frac{1 - F(t)}{1 - F(0)} = -\int_0^t h(x) \, dx$$

or

$$1 - F(t) = (1 - F(0)) \exp\left[-\int_0^t h(x) \, dx\right] \tag{6.93}$$

Also,

$$f(t) = (1 - F(0)) h(t) \exp\left[-\int_0^t h(x) \, dx\right] \tag{6.94}$$

We see from Eq. (6.93) that as $t \to \infty$, we must have

$$\int_0^t h(x) \, dx \to \infty, \quad \text{so that } F(\infty) = 1$$

(otherwise $F(t)$ would not be a proper distribution function). The value of $F(0)$ could be any number between zero and one; i.e., $F(0)$ is the probability of initial failure at time $t = 0$.* For the discussion which

---

* Strictly speaking, we could have used the notation $F(0+)$ throughout, which means $\lim F(t)$ as $t \to 0$ for $t > 0$. Then $F(0-)$ could always be defined identically equal to zero, but $F(0+)$ would not necessarily equal zero. Continuity of $F(t)$ at $t = 0$ is not a necessary condition for the solution given by Eq. (6.93).

DISCRETE AND CONTINUOUS DISTRIBUTION MODELS   137

follows, we will consider $F(0) = 0$ and write Eqs. (6.93) and (6.94) as

$$F(t) = 1 - \exp\left[-\int_0^t h(x)\,dx\right] \qquad (6.93')$$

and

$$f(t) = h(t)\exp\left[-\int_0^t h(x)\,dx\right] \qquad (6.94')$$

6.8.1 THE EXPONENTIAL DISTRIBUTION

The simplest form of $h(t)$ is

$$h(t) = \lambda, \qquad \lambda > 0 \qquad (6.95)$$

where λ is a constant. In this case, using Eq. (6.94') we easily find

$$f(t) = \lambda e^{-\lambda t} \qquad (6.96)$$

the exponential probability density function. The constant λ is interpreted as the *failure rate* and is the reciprocal of the *mean time-to-failure*.

6.8.2 THE WEIBULL DISTRIBUTION

As a further example, let us consider another simple form of $h(t)$, namely

$$h(t) = \alpha\lambda t^{\alpha-1} \qquad \text{where } \alpha > 0, \lambda > 0 \qquad (6.97)$$

Then

$$\int_0^t h(x)\,dx = \lambda t^\alpha \qquad (6.98)$$

Therefore

$$f(t) = \alpha\lambda t^{\alpha-1}e^{-\lambda t^\alpha} \qquad (6.99)$$

This is the so-called two-parameter Weibull probability density function. It is evident that the exponential distribution is a special case of the Weibull distribution, by setting $\alpha = 1$ in Eq. (6.99).

Let us find the mean and variance when τ has the probability density function given by Eq. (6.99). We have

$$E(\tau^n) = \int_0^\infty \alpha\lambda t^{n-1+\alpha}e^{-\lambda t^\alpha}\,dt \qquad (6.100)$$

Let $t^\alpha = u$, $\alpha t^{\alpha-1}\,dt = du$, or $dt = \dfrac{1}{\alpha}u^{(1/\alpha)-1}\,du$. Hence

$$E(\tau^n) = \int_0^\infty \lambda u^{n/\alpha}e^{-\lambda u}\,du \qquad (6.101)$$

or
$$E(\tau^n) = \Gamma\left(\frac{n}{\alpha} + 1\right)\lambda^{-n/\alpha} \quad (6.102)$$

When $n = 1$
$$E(\tau) = \alpha_1 = \Gamma\left(\frac{1}{\alpha} + 1\right)\lambda^{-1/\alpha} \quad (6.103)$$

When $n = 2$
$$E(\tau^2) = \mu_2 + \alpha_1^2 = \Gamma\left(\frac{2}{\alpha} + 1\right)\lambda^{-2/\alpha} \quad (6.104)$$

Hence
$$\text{Var}(\tau) = \mu_2 = \Gamma\left(\frac{2}{\alpha} + 1\right)\lambda^{-2/\alpha} - \left\{\Gamma\left(\frac{1}{\alpha} + 1\right)\right\}^2 \lambda^{-2/\alpha}$$

or
$$\mu_2 = \lambda^{-2/\alpha}\left[\Gamma\left(\frac{2}{\alpha} + 1\right) - \left\{\Gamma\left(\frac{1}{\alpha} + 1\right)\right\}^2\right] \quad (6.105)$$

We will return to the Weibull distribution in Sec. 7.5.5, where methods of estimating the parameters are given. Figure 6.6 illustrates the probability density functions, for several selected values of α and $\lambda = 1$. (Changing the value λ merely "squeezes" or broadens the curve; thus λ is a "scale" parameter.)

EXERCISE: Find $h(t)$ for the probability distribution whose distribution function is given by $F(t) = 1 - e^{-\mu t}[1 - P(2, \lambda t)]$, where

$$P(N, \lambda t) = \sum_{j=N}^{\infty} \frac{(\lambda t)^j e^{-\lambda t}}{j!} = \int_0^{\lambda t} \frac{u^{N-1} e^{-u}}{\Gamma(N)} du$$

Verify the last relationship.

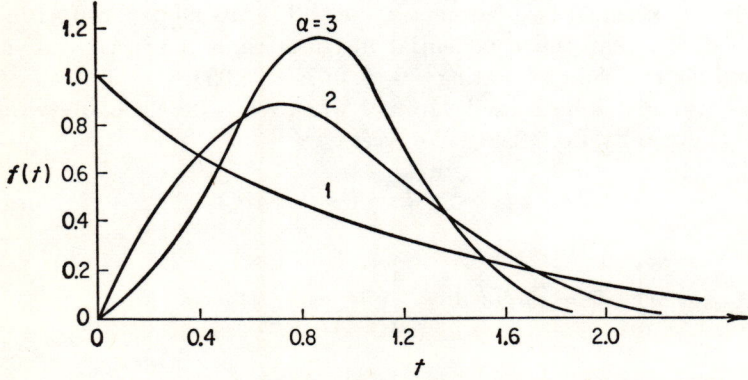

Fig. 6.6 The Weibull distribution (Eq. (6.99)) ($\lambda = 1$)

6.8.3 THE EXTREME VALUE DISTRIBUTION

One other form for the hazard function will be considered here. This is

$$h(t) = \alpha \gamma e^{\gamma t} \qquad (6.106)$$

We easily find from Eq. (6.94')

$$f(t) = \alpha \gamma e^{\gamma t} e^{-\alpha(e^{\gamma t}-1)} \qquad (6.107)$$

This probability density function can be converted by a change of variable into the *standard extreme-value* probability density function, as

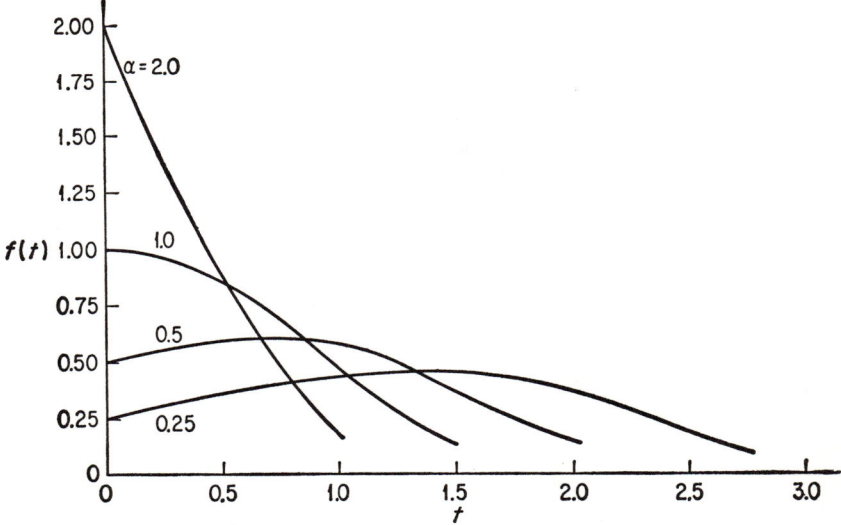

Fig. 6.7 The extreme-value distribution (Eq. (6.107)) ($\gamma = 1$)

will be shown below. In Sec. 6.8.3.2 a different approach and an application are given. Figure 6.7 illustrates the probability density function (Eq. (6.107)) for several selected values of α and $\gamma = 1$ (γ is a scale parameter).

A standard form of the two-parameter extreme value distribution is given in Ref. 7 as

$$P(\xi \leq x) = e^{-e^{-(x-u)/\beta}}, \qquad -\infty < x < \infty \qquad (6.108)$$

Since
$$P(\tau \leq t) = 1 - e^{-\alpha(e^{\gamma t}-1)}, \qquad 0 < t < \infty \qquad (6.109)$$

we can relate the random variables ξ and τ by the equation

$$e^{\gamma \tau} - 1 = e^{-\xi/\beta} \qquad (6.110)$$

where $\alpha = e^{u/\beta}$.

Furthermore if
$$\eta = (\xi - u)/\beta \qquad (6.111)$$
then
$$P(\eta \leq y) = P(\xi \leq \beta y + u) = e^{-e^{-y}}, \quad -\infty < y < \infty \qquad (6.112)$$

The random variable η is known as the *reduced* or standardized extreme value random variable (analogous to the standardized normal random variable). The distribution of η is tabulated in Ref. 8.

6.8.3.1 A Particular Form of the Hazard Function. In those cases when $h(t)$ is of the form
$$h(t) = \alpha h_1(t; \beta_1, \beta_2, \cdots) \qquad (6.113)$$
where α is considered as an unknown parameter, but where $\beta_1, \beta_2, \cdots$ are parameters whose values are known (or assumed known), then the hazard function models reduce to the simple exponential distribution model with one unknown parameter α. What this in effect means is that instead of using the sample space of real time τ, we use the pseudo-time
$$\xi \equiv \int_0^\tau h_1(x; \beta_1, \beta_2, \cdots)\, dx$$
where $\xi > 0$, to define the sample space. To show that this transformation results in the exponential distribution of ξ, we have
$$P(\xi \leq u) = P(H_1(\tau) \leq u) \qquad (6.114)$$
where
$$H_1(\tau) = \int_0^\tau h_1(x; \beta_1, \beta_2, \cdots)\, dx \qquad (6.115)$$
Since $H_1(\tau)$ increases in τ, the right side of Eq. (6.114) is equal to $P(\tau \leq H_1^{-1}(u))$ where $H_1(H_1^{-1}(u)) = u$. Thus
$$P(\xi \leq u) = 1 - e^{-\alpha u}, \quad 0 < u < \infty \qquad (6.116)$$
the distribution function for the exponential distribution.

In Sec. 8.5.1 we shall make use of this simplification to obtain exact confidence limits on reliability when the Weibull distribution model is applied to time-to-failure data. In general, however, one is not fortunate enough to know the parameters $\beta_1, \beta_2, \cdots$; and these must also be estimated from observed data. However in this case approximate methods based on the normal distribution can be used to obtain confidence limits on reliability as will be shown in Secs. 8.5 and 8.6.

6.8.3.2 A Particular Application of the Extreme Value Distribution. A failure model leading to the extreme value distribution has been described in Ref. 9. The type of failure considered is associated with corrosion. An example of this model is provided by the use of metal tubing as a lightweight construction material of a liquid rocket engine combustion chamber. The tubes themselves carry one of the propellants (the fuel) as a combustion chamber coolant. The tubes must be of minimum thickness to provide sufficient heat transfer to the coolant liquid. A failure would be detected if "pinholes" burned through the thin-walled tubes owing to the action of the hot combustion gases, causing leakage of the fuel into the combustion chamber. Generally, in this case, the rocket engine performance would drop off, with undesirable results.

The model considers that "microscopic" pits of various depths are present on the tube surfaces initially and each pit can increase in depth until a hole develops in the tube. To be precise we would define a failure to occur as soon as *one* pit has penetrated the thickness of the tube. If the time to penetration is proportional to the difference between the tube thickness and the initial depth of the pit, and the initial pit depths have an *exponential* probability distribution, then we are led to one form of the extreme value distribution as a time-to-failure distribution. We can derive this distribution as follows:

Let D be the thickness of a tube and d_i the initial depth of the ith pit (at time zero). Let $\tau_i = k(D - d_i)$ be the time-to-failure for the ith pit. Now, if the d_i ($i = 1, 2, \cdots, N$) are considered as a random sample from a (truncated) exponential distribution

$$1 - F(d) \equiv P(d_i \geq d) = \frac{e^{-\lambda d} - e^{\lambda D}}{1 - e^{-\lambda D}}, \qquad 0 \leq d \leq D \quad (6.117)$$

where $F(d)$ is the distribution function of d, then

$$G(t) = P(\tau_i \leq t) = P\left(d_i \geq D - \frac{t}{k}\right) = \frac{e^{\lambda t/k} - 1}{e^{\lambda D} - 1}, \qquad 0 \leq t \leq kD \quad (6.118)$$

Now let τ = time-to-failure; then

$$\tau = \min_i \tau_i \qquad (i = 1, 2, \cdots, N) \quad (6.119)$$

according to our assumption that failure occurs as soon as one pinhole burns through. It is easily shown that the distribution function for the smallest sample value of a random sample of size N is given by

$$P(\tau \leq t) = H(t) = 1 - (1 - G(t))^N \quad (6.120)$$

where $G(t)$ is the distribution function of the population from which the sample was taken. Now, it is reasonable to assume that the number of pits, N, is very large; and it is also easy to show that as $N \to \infty$

$$H(t) \sim 1 - e^{-NG(t)} \quad (6.121)$$

Thus from Eq. (6.118)

$$H(t) \sim 1 - \exp\left[-\frac{N}{e^{\lambda D} - 1}(e^{\lambda t/k} - 1)\right] \quad (6.122)$$

If we set

$$\alpha \equiv \frac{N}{e^{\lambda D} - 1} \quad (6.123)$$

$$\gamma \equiv \frac{\lambda}{k} \quad (6.124)$$

then

$$H(t) \sim 1 - e^{-\alpha(e^{\gamma t}-1)}, \quad t > 0 \quad (6.125)$$

and

$$H'(t) \sim \alpha\gamma e^{\gamma t}e^{-\alpha(e^{\gamma t}-1)} \quad (6.125')$$

which is the form of the extreme value distribution we obtained from the hazard function model in Sec. 6.8.3.

In applying the model to the rocket engine combustion chamber the following parameter values might be representative.

$$D = 2 \times 10^{-2} \text{ in.}, \quad k = 10^7 \text{ sec in.}^{-1}$$
$$\lambda = 4 \times 10^2 \text{ in.}^{-1}, \quad N = 3 \times 10^3$$

The quantity $\lambda D = 8$; thus in Eq. (6.117) we can almost consider $e^{-\lambda D}$ as negligible, or in other words, $P(d_i \geq d) \simeq e^{-\lambda d}$. The quantity $1/\lambda$ is then approximately the mean pit depth, which in this case would be 0.0025 in. We now determine the duration of combustion at which the survival probability is (say) $R = 0.50$.

$$H(t) = 1 - R = 1 - \exp\left[-\frac{3 \times 10^3}{3 \times 10^3}(e^{4 \times 10^{-5}t} - 1)\right] \quad (6.126)$$

or, for $R = 0.50$,

$$0.693 = e^{4 \times 10^{-5}t} - 1 \quad (6.127)$$

Thus

$$t \simeq 13{,}200 \text{ sec} \quad (6.128)$$

For $R = 0.95$, a similar calculation yields

$$t \simeq 1260 \text{ sec} \quad (6.129)$$

6.9 DEATH PROCESS MODELS

6.9.1 DERIVATION OF THE EXPONENTIAL DISTRIBUTION

In order to indicate the extension to a more general probability distribution of time-to-failure based on a many-state death process* the expo-

———
* See Ref. 1, Chap. 17, for a discussion of "birth and death" processes.

nential distribution will be derived by considering a two-state death process. The meaning of the term "death process" will become clear in the following paragraphs.

First consider that a device can be in one of two states: E_1, a successful state, and E_0, a failed state. Initially (at time $t = 0$) the device is assumed to be in state E_1. Let $P_1(t)$, $P_0(t)$ denote the respective probabilities that the device is in state E_1, E_0 at time t. A transition is allowed only from state E_1 to E_0. Once the device is in state E_0, it can only stay there. The transition is assumed to take place under the following condition: if the device is in E_1 at time t, the probability that it will make a change to state E_0 within the time interval $(t, t + h)$ is essentially proportional to h.* Now, the probability $P_0(t + h)$ that the device is in state E_0 at time $t + h$ is the sum of the following probabilities:

1. The probability that the device is in state E_0 at time t.
2. The probability that the device is in state E_1 at time t and a change to E_0 takes place during the time interval $(t, t + h)$.

Then we have

$$P_0(t + h) = P_0(t) + P_1(t)(\lambda h + o(h)) \qquad (6.130)$$

where λ is the proportionality constant and $o(h)$ denotes a function of h such that

$$\frac{o(h)}{h} \to 0 \quad \text{as} \quad h \to 0$$

We can rewrite Eq. (6.130) as

$$\frac{P_0(t + h) - P_0(t)}{h} = \lambda P_1(t) + \frac{o(h)}{h} \qquad (6.131)$$

Hence, as $h \to 0$, we have

$$P_0'(t) = \lambda P_1(t) \qquad (6.132)$$

Similarly, the probability $P_1(t + h)$ that the device is in state E_1 at time $t + h$ is the probability that it is in state E_1 at time t, and that no change takes place in the time interval $(t, t + h)$. Thus

$$P_1(t + h) = P_1(t)(1 - \lambda h - o(h)) \qquad (6.133)$$

or

$$\frac{P_1(t + h) - P_1(t)}{h} = -\lambda P_1(t) + \frac{o(h)}{h} \qquad (6.134)$$

and letting $h \to 0$,

$$P_1'(t) = -\lambda P_1(t) \qquad (6.135)$$

* The last term on the right side of Eq. (6.130) defines what is meant by "essentially proportional" to h; i.e., $\lambda h + o(h)$.

Since the device is in state E_1 at time $t = 0$, the initial conditions are:

$$P_1(0) = 1$$
$$P_0(0) = 0 \quad (6.136)$$

Solving Eqs. (6.132) and (6.135) together with the initial conditions (6.136), we obtain

$$P_0(t) = 1 - e^{-\lambda t} \quad (6.137a)$$
$$P_1(t) = e^{-\lambda t} \quad (6.137b)$$

The interpretation of this solution is as follows: $P_0(t)$ is the probability that the device is in the failed state at time t; hence, the quantity $1 - P_0(t)$ is the survival probability at time t. Also, $P_0(t) + P_1(t) = 1$, since the device is in either one of the two (mutually exclusive) states at time t. If we let τ be a random variable defined as the (waiting) time-to-failure, then clearly

$$G_0(t) \equiv P(\tau \leq t) = P_0(t) \quad (6.138)$$

i.e., $G_0(t)$ is the distribution function of the time-to-failure τ. Note that we had previously been considering the state number as the random variable, with time t as a parameter of the distribution of states.

We can get to Eq. (6.138) in still another way by considering the random variable τ as follows: let $g_0(t) \, dt$ denote the probability that the device just arrives in state E_0 within the time interval $(t, t + dt)$. Then $g_0(t)$ represents the probability density function of time-to-failure. We have

$$g_0(t) \, dt = P_1(t) \lambda \, dt + o(dt) \quad (6.139)$$

since the device must be in state E_1 at time t in order to just arrive in state E_0 within the time interval $(t, t + dt)$. Dividing Eq. (6.139) by dt and letting $dt \to 0$, we obtain

$$g_0(t) = \lambda P_1(t) \quad (6.140)$$

Using the previous result given by Eq. (6.137b)

$$g_0(t) = \lambda e^{-\lambda t} \quad (6.141)$$

which is recognized as the exponential probability density function of time-to-failure. Obviously

$$P_0(t) = \int_0^t g_0(u) \, du \quad (6.142)$$

corresponding to Eq. (6.138), is the distribution function of the random variable τ.

6.9.2 DERIVATION OF A MORE GENERAL DISTRIBUTION OF TIME-TO-FAILURE

In this generalization, we will allow the device to be in any one of a finite number N of states at time t. As before E_0 will denote the failed state, while $E_1, \cdots, E_N$ denote states of nonfailure. Transitions can take place only in the following manner: a state $E_k (1 \leq k < N)$ can be reached only by a transition from state E_{k+1}; and the state E_0 can be reached from any state. Once the device is in state E_0, it can only stay there. This (stochastic) process is illustrated in Fig. 6.8.* The arrows in the diagram denote the

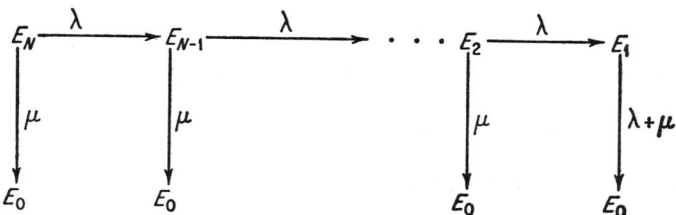

Figure 6.8

directions of the transitions; no transition is possible in the opposite directions. The quantities λ and μ are related to the transition probabilities, and will be defined below. More precisely, the postulates for this process are

1. If the device is in E_k $(2 \leq k \leq N)$ at time t, the probability that it makes a change to state E_{k-1} in the time interval $(t, t + h)$ is equal to $\lambda h + o(h)$. Also the probability that it makes a change to state E_0 in the time interval $(t, t + h)$ is equal to $\mu h + o(h)$. When $k = 1$, the probability of a single change (the only possible change is to E_0) in the time interval $(t, t + h)$ is $(\lambda + \mu)h + o(h)$.
2. The probability of more than one change (e.g., from E_k at time t to E_{k-1} to $E_{k-2} \cdots$) in the time interval $(t, t + h)$ is just $o(h)$, where $2 \leq k \leq N$.

The probability that the device is in state E_k at time t is denoted by the symbol $P_k(t)$. For $1 \leq k \leq N - 1$, the probability that the device is in E_k at time $t + h$ is equal to the sum of the following probabilities:

1. The probability that the device is in E_k at time t and that no change takes place in the time interval $(t, t + h)$. This probability is

$$P_k(t) \cdot (1 - \lambda h - \mu h - o(h))$$

---
* See Ref. 1, Chap. 17, for a derivation similar to that given in the following paragraphs.

2. The probability that the device is in E_{k+1} at time t and that a single change to E_k takes place in the time interval $(t, t+h)$. This probability is

$$P_{k+1}(t) \cdot (\lambda h + o(h))$$

3. The probability that the device is in E_{k+r} at time t ($r > 1$, $1 < k + r \leq N$) and that r changes take place in the time interval $(t, t+h)$. This probability is

$$P_{k+r}(t) \cdot o(h)$$

Therefore,

$$P_k(t+h) = P_k(t)(1 - \lambda h - \mu h) + P_{k+1}(t)\lambda h + o(h) \quad (6.143)$$

If we divide through by h and let $h \to 0$, the result is, for $k = 1, 2, \cdots, N-1$,

$$P'_k(t) = -(\lambda + \mu) P_k(t) + \lambda P_{k+1}(t) \quad (6.144)$$

By the same reasoning, the following is true:

$$P'_N(t) = -(\lambda + \mu) P_N(t) \quad (6.145)$$

A great variety of initial conditions are evidently possible; however, if it is assumed that the device is in state E_N at time $t = 0$, then the initial conditions are:

$$P_N(0) = 1$$
$$P_k(0) = 0, \quad 1 \leq k \leq N - 1 \quad (6.146)$$

Before considering the solution of Eqs. (6.144) and (6.145) for $P_k(t)$ let us now relate the random variable τ, which as before denotes the time-to-failure, to these quantities. Let $g_0(t) dt$ be the probability density function of τ. Then the probability that τ (the waiting time to state E_0) falls in the interval $(t, t + dt)$ is the sum of the following probabilities:

1. The probability that the device is in E_k at time t, followed by a single change to E_0 in the time interval $(t, t + dt)$. This probability is

$$P_k(t) \cdot (\mu \, dt + o(dt)), \quad k = 2, \cdots, N$$

2. The probability that the device is in E_1 at time t, followed by a single change to E_0 in the time interval $(t, t + dt)$. This probability is

$$P_1(t) \cdot ((\lambda + \mu) \, dt + o(dt))$$

3. The probability that the device is in E_r at time t ($2 \leq r \leq N$) and that m ($2 \leq m \leq r$) changes take place, putting the device into E_0 in the time interval $(t, t + h)$. This probability is

$$P_r(t) \cdot (o(dt))$$

Therefore,

$$g_0(t)\, dt = (\lambda + \mu) P_1(t)\, dt + \mu \sum_{k=2}^{N} P_k(t)\, dt + o(dt) \quad (6.147)$$

or, dividing Eq. 6.147 through by dt and letting $dt \to 0$,

$$g_0(t) = \lambda P_1(t) + \mu \sum_{k=1}^{N} P_k(t) \quad (6.148)$$

The probabilities $P_k(t)$, $k = 1, \cdots, N$ may be most easily obtained by using the Laplace transform on Eqs. (6.144) and (6.145) together with the initial conditions (6.146). These expressions, when put into Eq. (6.148) yield:

$$g_0(t) = e^{-\mu t} \left[\frac{\lambda(\lambda t)^{N-1} e^{-\lambda t}}{(N-1)!} + \mu \sum_{k=0}^{N-1} \frac{(\lambda t)^k e^{-\lambda t}}{k!} \right] \quad (6.149)$$

which is the probability density function of the time-to-failure τ. In addition, Eqs. (6.138) and (6.142) continue to hold true in the general model.

From the exercise in Sec. 6.8.2, using the fact that

$$\sum_{j=N}^{\infty} \frac{(\lambda t)^j e^{-\lambda t}}{j!} = \int_0^{\lambda t} \frac{u^{N-1} e^{-u}}{\Gamma(N)}\, du \equiv P(N, \lambda t)$$

Equation (6.149) can be rewritten

$$g_0(t) = e^{-\mu t} \left[\frac{\lambda(\lambda t)^{N-1} e^{-\lambda t}}{(N-1)!} + \mu \left(1 - \int_0^{\lambda t} \frac{u^{N-1} e^{-u}}{\Gamma(N)}\, du \right) \right] \quad (6.149')$$

Therefore, the distribution function $G_0(t)$ (changing the variable of integration by letting $u = \lambda x$), is

$$G_0(t) = 1 - e^{-\mu t} \left[1 - \int_0^t \frac{\lambda^N x^{N-1} e^{-\lambda x}}{\Gamma(N)}\, dx \right] \quad (6.150)$$

which can be verified by differentiating $G_0(t)$ to obtain $g_0(t)$.

Hence the survival probability at time t is

$$R(t) = e^{-\mu t} \left[1 - \int_0^t \frac{\lambda^N x^{N-1} e^{-\lambda x}}{\Gamma(N)}\, dx \right] \quad (6.151)$$

Thus the general model results in a combined chance failure (exponential) and *wearout* failure distribution. Equation (6.151) shows that the probabilities of the events "no chance failure prior to time t" and "no wearout failure prior to time t" multiply to give the probability of no failure (of either kind) prior to time t.

6.9.2.1 The Gamma Distribution. Before concluding the discussion of this general model let us note the integral in Eq. (6.151) for which we have used the notation $P(N, \lambda t)$. This is the well-known gamma distribution function, which is usually written as

$$P(N, \lambda t) = \frac{1}{\Gamma(N)} \int_0^t \lambda^N x^{N-1} e^{-\lambda x} \, dx \qquad (6.152)$$

Figure 6.9 illustrates the shape of the probability density function $dP(N, \lambda t)/dt$ for various values of N and λ. One can readily verify that the *mean* is given by $\mu = N/\lambda$ and the *variance* by $\sigma^2 = N/\lambda^2$. We shall return to the gamma distribution again in Chaps. 7 and 8 in its application to reliability estimation.

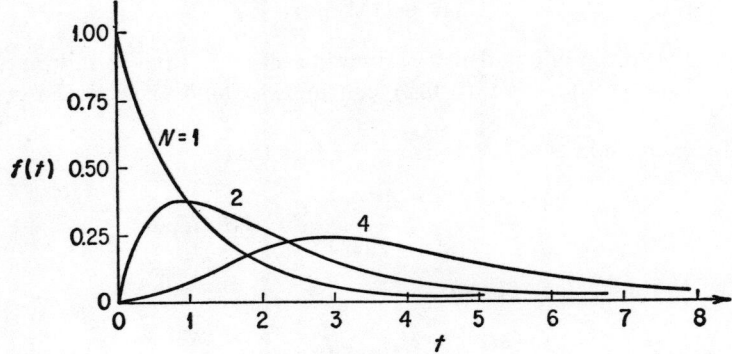

Fig. 6.9 The gamma probability density function ($\lambda = 1$)

Returning to the general model, the expression for $g_0(t)$ in Eq. (6.149') appears to be unwieldy, but it turns out that both $g_0(t)$ and the distribution function $G_0(t)$ can be calculated easily by using Molina's tables (Ref. 10). If in Molina's notation

$$q(x, a) = \frac{a^x e^{-a}}{x!} \qquad (6.153)$$

and

$$P(c, a) = \sum_{x=c}^{\infty} \frac{a^x e^{-a}}{x!} \qquad (6.154)$$

then from our notation, it can be shown that the generalized probability density function of time-to-failure given in Eq. (6.149) becomes

$$g_0(t) = e^{-\mu t}\{\lambda q(N - 1, \lambda t) + \mu[1 - P(N, \lambda t)]\} \qquad (6.155)$$

Also, the distribution function (Eq. (6.150)) becomes

$$G_0(t) = 1 - e^{-\mu t}[1 - P(N, \lambda t)] \qquad (6.156)$$

Hence, the survival probability at time t (Eq. (6.151)) is

$$R(t) = e^{-\mu t}[1 - P(N, \lambda t)] \qquad (6.157)*$$

We see that the general model results in an expression containing the combined exponential and gamma distribution for which probabilities can be calculated by means of Eqs. (6.157) and (6.154).

For the more general time-to-failure distribution, some examples of probability density curves (Eq. (6.155)) are given in Figs. 6.10, 6.11, 6.12. The curves are standardized by the following relation:

$$ag_0(at; \lambda/a, \mu/a, N) = g_0(t; \lambda, \mu, N)$$

$$(6.155a)$$

Thus, λ can always be chosen equal to unity, and the probability density functions can be plotted as a two-parameter system, since if the time scale is multiplied by a factor, then the parameters and the frequency curve change as indicated by Eq. (6.155a).

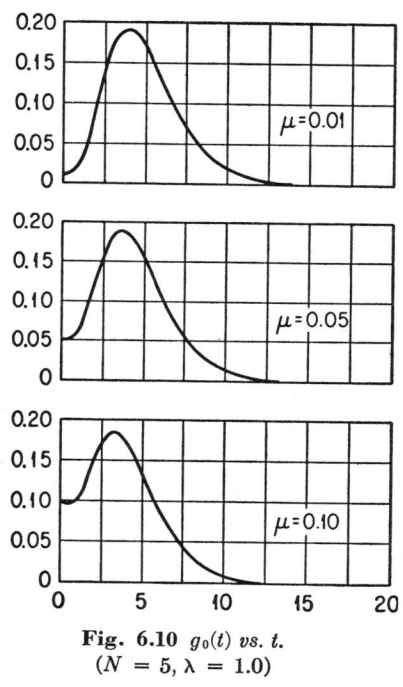

Fig. 6.10 $g_0(t)$ vs. t.
($N = 5, \lambda = 1.0$)

---

* It is interesting to note at this point that in Ref. 11 it was proposed that an expression for $R(t)$ be of the form

$$R(t) = e^{-\mu t}\left[1 - \Phi\left(\frac{t - m}{\sigma}\right)\right] \qquad (6.157a)$$

where $\Phi(X)$ is the standard normal distribution function

$$\Phi(X) = \frac{1}{(2\pi)^{1/2}} \int_{-\infty}^{X} \exp\left[-\frac{1}{2}v^2\right] dv \qquad (6.157b)$$

The similarity of Eqs. (6.157) and (6.157a) lies not only in the form but also in the fact that $P(N, \lambda t)$, the gamma distribution function, can be approximated by the normal distribution function in a certain limiting sense (see also Eq. (6.159) and Sec. 6.10). The bothersome point about Eq. (6.157a) is that it does not represent a proper probability distribution over the range $0 < t < \infty$, since the normal distribution has a "tail" left over on the negative t-axis. However, this is of negligible importance if the parameter m is sufficiently large compared to the parameter σ.

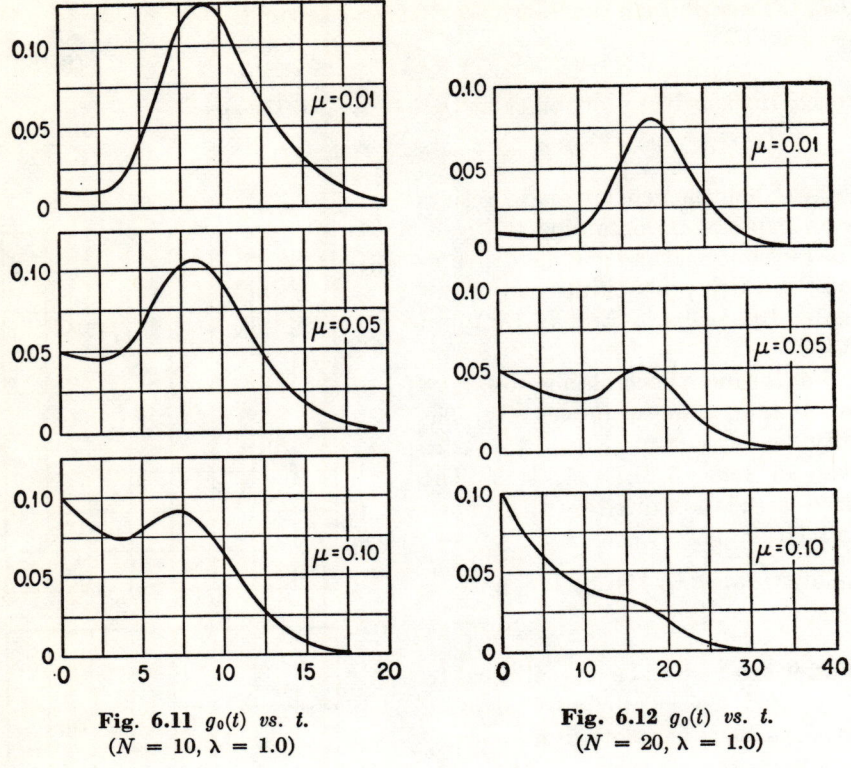

Fig. 6.11 $g_0(t)$ vs. t.
($N = 10, \lambda = 1.0$)

Fig. 6.12 $g_0(t)$ vs. t.
($N = 20, \lambda = 1.0$)

6.9.2.2 Limiting Cases of the General Time-to-Failure Distribution.
There are several special and limiting cases of the probability density function of the time-to-failure distribution (Eq. (6.149)):

1. $\mu > 0, \lambda = 0, N$ finite
2. $\mu = 0, \lambda > 0, N$ finite
3. $\mu > 0, \lambda > 0$, limit as $N \to \infty$
4. $\mu = 0, \lambda = \beta N$, large but fixed N
5. $\mu > 0, \lambda = \beta N$, limit as $N \to \infty$

The results for these five cases are presented below:

In Case 1, we have the exponential time-to-failure (chance) probability density function

$$g_0(t) = \mu e^{-\mu t} \qquad (6.158)*$$

* Since $\lambda = 0$, no transition is possible from state E_N except to state E_0. This means we can simply relabel E_N as E_1, and we have the two-state process derived previously.

DISCRETE AND CONTINUOUS DISTRIBUTION MODELS   151

In Case 2, we have the gamma time-to-failure (wearout) probability density function

$$g_0(t) = \frac{\lambda^N t^{N-1}}{(N-1)!} e^{-\lambda t} \qquad (6.159)$$

In Case 3, as $N \to \infty$ the probability density function tends to the exponential time-to-failure (chance) distribution given by Eq. (6.158), since with increasing N the "wearout peak" of the distribution travels off to infinity. This also agrees with our intuition concerning the model, since as $N \to \infty$ it becomes virtually impossible for failure to occur by passing through the whole sequence of states.

In Case 4, we let λ be proportional to N, in order to give the device a chance (so to speak) of moving down the sequence of states even when N is large. In this case (assuming $\mu = 0$, $\lambda = \beta N$), the probability density function $g_0(t)$ approaches a normal probability density function with mean $1/\beta$ and standard deviation $1/\beta N^{1/2}$. Thus for large but fixed N, $g_0(t)$ is approximately a normal p.d.f.

In Case 5, the frequency function $g_0(t)$ approaches a limiting distribution equal to $\mu e^{-\mu t}$ for $t < 1/\beta$, with a sharp "spike" at $t = 1/\beta$. The probability contained in the interval $0 < t < 1/\beta$ is $1 - e^{-\mu/\beta}$; and at $t = 1/\beta$, the spike contains probability $e^{-\mu/\beta}$.

It is almost intuitively evident that we cannot expect to obtain a limiting probability density function which looks like some of the selected examples given in Figs. 6.10–6.12, if we let $N = \infty$; that is, at least three parameters are needed to describe such a distribution.

6.9.2.3 Initial Failure.
Up to now, we have not incorporated the initial failure concept in the derivation of the model. However, the first approach is to assume that there is a probability α that the device is in state E_N at time $t = 0$. Thus, the set of initial conditions on the process corresponding to Eqs. (6.146) become

$$\begin{aligned} P_N(0) &= \alpha, & 0 < \alpha < 1 \\ P_k(0) &= 0, & 1 \leq k \leq N-1 \end{aligned} \qquad (6.146')$$

The result is to multiply the value of $g_0(t)$, given by Eq. (6.149) or (6.155), by α; therefore, we simply define $P_0(0) = 1 - \alpha$.

The "delayed" type initial failure discussed previously can be incorporated into the model by assuming that the proportionality constant corresponding to a jump from state E_N to E_0 is $\mu + \delta$, where δ is positive. This is illustrated in Fig. 6.13. The result would be a failure rate curve looking essentially like that shown in Fig. 6.14. However, the derivation becomes fairly complicated, and is not worked out here.

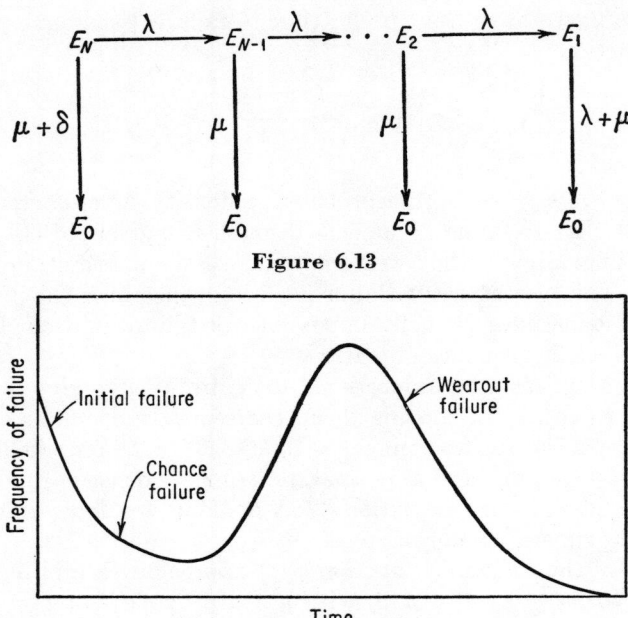

Figure 6.13

Fig. 6.14 Frequency of failure *vs.* time. (Delayed initial failure.)

6.9.2.4 Physical Interpretation of the Model Parameters. In order to interpret the parameters N, λ and μ in a physical sense, two examples are discussed below.

First, consider the example of an automobile tire. We might imagine that the tread of the tire consists of a large number of thin layers of rubber. As the tire rolls on the road, a layer can suddenly leave the tire. If the number of layers left on the tire at time t is called the state number of the tire at time t, then it is easy to imagine the transitions from one state to the next as satisfying the conditions of the model. When the last layer leaves the tire, it has worn out. However, we also realize that the tire can run over a nail, which could cause a blowout no matter how many layers of rubber are left. The latter situation corresponds to the allowed transitions from any state to the state of failure, in the model.

Another example, which could be interpreted as generally applicable to all types of equipment which are subject to severe environment—e.g., handling, transport, dropping, heat, cold, or even interplanetary space environment, etc.—is as follows: The various environments might be classified into two categories: (1) those which are severe enough so that if applied once, the equipment ceases to operate; (2) those which are mild enough that in one exposure the equipment will not fail, but the cumulative effect of several or a large number of such exposures would also cause the equipment to fail (possibly in a different manner). The parameter N could then be interpreted as the "resistance" of the equipment to mild exposures or shocks; i.e., failure of the equipment occurs only when a finite number N of such shocks have occurred. The parameter λ specifies the rate of occurrence or "arrival rate" of such shocks. The parameter μ is the occurrence rate of the type of shock that would immediately cause failure.

A good example is given by the power output equipment on several of the space probe satellites launched in the past few years. Power for the satellite is stored by several dozen batteries (the energy source consists of several thousand solar cells on "paddle wheels"). Hypothetically, one battery at a time could be put out of action by exposure to micrometeoroids (this might in itself be a cumulative type of damage). When enough batteries are not operating, the power supply is reduced to a sufficiently low level to prevent effective data transmission from the satellite. On the other hand, irrespective of the number of active batteries, the command transmit switch could fail completely, giving the same result.

6.10 THE NORMAL DISTRIBUTION

One of the most important continuous probability distribution models is the *normal* (Gaussian) distribution. We have already encountered this distribution as a limiting case of the binomial (a discrete distribution) in Sec. 6.2.3, and of the gamma distribution (a continuous distribution) in Sec. 6.9.2.2. For convenience we redefine it here: A (one-dimensional) random variable ξ is said to be normally distributed (or have a normal probability distribution) with mean μ, standard deviation σ when its probability density function is given by

$$f(x) = \frac{1}{\sigma\sqrt{2\pi}} \exp\left[-\frac{1}{2}\left(\frac{x-\mu}{\sigma}\right)^2\right], \quad \begin{array}{l} -\infty < x < \infty \\ \sigma > 0 \\ -\infty < \mu < \infty \end{array} \quad (6.160)$$

Correspondingly the distribution function of ξ is given as

$$P(\xi \leq x) \equiv F(x) = \frac{1}{\sigma\sqrt{2\pi}} \int_{-\infty}^{x} \exp\left[-\frac{1}{2}\left(\frac{t-\mu}{\sigma}\right)^2\right] dt \quad (6.161)$$

When $\mu = 0$ and $\sigma = 1$, then we call ξ a *standard* normal random variable with *standard* normal probability density and distribution functions.

A special notation is used to denote these functions, respectively, as

$$\phi(x) = \frac{1}{\sqrt{2\pi}} e^{-x^2/2}, \qquad -\infty < x < \infty \tag{6.162}$$

and
$$\Phi(x) = \frac{1}{\sqrt{2\pi}} \int_{-\infty}^{x} e^{-t^2/2}\, dt \tag{6.163}$$

The function $\phi(x)$ is a "bell-shaped" curve when graphed as a function of x, symmetric about the line $x = 0$ in the plane, with maximum value at $x = 0$ equal to $1/\sqrt{2\pi} = 0.399$ approximately, as shown in Fig. 6.15.

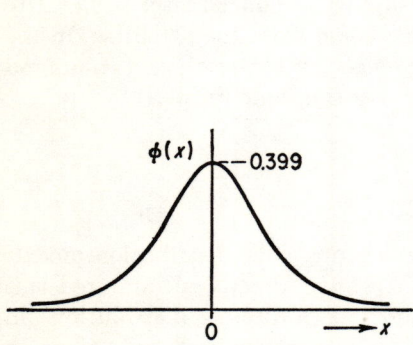

Fig. 6.15 The standard normal probability density function, $\phi(x)$.

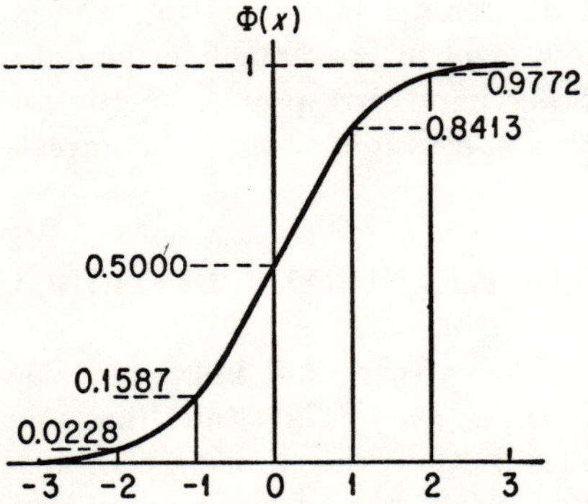

Fig. 6.16 The standard normal distribution function, $\Phi(x)$.

Evidently the standard normal distribution function $\Phi(x)$ has the necessary properties of a distribution function (Sec. 5.5), namely that

$$\Phi(-\infty) = 0 \tag{6.164}$$

$$\Phi(+\infty) = 1 \tag{6.165}$$

and owing to the symmetrical nature of $\phi(x)$,

$$\Phi(-x) = 1 - \Phi(x) \tag{6.166}$$

Directly from Eq. (6.166) we find that, when $x = 0$, $\Phi(0) = 1 - \Phi(0)$ which yields $\Phi(0) = \frac{1}{2}$. Figure 6.16 shows a graph of the essential features of $\Phi(x)$.

The property (6.165) can be shown by writing

$$\Phi(+\infty) = \frac{2}{\sqrt{2\pi}} \int_{0}^{\infty} e^{-x^2/2}\, dx \tag{6.167}$$

changing variables by letting $x^2/2 = y$, whereby $dx = dy/\sqrt{2y}$, then obtaining

$$\Phi(+\infty) = \frac{1}{\sqrt{\pi}} \int_0^\infty e^{-y} y^{-1/2} \, dy = \frac{\Gamma(\frac{1}{2})}{\sqrt{\pi}} = \frac{\sqrt{\pi}}{\sqrt{\pi}} = 1 \quad (6.168)$$

where we have used a well-known theorem of Laplace transforms:

$$\int_0^\infty e^{-st} t^\alpha \, dt = \frac{\Gamma(\alpha + 1)}{s^{\alpha+1}}, \quad \alpha > -1 \quad (6.169)$$

and the fact that $\Gamma(\frac{1}{2}) = \sqrt{\pi}$.

The probability contained in an interval (a, b) where ξ has any normal distribution can always be found by using tables of the standard normal distribution function, as follows:

If ξ has mean μ and standard deviation σ then $(\xi - \mu)/\sigma$ is a standard normal variable (Sec. 5.7). Hence

$$P(a < \xi < b) = P\left(\frac{a - \mu}{\sigma} < \frac{\xi - \mu}{\sigma} < \frac{b - \mu}{\sigma}\right) \quad (6.170)$$

is equal to $\Phi[(b - \mu)/\sigma] - \Phi[(a - \mu)/\sigma]$. The National Bureau of Standards Tables (Ref. 2) give values of $\phi(x)$ defined by Eq. (6.162) and values of $2\Phi(x) - 1$, where $\Phi(x)$ is defined by Eq. (6.163), to 15 decimal places for $x = 0$ (0.0001) 1 (0.001) 7.800 (various numbers of places) 8.285.

6.10.1 THE CENTRAL-LIMIT THEOREM

The normal distribution appears, as we have indicated previously in two specific cases, as a limiting form of many probability distributions. Furthermore, and perhaps most important, the limiting distribution of any reasonably behaving function of a sample mean or of sample central moments is asymptotically normal as the sample size tends to infinity irrespective of the probability distribution of the underlying population. We shall make use of this fact in Sec. 8.3 when we wish to find approximate confidence intervals on reliability estimators.

The reader probably accepts the fact that many kinds of measurements are approximately normally distributed. The theoretical justification for this is contained in the remarkable *central-limit theorems*. In general, a central-limit theorem applies when the measurement, or random variable we are considering is the *sum* of a large number of other random variables or "chance effects." The simplest form of central-limit theorem is as follows (Ref. 12, p. 214).

Let $\xi_1, \xi_2, \cdots, \xi_n$ be identically distributed, mutually independent random variables for any value of n, with common mean μ and standard deviation σ. Then if

$$\xi = \sum_{j=1}^{n} \xi_j$$

we have as $n \to \infty$

$$P(\xi \leq x) \sim \Phi\left(\frac{x - n\mu}{\sigma\sqrt{n}}\right) \qquad (6.171)$$

which states that ξ is asymptotically normal $(n\mu, \sigma\sqrt{n})$. In particular the mean $\bar{x}$ of a random sample of size n is asymptotically normal $(u, \sigma/\sqrt{n})$.*

EXERCISE: Show that the last statement is true.

Actually as we have previously stated, any "reasonably behaving" function of $\bar{x}$, say $f(\bar{x})$, is asymptotically normal; i.e.,

$$P(f(\bar{x}) \leq x) \sim \Phi\left[\frac{x - f(\mu)}{\left|\dfrac{df}{d\bar{x}}\right|_{\bar{x}=\mu}\left(\dfrac{\sigma}{\sqrt{n}}\right)}\right] \qquad (6.172)$$

(see Sec. 8.3).

We can infer from (6.172) for example that the arcsine transformation (Sec. 8.3.3) results in an asymptotically normal variable arcsine $\sqrt{\hat{p}}$; i.e., normal (arcsine $\sqrt{p}$, $1/(2\sqrt{N})$). Of course, $\hat{p} = f/N$ is itself asymptotically normal $(p, \sqrt{pq/N})$. Similarly the estimate $\hat{\lambda}$ of the parameter λ of the Poisson distribution (Eq. (6.86)), being the sample mean, is asymptotically normal $(\lambda, \sqrt{\lambda/N})$.

For more complete discussions of the normal distribution and various related distributions, the reader is urged to consult many of the references, particularly Refs. 12, 13, and 14. Reference 14 contains an elementary discussion of the properties of the multivariate normal distribution. Reference 15 is a thorough account of application of the lognormal distribution.

REFERENCES

1. W. Feller, *An Introduction to Probability Theory and Its Applications*, 2d ed., John Wiley & Sons, New York, 1957, vol. I, p. 434.

---

* What is meant by "asymptotically normal (a, b)" is only that the limiting distribution function is asymptotically $\Phi((x - a)/b)$. It does not necessarily imply that the mean and standard deviation of ξ tend to a and b.

2. *Tables of Probability Functions*, National Bureau of Standards, Applied Mathematics Series 23, 1953.
3. *Tables of the Cumulative Binomial Probability Distribution*, Harvard University Press, Cambridge, Mass., 1953.
4. M. A. Girshick, F. Mosteller, and L. J. Savage, "Unbiased Estimates for Certain Binomial Sampling Problems With Applications," *Ann. Math. Stat.*, **17,** 13–23 (1946).
5. E. C. Titchmarsh, *The Theory of Functions*, 2d ed., Oxford University Press, 1949, pp. 57–58.
6. A. Wald, *Sequential Analysis*, John Wiley & Sons, Inc., New York, 1947.
7. E. J. Gumbel, *Statistics of Extremes*, Columbia University Press, New York, 1958.
8. *Probability Tables for the Analysis of Extreme Value Data*, National Bureau of Standards, Applied Mathematics Series 22, 1953.
9. B. Epstein, "The Exponential Distribution and its Role in Life Testing," *Ind. Qual. Control*, **XV:**6 (December 1958).
10. E. C. Molina, *Poisson's Exponential Binomial Limit*, D. Van Nostrand Company, Inc., Princeton, N.J., 1942.
11. W. A. Gunn, "The Reliability of Complex Systems," *Western Conference Proceedings, ASQC, August 20–21, 1956.*
12. H. Cramér, *Mathematical Methods of Statistics*, Princeton University Press, Princeton, N.J., 1946.
13. S. S. Wilks, *Mathematical Statistics*, Princeton University Press, Princeton, N.J., 1943.
14. A. M. Mood, *Introduction to the Theory of Statistics*, McGraw-Hill Book Company, Inc., New York, 1950.
15. J. Aitchison and J. A. Brown, *The Lognormal Distribution, With Special Reference to its Use in Economics*, Cambridge University Press, 1957.

ADDITIONAL READING

Acheson, M. A., "Life Factors Affecting Acceptance Procedures," *Proc. 2d National Symposium on Quality Control and Reliability in Electronics, Washington, D.C., January 9–10, 1956*, 156–164.

Allen, W. R., "Inference From Tests with Continuously Increasing Stress," *Operations Research*, **7:**3, 303–12 (June 1959).

Bartholomew, D. J., "Testing for Departure from the Exponential Distribution," *Biometrika*, **44,** Parts 1 and 2, 253–57 (June 1957).

Birnbaum, Z. W., "A Statistical Model for Life-Length of Materials," *J. Am. Stat. Assoc.*, **53:**281, 151–60 (March 1958).

Broadbent, S., "Simple Mortality Rates," *Applied Statistics*, **7:**2, 86–95 (June 1958).

Cohen, A. C., "Estimating the Parameters of a Modified Poisson Distribution," *J. Am. Stat. Assoc.*, **55:**289. 139–43 (March 1960).

Cox, D. R., "The Analysis of Exponentially Distributed Life-Times with Two Types of Failure," *J. Roy. Stat. Soc.*, **21B**:2, 411–21 (1959).

Davis, D. J., "An Analysis of Some Failure Data," *J. Am. Stat. Assoc.*, **47**, 113–150 (1952).

Eldredge, G. G., "Analysis of Corrosion Pitting by Extreme-Value Statistics and Its Application to Oil Well Tubing Caliper Surveys," *Corrosion*, **13**, 51t–60t (January 1957).

Flehinger, B. J., and Lewis, P. A., "Two-Parameter Lifetime Distributions for Reliability Studies of Renewal Processes," *IBM Journal Research and Development*, **5**, 58–73 (January 1959).

Godfrey, M. L., "Theory of Extremal Values Applied to Tests," *Ind. Labs.*, **9**, 9–12 (July 1958).

————, "Theory of Extremal Values Applied to Tests," *Ind. Labs.*, **9**, 74–79 (August 1958).

Gumbel, E. J., "Statistician Attacks Extreme Values in Technical Problems," *Ind. Labs.*, **7**, 22–30 (December 1956).

Kao, J. H. K., "A New Life-Quality Measure for Electron Tubes," *I.R.E. Trans. Reliability and Quality Control*, **PGRQC–7**, 1–11 (April 1956).

Raff, M. S., "On Approximating the Point Binomial," *J. Am. Stat. Assoc.*, **51**, 293–303 (June 1956).

Saito, Kin-ichiro, "Maximum-Likelihood Estimate of Proportion Using Supplementary Information," *Bull. Math. Statistics (Japan)*, **7**:1/2, 11–17 (December 1956).

Stoller, D. S., "A Failure Model for Equipment Undergoing Complex Operation," *Operations Research*, **6**:5, 723–28 (September–October 1958).

Wohl, J. G., "Dependability of Military Equipment: A Systems Approach," *Elec. Mfg.*, **63**, 93–100, 153 (March 1959).

CHAPTER SEVEN
RELIABILITY ESTIMATION: PART I

7.1. INTRODUCTION: THE RELIABILITY FUNCTION, R

In Chap. 6, the important discrete and continuous distribution models with which we shall be concerned were presented.

The problem taken up in this and the following chapter is to estimate R, where, as we shall see, R is some specified function of the parameters of the distribution and is obtained by summation or integration of the probability density function over the sample space. Thus, if $P(\xi \leq x) \equiv F(x; \theta_1, \theta_2, \cdots)$ is the (cumulative) distribution function and $\theta_1, \theta_2, \cdots$ are the parameters of the distribution, then if x_1, x_2 are the limits which define the event: Success, R is given by

$$R = P(x_1 \leq \xi \leq x_2) = F(x_2; \theta_1, \theta_2, \cdots) - F(x_1; \theta_1, \theta_2, \cdots)$$

$$= R(\theta_1, \theta_2, \cdots) \qquad (7.1)$$

which we shall refer to as the *reliability function*.

7.1.1 EXAMPLE FOR THE NORMAL DISTRIBUTION

For instance, if x_1, x_2 represent the lower and upper specification limits on a given performance parameter, ξ, normally distributed, with mean θ_1 and standard deviation θ_2, we can define success as occurring when the performance lies within specification limits. Here

$$R = P(x_1 \leq \xi \leq x_2)$$

$$= \int_{x_1}^{x_2} \frac{1}{\theta_2 \sqrt{2\pi}} \exp\left[-\frac{(x - \theta_1)^2}{2\theta_2^2}\right] dx$$

$$= R(\theta_1, \theta_2) \qquad \text{since } x_1 \text{ and } x_2 \text{ are known}$$

7.1.2 EXAMPLE FOR THE EXPONENTIAL DISTRIBUTION

In the case of the exponential distribution, for example, the reliability would be defined as the probability that the device operates for a period greater than or equal to time T. Thus

$$R = P(\tau > T) = \int_T^\infty \frac{1}{\theta} e^{-t/\theta}\, dt$$

$$= [-e^{-t/\theta}]_T^\infty = e^{-T/\theta} \quad (7.2)$$

which can be written $R(\theta)$ since T is known.

7.1.3 EXAMPLE FOR THE BINOMIAL DISTRIBUTION

The probability distribution need not be a continuous distribution. Thus, for the binomial distribution, if θ is the probability of any one component functioning successfully and reliability is defined in this particular example as the ability, say, for at least r components out of a total of n to function successfully, then

$$R = P(r \leq \zeta \leq n)$$

$$= \binom{n}{n}\theta^n + \binom{n}{n-1}\theta^{n-1}(1-\theta) + \cdots + \binom{n}{r}\theta^r(1-\theta)^{n-r}$$

$$= R(\theta) \qquad \text{since } r \text{ and } n \text{ are known}$$

7.1.4 THE PROBLEM OF ESTIMATING R

Thus again we see that reliability can be expressed as some function of the parameter(s) of the distribution. Generally the quantity R does not appear explicitly in the mathematical forms of the probability density functions, and also, the true values of $\theta_1, \theta_2, \cdots$ are not known. It is convenient, therefore, in approaching the problem of estimating R to first consider the problem of estimating the parameters of the distribution model which do appear explicitly. Once such estimators $\hat{\theta}_1, \hat{\theta}_2, \cdots$ of $\theta_1, \theta_2, \cdots$ are obtained we can then consider how to utilize the estimates of the parameters to obtain estimates of reliability.

In this chapter, therefore, we will cover the basic aspects of estimation of distribution model *parameters*. The material is "classical," in the sense that it is the same subject, covered much more thoroughly, in sections on estimation theory of the referenced statistical texts. However, in this chapter the orientation will be to the specific distribution models considered of importance in reliability analysis, as well as to the material in Chap. 8.

RELIABILITY ESTIMATION: PART I   161

7.2. PROPERTIES OF ESTIMATORS

In Sec. 5.10 we discussed the sample mean and variance, which are examples of the most universally used estimators. The sample mean is generally the best estimator for the true mean μ of the population from which the sample is taken. It provides an estimate of the *location* of the population distribution. The sample variance under the same conditions is an estimator for the "spread," or *dispersion* of the population distribution as given by the population variance, σ^2.

Generalizing, we say that an estimator is a function of the sample, which "estimates" in some sense a function of the parameters of the distribution from which the sample is taken.

Whether an estimator is "good" or "bad" depends upon what function of the sample is selected and what function of the parameters of the distribution it is used to estimate. A "good" estimator should have the property that it is "close" to the true value of the function it is estimating.

Recalling that the sample values are themselves random variables (Sec. 5.10) which implies that an estimator is a random variable, we now define some optimum properties of estimators.

7.2.1 UNBIASEDNESS

An estimator $\hat{\theta}$ is said to be *unbiased* if

$$E(\hat{\theta}) = \theta \tag{7.3}$$

That is to say, if $x_1, \cdots, x_n$ are the sample values taken from a population distribution whose p.d.f. (or probability distribution) is a function of θ (and possibly other parameters), then $\hat{\theta}$ is a function of $x_1, \cdots, x_n$ such that the *expected* value of $\hat{\theta}$ is the parameter θ. We say that $\hat{\theta}$ is an unbiased estimator of θ.

EXAMPLE: $\bar{x} \equiv \sum x_i/n$ is an unbiased estimator of μ. For, $E(x_i) = \mu$; and $E(\bar{x}) = 1/n \sum E(x_i) = n\mu/n = \mu$.

EXERCISE: The sample variance defined by

$$s^2 \equiv m_2 \equiv \frac{1}{n} \sum (x_i - \bar{x})^2$$

is not an unbiased estimator of σ^2. To show this, use the fact that $E(x_i^2) = \sigma^2 + \mu^2$. What *is* an unbiased estimator for σ^2? (See Sec. 5.10.)

EXAMPLE: Interpret the previous example for the *binomial* probability distribution (parameter $n = 1$):

$$P(\zeta = 0) = q$$
$$P(\zeta = 1) = p \equiv 1 - q$$

The interpretation is as follows: We select a random sample, $x_1, \cdots, x_n$ of size n from this population distribution; i.e., the x_i's are zeros and/or ones. Then $\bar{x} \equiv \sum x_i/n$ is equal to the number of ones, say, f ones, in the sample, divided by n. Since $E(x_i) = p$, then $\bar{x} = f/n$ is an unbiased estimator for p. With our usual interpretation, zeros denote successes, ones denote failures, where q and p are the probabilities for success and failure, respectively. Thus the observed ratio of failures to sample size is an unbiased estimator for the true probability of failure, p. Note that drawing a random sample of size n from the above binomial distribution is formally the same as making n independent trials with constant probability p of failure on each trial.

7.2.2 CONSISTENCY

An estimator $\hat{\theta}$ is said to be *consistent* if

$$\lim_{n \to \infty} P(|\hat{\theta} - \theta| > \epsilon) = 0$$

for arbitrary $\epsilon > 0$, where n is the sample size. Thus, by taking a large enough sample size, we can make the probability as small as we please that a consistent estimator differs from the parameter it estimates by more than a previously fixed amount, no matter how small the amount.

EXAMPLE: $\bar{x}$ is a consistent estimator of μ. To show this, we assume that the variance σ^2 of the population distribution exists (this assumption is not necessary, as the theorem is true if only the mean μ exists) (see Ref. 1, p. 254). We have by Tchebycheff's inequality (Sec. 5.7.1)

$$P(|\bar{x} - \mu| > \epsilon) \leq \frac{\operatorname{Var} \bar{x}}{\epsilon^2} \tag{7.4}$$

But $\operatorname{Var} \bar{x} = \sigma^2/n$ (Sec. 5.10). Therefore

$$P(|\bar{x} - \mu| > \epsilon) \leq \frac{\sigma^2}{n\epsilon^2}$$

which tends to zero as $n \to \infty$.

EXAMPLE:

$$\bar{x}' \equiv \frac{1}{n-1} \sum_{i=1}^{n} x_i$$

is a consistent estimator for μ (it is biased however). Note that it is *asymptotically* unbiased; i.e., as $n \to \infty$, the bias approaches zero.

7.2.3 MINIMUM VARIANCE

A third important property of a desirable estimator is that it have as small as possible a variance for any sample size n. This leads us to more complicated aspects of the theory of estimation, and it will suffice here to state that the *method of maximum likelihood* will generally find estimators of this kind when they exist.*

7.3. METHODS FOR FINDING ESTIMATORS

For the purposes of this chapter, the procedures for finding estimators will be merely stated, as the theory is rather complicated. However, with the present tools we can examine some of the important properties of the estimators obtained.

The *method of maximum likelihood* for obtaining estimators is generally easy to apply and also tends to yield better estimates than other methods; e.g., an alternative method is to equate moments of the sample with population moments. First we shall discuss the method of maximum likelihood in general terms, then apply it to a few examples. In many cases this method is equivalent to the above-mentioned alternative method. The method of matching moments, however, may in some cases be easier to use computationally. Whichever method is used, the estimators obtained will generally be *asymptotically normal*; i.e., as the sample size n tends to infinity the distribution of the estimator will approach the normal distribution.

7.4. THE METHOD OF MAXIMUM LIKELIHOOD (ONE UNKNOWN PARAMETER)

Suppose we have a random sample $x_1, \cdots, x_n$ from the population distribution whose p.d.f. is $f(x; \theta)$. Here, θ denotes a parameter of the distribution, which might be the mean of the distribution, but not necessarily. The problem is to find some function of the sample values $\hat{\theta}(x_1, \cdots, x_n)$, such that $\hat{\theta}$ is a good estimator for θ. To do this, we consider the function

$$L(x_1, \cdots, x_n; \theta) = f(x_1; \theta) \cdots f(x_n; \theta) \qquad (7.5)$$

The function L, as defined by Eq. (7.5), is called the *likelihood function of the sample*. Let us see how this definition leads to a good estimator for θ. First of all, the righthand side of Eq. (7.5) is identical in form with the joint probability density function of n independent random variables, each with the same probability distribution. Thus L can be considered as the *a priori* probability of obtaining the result $x_1, \cdots, x_n$. However, since we regard the values of $x_1, \cdots, x_n$ as already observed from an experiment, then L is simply a function of the unknown parameter θ which we wish to estimate. The method of maximum likelihood consists of finding the value

* See Ref. 1, Chaps. 32, 33.

of θ in Eq. (7.5) which maximizes L. The idea behind this is that this value of θ maximizes the "probability of obtaining the observed result"—making use of the formal identity stated above. Thus, in this analogy we attempt to find the value of θ which would make it "most likely" that we observe the (already given) result of the experiment.

Since $\mathcal{L} \equiv \log L$ is an increasing function of L, then the value of θ which makes $\log L$ as large as possible will also make L as large as possible. In general, this value of θ can be obtained by solving the *likelihood equation*

$$\frac{\partial \mathcal{L}}{\partial \theta} = 0$$

for θ. The value of θ thus obtained will be denoted by $\hat{\theta}$, and is a function of the $x_1, \cdots, x_n$.

7.4.1 APPLICATIONS TO CONTINUOUS DISTRIBUTIONS

EXAMPLE: Suppose n electron tubes are put on life test. The experiment is stopped when all tubes have failed, the times of failure being $t_1, \cdots, t_n$. Assume that the time-to-failure has an exponential distribution with p.d.f.

$$f(t; \theta) = \theta^{-1} e^{-t/\theta}, \quad t > 0$$
$$= 0, \quad t \leq 0$$

where the parameter θ is the mean of the distribution. The likelihood function L is then

$$L = \theta^{-n} \exp\left[-\frac{1}{\theta} \sum_{i=1}^{n} t_i\right] \tag{7.6}$$

therefore

$$\mathcal{L} = -n \log \theta - \frac{1}{\theta} \sum t_i \tag{7.7}$$

$$\frac{\partial \mathcal{L}}{\partial \theta} = -\frac{n}{\theta} + \frac{1}{\theta^2} \sum t_i \tag{7.8}$$

Setting $\partial \mathcal{L}/\partial \theta = 0$ and solving Eq. (7.8) for θ, the solution being denoted by $\hat{\theta}$, we obtain

$$\hat{\theta} = \frac{1}{n} \sum_{i=1}^{n} t_i \tag{7.9}$$

EXERCISE: Show that the value of θ given by Eq. (7.9) gives a *maximum* value for $\mathcal{L}$, by differentiating Eq. (7.8) and using Eq. (7.9). Note that $\hat{\theta}$ obtained above is just the mean of the sample $t_1, \cdots, t_n$. Thus we would have obtained the same estimator, $\hat{\theta}$, by matching the *sample mean* with the *population mean*.

Once we have obtained the estimator, by any means whatever, we then consider it as a random variable, since it is really a certain function of the outcome of the experiment $(x_1, \cdots, x_n)$, which was also stated in Sec. 5.10. Thus the estimator has a probability distribution of its own, with mean, variance, and so on. For example, from Eq. (5.54) we find that $\hat{\theta}$ in the previous example has a gamma distribution. It is also important to note here that the gamma distribution is related to the well-known χ^2 distribution. To show this, we have (from Eq. (5.54))

$$P(\hat{\theta} \leq x) = \frac{1}{\Gamma(n)} \int_0^x \left(\frac{n}{\theta}\right)^n t^{n-1} e^{-tn/\theta} \, dt \tag{7.10}$$

Now define $\chi^2 = 2n\hat{\theta}/\theta$. Then

$$P(\chi^2 \leq y) = P\left(\hat{\theta} \leq \frac{\theta y}{2n}\right) \tag{7.11}$$

$$= \frac{1}{\Gamma(n)} \int_0^{\theta y/2n} \left(\frac{n}{\theta}\right)^n t^{n-1} e^{-tn/\theta} \, dt \tag{7.12}$$

By transforming the variable of integration in Eq. (7.12) we obtain

$$P(\chi^2 \leq y) = \frac{1}{2^n \Gamma(n)} \int_0^y t^{n-1} e^{-t/2} \, dt \tag{7.13}$$

which is identified as the χ^2 distribution function with $2n$ "degrees of freedom." We will hereafter use χ^2_{2n} as notation for the random variable to show the dependence on the parameter $2n$ (Ref. 1, p. 234).

EXERCISE: In the previous example $E(\hat{\theta}) = \theta$; i.e., $\hat{\theta}$ is an unbiased estimator for θ. Also Var $\hat{\theta} = \theta^2/n$. Show these facts in two ways.

7.4.1.1. A Special Example. The following example is quite different from the previous examples in that the maximum likelihood estimator cannot be obtained by differentiation of the likelihood function. Let

$$f(x; \theta) = \frac{1}{\theta} \quad \text{for } 0 < x < \theta \tag{7.14}$$

$$= 0 \text{ elsewhere}$$

(It is easy to see that $f(x; \theta)$ is a p.d.f., since

$$\int_{-\infty}^{\infty} f(x; \theta) \, dx = \int_0^\theta \frac{dx}{\theta} = 1)$$

We have
$$L(x_1, \cdots, x_n; \theta) = \left(\frac{1}{\theta}\right)^n, \quad 0 < x_i < \theta$$
$$i = 1, 2, \cdots, n \qquad (7.15)$$

By inspection of Eq. (7.15), L is maximized when θ takes on its smallest possible value. Also by (7.15), we see that θ can be no smaller than the *largest* of the x_i, $i = 1, 2, \cdots, n$. Thus the maximum likelihood estimator for θ is
$$\hat{\theta} = \max_i x_i \qquad (7.16)$$

A maximum likelihood estimator is not necessarily unbiased.* Again consider the preceding example. Since $\hat{\theta}$ is the maximum value of the sample, it has the probability density element
$$P(y < \hat{\theta} < y + dy) = g(y)\, dy = n\overline{F(y; \theta)}^{n-1} f(y; \theta)\, dy$$
where $f(y; \theta)$ is given by Eq. (7.14) and $F(y; \theta)$ as usual denotes the distribution function of the distribution whose p.d.f. is $f(y; \theta)$.† Thus
$$g(y) = n\left(\frac{y}{\theta}\right)^{n-1} \frac{1}{\theta} = \frac{n}{\theta^n} y^{n-1} \qquad (7.17)$$

Now
$$E(\hat{\theta}) = \frac{n}{\theta^n} \int_0^\theta y^n\, dy \qquad (7.18)$$

or
$$E(\hat{\theta}) = \frac{n}{n+1} \theta \qquad (7.19)$$

Thus $\hat{\theta}$, defined by Eq. (7.16), is not an unbiased estimator of θ; but
$$\hat{\theta}' \equiv \frac{n+1}{n} \hat{\theta}$$
would be unbiased.

EXERCISE: Show that
$$\operatorname{Var} \hat{\theta} = \frac{\theta^2 n}{(n+1)^2(n+2)}$$

---

* It is not necessarily even consistent (cf. Sec. 7.2.2). (See also J. Neyman, *Lectures and Conferences on Mathematical Statistics and Probability*, Graduate School, U. S. Department of Agriculture, Washington, Second Edition (1952) pp. 189–190.)

† This is an application of the multinomial distribution (*cf.* Sec. 6.3).

RELIABILITY ESTIMATION: PART I   167

7.4.2 APPLICATION TO DISCRETE DISTRIBUTIONS

The maximum likelihood method is applied to discrete probability distributions by writing, instead of Eq. (7.5), the following:

$$L(x_1, \cdots, x_n; \theta) = P(\zeta = x_1) \cdots P(\zeta = x_n) \qquad (7.20)$$

where $P(\zeta = x_i)$ involves the parameter θ. For example, let ζ have the binomial distribution

$$\begin{aligned} P(\zeta = 0) &= q \\ P(\zeta = 1) &= p \equiv 1 - q \end{aligned} \qquad (7.21)$$

and consider p as the parameter to be estimated. Then,

$$\begin{aligned} L(x_1, \cdots, x_n; p) &= \prod_{i=1}^{n} p^{x_i}(1-p)^{1-x_i} \\ &= p^{\Sigma x_i}(1-p)^{n-\Sigma x_i} \end{aligned} \qquad (7.22)$$

(Recall that the x_i are zeros or ones.) Hence

$$\mathcal{L} = \left(\sum x_i\right) \log p + \left(n - \sum x_i\right) \log (1-p)$$

Set
$$\frac{\partial \mathcal{L}}{\partial p} = 0 = \frac{\sum x_i}{p} - \frac{n - \sum x_i}{1-p}. \qquad (7.23)$$

Solving for p in Eq. (7.23) and denoting the solution by $\hat{p}$, we have

$$\hat{p} = \frac{\sum x_i}{n} \qquad (7.24)$$

Again, as in the example using the exponential distribution (Eq. 7.9), the mean of the sample turns out to be the maximum likelihood estimator for the parameter (in this case p) which is the true mean of the population distribution given by Eqs. (7.21).

EXERCISE: Find the maximum likelihood estimator for the parameter λ of the Poisson distribution:

$$P_k \equiv P(\zeta = k) = \frac{\lambda^k e^{-\lambda}}{k!}, \qquad k = 0, 1, 2, \cdots$$

Answer: $\hat{\lambda} = \sum k_i/n \qquad (7.25)$

where n trials are made and k_i events occur in the ith trial; $i = 1, 2, \cdots, n$.

7.4.3 THE VARIANCE OF THE MAXIMUM LIKELIHOOD ESTIMATOR

The likelihood equation also provides a convenient method of obtaining the variance of the estimator, $\hat{\theta}$, which in general is easier to compute than $E(\hat{\theta} - E(\hat{\theta}))^2$, the exact variance. Thus the formula

$$\operatorname{Var} \hat{\theta} \simeq -\left(\frac{\partial^2 \mathcal{L}}{\partial \theta^2}\right)^{-1}$$

may be used under certain conditions when the sample values are replaced by their expected values after differentiation. Since a maximum likelihood estimator is not necessarily unbiased, this method may provide the exact variance or only its limiting form as n becomes large. We shall see in Sec. 7.5.4 that the above expression is a particular case of a formula to obtain variances and covariances for two or more parameter estimators.

EXAMPLE: Consider the exponential distribution. We differentiate Eq. (7.8) with respect to θ, obtaining:

$$\frac{\partial^2 \mathcal{L}}{\partial \theta^2} = \frac{n}{\theta^2} - \frac{2}{\theta^3} \sum t_i \qquad (7.26)$$

We then replace $\sum t_i$ by $E(\sum t_i) = n\theta$, obtaining

$$\frac{\partial^2 \mathcal{L}}{\partial \theta^2} = -\frac{n}{\theta^2} \qquad (7.27)$$

and therefore

$$\operatorname{Var} \hat{\theta} = \frac{\theta^2}{n} \qquad (7.28)$$

which is also the exact variance.

7.4.4 CONFIDENCE LIMITS FOR PARAMETERS

As stated previously, maximum likelihood estimators tend generally to be distributed normally for large sample sizes. Thus if $\hat{\theta}$ is the maximum likelihood estimator for θ, based on a sample size n, then if

$$\zeta \equiv \frac{\hat{\theta} - \theta}{\sqrt{\operatorname{Var} \hat{\theta}}}$$

we have

$$P(\zeta \leq x) \to \Phi(x) = \frac{1}{\sqrt{2\pi}} \int_{-\infty}^{x} e^{-t^2/2} \, dt \qquad (7.29)$$

as n becomes large.*

This fact enables us to find approximate *confidence limits* for probability distribution parameters, as follows:

Let γ be the stated *confidence coefficient*, and suppose we wish to find an interval about θ whose end points C_1, C_2 are functions of the sample

* The example of Sec. 7.4.1.1 is an exception, a reason being that $\operatorname{Var} \hat{\theta} = O(1/n^2)$ rather than $O(1/n)$.

values (i.e., C_1, C_2 are random variables) such that

$$P(C_1 < \theta < C_2) = \gamma \quad (7.30)$$

no matter what the true value of θ is.

From Eq. (7.29)

$$P\left(-K_{(1-\gamma)/2} < \frac{\hat{\theta} - \theta}{\sqrt{\operatorname{Var} \hat{\theta}}} < K_{(1-\gamma)/2}\right) \simeq \gamma \quad (7.31)$$

where K_α is defined by

$$\alpha = \frac{1}{\sqrt{2\pi}} \int_{K_\alpha}^{\infty} e^{-t^2/2}\,dt = 1 - \Phi(K_\alpha)$$

i.e., K_α is the "normal deviate exceeded with probability α." Now if $\operatorname{Var} \hat{\theta}$, which is generally a function of θ and other parameters, is evaluated by replacing the parameters by their estimators, and then denoted by V, then the event in parenthesis on the lefthand side of Eq. (7.31) is approximately equivalent to the event

$$(\hat{\theta} - K_{(1-\gamma)/2} V^{1/2} < \theta < \hat{\theta} + K_{(1-\gamma)/2} V^{1/2})$$

Thus the interval $\hat{\theta} \pm K_{(1-\gamma)/2} V^{1/2}$ contains θ with probability approximately γ. Note that we could have used any pair of K_{ϵ_1}, K_{ϵ_2} in Eq. (7.31) such that $\epsilon_1 + \epsilon_2 = 1 - \gamma$.

EXAMPLE: Let $\bar{x}$ and s^2 be the sample mean and variance of a sample drawn from a normal population. Find approximate confidence limits for μ. Since $\operatorname{Var} \bar{x} = \sigma^2/n$, then $V = s^2/(n-1)$ (using the unbiased estimator of σ^2). Hence the interval

$$\bar{x} \pm K_{(1-\gamma)/2} \frac{s}{\sqrt{n-1}}$$

contains the population mean μ with probability γ approximately. If σ^2 were a *known* parameter then the probability that the interval

$$\bar{x} \pm K_{(1-\gamma)/2} \frac{\sigma}{\sqrt{n}}$$

contains μ would be *exactly* γ.*

EXERCISE: Find a *one-sided* (lower) confidence limit for the parameter $R \equiv 1 - p$ of a binomial distribution; i.e., find $\hat{R}_L$, such that

$$P(\hat{R}_L < R < 1) \simeq \gamma$$

---

* The Student t-distribution would be used to determine exact confidence limits on μ when σ^2 is *unknown*. Thus the limits $\bar{x} \pm t_{(1-\gamma)/2} s/\sqrt{n-1}$ will contain μ with probability *exactly* γ, where t_α is the Student t-deviate with $n-1$ degrees of freedom exceeded with probability α (see Table 3, p. 588, A. H. Bowker and G. J. Lieberman, *Engineering Statistics*, Prentice-Hall, Inc., Englewood Cliffs, N. J., 1959).

Use the normal approximation given previously; but do not replace Var $\hat{R} = R(1 - R)/n$ by $(n - f)f/n^3$ where f is the observed number of failures.

Answer:

$$\hat{R}_L = \frac{n}{n + K_{1-\gamma}^2}\left(\hat{R} + \frac{K_{1-\gamma}^2}{2n} - K_{1-\gamma}\sqrt{\frac{\hat{R}(1-\hat{R})}{n} + \frac{K_{1-\gamma}^2}{4n^2}}\right) \quad (7.32)$$

7.5. THE APPLICATION OF MAXIMUM LIKELIHOOD TO DISTRIBUTIONS WITH TWO OR MORE PARAMETERS

7.5.1 ESTIMATION OF PARAMETERS OF THE NORMAL DISTRIBUTION

The maximum likelihood method easily generalizes to the case of two unknown parameters. The best-known application is to the parameters of the normal distribution. Since

$$f(x; \mu, \sigma^2) = (2\pi\sigma^2)^{-\frac{1}{2}} \exp\left[-\frac{(x-\mu)^2}{2\sigma^2}\right], \quad -\infty < x < \infty$$
$$-\infty < \mu < \infty \quad (7.33)$$
$$\sigma^2 > 0$$

then

$$L(x_1, \cdots, x_n; \mu, \sigma^2) = (2\pi\sigma^2)^{-n/2} \exp\left[-\frac{1}{2\sigma^2}\sum(x_i - \mu)^2\right] \quad (7.34)$$

$$\mathcal{L} \equiv \log L = \text{constant} - \frac{n}{2}\log\sigma^2 - \frac{1}{2\sigma^2}\sum(x_i - \mu)^2$$

Now we set $\partial\mathcal{L}/\partial\mu = 0$, $\partial\mathcal{L}/\partial(\sigma^2) = 0$. The first equation yields

$$\frac{1}{\sigma^2}\sum(x_i - \mu) = 0 \quad (7.35)$$

Since $\sigma^2 > 0$, then the solution of Eq. (7.35), denoted by $\hat{\mu}$, is

$$\hat{\mu} = \bar{x} = \frac{1}{n}\sum x_i \quad (7.36)$$

The second equation yields

$$-\frac{n}{2\sigma^2} + \frac{1}{2(\sigma^2)^2}\sum(x_i - \mu)^2 = 0 \quad (7.37)$$

whose solution, denoted by s^2 (or m_2), is

$$s^2 = \frac{1}{n} \sum (x_i - \hat{\mu})^2 \qquad (7.38)$$

(Note that $\hat{\mu}$ appears in Eq. (7.38) instead of μ since we are solving Eqs. (7.35) and (7.37) simultaneously.)

We have already seen that s^2 is a biased estimator for σ^2, but that $[n/(n-1)]s^2$ is unbiased. It should be mentioned that neither s, nor even $s\sqrt{n/(n-1)}$, is an unbiased estimator for σ (see Ref. 1, p. 484).

7.5.2 ESTIMATION OF PARAMETERS OF THE GAMMA DISTRIBUTION

Another example in which two parameters are to be estimated is the case in which the distribution is a gamma p.d.f. (Eq. (6.159)) given by

$$f(x; \lambda, \alpha) = \frac{\lambda^\alpha x^{\alpha-1} e^{-\lambda x}}{\Gamma(\alpha)}, \qquad x > 0$$

$$\binom{\alpha > 0}{\lambda > 0} \qquad (7.39)$$

$$= 0, \qquad x \leq 0$$

We have

$$\mathcal{L} = \alpha n \log \lambda - n \log \Gamma(\alpha) + (\alpha - 1) \sum \log x_i - \lambda \sum x_i \qquad (7.40)$$

$$\frac{\partial \mathcal{L}}{\partial \alpha} = n \log \lambda - n \Psi(\alpha) + \sum \log x_i = 0 \qquad (7.41)^*$$

$$\frac{\partial \mathcal{L}}{\partial \lambda} = \frac{\alpha n}{\lambda} - \sum x_i = 0 \qquad (7.42)$$

By inspection of Eqs. (7.41) and (7.42) we see that the simultaneous solution for $\hat{\alpha}$, $\hat{\lambda}$ can only be accomplished by trial and error. A good approximation to $\Psi(\alpha)$, when α is not too small, say $\alpha \geq 2$, is

$$\Psi(\alpha) \approx \log(\alpha - \tfrac{1}{2}) + \frac{1}{24(\alpha - \tfrac{1}{2})^2} \qquad (7.43)$$

For example, when $\alpha = 2, 3$, the exact values of $\Psi(\alpha)$ are 0.423 and 0.923 respectively; the values given by Eq. (7.43) are 0.424 and 0.923

---

*$\Psi(Z) \equiv d/dZ \log \Gamma(Z)$ is called "the logarithmic derivative of the Γ-function," or simply "psi-function."

respectively.* Equations (7.42) and (7.41), respectively, become

$$\hat{\alpha} = \frac{\hat{\lambda}}{n} \sum x_i \tag{7.44}$$

$$\hat{\lambda} = \exp\left[\Psi(\hat{\alpha}) - (1/n) \sum \log x_i\right] \tag{7.45}$$

Equation (7.45) can be simplified by writing

$$\exp \Psi(\hat{\alpha}) \simeq \left(\alpha - \frac{1}{2}\right)\left(1 + \frac{1}{24(\alpha - \frac{1}{2})^2}\right) \tag{7.46}$$

$$= \alpha - \frac{1}{2} + \frac{1}{24(\alpha - \frac{1}{2})} \tag{7.47}$$

where we have used the approximation $e^x \approx 1 + x$ for x small, in Eq. (7.46). If we define $\hat{w} \equiv \hat{\alpha} - \frac{1}{2}$, then Eqs. (7.44) and (7.45) become

$$\hat{w} = -\frac{1}{2} + \frac{\hat{\lambda}}{n} \sum x_i \tag{7.48}$$

$$\hat{\lambda} = \left(\hat{w} + \frac{1}{24\hat{w}}\right)(\prod x_i)^{-1/n} \tag{7.49}$$

EXAMPLE: It is assumed that the time-to-failure distribution for an electrical generator is represented by Eq. (7.39). Seven such generators were observed to fail at 100, 110, 150, 175, 185, 200, and 220 hours.† Estimate the parameters α and λ. We have

$$\frac{\sum x_i}{n} = 162.9 \quad \text{and} \quad \exp\left[-\frac{1}{n}\sum \log x_i\right] = (\prod x_i)^{-1/n} = 0.00637$$

Hence

$$\hat{w} = -\frac{1}{2} + 162.9\hat{\lambda} \tag{7.48'}$$

$$\hat{\lambda} = \left(\hat{w} + \frac{1}{24\hat{w}}\right)(0.00637) \tag{7.49'}$$

*See Ref. 2, p. 16, for exact values (note that Ref. 2 defines their $\Psi^*(x) \equiv$ our $\Psi(x + 1)$; so that $\Psi(2) = \Psi^*(1)$, etc.).

† The data of this example will be used to make numerical calculations of confidence limits (cf. Secs. 8.5, 8.6) wherein normal distribution approximations are involved. It must not be supposed however that a sample size as small as seven is sufficient for accurate use of the normal approximation; a sample size of 50–100 is probably the minimum for this requirement. The small sample size is used here merely for illustrative purposes to allow the reader to follow conveniently the detailed calculations.

RELIABILITY ESTIMATION: PART I   173

Using the technique given in Appendix 7A or 7B,* the estimators obtained are

$$\hat{\lambda} = 0.0841, \qquad \hat{w} = 13.20 \quad \text{or} \quad \hat{\alpha} = 13.70 \qquad (7.50)$$

7.5.3 METHOD OF MATCHING MOMENTS

An alternative method, "by matching moments," was mentioned in Sec. 7.3. Let us see how this method works using the data of the previous example. The mean of the distribution whose p.d.f. is given by Eq. (7.39) is

$$\mu = \frac{\alpha}{\lambda} \qquad (7.51)$$

and the variance is

$$\sigma^2 = \frac{\alpha}{\lambda^2} \qquad (7.52)$$

The procedure is to equate the sample mean and (unbiased) sample variance with the population mean and variance given by Eqs. (7.51) and (7.52), respectively. The sample mean has already been obtained as $\bar{x} = 162.9$. The sample variance multiplied by a factor $n/(n-1)$ to make it unbiased is

$$\frac{ns^2}{n-1} = \frac{\sum (x_i - \bar{x})^2}{n-1} = 2032 \qquad (7.53)$$

Thus we have two equations in two unknowns, α and λ:

$$\frac{\alpha}{\lambda} = 162.9, \qquad \frac{\alpha}{\lambda^2} = 2032$$

from which we easily obtain

$$\hat{\lambda} = \frac{162.9}{2032} = 0.0802$$

$$\hat{\alpha} = 13.06 \qquad (7.54)$$

It is evident that very little computation was necessary to obtain the estimators given by Eq. (7.54), whereas to obtain the maximum likelihood

---

* Note that it is simpler to treat Eqs. (7.48') and (7.49') above as one equation. Thus assume $\hat{\lambda}$, solve for $\hat{w}$, substitute in (7.49') and obtain the next value of $\hat{\lambda}$, etc.

estimators (7.50) was relatively difficult. The numerical values obtained by the two methods differ somewhat, but so far there is nothing to tell us which set of estimators is better. The theory tells us that the maximum likelihood estimators are better for *large* samples, in that their variances tend to the minimum attainable value* as the sample size gets larger, but for small samples there is apparently no easy way of choosing between the two ways of estimation for this particular problem.

7.5.4 VARIANCES AND COVARIANCES OF MAXIMUM LIKELIHOOD ESTIMATORS

Approximate values for variances and covariances for two or more estimators can be found directly from the likelihood equations. The method will be illustrated using the previous example on the gamma distribution.

First calculate the second partial derivatives of $\mathcal{L}$ (from Eqs. (7.41) and (7.42)):

$$\mathcal{L}_{\alpha\alpha} \equiv \frac{\partial^2 \mathcal{L}}{\partial \alpha^2} = -n\Psi'(\alpha) \tag{7.55}$$

$$\mathcal{L}_{\lambda\lambda} \equiv \frac{\partial^2 \mathcal{L}}{\partial \lambda^2} = -\frac{\alpha n}{\lambda^2} \tag{7.56}$$

$$\mathcal{L}_{\lambda\alpha} \equiv \frac{\partial^2 \mathcal{L}}{\partial \alpha \, \partial \lambda} = \frac{n}{\lambda} = \mathcal{L}_{\alpha\lambda} \tag{7.57}$$

(If any of the second partial derivatives had involved sample values, e.g., $\sum x_i$, these would be replaced by expected values, e.g., $n\alpha/\lambda$.)

Next, form the matrix **A**:

$$\mathbf{A} = \begin{bmatrix} \mathcal{L}_{\alpha\alpha} & \mathcal{L}_{\alpha\lambda} \\ \mathcal{L}_{\alpha\lambda} & \mathcal{L}_{\lambda\lambda} \end{bmatrix} \tag{7.58}$$

Then the matrix **B** defined as

$$\mathbf{B} = \begin{bmatrix} \text{Var } \hat{\alpha} & \text{Cov } (\hat{\alpha}, \hat{\lambda}) \\ \text{Cov } (\hat{\alpha}, \hat{\lambda}) & \text{Var } \hat{\lambda} \end{bmatrix} \tag{7.59}$$

* It can be shown under general circumstances that there is a minimum attainable value for the variance of an estimator, and that the variances of maximum likelihood estimators will approach the minimum attainable as the sample size n gets large (Ref. 3, pp. 130, 157).

RELIABILITY ESTIMATION: PART I 175

is the negative of the inverse of **A**; i.e.,

$$\mathbf{B} = -\mathbf{A}^{-1} \qquad (7.60)\,*\dagger$$

Thus, in our example of the gamma distribution:

$$\mathbf{A} = \begin{bmatrix} -n\Psi'(\alpha) & \dfrac{n}{\lambda} \\[2ex] \dfrac{n}{\lambda} & \dfrac{-\alpha n}{\lambda^2} \end{bmatrix}$$

To obtain the inverse, we first calculate the determinant of **A**:

$$\text{Det } \mathbf{A} = \frac{n^2 \alpha \Psi'(\alpha)}{\lambda^2} - \frac{n^2}{\lambda^2} = \frac{n^2}{\lambda^2}(\alpha \Psi'(\alpha) - 1) \qquad (7.61)$$

Then

$$\mathbf{A}^{-1} = \frac{1}{\text{Det } \mathbf{A}} \begin{bmatrix} \dfrac{-\alpha n}{\lambda^2} & \dfrac{-n}{\lambda} \\[2ex] \dfrac{-n}{\lambda} & -n\Psi'(\alpha) \end{bmatrix} \qquad (7.62)$$

where the matrix in Eq. (7.62) is obtained by replacing each element of the matrix **A** by its cofactor.

---

* The preceding method for obtaining variances and covariances is somewhat different from that usually given (see Ref. 4, p. 212). In the usual method, one starts with the original density function f and finds expected values of

$$\frac{\partial^2 \log f}{\partial \alpha^2}, \quad \frac{\partial^2 \log f}{\partial \lambda^2}, \quad \frac{\partial^2 \log f}{\partial \alpha\, \partial \lambda}$$

with respect to the probability density function f. The expected values are then put into the matrix **A**, and the negative of the inverse found; then each of the elements is divided by n, the sample size. The reason for using the method given here is that in more complicated problems where samples of size $n_1, n_2, \cdots$, etc. are taken from several populations with common parameters, it is not easy to find what n to use in the usual method. On the other hand, if one already has the likelihood equations, it is a simple matter to differentiate the expressions again, and proceed as above. In the example we are discussing, the two methods lead to exactly the same results. In other examples, such as estimating the parameters of the normal or Weibull distributions, one must replace quantities such as $\Sigma (x_i - \mu)^2$, which appear in the second derivatives of the likelihood function, by expected values (e.g., $n\sigma^2$), or by solving one or more of the likelihood equations with respect to the above quantity in terms of the other parameters.

† When only one parameter (say α) is involved $\mathbf{B} = -\mathbf{A}^{-1} = -(\partial^2 \mathcal{L}/\partial \alpha^2)^{-1}$, again with the proviso to replace sample values by their expected values (see Sec. 7.4.3).

Hence, from (7.60)

$$B = \frac{1}{n[\alpha\Psi'(\alpha) - 1]} \begin{bmatrix} \alpha & \lambda \\ \lambda & \lambda^2\Psi'(\alpha) \end{bmatrix} \quad (7.63)$$

therefore

$$\text{Var } \hat{\alpha} = \frac{\alpha}{n[\alpha\Psi'(\alpha) - 1]} \quad (7.64)$$

$$\text{Var } \hat{\lambda} = \frac{\lambda^2\Psi'(\alpha)}{n[\alpha\Psi'(\alpha) - 1]} \quad (7.65)$$

$$\text{Cov } (\hat{\alpha}, \hat{\lambda}) = \frac{\lambda}{n[\alpha\Psi'(\alpha) - 1]} \quad (7.66)$$

The function $\Psi'(\alpha)$ can be shown to be $\simeq 1/(\alpha - \frac{1}{2})$ (for example, $\Psi'(2) = 0.6449$ and $1/(2 - \frac{1}{2}) = 0.6667$; $\Psi'(3) = 0.3949$ and $1/(3 - \frac{1}{2}) = 0.40000$; $\Psi'(4) = 0.2838$, $1/(4 - \frac{1}{2}) = 0.2857$, etc.); therefore $\alpha\Psi'(\alpha) - 1 \simeq 1/(2\alpha - 1)$. The expressions (7.64), (7.65), and (7.66) then become approximately

$$\text{Var } \hat{\alpha} \simeq \frac{\alpha(2\alpha - 1)}{n} \quad (7.64')$$

$$\text{Var } \hat{\lambda} \simeq \frac{2\lambda^2}{n} \quad (7.65')$$

$$\text{Cov } (\hat{\alpha}, \hat{\lambda}) \simeq \frac{\lambda(2\alpha - 1)}{n} \quad (7.66')$$

EXERCISE: Find the approximate variances and covariances of $\bar{x}$ and s^2 defined by Eqs. (7.36) and (7.38). Remember to replace $\sum (x_i - \mu)$ and $\sum (x_i - \mu)^2$ by expected values *after* differentiating.

Going back to the example of the seven electrical generators, we can get approximate values for the standard deviations of the estimators $\hat{\alpha}$ and $\hat{\lambda}$ by inserting in Eqs. (7.64') and (7.65') the numerical estimates (7.50) obtained by the maximum likelihood method. Thus $\alpha \simeq \hat{\alpha} = 13.70$ and $\lambda \simeq \hat{\lambda} = 0.0841$. Therefore*

$$\langle \text{Var } \hat{\alpha} \rangle^{\frac{1}{2}} \simeq \sqrt{\frac{(13.70)(26.4)}{7}} = 7.19$$

$$\langle \text{Var } \hat{\lambda} \rangle^{\frac{1}{2}} \simeq \sqrt{\frac{2(0.0841)^2}{7}} = 0.0450$$

* $\langle \ \rangle$ will denote a numerical estimate of the quantity in the brackets.

RELIABILITY ESTIMATION: PART I 177

Also
$$\langle \text{Cov}(\hat{\alpha}, \hat{\lambda}) \rangle = \frac{(0.0841)(26.4)}{7} = 0.317$$

Therefore the correlation coefficient of $\hat{\alpha}, \hat{\lambda}$ is approximately
$$\langle \hat{\rho}(\hat{\alpha}, \hat{\lambda}) \rangle \simeq \frac{0.317}{(7.19)(0.0450)} = 0.98$$

In general,
$$\hat{\rho} \simeq \frac{2\alpha - 1}{\sqrt{2\alpha(2\alpha - 1)}} \qquad (7.67)$$

which is very close to one for even moderately small values of α. It is re-emphasized that the formulas for variances are valid only for large sample sizes. In the case of the gamma distribution (as well as the Weibull) there is also an unknown error due to estimator bias (which, however, disappears as $n \to \infty$). The result is, with the given procedure, to underestimate variances when the sample size is small.

7.5.5 ESTIMATION OF PARAMETERS OF THE WEIBULL DISTRIBUTION

In Sec. 6.8.2 the Weibull probability density function was given as
$$f(t) = \alpha \lambda t^{\alpha-1} e^{-\lambda t^{\alpha}}, \qquad t > 0, \alpha > 0, \lambda > 0 \qquad (7.68)$$

We will now apply the maximum likelihood method to this problem, as we did previously for the gamma distribution. The logarithm of the likelihood function is
$$\mathcal{L} \equiv \log L = n \log \alpha + n \log \lambda + (\alpha - 1) \sum_{i=1}^{n} \log t_i - \lambda \sum_{i=1}^{n} t_i^{\alpha} \qquad (7.69)$$

Therefore,
$$\frac{\partial \mathcal{L}}{\partial \alpha} = \frac{n}{\alpha} + \sum \log t_i - \lambda \sum t_i^{\alpha} \log t_i = 0 \qquad (7.70)$$

$$\frac{\partial \mathcal{L}}{\partial \lambda} = \frac{n}{\lambda} - \sum t_i^{\alpha} = 0 \qquad (7.71)$$

Denoting the solutions by $\hat{\alpha}$ and $\hat{\lambda}$, Eqs. (7.70) and (7.71) can be rewritten as
$$\hat{\lambda} = \frac{n}{\sum t_i^{\hat{\alpha}}} \qquad (7.70')$$

and
$$\hat{\alpha} = \frac{n}{\hat{\lambda} \sum t_i^{\hat{\alpha}} \log t_i - \sum \log t_i} \tag{7.71'}$$

The technique given in Appendix 7A or 7B of this chapter is useful for finding the above estimators. Again it is simpler to assume $\hat{\alpha}$, compute $\hat{\lambda}$ from Eq. (7.70'); then recompute $\hat{\alpha}$ from Eq. (7.71') into which the previously obtained value of $\hat{\lambda}$ has been substituted. Some experience is probably necessary in choosing a good initial value of $\hat{\alpha}$; although the process of iteration as given in Appendixes 7A and 7B almost assuredly will converge to the correct values, no matter what the initial value chosen for $\hat{\alpha}$. In the example below, we will make use of the known expression for the mean of the distribution to obtain an initial value for $\hat{\alpha}$.

As an example let us use the time-to-failure data of the seven electrical generators, given previously. We have $t_1, \cdots, t_n = 100, 110, 150, 175, 185, 200$, and 220 hours, respectively. To obtain a reasonable initial value for $\hat{\alpha}$ we equate the mean of the sample to the previously determined mean of the distribution (Eq. (6.103)).

$$162.9 = \lambda^{-1/\alpha} \Gamma\left(1 + \frac{1}{\alpha}\right) \tag{7.72}$$

By using Eqs. (7.72) and (7.70') together with tables of $\Gamma(x)$ (Ref. 5, p. 193) one can quickly obtain an initial value of $\hat{\alpha} = 2.2$. Then by applying the iteration procedure of Appendixes 7A and 7B, we obtain

$$\hat{\alpha} = 4.6690, \quad \hat{\lambda} = 3.04238 \times 10^{-11} \tag{7.73}$$

To obtain estimates of the variances and covariances, we proceed as we did for the gamma distribution. From Eqs. (7.70) and (7.71)

$$\frac{\partial^2 \mathcal{L}}{\partial \alpha^2} = -\frac{n}{\alpha^2} - \lambda \sum t_i^\alpha \log^2 t_i \tag{7.74}$$

$$\frac{\partial^2 \mathcal{L}}{\partial \lambda^2} = -\frac{n}{\lambda^2} \tag{7.75}$$

$$\frac{\partial^2 \mathcal{L}}{\partial \alpha\, \partial \lambda} = -\sum t_i^\alpha \log t_i \tag{7.76}$$

This time, we will not replace the quantities involving t_i by their expected values, since this may tend to cause underestimation of the variances (see last paragraph Sec. 7.5.4).

RELIABILITY ESTIMATION: PART I

Using the estimators (7.73) in Eqs. (7.74)–(7.76), we obtain

$$\left\langle \frac{\partial^2 \mathcal{L}}{\partial \alpha^2} \right\rangle = -194.9062$$

$$\left\langle \frac{\partial^2 \mathcal{L}}{\partial \lambda^2} \right\rangle = -0.75626 \times 10^{22}$$

$$\left\langle \frac{\partial^2 \mathcal{L}}{\partial \alpha \, \partial \lambda} \right\rangle = -12.12624 \times 10^{11}$$

Thus our matrix $\mathbf{A}$ becomes

$$\mathbf{A} = \begin{bmatrix} -194.9062 & -12.12624 \times 10^{11} \\ -12.12624 \times 10^{11} & -0.75626 \times 10^{22} \end{bmatrix}$$

Det $\mathbf{A} = 0.35406 \times 10^{22}$. Thus,

$$-\mathbf{A}^{-1} = \frac{1}{0.35406 \times 10^{22}} \begin{bmatrix} 0.75626 \times 10^{22} & -12.12624 \times 10^{11} \\ -12.12624 \times 10^{11} & 194.9062 \end{bmatrix}$$

Thus

$$\langle \text{Var } \hat{\alpha} \rangle \simeq 2.1360 \simeq 2.14$$

$$\langle \text{Var } \hat{\lambda} \rangle \simeq 550.49 \times 10^{-22} \simeq 550 \times 10^{-22}$$

$$\langle \text{Cov } (\hat{\alpha}, \hat{\lambda}) \rangle \simeq -34.249 \times 10^{-11} \simeq -34.2 \times 10^{-11}$$

In this case we also find

$$\langle \hat{\rho}(\hat{\alpha}, \hat{\lambda}) \rangle = \frac{-34.249 \times 10^{-11}}{10^{-11} \sqrt{(2.1360)(550.49)}} \simeq -1.00$$

Thus the estimators $\hat{\alpha}$ and $\hat{\lambda}$ have a high negative correlation. It is interesting to check the values of the minimum possible variances, assuming that the estimators $\hat{\alpha}$ and $\hat{\lambda}$ were actually unbiased. This is done by replacing the expressions involving the sample values in Eqs. (7.74) and (7.76) by their expected values. We must find, for example, from Eq. (7.74) the expected value of $\sum t_i^\alpha \log^2 t_i$. Since the t_i are independent random variables, we have

$$E(\sum t_i^\alpha \log^2 t_i) = nE(\tau^\alpha \log^2 \tau) \tag{7.77}$$

where τ has the p.d.f. given by Eq. (7.68). Thus

$$E(\tau^\alpha \log^2 \tau) = \int_0^\infty \alpha \lambda t^{2\alpha-1} \log^2 t \, e^{-\lambda t^\alpha} \, dt \tag{7.78}$$

This integral can be evaluated by replacing the quantity $t^{2\alpha-1}$ by $t^{\beta-1}$; by evaluating the expression (using the Laplace transform)

$$\int_0^\infty \alpha\lambda t^{\beta-1} e^{-\lambda t^\alpha}\, dt$$

then by differentiating the result twice with respect to β, and finally replacing β by 2α. This process yields

$$E(\tau^\alpha \log^2 \tau) = \frac{1}{\alpha^2 \lambda}\left[\Psi'(2) + (\Psi(2) - \log \lambda)^2\right] \quad (7.79)$$

Also, the same method is used for $E(\tau^\alpha \log \tau)$, except that one differentiates only once with respect to β. Thus

$$E(\tau^\alpha \log \tau) = \frac{1}{\alpha\lambda}(\Psi(2) - \log \lambda) \quad (7.80)$$

EXERCISE: Work through the details of the above method. Note that $\Psi(2) = 1 - E$, where E is Euler's constant, $E = 0.577215665$. Also

$$\Psi'(2) = \frac{\pi^2}{6} - 1 = 0.644934067$$

If we denote the quantity $\Psi(2) - \log \lambda$ by Δ for the moment, the variance-covariance matrix turns out to be (taking α and λ in that order)

$$-\mathbf{A}^{-1} = \begin{bmatrix} \dfrac{\alpha^2}{n}\left(\dfrac{\pi^2}{6}\right)^{-1} & \dfrac{-\alpha\lambda\Delta}{n}\left(\dfrac{\pi^2}{6}\right)^{-1} \\[2ex] \dfrac{-\alpha\lambda\Delta}{n}\left(\dfrac{\pi^2}{6}\right)^{-1} & \dfrac{\lambda^2}{n}\left(1 + \Delta^2\left(\dfrac{\pi^2}{6}\right)^{-1}\right) \end{bmatrix} \quad (7.81)$$

The correlation coefficient $\hat{\rho}(\hat{\alpha}, \hat{\lambda})$ is

$$\hat{\rho} = \frac{-\Delta}{\sqrt{\pi^2/6 + \Delta^2}} \quad (7.82)$$

If we now *estimate* the variances and covariances by inserting the estimators (7.73) into the matrix $-\mathbf{A}^{-1}$ we obtain

$$\langle \text{Var }\hat{\alpha}\rangle \simeq 1.893$$

$$\langle \text{Var }\hat{\lambda}\rangle \simeq 489.3 \times 10^{-22} \quad (7.83)$$

$$\langle \text{Cov }(\hat{\alpha}, \hat{\lambda})\rangle \simeq -30.40 \times 10^{-11}$$

We also note that when λ is extremely small, as indicated by the estimator $\hat{\lambda}$, Δ will be extremely large; and thus from (7.82) $\langle \hat{\rho} \rangle \simeq -1$. The above estimated values of the variances and covariances represent the smallest attainable values, assuming that the biases in $\hat{\alpha}$ and $\hat{\lambda}$ be zero. Thus the previously obtained estimates are reasonable and are perhaps fairly accurate. Note that one must retain several extra decimal places throughout the entire computation. This is made necessary by the high negative correlation of $\hat{\alpha}$ and $\hat{\lambda}$.

EXERCISE: Apply the preceding methods of estimation to the extreme value distribution (Sec. 6.8.3).

REFERENCES

1. H. Cramér, *Mathematical Methods of Statistics*, Princeton University Press, Princeton, N.J., 1946.
2. E. Jahnke and F. Emde, *Tables of Functions*, 4th ed., Dover Publications, Inc., New York, 1945.
3. C. R. Rao, *Advanced Statistical Methods for Biometric Research*, John Wiley & Sons, Inc., New York, 1952.
4. A. M. Mood, *Introduction to the Theory of Statistics*, McGraw-Hill Book Company, Inc., New York, 1950.
5. H. B. Dwight, *Tables of Integrals and Other Mathematical Data*, The Macmillan Company, New York, 1934.
6. F. A. Willers, *Practical Analysis*, Dover Publications, Inc., 1948.

ADDITIONAL READING

Bartholomew, D. J., "A Problem of Life Testing," *J. Am. Stat. Assoc.*, **52:279**, 350–54 (September 1957).

Chapman, D. G., "Estimating the Parameters of a Truncated Gamma Distribution," *Ann. Math. Stat.*, **27:2**, 493–505 (June 1956).

Cohen, A. C., Jr., "Simplified Estimators for the Normal Distribution When Samples Are Singly Censored or Truncated," *Technometrics*, **1:3**, 217–37 (August 1959).

Durand, D., and Greenwood, J. A., "Aids for Fitting the Gamma Distribution by Maximum Likelihood," *Technometrics*, **2:1**, 55–56 (February 1960).

Hartley, H. O., "The Modified Gauss-Newton Method for the Fitting of Non-Linear Regression Functions by Least Squares," *Technometrics*, **3:4**, 269–280 (May 1961).

Herd, G. R., "Estimation of Reliability Functions," *Proc. 3rd National Symposium on Reliability and Quality Control in Electronics (I.R.E.), Washington, D. C., January 14–16, 1957*, 113–22.

Kao, J. H. K., "A Graphical Estimation of Mixed Weibull Parameters in Life-Testing of Electron Tubes," *Technometrics*, **1**:4, 389–407 (November 1959).

Kao, J. H. K., "Computer Methods for Estimating Weibull Parameters in Reliability Studies," *I.R.E. Trans. Reliability and Quality Control*, **13,** 15–22 (July 1958).

Moore, P. G., "The Transformation of a Truncated Poisson Distribution," *Skand. Akt.*, **39**:1–2, 19–25 (1956).

APPENDIX 7A

SOLUTION OF EQUATIONS OF THE FORM $x = f(x)$

In certain cases the equation obtained for the maximum likelihood estimator is not solvable directly for the estimator but must be obtained by a trial-and-error technique. When the equation can be thrown into the form $x = f(x)$ where f is a known function of x, there is a particularly efficient way of proceeding to the solution, which is also capable of generalization to the case of two or more equations put into similar form; e.g., for two equations in two unknowns the equations are

$$x = f_1(x, y)$$

$$y = f_2(x, y)$$

We will consider the one-variable case first, then generalize to two or more variables in Appendix 7B.

Let $x = \zeta$ be the root of the above one-variable equation; i.e.,

$$\zeta = f(\zeta)$$

Let x_0 be an initial trial value for the root ζ and define $x_1 = f(x_0)$. Suppose that

$$x_0 = \zeta + \epsilon \tag{7A.1}$$

then $\quad x_1 = f(x_0) = f(\zeta + \epsilon) = f(\zeta) + \epsilon f'(\zeta) + O(\epsilon^2)$

or $\quad x_1 = \zeta + \epsilon f'(\zeta) + O(\epsilon^2) \tag{7A.2}$

Now, let x_0^1 denote the next initial value to use, and define it as a linear combination of the first initial trial value x_0 and the derived value x_1

$$x_0^1 = Ax_0 + Bx_1 \tag{7A.3}$$

However, using (7A.1) and (7A.2) in (7A.3),

$$x_0^1 = A(\zeta + \epsilon) + B(\zeta + \epsilon f'(\zeta) + O(\epsilon^2)) \tag{7A.4}$$

or
$$x_0^1 = \zeta(A + B) + \epsilon(A + Bf'(\zeta)) + O(\epsilon^2) \tag{7A.5}$$

The quantities A and B are at our disposal. Since we wish x_0^1 to be as close to the root ζ as possible, we set

$$A + B = 1 \tag{7A.6}$$

and also the coefficient of ϵ equal to zero:

$$A + Bf'(\zeta) = 0 \tag{7A.7}$$

Solving Eqs. (7A.6) and (7A.7) for A and B:

$$B = \frac{1}{1 - f'(\zeta)} \tag{7A.8}$$
$$A = 1 - B$$

The derivative $f'(\zeta)$ is not known, however; but it can be approximated by, say, $f'(x_0)$. To avoid differentiating $f(x)$, we can define $x_2 \equiv f(x_1)$, then in turn approximate $f'(x_0)$ by $(x_2 - x_1)/(x_1 - x_0)$. Using this expression together with Eq. (7A.8) in Eq. (7A.3), we finally obtain

$$x_0^1 = \frac{x_1^2 - x_0 x_2}{2x_1 - x_0 - x_2} = x_2 + \frac{(x_1 - x_2)^2}{2x_1 - x_0 - x_2} \tag{7A.9}$$

Thus x_0^1 can be used as a new initial trial value and the process repeated. Convergence is usually very rapid, since the error for each successive trial value is $O(\epsilon^2)$, when the error for the previous trial value is ϵ. However, if $f'(\zeta) = 1$, the process, if it converges at all, converges more slowly, since from Eq. (7A.5) we cannot then make both $A + B = 1$ and the error of order ϵ vanish also.*

EXAMPLE: Find the square root of any number.
Let $N > 0$ be the number. We attempt to solve the equation

$$x^2 = N \tag{7A.10}$$

or
$$x = \frac{N}{x} \tag{7A.11}$$

---

* This result is known as Aitken's δ^2 process (see F. B. Hildebrand, *Introduction to Numerical Analysis*, New York: McGraw-Hill Book Co., (1956), p. 445).

which is the form required for the method. Let x_0 be any initial trial value; then $x_1 = N/x_0$; $x_2 = N/x_1 = x_0$. Therefore from Eq. (7A.9)

$$x_0^1 = \frac{\left(\dfrac{N}{x_0}\right)^2 - x_0^2}{2\dfrac{N}{x_0} - 2x_0} = \frac{1}{2}\left(\frac{N}{x_0} + x_0\right) \tag{7A.12}$$

In general, the nth iteration $x_0^{(n)}$ is

$$x_0^{(n)} = \frac{1}{2}\left(\frac{N}{x_0^{(n-1)}} + x_0^{(n-1)}\right) \tag{7A.13}*$$

EXERCISE: Solve the equation

$$x = 2 \sin x \tag{7A.14}$$

(from Ref. 6, p. 223, the root $\zeta = 1.8954942$). Start with $x_0 = 2$.

* Equation (7A.13) is a "lucky accident" since one would not in general obtain so simple an iteration formula.

APPENDIX 7B
SOLUTION OF n EQUATIONS IN n UNKNOWNS

Suppose that we have a set of n equations with n unknowns to be found; and that the equations can be written in the form

$$x_1 = f_1(x_1, \cdots, x_n)$$
$$\cdots \cdots \cdots$$
$$x_n = f_n(x_1, \cdots, x_n) \quad (7B.1)$$

Equation (7B.1) can be written in matrix form

$$\mathbf{X} = f(\mathbf{X}) \quad (7B.2)$$

where $\mathbf{X}$ and $f(\mathbf{X})$ are $n \times 1$ matrices (column vectors).

Let $\boldsymbol{\zeta}$, an $n \times 1$ matrix, be the solution; i.e.

$$\boldsymbol{\zeta} \equiv \begin{bmatrix} \zeta_1 \\ \cdot \\ \cdot \\ \cdot \\ \zeta_n \end{bmatrix} = \begin{bmatrix} f_1(\zeta_1, \zeta_2, \cdots, \zeta_n) \\ \\ \\ f_n(\zeta_1, \zeta_2, \cdots, \zeta_n) \end{bmatrix} \quad (7B.3)$$

Let $\mathbf{X}_0$ denote the $n \times 1$ matrix of trial values, such that

$$\mathbf{X}_0 = \boldsymbol{\zeta} + \boldsymbol{\varepsilon} \quad (7B.4)$$

where

$$\boldsymbol{\varepsilon} = \begin{bmatrix} \epsilon_1 \\ \epsilon_2 \\ \cdot \\ \cdot \\ \cdot \\ \epsilon_n \end{bmatrix} \quad (7B.5)$$

Also, let
$$\mathbf{X}_1 = f(\mathbf{X}_0) \quad (7B.6)$$

Now
$$\mathbf{X}_1 = f(\boldsymbol{\zeta} + \boldsymbol{\varepsilon}) = \boldsymbol{\zeta} + \mathbf{J}\boldsymbol{\varepsilon} + \boldsymbol{O}(\epsilon^2) \quad (7B.7)*$$

* $\boldsymbol{O}(\epsilon^2)$ is an $n \times 1$ matrix, each of whose elements is $O(\epsilon^2)$, where $\epsilon = \max(\epsilon_i)$.

RELIABILITY ESTIMATION: PART I

where $\mathbf{J}$ is the $n \times n$ matrix of the partial derivatives of the f_i with respect to the x_j, evaluated at $x_j = \zeta_j$; thus $\mathbf{J}$ is the usual "Jacobian" matrix of partial derivatives. (Note that $\mathbf{J}\boldsymbol{\varepsilon}$ is an $n \times 1$ matrix since $\boldsymbol{\varepsilon}$, an $n \times 1$ matrix, is premultiplied by $\mathbf{J}$, an $n \times n$ matrix). For example, when $n = 2$,

$$\mathbf{J} = \begin{bmatrix} \dfrac{\partial f_1}{\partial x_1} & \dfrac{\partial f_1}{\partial x_2} \\[2mm] \dfrac{\partial f_2}{\partial x_1} & \dfrac{\partial f_2}{\partial x_2} \end{bmatrix} \tag{7B.8}$$

where the derivatives are evaluated at $(x_1, x_2) = (\zeta_1, \zeta_2)$.

If we let
$$\mathbf{X}_0^1 = \mathbf{A}\mathbf{X}_0 + \mathbf{B}\mathbf{X}_1 \tag{7B.9}$$

where $\mathbf{A}$ and $\mathbf{B}$ are both $n \times n$ matrices, then

$$\mathbf{X}_0^1 = (\mathbf{A} + \mathbf{B})\boldsymbol{\zeta} + (\mathbf{A} + \mathbf{B}\mathbf{J})\boldsymbol{\varepsilon} + O(\epsilon^2)$$

Thus, we set
$$\mathbf{A} + \mathbf{B} = \mathbf{I}, \text{ the unit matrix}$$
and
$$\mathbf{A} + \mathbf{B}\mathbf{J} = \mathbf{O}, \text{ the zero matrix} \tag{7B.10}$$

Hence, subtracting,
$$\mathbf{B}(\mathbf{I} - \mathbf{J}) = \mathbf{I}$$
or
$$\mathbf{B} = (\mathbf{I} - \mathbf{J})^{-1} \tag{7B.11}$$

where $(\mathbf{I} - \mathbf{J})^{-1}$ is the *inverse* of the matrix $\mathbf{I} - \mathbf{J}$. Also
$$\mathbf{A} = \mathbf{I} - (\mathbf{I} - \mathbf{J})^{-1}$$
thus
$$\mathbf{X}_0^1 = [\mathbf{I} - (\mathbf{I} - \mathbf{J})^{-1}]\mathbf{X}_0 + (\mathbf{I} - \mathbf{J})^{-1}\mathbf{X}_1 \tag{7B.12}$$

As in the case of the single equation with one unknown, the partial derivatives in the matrix $\mathbf{J}$ may be approximated by finding $\mathbf{X}_2 \equiv f(\mathbf{X}_1)$; then by using the appropriate ratios of differences to estimate $\partial f_1/\partial x_1$, etc.

EXAMPLE: Solve the equations

$$x = -\frac{1}{2} + 162.9y$$

$$y = \left(x + \frac{1}{24x}\right)(0.00637)$$

This is the set of approximate equations that are used to obtain maximum likelihood estimators for α and λ in the gamma distribution problem, Sec.

7.5.2. Although it is easier to treat these as one equation, by substituting the first relation into the second, we will now for purposes of illustration keep the equations separate. As the initial solution vector, choose $\mathbf{X_0} = \begin{bmatrix} 1 \\ 1 \end{bmatrix}$. Then

$$\mathbf{X_1} = \begin{bmatrix} 162.4 \\ 0.0066354 \end{bmatrix}$$

$$\mathbf{X_2} = \begin{bmatrix} 0.58091 \\ 1.03449 \end{bmatrix}$$

To obtain the matrix $\mathbf{J}$, we now calculate estimates of the partial derivatives $\partial f_1/\partial x$, $\partial f_1/\partial y$, $\partial f_2/\partial x$, $\partial f_2/\partial y$:

$$\frac{\partial f_1}{\partial x} \cong \frac{f_1(x_1, y_0) - f_1(x_0, y_0)}{x_1 - x_0} = 0$$

$$\frac{\partial f_1}{\partial y} \cong \frac{f_1(x_0, y_1) - f_1(x_0, y_0)}{y_1 - y_0} = 162.9$$

$$\frac{\partial f_2}{\partial x} \cong \frac{f_2(x_1, y_0) - f_2(x_0, y_0)}{x_1 - x_0} = 0.0063683$$

$$\frac{\partial f_2}{\partial y} \cong \frac{f_2(x_0, y_1) - f_2(x_0, y_0)}{y_1 - y_0} = 0$$

Thus
$$\mathbf{J} = \begin{bmatrix} 0 & 162.9 \\ 0.0063683 & 0 \end{bmatrix}$$

Hence
$$\mathbf{I} - \mathbf{J} = \begin{bmatrix} 1 & -162.9 \\ -0.0063683 & 1 \end{bmatrix}$$

$$\mathbf{B} = (\mathbf{I} - \mathbf{J})^{-1} = \begin{bmatrix} -26.738 & -4355.6 \\ -0.17028 & -26.738 \end{bmatrix}$$

$$\mathbf{A} = \mathbf{I} - (\mathbf{I} - \mathbf{J})^{-1} = \begin{bmatrix} 27.738 & 4355.6 \\ 0.17028 & 27.738 \end{bmatrix}$$

Hence

$$\mathbf{X_0^1} = \begin{bmatrix} 27.738 & 4355.6 \\ 0.17028 & 27.738 \end{bmatrix} \begin{bmatrix} 1 \\ 1 \end{bmatrix} + \begin{bmatrix} -26.738 & -4355.6 \\ -0.17028 & -26.738 \end{bmatrix} \begin{bmatrix} 162.4 \\ 0.0066354 \end{bmatrix}$$

or
$$\mathbf{X_0^1} = \begin{bmatrix} 4383.34 \\ 27.90828 \end{bmatrix} + \begin{bmatrix} -4371.15 \\ -27.83089 \end{bmatrix}$$

Thus the new approximation is

$$\mathbf{X_0^1} = \begin{bmatrix} 12.19 \\ 0.07739 \end{bmatrix}$$

RELIABILITY ESTIMATION: PART I 189

Recalling that the solution already obtained was $\begin{bmatrix} 13.20 \\ 0.0841 \end{bmatrix}$, we see that the first approximation gives a remarkably close result, considering that we started with the initial value $\mathbf{X_0} = \begin{bmatrix} 1 \\ 1 \end{bmatrix}$.

We will now carry through the second approximation, starting with $\mathbf{X_0} = \begin{bmatrix} 12.2 \\ 0.077 \end{bmatrix}$. Therefore

$$\mathbf{X_1} = \begin{bmatrix} 12.0433 \\ 0.0777358 \end{bmatrix} \quad \text{and} \quad \mathbf{X_2} = \begin{bmatrix} 12.163162 \\ 0.0767379 \end{bmatrix}$$

$$\mathbf{J} = \begin{bmatrix} 0 & 162.90024 \\ 0.00636822 & 0 \end{bmatrix}$$

$$\mathbf{I} - \mathbf{J} = \begin{bmatrix} 1 & -162.90024 \\ -0.00636822 & 1 \end{bmatrix}$$

$$\mathbf{B} = (\mathbf{I} - \mathbf{J})^{-1} = \begin{bmatrix} -26.7490 & -4357.41852 \\ -0.1703435 & -26.7490 \end{bmatrix}$$

$$\mathbf{A} = \mathbf{I} - (\mathbf{I} - \mathbf{J})^{-1} = \begin{bmatrix} 27.7490 & 4357.41852 \\ 0.1703435 & 27.7490 \end{bmatrix}$$

Thus
$$\mathbf{X_0^1} = \mathbf{A} \begin{bmatrix} 12.2 \\ 0.077 \end{bmatrix} + \mathbf{B} \begin{bmatrix} 12.0433 \\ 0.0777358 \end{bmatrix}$$

$$= \begin{bmatrix} 674.05903 \\ 4.21486 \end{bmatrix} + \begin{bmatrix} -660.87365 \\ -4.13085 \end{bmatrix}$$

$$= \begin{bmatrix} 13.18538 \\ 0.08401 \end{bmatrix}$$

The next approximation can be obtained using the previous $\mathbf{A}$ and $\mathbf{B}$ matrices, since they have become fairly stable. The result is

$$\mathbf{X} = \begin{bmatrix} 13.1956 \\ 0.084074 \end{bmatrix}$$

which checks very closely when substituted into the original equations.

CHAPTER EIGHT
RELIABILITY ESTIMATION: PART II

8.1. RELIABILITY ESTIMATES AND CONFIDENCE LIMITS FROM THE RELIABILITY FUNCTION

We have seen that reliability can be expressed as a function of the parameters $\theta_1, \theta_2, \cdots$, of the distribution. Methods of obtaining estimators $\hat{\theta}_1, \hat{\theta}_2, \cdots$, of these parameters were presented in Chap. 7. We shall now consider how to relate these two sets of information so that reliability confidence limits as well as point estimates can be obtained. That is, the construction and properties of a reliability function or estimator $\hat{R} = R(\hat{\theta}_1, \hat{\theta}_2, \cdots)$ will be examined.

Consider first a single parameter distribution, such as the exponential distribution. We can state that

$$\hat{R} = e^{-T/\hat{\theta}} \qquad (8.1)$$

is an estimator for

$$R = e^{-T/\theta} \qquad (8.2)$$

where $\hat{\theta}$ is an estimator for θ. In Eq. (8.1) if $\hat{\theta}$ is the maximum likelihood estimator, it can be shown* that $\hat{R}$ is also the maximum likelihood estimator for R. The reader will have noticed that the reliability estimate in Eq. (8.1) is obtained simply by substituting the estimator $\hat{\theta}$ for θ in Eq. (8.2). If we further wish to associate a confidence statement with the reliability estimate, then this is done by computing a confidence limit for θ, i.e., $\hat{\theta}_c$, and substituting this value in Eq. (8.2) to give the reliability confidence limit $\hat{R}_c = e^{-T/\hat{\theta}_c}$. This can always be done provided the reliability function is monotonic in a single parameter. To see this, we note that the events $\hat{\theta}_c < \theta$ and $R(\hat{\theta}_c) < R(\theta)$ are equivalent if $R(\theta)$ is an increasing function of θ where $\hat{\theta}_c$ is a *lower* confidence limit on θ. However, if R were a decreasing function of its argument, then we should have to

---
* See exercise below, Sec. 8.2.

find an *upper* confidence limit on θ, i.e., $\hat{\theta}'_c > \theta$, in order to obtain a *lower* confidence limit on $R(\theta)$; i.e., $R(\hat{\theta}'_c) < R(\theta)$.

When the reliability function contains two or more unknown parameters, we can obtain a point estimate $\hat{R}$ of R simply by substituting the point estimates of the unknown parameters in the reliability function. *However, we cannot in general substitute confidence limits on the given parameters directly into the reliability function to obtain a confidence limit on reliability, where there is more than one unknown parameter.* We shall return to this important topic in Sec. 8.4.

8.2. A METHOD OF RELIABILITY ESTIMATION FOR SINGLE-PARAMETER DISTRIBUTIONS

First, let us consider two examples of reliability estimation for single-parameter failure distributions.

EXERCISE: Show directly that the maximum likelihood estimator for R is given by $e^{-T/\hat{\theta}}$, where $\hat{\theta}$ is the maximum likelihood estimator for θ (Eq. (7.9)) for the exponential time-to-failure distribution. (*Hint*: Express the probability density function in terms of R.)

EXAMPLE: Ten failures are observed in 100 trials. Assuming binomial trials, compute a lower 90 per cent confidence limit on $R = 1 - p$ where p is the probability of failure in any one trial (a) when it was determined in advance that 100 trials would be made; (b) when it was determined in advance that trials would stop as soon as ten failures were observed.

Answer: (a) Use the approximate equation (7.32) to obtain $\hat{R}_L = 0.862$. In Appendix 8A there are constructed upper confidence limits on $p \equiv 1 - R$ which have *at least* confidence γ (but no approximations are involved). Figure A.3, p. 558, yields $\hat{R}_L = 0.85$ (one minus the upper confidence limit on p) using the method of Appendix 8A. It appears that the former limit is optimistically high; however, one cannot easily be sure about this since different "ground rules" are used for deriving the two methods of finding confidence limits. It is recommended that the latter method be used if one needs to be on the "safe side."

(b) Here one can use the results of Appendix 8C and Fig. A.3 to obtain $\hat{R}_L \simeq 0.865$ approximately. To find the analogue of Eq. (7.32) is a difficult problem since Var $\hat{R}$ is a very complicated function of p (or R). See the related Prob. 33, Chap. 9 of Ref. 1.

8.3. A GENERAL METHOD FOR OBTAINING APPROXIMATE RELIABILITY ESTIMATES AND CONFIDENCE LIMITS

An alternative method is available for reliability estimation which is important because it is not limited to use with single-parameter distributions but can be extended to apply to multiparameter distributions, as shown in subsequent sections. To begin with we deal with the alternative method for single-parameter distributions only.

If we can find the mean and variance of an estimator $\hat{R} = R(\hat{\theta})$, then *approximate confidence limits on $R(\theta)$ may be found by using the general rule that $\hat{R}$ is approximately normally distributed when the sample size n is large*. First, we give general formulas for the mean and variance with several examples of their use.

8.3.1 THE MEAN OF THE RELIABILITY FUNCTION

In general, the expected value of a function of a sample moment, e.g., $\bar{x}$, s^2 (or, in general, estimators which are functions of sums of functions of all the sample values*), can be found by the following formula (Ref. 2, p. 345):

$$E(G(\hat{\theta})) = G(\theta) + O\!\left(\frac{1}{n}\right) \quad (8.3)$$

where $G(\theta)$ is some function of θ (in our case the reliability function) and θ is the corresponding population moment or parameter value such that $E(\hat{\theta}) \to \theta$ as $n \to \infty$.† The quantity $O(1/n)$ is a function of n, the sample size, which tends to zero as fast as $1/n$ when $n \to \infty$. For example, if $\hat{\theta} = \bar{x}$, the sample mean, and $G(x) = x^2$, we have for large n,

$$E(\bar{x}^2) \simeq \mu^2 \quad (8.4)$$

(In actuality, $E(\bar{x}^2) = \mu^2 + \sigma^2/n$ which evidently $\to \mu^2$ as $n \to \infty$.)

As a second example, if

$$G(\bar{x}) = \sin^{-1}\sqrt{\bar{x}}$$

then
$$E(\sin^{-1}\sqrt{\bar{x}}) \simeq \sin^{-1}\sqrt{\mu} \quad (8.5)$$

8.3.2 THE VARIANCE OF THE RELIABILITY FUNCTION

Correspondingly, the variance of the function $G(\hat{\theta})$ can generally be found by (Ref. 2, p. 345)

$$\operatorname{Var}(G(\hat{\theta})) = \left(\frac{\partial G}{\partial \hat{\theta}}\right)^2_{\hat{\theta}=\theta} \operatorname{Var}\hat{\theta} + O\!\left(\frac{1}{n^{3/2}}\right) \quad (8.6)$$

---

*There is some conjecture here. What is meant is that $G = G(\Sigma f(x_i))$; e.g., $G = (\Sigma \log x_i)^2$; etc.

† In this case we must have either $E(\hat{\theta}) = \theta$ or $E(\hat{\theta}) = \theta f(n)$ where $f(n) = 1 + O(1/n)$ as $n \to \infty$; i.e., $f(n) \to 1$ no slower than indicated, otherwise the term $O(1/n)$ in Eq. (8.3) may not be correct.

Thus, for the previous examples, when n is large, the approximate expression for the variances are, respectively

$$\text{Var}(\bar{x}^2) \simeq \frac{4\mu^2\sigma^2}{n} \tag{8.7}$$

and
$$\text{Var}(\sin^{-1}\sqrt{\bar{x}}) \simeq \frac{1}{4\mu(1-\mu)}\frac{\sigma^2}{n} \tag{8.8}$$

8.3.3 EXAMPLES

EXAMPLE: Let $\bar{x} = f/n$ where f is the total number of failures (ones) in a sample of size n drawn from the binomial distribution (Eq. (7.21)). Then, if

$$G(\bar{x}) \equiv \sin^{-1}\sqrt{\bar{x}} = \sin^{-1}\sqrt{\frac{f}{n}}$$

find the approximate variance of G from Eq. (8.8).
We have $\mu = p$ and $\sigma^2 = pq = p(1-p)$; thus

$$\text{Var}\left[\sin^{-1}\sqrt{\frac{f}{n}}\right] \simeq \frac{1}{4n} \tag{8.9}$$

In the example above the variance is independent of the parameter p except for terms of order equal to or higher than $n^{-3/2}$. The use of the function $\sin^{-1}\sqrt{f/n}$ is known as the *arcsine transformation of a binomial proportion*. Its purpose is to produce an estimator (of $\sin^{-1}\sqrt{p}$) with "stable" variance, i.e., variance essentially independent of p.

EXAMPLE: Find Var $s = \text{Var}\sqrt{s^2}$, where s^2 is defined by Eq. (7.38). Use the fact that

$$\text{Var}(s^2) = \frac{2(n-1)}{n^2}\sigma^4$$

(when the sample is taken from a *normally* distributed population) (see Eq. (5.99) *et seq.*).

Answer:

$$\text{Var } s = \frac{\sigma^2}{2n} + O\left(\frac{1}{n^{3/2}}\right)$$

In actuality

$$\text{Var } s = \frac{\mu_4 - \sigma^4}{4\sigma^2 n} + O\left(\frac{1}{n^2}\right)$$

for *any* underlying population distribution (Ref. 2, p. 353). Since $\mu_4 = 3\sigma^4$ for a normal distribution, then actually

$$\text{Var } s = \frac{\sigma^2}{2n} + O\!\left(\frac{1}{n^2}\right)$$

Thus in a specific case, the error term may be less than that given by the more general Eq. (8.6).

EXAMPLE: Find the approximate expected value and standard deviation of $\hat{R}$, defined by Eq. (8.1), where $\hat{\theta}$ is given by Eq. (7.9).

Answer: $E(\hat{R}) \simeq e^{-T/\theta} = R$

$$\sigma_{\hat{R}} \equiv +\sqrt{\text{Var } \hat{R}} \simeq \frac{T}{\theta\sqrt{n}} e^{-T/\theta} = \frac{R \log 1/R}{\sqrt{n}}$$

The result of the previous example and the result following Eq. (7.31) can now be used to obtain *approximate* confidence limits on $R = e^{-T/\theta}$, for example. We have as the result:

$$P(C_1 < e^{-T/\theta} < C_2) \simeq \gamma$$

where

$$C_1 = e^{-T/\hat{\theta}}\!\left(1 - K_{(1-\gamma)/2}\frac{T}{\hat{\theta}\sqrt{n}}\right)$$

$$C_2 = e^{-T/\hat{\theta}}\!\left(1 + K_{(1-\gamma)/2}\frac{T}{\hat{\theta}\sqrt{n}}\right)$$

(8.10)

and γ is the confidence coefficient.

If we only wish to find a *lower* confidence limit on $R = e^{-T/\theta}$, with confidence coefficient γ, then $C_2 \equiv 1$ and

$$C_1 = e^{-T/\hat{\theta}}\!\left(1 - K_{1-\gamma}\frac{T}{\hat{\theta}\sqrt{n}}\right) \tag{8.11}$$

EXAMPLE: *Exact* lower confidence limits on $R = e^{-T/\theta}$ can be obtained by making use of Eqs. (7.11)–(7.13). From Eq. (7.11) we know that the random variable $2n\hat{\theta}/\theta$ has a χ^2_{2n} distribution. If $\chi^2_{2n;1-\gamma}$ denotes the χ^2 deviate with $2n$ degrees of freedom exceeded with probability $1-\gamma$, then

$$P\!\left[\frac{2n\hat{\theta}}{\theta} < \chi^2_{2n;1-\gamma}\right] = \gamma \tag{8.12}$$

or

$$P\!\left[\frac{2n\hat{\theta}}{\chi^2_{2n;1-\gamma}} < \theta\right] = \gamma \tag{8.13}$$

Thus, $2n\hat{\theta}/\chi^2_{2n;1-\gamma} \equiv \hat{\theta}_c$ is a lower confidence limit for θ. It follows from the remarks subsequent to Eq. (8.2) that

$$R(\hat{\theta}_c) = \exp\left(-\frac{T\chi^2_{2n;1-\gamma}}{2n\hat{\theta}}\right) \tag{8.14}$$

is a lower confidence limit for R, with confidence γ. We will apply the same method later in Sec. 8.5.1 to the Weibull and in 8.6.1 to the gamma distribution.

EXERCISE: Compare the lower confidence limit as given by (8.11) to that given by (8.14) (make up a numerical example). Use Table A.8, p. 552 for values of $\chi^2_{2n;1-\gamma}/2n$.

8.4. EXTENSION OF THE APPROXIMATE METHOD TO TWO-PARAMETER DISTRIBUTIONS

When G is a function of two parameter estimators, say $G(\hat{\theta}, \hat{\lambda})$, then it is generally true that

$$E[G(\hat{\theta}, \hat{\lambda})] = G(\theta, \lambda) + O\left(\frac{1}{n}\right) \tag{8.15}$$

$$\text{Var}\,[G(\hat{\theta}, \hat{\lambda})] = \left(\frac{\partial G}{\partial \hat{\theta}}\right)^2 \text{Var}\,\hat{\theta} + \left(\frac{\partial G}{\partial \hat{\lambda}}\right)^2 \text{Var}\,\hat{\lambda}$$

$$+ 2\left(\frac{\partial G}{\partial \hat{\theta}}\right)\left(\frac{\partial G}{\partial \hat{\lambda}}\right) \text{Cov}\,(\hat{\theta}, \hat{\lambda}) + O\left(\frac{1}{n^{3/2}}\right) \tag{8.16}$$

The derivatives in Eq. (8.16) are evaluated at $\hat{\theta} = \theta$, $\hat{\lambda} = \lambda$, where $E(\hat{\theta}) \simeq \theta$ and $E(\hat{\lambda}) \simeq \lambda$, as indicated previously for the single-parameter case.

8.4.1 EXAMPLES

EXERCISE: Let $\bar{x}$ be the sample mean, and s^2 the sample variance of a random sample from a normally distributed population. Define

$$\hat{R} = 1 - \Phi\left(\frac{T - \bar{x}}{s}\right) \tag{8.17}$$

where $\Phi(Z)$ is the standard normal distribution function

$$(2\pi)^{-1/2} \int_{-\infty}^{Z} e^{-t^2/2}\,dt$$

Find an approximate expression for Var $\hat{R}$. Use the fact that $\bar{x}$ and s are statistically independent (Ref. 2, pp. 381–82).

Answer:
$$\operatorname{Var} \hat{R} \simeq \frac{\phi^2}{n}\left[1 + \frac{1}{2}\left(\frac{T-\mu}{\sigma}\right)^2\right]$$

where

$$\phi \equiv \phi\left(\frac{T-\mu}{\sigma}\right) = \frac{1}{\sqrt{2\pi}} \exp\left[-\frac{1}{2}\left(\frac{T-\mu}{\sigma}\right)^2\right]$$

EXAMPLE: Fifty 200-watt light bulbs of a certain brand are tested until each fails. It is assumed that the time-to-failure distribution is approximately normal with unknown mean μ and standard deviation σ. The times-to-failure are observed and the data are reduced to the sample mean

$$\bar{x} = \tfrac{1}{50} \sum t_i = 1500 \text{ hours}$$

and standard deviation

$$s = \left[\tfrac{1}{50} \sum (t_i - 1500)^2\right]^{1/2} = 100 \text{ hours}$$

What is a 90 per cent lower confidence limit for the probability R that a light bulb will not fail before 1200 hours?

Answer: As an estimator for this probability we have from (8.17)

$$\hat{R} = 1 - \Phi\left(\frac{1200 - 1500}{100}\right)$$

$$= 1 - \Phi(-3)$$

$$= 0.99865$$

From the result of the previous exercise, and replacing the quantities μ and σ by their estimators $\bar{x}$ and s, we obtain (the angle brackets denote such an "estimate" of $\operatorname{Var} \hat{R}$)

$$\langle \operatorname{Var} \hat{R} \rangle^{1/2} \equiv V^{1/2} = 0.00147$$

Finally from Sec. 7.4.4 we obtain as an approximate 90 per cent lower confidence limit on R:

$$\hat{R}_L = 0.99865 - 1.282(0.00147) = 0.9968$$

More accurate methods for dealing with this problem when the normal distribution is involved will be given later in Sec. 8.7. The following sections apply both the approximate and exact methods that we have been discussing to the Weibull, gamma, and extreme value distributions (the latter application is left as an exercise). The example given in Sec. 7.5.2 on the results of seven electrical generator lifetimes is used for illustrative purposes throughout.*

---

* See footnote, p. 172.

8.5. ESTIMATOR AND CONFIDENCE LIMIT FOR R IN THE CASE OF THE WEIBULL DISTRIBUTION

We have, as an estimator for the probability R that $\tau > T$:

$$\hat{R} = \exp(-\hat{\lambda} T^{\hat{\alpha}}) \tag{8.18}$$

Thus, using (8.15) and (8.16)

$$E(\hat{R}) \simeq \exp(-\lambda T^{\alpha}) \tag{8.19}$$

and

$$\text{Var } \hat{R} \simeq T^{2\alpha} e^{-2\lambda T^{\alpha}} \{\lambda^2 \log^2 T \text{ Var } \hat{\alpha} + \text{Var } \hat{\lambda} + 2\lambda \log T \text{ Cov } (\hat{\alpha}, \hat{\lambda})\} \tag{8.20}$$

Estimated values of $E(\hat{R})$ and $\sqrt{\text{Var } \hat{R}}$ can be obtained by replacing α and λ wherever they appear in (8.19) and (8.20) by the estimators (Eq. (7.73)) and also by using the values of the variances and covariances as given in Sec. 7.5.5. For example, for $T = 100$ hours

$$\langle E(\hat{R}) \rangle \simeq 0.936$$
$$\langle \sqrt{\text{Var } \hat{R}} \rangle \simeq 0.065 \tag{8.21}$$

Thus a 90 per cent lower confidence limit for R is given by

$$0.936 - K_{0.10}(0.065) = 0.936 - 1.282(0.065) = 0.853$$

We would then state that if 100 hours were the required life, then the probability of meeting this requirement (reliability) would be at least 0.853, with 90 per cent confidence.

8.5.1 EXACT CONFIDENCE LIMIT FOR R IN THE CASE OF THE WEIBULL DISTRIBUTION

When the parameter α of the Weibull distribution is known, it is possible to obtain exact confidence limits on R, because the exact distribution of the maximum likelihood estimator

$$\frac{1}{\hat{\lambda}} = \frac{\sum t_j^{\alpha}}{n} \tag{8.22}$$

can be found. To do this, we have first of all

$$P(\tau \leq t) = 1 - e^{-\lambda t^{\alpha}} \tag{8.23}$$

i.e., τ has the Weibull distribution. Let $\zeta = \tau^{\alpha}$; then

$$P(\zeta \leq y) = P(\tau^{\alpha} \leq y) = P(\tau \leq y^{1/\alpha}) \tag{8.24}$$

Hence from (8.23)

$$P(\zeta \leq y) = 1 - e^{-\lambda y} \qquad (8.25)$$

Thus ζ or τ^α has the ordinary exponential distribution. From Eq. (7.11) we already know that the mean of a sample of size n from an exponential distribution has a distribution related to that of χ^2; thus

$$P\left(\frac{2n\lambda}{\hat{\lambda}} \leq y\right) = P(\chi^2_{2n} \leq y) \qquad (8.26)$$

Hence $2n\lambda/\hat{\lambda}$ has a χ^2 distribution with $2n$ degrees of freedom.

From the last remark we can obtain confidence limits on λ and therefore on $R = e^{-\lambda T^\alpha}$. Let $\chi^2_{2n;\alpha}$ denote the value of a chi-square random variable exceeded with probability α. Thus

$$P(\chi^2_{2n} \geq \chi^2_{2n;1-\gamma}) = 1 - \gamma \qquad (8.27)$$

For example, when $2n = 14$ and $\gamma = 0.90$, then from tables of the χ^2 distribution (Ref. 3, p. 557), $\chi^2_{14;0.10} = 21.064$. From the remark immediately following (8.26), therefore,

$$P\left(\frac{2n\lambda}{\hat{\lambda}} \leq \chi^2_{2n;1-\gamma}\right) = \gamma \qquad (8.28)$$

or

$$P\left(\lambda \leq \frac{\hat{\lambda}\chi^2_{2n;1-\gamma}}{2n}\right) = \gamma \qquad (8.29)$$

Equation (8.29) states that the quantity $\hat{\lambda}\chi^2_{2n;1-\gamma}/2n$ is an upper confidence limit on λ with confidence coefficient γ. Since the event in brackets in Eq. (8.29) is equivalent to the event

$$\exp(-\lambda T^\alpha) \geq \exp\left(-\frac{\hat{\lambda}T^\alpha \chi^2_{2n;1-\gamma}}{2n}\right) \qquad (8.30)$$

then we have immediately

$$P\left[R \geq \exp\left(-\frac{\hat{\lambda}T^\alpha \chi^2_{2n;1-\gamma}}{2n}\right)\right] = \gamma \qquad (8.31)$$

The example previously given can be considered again by now assuming that the parameter α is known already to be $\alpha = 4.669$ (from Eq. (7.73)). Since $\hat{\lambda} = 3.04238 \times 10^{-11}$, $T = 100$; $n = 7$, $\chi^2_{14;0.10} = 21.064$, a lower confidence limit $\hat{R}_L$ (confidence 0.90) for the probability

RELIABILITY ESTIMATION: PART II   199

R of surviving $T = 100$ hours is

$$\hat{R}_L = \exp\left[-\frac{(3.04238 \times 10^{-11})(100^{4.669})(21.064)}{14}\right]$$

$$= e^{-0.0997}$$
(8.32)

Thus
$$\hat{R}_L = 0.905 \tag{8.33}$$

Comparison of the value of $\hat{R}_L$ with the value 0.853 obtained previously for the two-parameter Weibull distribution illustrates a fact which is generally true, namely that with additional knowledge (knowledge of the parameter α) a tighter lower bound to R can be obtained.

EXERCISE: Apply the results of this section and Sec. 7.5.5 to the *extreme value* distribution, with distribution function:

$$P(\tau \le t) = 1 - e^{-\alpha(e^{\gamma t}-1)} \tag{8.34}$$

to obtain approximate and exact confidence limits on R. Use the same data.

8.6. ESTIMATOR AND CONFIDENCE LIMIT FOR R IN THE CASE OF THE GAMMA DISTRIBUTION

An estimate of the probability that time-to-failure exceeds T is given by

$$\hat{R} = \int_T^\infty \frac{\hat{\lambda}^{\hat{\alpha}} x^{\hat{\alpha}-1} e^{-\hat{\lambda}x}}{\Gamma(\hat{\alpha})} \, dx \tag{8.35}$$

or, changing variables,

$$\hat{R} = \int_{\hat{\lambda}T}^\infty \frac{u^{\hat{\alpha}-1} e^{-u}}{\Gamma(\hat{\alpha})} \, du \tag{8.36}$$

As we saw in Sec. 6.8.2, the above integral can be written as a sum:*

$$\hat{R} = 1 - \sum_{k=\hat{\alpha}}^{\infty} \frac{(\hat{\lambda}T)^k e^{-\hat{\lambda}T}}{k!}$$

Thus we can use Molina's table II (Ref. 4) which tabulates

$$P(c, a) = \sum_{x=c}^{\infty} \frac{a^x e^{-a}}{x!}$$

Suppose $T = 100$ hours, $a \equiv \hat{\lambda}T = (0.0841)(100) = 8.41$, $c \equiv \hat{\alpha} = 13.7$.

* This is not strictly correct since the index in the summation must be an integer; nevertheless, interpolation to non-integral values of c in Molina's table II will give a good approximation.

From the referenced tables, we then have (by interpolation)

$$\hat{R} = 1 - 0.05924 = 0.941 \tag{8.37}$$

To obtain the variance of $\hat{R}$ given by Eq. (8.35), Eq. (8.16) could be used directly; however, the resulting expressions are too difficult to evaluate. To avoid this difficulty, we use the *Edgeworth* expansion of the gamma distribution function in terms of the normal distribution function (Ref. 2, p. 229). Thus

$$\hat{R} \approx 1 - \Phi\left(\frac{\hat{\lambda}T - \hat{\alpha}}{\sqrt{\hat{\alpha}}}\right)$$

$$+ \frac{\left[\left(\frac{\hat{\lambda}T - \hat{\alpha}}{\sqrt{\hat{\alpha}}}\right)^2 - 1\right]}{3\sqrt{\hat{\alpha}}} \frac{1}{\sqrt{2\pi}} \exp\left[-\frac{1}{2}\left(\frac{\hat{\lambda}T - \hat{\alpha}}{\sqrt{\hat{\alpha}}}\right)^2\right] - \cdots \tag{8.38}$$

The term $1 - \Phi[(\hat{\lambda}T - \hat{\alpha})/\sqrt{\hat{\alpha}}]$ can be recognized as the normal approximation to the "tail" of a gamma distribution function, wherein the normal distribution is made to have the same mean, $\hat{\alpha}/\hat{\lambda}$, and standard deviation, $\sqrt{\hat{\alpha}}/\hat{\lambda}$, as does the gamma distribution. The last correction term in (8.38) takes into account the "skewness" of the gamma distribution.

EXAMPLE: Using (8.38) as a representation of $\hat{R}$, calculate $\hat{R}$ by substituting in the values of $\hat{\alpha} = 13.70$, $\hat{\lambda} = 0.0841$, with $T = 100$, and using appropriate tables of the normal distribution and probability density functions. Then, applying Eq. (8.16) to (8.38) find a formula for Var $\hat{R}$ and also an estimate of Var $\hat{R}$, using only the normal approximation term of (8.38).

If we define $\hat{\beta} \equiv (\hat{\lambda}T - \hat{\alpha})/\sqrt{\hat{\alpha}}$, then using the given estimates of $\hat{\alpha}$ and $\hat{\lambda}$, we obtain $\hat{\beta} = -1.43$. Since

$$\hat{R} = 1 - \Phi(\hat{\beta}) + \frac{\hat{\beta}^2 - 1}{3\sqrt{\hat{\alpha}}} \phi(\hat{\beta}) \tag{8.39}$$

we have from tables $1 - \Phi(\hat{\beta}) = 0.9236$, $\phi(\hat{\beta}) = 0.1435$; thus $\hat{R} = 0.9236 + 0.0135 = 0.937$, which is fairly close to the (interpolated) value given by (8.37). We find (using Eqs. (7.64')–(7.66'))

$$\langle \text{Var } \hat{R} \rangle \simeq \frac{\phi^2(\hat{\beta})}{n} \left[\frac{2\hat{\lambda}^2 T^2}{\hat{\alpha}} - (2\hat{\alpha} - 1)\left(1 + \frac{\hat{\beta}}{2\sqrt{\hat{\alpha}}}\right)\left(1 + \frac{3\hat{\beta}}{2\sqrt{\hat{\alpha}}}\right)\right] \tag{8.40}$$

and using the previously given numerical values of $\hat{\alpha}$, $\hat{\lambda}$, and $\hat{\beta}$, we obtain

$$V \equiv \langle \text{Var } \hat{R} \rangle \simeq 0.00403 \quad \text{and} \quad V^{1/2} \simeq 0.0635 \tag{8.41}$$

Thus a 90 per cent lower confidence limit for R is given by

$$0.937 - K_{0.10}(0.0635) = 0.937 - 1.282(0.0635) = 0.856$$

Compare these results with those obtained previously where the Weibull distribution was assumed to be the correct model for time-to-failure (Sec. 8.5).

8.6.1 EXACT CONFIDENCE LIMIT FOR R IN THE CASE OF THE GAMMA DISTRIBUTION

When the parameter α in the gamma distribution is known, one can determine exact confidence limits for the parameter λ, and also for the probability R of surviving time T. The probability density function for the gamma distribution is repeated here for convenience:

$$f(t; \lambda, \alpha) = \frac{\lambda^{\alpha} t^{\alpha-1} e^{-\lambda t}}{\Gamma(\alpha)}; \quad t > 0 \quad \begin{pmatrix} \alpha > -1 \\ \lambda > 0 \end{pmatrix} \quad (8.42)$$

$$= 0 \quad t \leq 0$$

The maximum likelihood estimator for the parameter $\alpha/\lambda \equiv \beta$ is then (from Eq. (7.42))

$$\hat{\beta} = \frac{\sum t_i}{n} \quad (8.43)$$

It can be shown* that the random variable $2n\hat{\beta}\lambda$ has a χ^2 distribution with $2n\alpha$ "degrees of freedom." Thus

$$P(2n\hat{\beta}\lambda \leq \chi^2_{2n\alpha; 1-\gamma}) = \gamma \quad (8.44)$$

or

$$P\left(\lambda \leq \frac{\chi^2_{2n\alpha; 1-\gamma}}{2n\hat{\beta}}\right) = \gamma \quad (8.45)$$

Thus the quantity in the righthand side of the brackets is an upper confidence limit for λ with confidence coefficient γ.

We now wish to convert Eq. (8.45) into a confidence limit statement on R, where

$$R \equiv R(\lambda T) = \int_{\lambda T}^{\infty} \frac{u^{\alpha-1} e^{-u}}{\Gamma(\alpha)} du \quad (8.46)$$

By inspecting (8.46), we see that $R(\lambda T)$ is a decreasing function of λ; this means that if U is an upper bound on λ, then $R(UT)$ is a lower bound

* Ref. 2, p. 505.

on R. From (8.45) and (8.46) we then have immediately

$$P\left[R \geq R\left(\frac{T\chi^2_{2n\alpha;1-\gamma}}{2n\hat{\beta}}\right)\right] = \gamma \tag{8.47}$$

Thus the right side of the bracketed quantity gives a lower confidence limit on the probability R of surviving time T, for the gamma distribution.

As an example, we again use the data on the seven electrical generators; however, we assume α is *known* equal to 13.7. From the example previously referred to, we then have $T = 100$, $n = 7$, $\hat{\beta} = 162.9$. To obtain the value of $\chi^2_{2n\alpha;0.10} = \chi^2_{191.8;0.10}$, we use the fact that for large "degrees of freedom" m, $\sqrt{2\chi^2_m}$ is approximately normally distributed with mean $\sqrt{2m-1}$ and standard deviation one (Ref. 2, p. 251).* Thus if $K_{1-\gamma}$ denotes a standard normal deviate exceeded with probability $1 - \gamma$, then

$$K_{1-\gamma} \simeq \sqrt{2\chi^2_{m;1-\gamma}} - \sqrt{2m-1} \tag{8.48}$$

or

$$\chi^2_{m;1-\gamma} \simeq \tfrac{1}{2}(K_{1-\gamma} + \sqrt{2m-1})^2 \tag{8.49}$$

For $\gamma = 0.90$, $m = 191.8$, then Eq. (8.49) gives

$$\chi^2_{191.8;0.10} = \tfrac{1}{2}(1.282 + \sqrt{382.6})^2 = 217.2 \tag{8.50}$$

Hence

$$\frac{T\chi^2_{2n\alpha;1-\gamma}}{2n\hat{\beta}} = \frac{100\,(217.2)}{14\,(162.9)} = 9.52$$

Now we must evaluate R from Eq. (8.46). To do this, we note the remarks following Eq. (8.36), which state that one should enter Molina's table II† with $a = 9.52$, $c = 13.7$. We obtain (by interpolation) $P(c, a) = 0.122$. Thus the lower confidence limit on R with confidence coefficient 0.90 is

$$\hat{R}_L = 1 - 0.122 = 0.878 \tag{8.51}$$

This value can be compared with the result following Eq. (8.41). It would be expected that the lower confidence limit (0.856) on R when two parameters must be estimated would be smaller than the value of $\hat{R}_L = 0.878$ given by Eq. (8.51); i.e., a "more precise" estimate is obtained when additional knowledge of the underlying distribution of time-to-failure is available. One can also compare the lower confidence limit obtained from the same data, assuming the Weibull distribution (Eq. (8.33)) when one parameter is assumed known.

* A better approximation (see Ref. 3, p. 556) is to assume that $(\chi^2_m/m)^{1/3}$ is normal with mean $1 - 2/9m$ and variance $2/9m$. In our particular example the difference between the two is negligible.

† Reference 4.

8.7. ESTIMATOR AND CONFIDENCE INTERVAL FOR R IN THE CASE OF THE NORMAL DISTRIBUTION

In Sec. 8.4.1 of this chapter an approximate method was given for finding an estimator and confidence interval for

$$R = 1 - \Phi\left(\frac{T-\mu}{\sigma}\right) = \Phi\left(\frac{\mu-T}{\sigma}\right)$$

where $\Phi(x)$ is the standard normal distribution function (Sec. 6.10). We shall now indicate how to obtain exact confidence intervals for $R = \Phi[(\mu - T)/\sigma]$ when only sample estimates of μ and σ are available. As noted previously, T will frequently represent the required time-of-operation although there is no reason to restrict the definition to these cases. For example, one could define "reliability" as the probability that a specified lower limit on total impulse is exceeded. We shall see from the later examples how the answers to such problems are obtained from the tables of *one-sided tolerance factors* (Table A.9, p. 554). These tables were generated to solve the converse problem of *calculating*, from an observed sample, a *tolerance limit* on a *specified* proportion of a distribution. The problem at hand requires that the *"tolerance" limit* be *specified* and that the proportion of a distribution exceeding (or falling short of) the limit be *calculated* from the observed sample. In each case a confidence coefficient is specified in advance.

We are also interested in estimating a function of the form

$$R = \Phi\left(\frac{T_2 - \mu}{\sigma}\right) - \Phi\left(\frac{T_1 - \mu}{\sigma}\right)$$

as this expression represents the probability that a normally distributed random variable lies between "specification limits" T_1 and T_2; i.e., T_1, T_2 will frequently represent the required limits on some performance parameters such as thrust, mixture ratio, etc. For this last problem we shall indicate how to obtain an approximate lower confidence limit on R.

Tolerance limits are obtained as follows: A random sample of size n is taken from a normally distributed population (time-to-failure, performance values, etc.). The mean $\bar{x}$ and (positive) square root of the unbiased estimator of the variance,* s', are then calculated. A *one-sided tolerance limit* is given by the quantity $\bar{x} - Ks'$ (or $\bar{x} + Ks'$) with the property that the probability is γ that *at least* a proportion $(1 - \alpha)$ of the distribution will be contained within the interval $\bar{x} - Ks'$ and plus infinity (or minus infinity and $\bar{x} + Ks'$). The quantity K is called a *tolerance factor* and is evidently a function of n, γ, and α. Table A.9 presents values of K

---

* The quantity $s' = s\sqrt{n/(n-1)}$, where s is the square root of the sample variance; i.e., $s^2 = (1/n) \Sigma(x_i - \bar{x})^2$, which was shown to be biased.

for $n = 3(1)25(5)50$; $\gamma = 0.75, 0.90, 0.95, 0.99$; $\alpha = 0.25, 0.10, 0.05, 0.01, 0.001$.*

8.7.1 EXAMPLES

The following examples will illustrate the use of the tolerance factor tables.

EXAMPLE: We use the data of the example of Sec. 8.4.1, in which $\bar{x} = 1500$ hours and

$$s' = \sqrt{\frac{n}{n-1}}\, s = \sqrt{\frac{50}{49}}\,(100) = 101 \text{ hours}$$

What is a lower tolerance limit of the form $\bar{x} - Ks'$ such that the probability is $\gamma = 0.90$ that the probability is at least $1 - \alpha = 0.95$ that the time-to-failure of this kind of light-bulb will exceed the calculated lower tolerance limit? From Table A.9 we find the tolerance factor $K = 1.965$ to calculate the $1 - \alpha = 0.95$ lower tolerance limit $\bar{x} - Ks' = 1302$ hours. This means that we would make the statement "the probability is at least 0.95 that the time-to-failure of this light-bulb exceeds 1302 hours," and we would be 90 per cent confident that the statement was correct. This further means that if the entire sampling procedure and calculation were repeated, the resulting statement would differ from the above statement *only in the numerical value of the lower tolerance limit*. For the converse problem, given in the following example, repetition of the sampling procedure would yield statements which differed *only in the numerical values of the probability* of exceeding the *fixed* lower limit. In either case the confidence coefficient (0.90) is the expected proportion of *correct statements* in a long series of repetitions of the procedure.

EXAMPLE: Using the same data as in the previous example, what is a 90 per cent lower confidence limit on the probability that the time-to-failure exceeds the (pre-specified) required life of 1302 hours?

Answer: 0.95. To obtain this answer, one calculates the quantity $K = (1500 - 1302)/101 = 1.965$; then from Table A.9, p.554, under $\gamma = 0.90$ one finds the value of α corresponding to the value of $K = 1.965$, for $n = 50$. We find $\alpha = 0.05$, and therefore $1 - \alpha = 0.95$, the desired lower confidence limit. If the quantity K were not directly available in the table, we would have to interpolate to find the correct value of α. The following example requires such interpolation.

EXAMPLE: Solve the problem of Sec. 8.4.1 by the method given in the previous example. In this case $K = (1500 - 1200)/101 = 2.97$. From Table A.9 we find under $\gamma = 0.90$, $n = 50$, that α is between 0.01 and 0.001. The following method of interpolation is preferred:

From the tables of the normal distribution function (Ref. 5), the values of K_α (standard normal deviate exceeded with probability α) for $\alpha = 0.01$

---

* Reprinted with the kind permission of the publishers, Prentice-Hall, Inc., and the authors, A. H. Bowker and G. J. Lieberman, from *Engineering Statistics*, Table 8.3, pp. 230–31 (1959).

and 0.001 are $K_{0.01} = 2.326$ and $K_{0.001} = 3.090$. The corresponding *tolerance factors* are 2.735 and 3.604. By usual linear interpolation methods on K_α we obtain $K_\alpha = 2.326\lambda + 3.090(1 - \lambda)$, where $\lambda = (3.604 - 2.97)/(3.604 - 2.735) = 0.73$. Thus $K_\alpha = 2.532$. Tables of the normal distribution function yield $\alpha = 0.00567$ and therefore $1 - \alpha = 0.99433$. Note that there is no significance in the use of normal deviates for interpolation purposes; it merely happens that K_α is reasonably linear in log α and is well tabulated. The number 0.99433 should be compared with the lower confidence limit 0.9968 obtained in Sec. 8.4.1.

EXERCISE: Use the same data as in the previous example, but with required life $T = 1350$ hours. Compare the lower 90 per cent confidence limits on $R = \Phi[(\mu - T)/\sigma]$ obtained by the method of this section and of Sec. 8.4.1.

Answer: 0.881 and 0.899, respectively. Note that the approximate method of Sec. 8.4.1 appears again to be slightly optimistic, as we saw in the previous example.

Methods of finding an exact confidence interval of the form $(\hat{R}_L < R < \hat{R}_U)$, where $R = \Phi[(\mu - T)/\sigma]$, are given in Ref. 6.

8.7.1.1. Confidence Limits for R (Two-sided Specification). We now turn to the problem of estimation of probabilities that a normally distributed variable is contained between *two* specification limits; i.e., we wish to place confidence limits on the reliability function

$$R = \Phi\left(\frac{T_2 - \mu}{\sigma}\right) - \Phi\left(\frac{T_1 - \mu}{\sigma}\right)$$

It can be shown using the methods of Sec. 8.4 that the following set of formulas can be derived to obtain a lower confidence limit on R. Thus, given

$$\hat{R} = \Phi\left(\frac{T_2 - \bar{x}}{s}\right) - \Phi\left(\frac{T_1 - \bar{x}}{s}\right)$$

one can derive the expression

$$\operatorname{Var} \hat{R} = \frac{1}{n}\left[\phi_1^2\left(1 + \frac{1}{2}\Delta_1^2\right) + \phi_2^2\left(1 + \frac{1}{2}\Delta_2^2\right) - 2\phi_1\phi_2\left(1 + \frac{1}{2}\Delta_1\Delta_2\right)\right]$$

where

$$\left.\begin{aligned}\phi_i &= (2\pi)^{-1/2} \exp\left[-\frac{1}{2}\left(\frac{T_i - \mu}{\sigma}\right)^2\right] \\ \Delta_i &= \frac{T_i - \mu}{\sigma}\end{aligned}\right\} \quad i = 1, 2$$

Therefore

$$\hat{R}_L = \hat{R} - K_{1-\gamma}V^{1/2}$$

is a lower confidence limit on R with confidence coefficient γ, where $K_{1-\gamma}$ is the standard normal deviate exceeded with probability $1 - \gamma$, and $V^{1/2} \equiv \langle \text{Var } \hat{R} \rangle^{1/2}$, where $\langle \text{Var } \hat{R} \rangle$ denotes the value of Var $\hat{R}$ when μ and σ are replaced by the estimators $\bar{x}$ and s, respectively.

EXAMPLE: The specification limits on the thrust to be delivered by a solid-propellant rocket engine are given as "20,000 lb$_f$ $\pm$ 3%"; i.e., 19,400 lb$_f$ and 20,600 lb$_f$ are the lower and upper specification limits, respectively. A random sample of twenty engines are test-fired and the data reduced to yield a mean thrust $\bar{x} = 19,900$ lb$_f$ and sample standard deviation $s = 250$ lb$_f$. What is a 90 per cent lower confidence limit on the probability that the rocket engine will meet its thrust requirements?

By the approximate method of the previous paragraph, we have

$$\Delta_2 \equiv \frac{T_2 - \bar{x}}{s} = \frac{20{,}600 - 19{,}900}{250} = 2.8$$

$$\Delta_1 \equiv \frac{T_1 - \bar{x}}{s} = \frac{19{,}400 - 19{,}900}{250} = -2.0$$

Thus

$$\hat{R} = \Phi(2.8) - \Phi(-2.0) = \Phi(2.8) + \Phi(2.0) - 1$$
$$= 0.99744 + 0.97725 - 1 = 0.97469$$

Also $\phi_1 = 0.05399$, $\phi_2 = 0.00792$, using tables of the standard normal probability density function. We can then compute

$$V = \langle \text{Var } \hat{R} \rangle = 0.000530 \qquad \text{and} \qquad V^{\frac{1}{2}} = 0.0230$$

Thus

$$\hat{R}_L = 0.9747 - 1.282(0.0230) = 0.9452$$

The reader is referred to Ref. 6 for a method of finding a confidence interval on

$$R = \Phi\left(\frac{T_2 - \mu}{\sigma}\right) - \Phi\left(\frac{T_1 - \mu}{\sigma}\right)$$

which is considered to be more accurate, but is considerably more difficult to use than the method illustrated.

8.8. CONFIDENCE LIMITS FOR PARAMETERS OF BINOMIAL, NEGATIVE BINOMIAL, AND POISSON DISTRIBUTIONS

Appendixes A–D of this chapter present methods for constructing one-sided upper confidence limits on probability of failure p, based on results of binomial trials, and on failure rate λ based on sampling results when the Poisson distribution applies.

The results can be used to obtain lower confidence limits on reliability R for the cases when R is any nonincreasing (generally strictly decreasing) function of p. For example, when reliability is defined as the probability of success of an *independent serial* system (Sec. 9.2) all of whose n subsystems have the *same* probability of failure p, then

$$R = (1 - p)^n \tag{8.52}$$

If $\hat{p}_c$ is an upper confidence limit on p with confidence coefficient γ, then

$$\hat{R}_c = (1 - \hat{p}_c)^n \tag{8.53}$$

is a lower confidence limit on R with confidence coefficient γ (see Sec. 8.1 for a general proof of this statement). The same result follows if the system is *independent parallel* (Sec. 9.4) with all n subsystems having the *same* probability of failure p, since then

$$R = 1 - p^n \tag{8.54}$$

and R is still a decreasing function of p. It is also evident that the result follows when the system is an *independent mixed serial-parallel* system of all of whose components have the *same* probability of failure p. Thus, in general

$$R = f(p) \tag{8.55}$$

will be a decreasing function of p; and

$$\hat{R}_c = f(\hat{p}_c) \tag{8.56}$$

will be a lower confidence limit on R with confidence coefficient γ, where $\hat{p}_c$ is an upper confidence limit on p with confidence coefficient γ.

Methods of finding lower confidence limits on R for independent serial and parallel systems, when the probabilities of failure of the subsystems are not all equal to the same value p, will be discussed in Secs. 9.2.1 and 9.4.

REFERENCES

1. W. Feller, *An Introduction to Probability Theory and Its Applications*, 2d ed., John Wiley & Sons, Inc., New York, 1957, vol. I.
2. H. Cramér, *Mathematical Methods of Statistics*, Princeton University Press, Princeton, N.J., 1946.
3. A. H. Bowker and G. J. Lieberman, *Engineering Statistics*, Prentice-Hall, Inc., Englewood Cliffs, N.J., 1959.
4. E. C. Molina, *Poisson's Exponential Binomial Limit*, D. Van Nostrand Company, Princeton, N.J., 1942.
5. *Tables of Normal Probability Functions*, National Bureau of Standards, Applied Mathematics Series 23, 1953.
6. G. J. Resnikoff and G. J. Lieberman, *Tables of the Non-Central t-Distribution*, Stanford University Press, Stanford, Calif., 1957.

7. C. R. Blythe and D. W. Hutchinson, "Table of Neyman—Shortest Unbiased Confidence Intervals for the Binomial Parameter," *Biometrika*, **47**, Parts 3 and 4, 381–391 (December 1960).

ADDITIONAL READING

Cohen, A. C., Jr., "On the Solution of Estimating Equations for Truncated and Censored Samples from Normal Populations," *Biometrika*, **44**, Parts 1 and 2, 225–36 (June 1957).

Connor, W. S., "Interpreting Reliability by Fitting Theoretical Distributions to Failure Data," *Ind. Eng. Chem.*, **52**, 75A–76A (February 1960); *52* 71A–72A (April 1960).

Crow, E. L., "Confidence Intervals for a Proportion," *Biometrika*, **43**, Parts 3 and 4, 423–35 (December 1956).

Crow, E. L., and Gardner, R. S., "Confidence Intervals for the Expectation of a Poisson Variable," *Biometrika*, **46**, Parts 3 and 4, 441–53 (December 1959).

Sarhan, A. E., and Greenberg, B. G., "Tables for Best Linear Estimates by Order Statistics of the Parameters of Single Exponential Distributions from Singly and Doubly Censored Samples," *J. Am. Stat. Assoc.*, **52**:277, 58–87 (March 1957)

Walsh, J. E., "Asymptotic Efficiencies of a Nonparametric Life Test for Smaller Percentiles of a Gamma Distribution," *J. Am. Stat. Assoc.*, **51**, 467–80 (September 1956).

Walsh, J. E., "Estimating Population Mean, Variance, and Percentage Points from Truncated Data," *Skand. Akt.* **39**: 1–2, 47–58 (1956).

APPENDIX 8A

ONE-SIDED BINOMIAL CONFIDENCE LIMITS

This appendix presents a proof that the numbers $C(F)$ defined by Eq. (8A.1) are upper confidence limits to a binomial parameter p. Figures A.1–A.5 (pp. 556-560) give values of $C(F)$ for various values of F and for sample sizes N ranging from 5–1000, for $\gamma = 0.50, 0.80, 0.90, 0.95,$ and 0.99. In this case, N independent trials are made of a "system," and a number of failures F are observed. A statement is then made: "The true probability of failure p of the system is not more than C." If a series of such statements is to be made, each after observing a number of failures in a sample of independent trials, we would like to guarantee in some sense that most of the statements are correct. This can be accomplished if we agree in advance on a "confidence coefficient" γ, $0 < \gamma < 1$ (usually 0.90 or higher), such that whatever the true value of the parameter p, the expected proportion of correct statements in a long series of such statements is to be equal to or greater than γ. If for any single statement we know that the probability that the statement is correct is at least equal to γ, independent of the true value of p, then by the long-run frequency interpretation of probability, the above type of guarantee will be achieved. Note that the value of p could be different from population to population from which each sample of trials was taken; also the sample size N of trials could be different from sample to sample (the proviso is that the sample size N be fixed in advance of taking each sample).

The above objectives are accomplished as follows: First consider that N independent trials are to be made from a population in which the probability of observing a failure is equal to some (unknown) constant p, $0 \leq p \leq 1$. The set of observed outcomes consists of all possible numbers of failures $(0, 1, \cdots, N)$ and is called the *sample space* (Sec. 6.6). Consider now the following equation which defines certain numbers $C(F)$; $F = 0, \cdots, N$:

$$\sum_{j=0}^{F} \binom{N}{j} [C(F)]^j [1 - C(F)]^{N-j} = 1 - \gamma \qquad (8A.1)$$

where $0 < \gamma < 1$. We will now show that the numbers $C(F)$ are a set of upper confidence limits for p.

First we see that $0 < C(0) < C(1) \cdots < C(F) < C(F+1) < \cdots < C(N) = 1$. To show this we must first show that Eq. (8A.1) has a unique solution (except when $F = N$, for which $C(N)$ is defined equal to unity). Consider the function:

$$g_F(x) = \sum_{j=0}^{F} \binom{N}{j} x^j (1-x)^{N-j}$$

We have $g_F(0) = 1$ and $g_F(1) = 0$; also

$$g'_F(x) = -\binom{N}{F}(N-F)x^F(1-x)^{N-F-1} \leq 0 \quad \text{for } 0 \leq x \leq 1$$

Hence $g_F(x) = 1 - \gamma > 0$ has exactly one root in the interval $0 < x < 1$. When $F = N$, however, the left side of Eq. (8A.1) is identically equal to unity; and so we will arbitrarily define $C(N) = 1$.

Now to show that $C(F) < C(F+1)$, consider

$$g_F(Z) = 1 - \gamma \tag{8A.2}$$

$$g_{F+1}(Y) = 1 - \gamma \tag{8A.3}$$

The problem is to show that $Y > Z$. We have

$$g_{F+1}(Y) = g_F(Y) + \binom{N}{F+1} Y^{F+1}(1-Y)^{N-F-1} = 1 - \gamma \tag{8A.4}$$

or

$$g_F(Y) = 1 - \gamma - \binom{N}{F+1} Y^{F+1}(1-Y)^{N-F-1} \tag{8A.5}$$

The right side of Eq. (8A.5) must be positive and less than $1 - \gamma$. If we assume $Y \leq Z$, then $g_F(Y) \geq g_F(Z)$, since $g_F(x)$ is nonincreasing in x (strictly decreasing for $0 < x < 1$). But the last inequality is contradicted by Eqs. (8A.2) and (8A.5) and the remark immediately following (8A.5). Thus $Y > Z$.

Now to prove the main theorem, suppose that p (the unknown probability of failure) lies in the interval $[C(F), C(F+1)]$; i.e., $C(F) \leq p \leq C(F+1)$, where $F = 0, 1, \cdots, N - 1$.

Let X denote the number of failures observed in N trials. It will now be shown that

$$P[C(X) \geq p] \geq \gamma \tag{8A.6}$$

To prove inequality (8A.6), we note that the event $X = F + 1$ or $F + 2$ or $\cdots$ or N is equivalent to the event $C(X) = C(F+1)$ or

RELIABILITY ESTIMATION: PART II   211

$C(F+2)$ or $\cdots$ or $C(N)$, which in turn implies the event $C(X) \geq p$. Thus (see example preceding Sec. 5.3.1)

$$P[X = F + 1 \text{ or } \cdots \text{ or } N] \leq P[C(X) \geq p] \qquad (8A.7)$$

The left side of (8A.7) can be written as

$$1 - \sum_{j=0}^{F} \binom{N}{j} p^j (1-p)^{N-j} \qquad (8A.8)$$

However, the latter function is increasing in p. Thus if we replace p by $C(F)$, expression (8A.8) is not increased. Therefore, using (8A.7)

$$1 - \sum_{j=0}^{F} \binom{N}{j} [C(F)]^j [1 - C(F)]^{N-j} \leq P[C(X) \geq p] \qquad (8A.9)$$

The left side of (8A.9) is equal to γ independently of the value of F, from (8A.1); hence the desired inequality (8A.6) follows. One step remains, since we have not yet considered the case when $0 \leq p \leq C(0)$. In this case, however, we immediately have

$$P[C(X) \geq p] = 1 \geq \gamma \qquad (8A.10)$$

Thus inequality (8A.6) is true independently of the true value of p, which concludes the proof.

Note that since reliability $R \equiv 1 - p$; then the numbers $C(F)$ defined by Eq. (8A.1) can be used to determine *lower* confidence limits for R. The event $C(X) \geq p$ is equivalent to the event $1 - C(X) \leq R$; thus the two events have the same probability. Hence

$$P[1 - C(X) \leq R] \geq \gamma \qquad (8A.11)$$

which states that the numbers $1 - C(F)$ are lower confidence limits for R (confidence coefficient γ).

APPENDIX 8B
EXACT ONE-SIDED BINOMIAL CONFIDENCE LIMITS

8B.1. INTRODUCTION

A method for computing exact upper confidence limits to a binomial parameter is shown in this appendix. The result is that upper confidence limits can be constructed with *exactly* confidence γ, instead of with confidence *at least* γ (where $\gamma = 0.90$, say). The latter confidence limits are given by the methods of Appendix 8A of this chapter. For either case, N independent trials are made, each with the same probability p that "failure" occurs. The number of failures F is noted. In order to find the inexact limits one then computes or finds in Figs. A.1–A.5 (pp. 556-560) the upper confidence limit $\bar{C}(F; \gamma)$, then states

$$\bar{C}(F; \gamma) \geq p \qquad (8B.1)*$$

The *a priori* probability that the statement is correct is *at least* equal to γ. In order to find the exact limits, an *additional sampling* is undertaken after F failures out of N trials are observed. This is accomplished by selecting a number r from a uniformly distributed population where $0 \leq r \leq 1$;† then one computes (or looks up in a table) $C(F + r; \gamma)$ and states that

$$C(F + r; \gamma) \geq p \qquad (8B.2)$$

The *a priori* probability that statement (8B.2) is correct is *exactly* equal to γ. Intuitively, one can see that C will always be equal to or less than $\bar{C}$, since statement (8B.2) utilizes no more "confidence" than necessary.

The basic theory will be given in this appendix, together with a chart that illustrates a simple case where $N = 2$; $\gamma = 0.50$ and $\gamma = 0.90$. For large sample sizes it makes very little difference which method is used. The method given here shows that improved inferences can be made for small samples.

---
* We are using $\bar{C}$ in this appendix to denote what was called C in appendix 8A.
† One can use a table of (say) three-digit random numbers.

8B.2. THEORY

Let X denote the number of failures observed in N independent trials with constant probability p of observing a failure in any single trial. Let r denote a random variable with a uniform distribution in the interval $(0, 1)$; i.e.,

$$P(r \leq z) = 0, \quad z < 0$$
$$= z, \quad 0 \leq z < 1 \quad (8B.3)$$
$$= 1, \quad z \geq 1$$

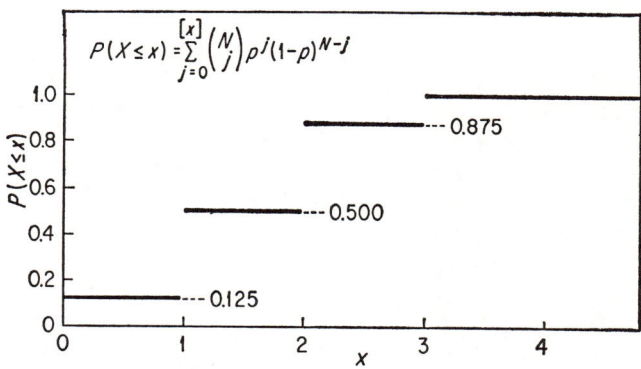

Fig. 8B.1 $P(X \leq x); p = 1/2, N = 3$.

It is not difficult to show that the cumulative distribution function for $X + r$ is given by:

$$P(X + r \leq x) = \sum_{j=0}^{[x-1]} \binom{N}{j} p^j (1-p)^{N-j} + \binom{N}{[x]} p^{[x]} (1-p)^{N-[x]} (x - [x]) \quad (8B.4)*$$

where $0 \leq x \leq N + 1$.

Figures 8B.1 and 8B.2 illustrate the difference in appearance between the distribution function of X and that of $X + r$.

Now we define a function $C(x)$ by the equation

$$\sum_{j=0}^{[x-1]} \binom{N}{j} C^j (1-C)^{N-j} + \binom{N}{[x]} C^{[x]} (1-C)^{N-[x]} (x - [x]) = 1 - \gamma$$

$$(8B.5)$$

where $0 < \gamma < 1$. It turns out that $C(x)$ is well defined by Eq. (8B.5) only for $1 - \gamma < x < N + 1 - \gamma$; it will be arbitrarily defined equal to

* $[x]$ = greatest integer $\leq x$; e.g., $[1.43] = 1$, $[2] = 2$, etc.

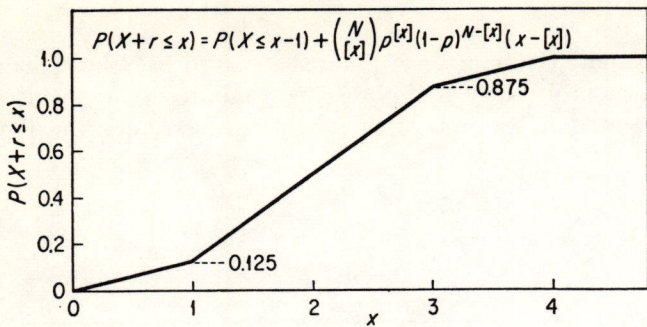

Fig. 8B.2 $P(X + r \leq x)$; $p = 1/2$, $N = 3$.

zero for $0 \leq x \leq 1 - \gamma$ and unity for $N + 1 - \gamma \leq x \leq N + 1$. To see why this is so, first consider Eq. (8B.5) when $1 - \gamma < x < N + 1 - \gamma$. The left side of Eq. (8B.5) when considered as a function of C is then greater than $1 - \gamma$ when $C = 0$; is zero when $C = 1$; and decreases when C increases from zero to unity. Accepting the fact that the left side of Eq. (8B.5) is a decreasing function of C for the moment, this means that Eq. (8B.5) has a unique root C such that $0 \leq C \leq 1$, for each x in the interval $1 - \gamma < x < N + 1 - \gamma$.

To prove that the left side of Eq. (8B.5) decreases, differentiate it with respect to C (note that all but one of the terms of the sum cancel when it is differentiated), obtaining:

$$\frac{d}{dC} = -\binom{N}{[x-1]}(N - [x-1])C^{[x-1]}(1 - C)^{N-[x]}$$

$$+ \binom{N}{[x]}[x]C^{[x-1]}(1 - C)^{N-[x]}(x - [x]) \quad (8B.6)$$

$$- \binom{N}{[x]}(N - [x])C^{[x]}(1 - C)^{N-[x]-1}(x - [x])$$

$$= \binom{N}{[x]}[x]C^{[x-1]}(1 - C)^{N-[x]}(x - [x] - 1)$$

$$- \binom{N}{[x]}(N - [x])C^{[x]}(1 - C)^{N-[x]-1}(x - [x]) \quad (8B.7)$$

< 0, since $(x - [x] - 1) < 0$. Note that the identity

$$\binom{N}{f-1}(N - f + 1) \equiv \binom{N}{f}f$$

RELIABILITY ESTIMATION: PART II

is used to obtain (8B.7) from (8B.6). This proves the desired result. When $0 \leq x \leq 1 - \gamma$ and $N + 1 - \gamma \leq x \leq N + 1$, the difference between the left and right sides of Eq. (8B.5) vanishe only at $x = 1 - \gamma$ and $x = N + 1 = \gamma$; so we arbitrarily define $C(x) \equiv 0$ for $0 \leq x \leq 1 - \gamma$ and $C(x) \equiv 1$ for $N + 1 - \gamma \leq x \leq N + 1$.

Now, it is asserted that $C(x)$ is nondecreasing in x (strictly increasing for $1 - \gamma < x < N + 1 - \gamma$). To show this, we use the previous result that the derivative of the left side of Eq. (8B.5) is < 0. Thus differentiating Eq. (8B.5) now with respect to x, we have

$$(<0) \frac{dC}{dx} + \binom{N}{[x]} (C(x))^{[x]} (1 - C(x))^{N-[x]} = 0 \quad (8B.8)$$

where (<0) denotes the right side of Eq. (8B.7). Thus

$$dC/dx > 0$$

since the second term on the left side of Eq. (8B.8) is positive, which proves the assertion.

Now, to prove the main result, that the function $C(x)$ gives upper confidence limits for p, with confidence exactly γ, let p be any number between 0 and 1. Then $p = C(x)$ for some x in the interval $1 - \gamma \leq x \leq N + 1 - \gamma$.

The event $X + r \geq x$ is equivalent to the event $C(X + r) \geq C(x)$; i.e., $C(X + r) \geq p$. Hence

$$P(X + r \geq x) = P[C(X + r) \geq p] \quad (8B.9)$$

Now,

$$P(X + r \geq x) = 1 - P(X + r \leq x)$$

$$= 1 - \sum_{j=0}^{[x-1]} \binom{N}{j} p^j (1-p)^{N-j}$$

$$+ \binom{N}{[x]} p^{[x]} (1-p)^{N-[x]} (x - [x]) \quad (8B.10)$$

and since $p = C(x)$, we see from Eq. (8B.4) that

$$P(X + r \geq x) = 1 - (1 - \gamma) = \gamma$$

Thus, from Eq. (8B.9),

$$P[C(X + r) \geq p] = \gamma \quad (8B.11)$$

The result given by Eq. (8B.11) shows that whatever the true value of p, the upper confidence limit calculated from Eq. (8B.4) will "cover" p with

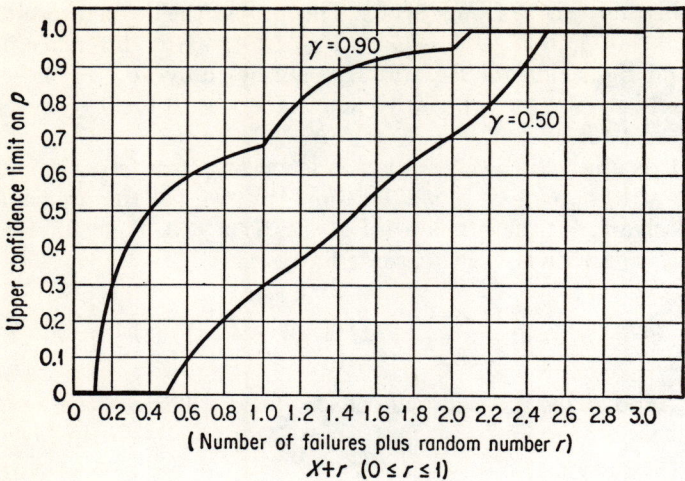

Fig. 8.3 Exact upper confidence limits on a binomial parameter p sample size $N = 2$, confidence coefficient $\gamma = 0.50$ and 0.90.

exactly probability γ. (If it turns out that $X + r \leq 1 - \gamma$ or $\geq N + 1 - \gamma$, then $C(X + r) \equiv 0$ or 1, respectively.)

Figure 8B.3 presents a plot of $C(x)$ versus x for $N = 2$; $\gamma = 0.50$ and 0.90. Reference 7 contains tables of *two-sided* exact confidence limits for $\gamma = 0.95$ and $\gamma = 0.99$ and $N = 2(1)24(2)50$.

APPENDIX 8C

UPPER CONFIDENCE LIMITS ON p FOR NEGATIVE BINOMIAL SAMPLING

In Sec. 6.4 the sample space was defined by c failures, n trials, where c is chosen in advance and the number of binomial trials is observed until c failures occur. For this sample space one can prove, in a similar manner to that for the case when the number of trials is fixed in advance, that upper confidence limits on p with confidence coefficient γ can be found by solving

$$\sum_{j=n}^{\infty} \binom{j-1}{c-1} C^c (1-C)^{j-c} = 1 - \gamma \qquad (8C.1)$$

for C. We can, however, use the results of Appendix 8A obtained for binomial trials with prechosen sample size by first writing (8C.1) as

$$1 - \sum_{j=n}^{\infty} \binom{j-1}{c-1} C^c (1-C)^{j-c} = \gamma \qquad (8C.2)$$

or

$$\sum_{j=c}^{n-1} \binom{j-1}{c-1} C^c (1-C)^{j-c} = \gamma \qquad (8C.3)$$

However, by Eq. (6.56), (8C.3) becomes

$$\sum_{j=c}^{n-1} \binom{n-1}{j} C^j (1-C)^{n-1-j} = \gamma \qquad (8C.4)$$

or

$$\sum_{j=0}^{c-1} \binom{n-1}{j} C^j (1-C)^{n-1-j} = 1 - \gamma \qquad (8C.5)$$

Now we see by Eq. (8A.1) that Figs. A.1–A.5 (pp. 556-560) can be used to obtain the upper confidence limit C if we identify $n - 1 \equiv N$, $c - 1 \equiv F$.

As an example, let $c = 2$. Suppose that the second failure occurs on the tenth trial. The upper confidence limit C, on p, with confidence coefficient $\gamma = 0.90$, from Fig. A.3, where $F = 1$, $N = 9$, is $C = 0.37$.

When $c = 1$ and a failure occurs on the first trial ($n = 1$), then we define $C = 1.00$.

APPENDIX 8D
UPPER CONFIDENCE LIMITS ON λ FOR THE POISSON DISTRIBUTION

When the Poisson distribution model applies (Sec. 6.5), we can obtain upper confidence limits on the parameter λ in a similar way to those given for binomial trials. Suppose the outcome of *one observation* is n events; e.g., n failures, where n is zero or a positive integer. Then upper confidence limits C on λ, with confidence coefficient γ, can be found by solving

$$\sum_{k=0}^{n} \frac{C^k e^{-C}}{k!} = 1 - \gamma \tag{8D.1}$$

Molina's tables (Ref. 4, Table II) can then be used since in Molina's notation

$$P(c, a) = \sum_{x=c}^{\infty} \frac{a^x e^{-a}}{x!} \tag{8D.2}$$

Since $$\sum_{k=0}^{n} \frac{C^k e^{-C}}{k!} = 1 - \sum_{k=n+1}^{\infty} \frac{C^k e^{-C}}{k!} = 1 - \gamma$$

we equate $a \equiv C$ and $c \equiv n + 1$, then find the value of a (interpolating as necessary) corresponding to $P(c, a) = \gamma$ in the body of the table.

As an example, if $\gamma = 0.90$ and $n = 5$, then the referenced tables yield $C = 9.275$. We can also use Figs. A.1–A.5 (pp. 556-560) to obtain C by recalling that the binomial distribution approaches the Poisson distribution when N, the sample size, gets large, and p is small such that $Np \sim \lambda$. Thus to obtain C for this example, we would use Fig. A.3 for $F = 5$ and read off the numerical value of the abscissa for N as large as possible; in this case the largest $N = 1000$. This yields an upper confidence limit on p if N were equal to 1000, namely 0.0093, approximately. Since $\lambda \sim Np$, then the upper confidence limit on λ would be approximately $1000(0.0093) = 9.3$ which agrees fairly closely with the value obtained from Molina's tables. In general, Figs. A.1–A.5 should not be used for this purpose if n is, say, larger than 5.

Now suppose K observations are made, yielding $n_1, n_2, \cdots, n_K$ events, respectively. Then the proper procedure is to solve the equation

$$\sum_{k=0}^{\sum_{1}^{K} n_j} \frac{(KC)^k e^{-KC}}{k!} = 1 - \gamma \qquad (8D.3)$$

since we know that

$$\sum_{1}^{K} n_j$$

has a Poisson distribution with parameter $K\lambda$. As an example we take the data of the example in Sec. 6.5. We have $K = 10$, $\sum n_j = 30$. If we choose $\gamma = 0.90$, then the upper confidence limit on λ can be obtained from Ref. 4 as $C = 3.83$. We might suppose that instead of this data we had obtained (say) $\sum n_j = 15$ and $K = 5$. The unbiased estimate of λ would still be $\hat{\lambda} = 3$, but the upper confidence limit with confidence coefficient 0.90 would be $C = 4.26$, which indicates, as expected, less precision than for the larger sample of data.

When $\sum n_j = 85$ or greater, and the confidence coefficient $\gamma \geq 0.90$, table II of Ref. 4 does not suffice for our purpose. It is not difficult to show, however, that

$$C = \frac{\chi^2_{2(f+1);1-\gamma}}{2K} \qquad (8D.4)$$

gives the correct upper confidence limit on λ, where $f = \sum n_j$; and therefore Table A.8, pp. 552-3 may be used.

EXERCISE: Show that Eq. (8D.4) is correct. *Hint:* see Sec. 6.8.2 and Eq. (7.13). Check the above examples.

EXERCISE: Derive the following approximate formula for C:

$$C \simeq \frac{f+1}{K} \left[1 + \frac{K_{1-\gamma}}{3\sqrt{f+1}} - \frac{1}{9(f+1)} \right]^3 \qquad (8D.5)$$

where $K_{1-\gamma}$ is the normal deviate exceeded with probability $1 - \gamma$. *Hint:* See bottom p. 202.

CHAPTER NINE
RELIABILITY STRUCTURE MODELS

9.1. INTRODUCTION

The reliability models discussed in this chapter are concerned with the relationship between the components of a system and their effect on the performance of that system. This relationship consists of two topics; we must consider, first, the output or performance of an individual component, and second, the functional relationship between the totality of components which together form the system's "structure."

The output of a component may be observed as attribute data or variables data. In the case of the attribute data we are concerned only with the result of the operation as success (S), or its complement, failure (F). However, a variable output might be used in several ways. First, we might observe whether the output falls inside or outside given limits, such as specification limits, and thus classify the test as a success or failure accordingly; i.e., using the output simply as attribute data. Here the output might be a single parameter which is indicative of reliability or it might consist of several parameters, some of which might be functionally related but no one of which could be completely identified with reliability. In this case the event, success, could be said to have occurred when each and all of the parameters had been evaluated and together found to satisfy the requirements for success. Second, we might use several observations of the output and, without classifying the individual observations as success or failure, use *variables analysis* to determine the probability of the output falling within certain limits (again, these limits might be specifications). Third, although each individual component might not have any output limits *per se*, the composite contribution of all the components' outputs, resulting in the system's output, would have limits of its own within which to perform. In turn the system's output might be attribute or variables data and analyzed accordingly.

The functional relationship of the system organizes the outputs of the components so that they will perform in a certain specified manner and order. The system might be designed so that for satisfactory operation *all* the components must themselves necessarily function satisfactorily. A system of this type is called a *serial system*. On the other hand, the system might be designed so that in the event of failure of any one of the components another is already available within the system to take over the operation. Systems which contain this type of provision are called *parallel systems*. Systems will frequently include both serial and parallel relationships. We are concerned with expressing the operation of systems composed of such functional structures in terms of probability relationships, when the components' performance is measured either as attribute or variables data.

The importance of understanding the reliability structure is that it permits us to recognize the weaknesses or potential weaknesses of the system from a reliability viewpoint. More important, it also provides us with knowledge and techniques enabling us to eliminate many unreliable areas. For example, we can compute how many redundant components we might need as back-ups to provide a system with a required level of reliability. Again, we can determine the probable reliability of a system composed of components with known or expected reliabilities. Or, inversely, we can establish how the system's reliability requirements can be apportioned into component reliability requirements and thereby determine the allocation of development time and funds to attain these goals. The techniques presented will also illustrate how we can determine reliability safety margins when there are fixed limits for successful operation (e.g., specification limits) or variable limits consisting of the neighboring component's variable input capability. This analysis is frequently referred to as *reliability stress versus strength analysis*. All of these studies provide an important contribution to the techniques of original and continuing reliability design review.

9.2. SERIAL SYSTEMS

Consider a *system* which consists of two or more parts, called *subsystems*. One of the simplest types of failure models is constructed by making use of the following assumptions: Let S denote the event that the system is successful. Let S_j, $j = 1, \cdots, n$, denote the events that the subsystems are successful. Assume that S occurs when, and only when, all the S_j occur; i.e., the system is successful if, and only if, all of the subsystems are successful. When this is the situation, we say that the system is *serial*. We express this assumption, using the ∩ notation given in Sec.

5.2.1, as
$$S = S_1 \cap S_2 \cap \cdots \cap S_n \tag{9.1}$$
and therefore
$$P(S) = P(S_1 \cap S_2 \cap \cdots \cap S_n) \tag{9.2}$$

Equation (9.2) can also be written (using the conditional probability relations of Sec. 5.3.1) as

$$P(S) = P(S_1 \mid S_2 \cap \cdots \cap S_n) P(S_2 \cap \cdots \cap S_n) \tag{9.3}$$
$$= P(S_1 \mid S_2 \cap \cdots \cap S_n) P(S_2 \mid S_3 \cap \cdots \cap S_n) P(S_3 \cap \cdots \cap S_n) \tag{9.4}$$

and in general

$$P(S) = \left[\prod_{j=1}^{n-1} P(S_j \mid S_{j+1} \cap \cdots \cap S_n) \right] P(S_n) \tag{9.5}$$

9.2.1 INDEPENDENT SERIAL SYSTEMS

If it is now assumed that the S_j are *mutually independent*, i.e., probability of success of the jth subsystem is the same irrespective of occurrence of any combination of events of the *other* subsystems, which implies that

$$P(S_j \mid S_{j+1} \cap \cdots \cap S_n) = P(S_j), \quad j = 1, 2, \cdots, n-1$$

then Eq. (9.5) takes the form

$$P(S) = \prod_{j=1}^{n} P(S_j) \tag{9.6}$$

or, denoting $P(S) \equiv R$, $P(S_j) \equiv R_j$, we obtain

$$R = \prod_{j=1}^{n} R_j \tag{9.7}$$

where R denotes "Reliability." Thus the over-all reliability of the system is the product of the reliabilities of the subsystems. This is the so-called "product rule" for *independent serial* systems.

When the assumption of independence is not met but the system is serial, then any of the relations (9.2)–(9.5) determine the reliability. We note that the subscripts in (9.1)–(9.5) may be permuted. For example, if $n = 2$,

$$P(S) = [P(S_1 \mid S_2)] P(S_2) \tag{9.8}$$

corresponding to Eq. (9.5), may be also written as

$$P(S) = [P(S_2 \mid S_1)] P(S_1) \tag{9.9}$$

EXERCISE: If $P(S_2 \mid S_1) < P(S_2)$, show that $P(S_1 \mid S_2) < P(S_1)$, and also that $P(F_2 \mid F_1) < P(F_2)$, where $F_j \equiv \bar{S}_j$.

EXERCISE: Define two random variables X_1 and X_2 such that $X_1 = 0(1)$ if subsystem no. 1 is a success (failure) and similarly $X_2 = 0(1)$ if subsystem no. 2 is a success (failure). Denote $P(X_1 = 1)$ by $p_1 \equiv 1 - R_1$ and $P(X_2 = 1)$ by $p_2 \equiv 1 - R_2$. Suppose $P(S_2 \mid S_1) = \delta$. What is the correlation coefficient of (X_1, X_2)?

Answer:

$$\rho(X_1, X_2) = \frac{R_1(\delta - R_2)}{\sqrt{R_1 R_2 (1 - R_1)(1 - R_2)}} \tag{9.10}$$

Assume that the system is serial. What is an expression for the reliability of the system $R = P(X_1 = 0, X_2 = 0)$* in terms of ρ?

Answer:

$$R = R_1 R_2 + \rho \sqrt{R_1 R_2 (1 - R_1)(1 - R_2)} \tag{9.11}$$

EXERCISE: Prove that

$$R_1 R_2 - \tfrac{1}{4} \le R \le R_1 R_2 + \tfrac{1}{4} \tag{9.12}$$

Thus, show that a serial system of two subsystems can never have reliability greater than $R_1 R_2 + \tfrac{1}{4}$ and for this upper limit to be attained, it is necessary that $R_1 \equiv R_2 = 0.5$.

EXAMPLE: Prove that

$$\min(R_1, R_2) \le R_1 R_2 + \tfrac{1}{4} \tag{9.13}$$

Suppose $R_1 = \min(R_1, R_2)$; i.e., $R_1 \le R_2$. We must prove $R_1 \le R_1 R_2 + \tfrac{1}{4}$ or, equivalently,

$$R_1(1 - R_2) \le \tfrac{1}{4} \tag{9.14}$$

But, for *any* R_2 such that $0 \le R_2 \le 1$,

$$R_2(1 - R_2) \le \tfrac{1}{4} \tag{9.15}$$

Since $R_1 \le R_2$, the lefthand side of (9.15) is not increased if we replace the first factor R_2 by R_1 and thus (9.14) is proved.

EXERCISE: The reliability of a serial system consisting of two subsystems is not greater than the minimum of the reliabilities of the two subsystems. *Hint:* With the preceding formulation we have $R = \delta R_1$. If we attempt to make δR_1 as large as possible we would choose δ equal to its maximum, one. However, since $P(X_1 = 0 \mid X_2 = 0) = \delta R_1 / R_2 \le 1$, then $\delta = 1$ implies $R_1 \le R_2$. Thus $R_1 = \min(R_1, R_2)$ and therefore $R \le \min(R_1, R_2)$.

EXERCISE: Use the fact that

$$S_1 \cap S_2 \cap \cdots \cap S_n \subset S_j, \quad j = 1, 2, \cdots, n$$

* Note that we are using a comma in place of $\cap$ (*cf.* Sec. 5.4).

to show that the reliability of any serial system is not greater than the minimum of the reliabilities of its subsystems.

In Sec. 9.3 we will consider the *"weakest-link"* model which is a special case of a *dependent serial system*. In this model the maximum possible serial system reliability is attained, i.e., the system reliability is the minimum of the subsystem reliabilities.

9.2.1.1. An Example of a Serial System.
Perhaps the best example of a serial system is that of a complex missile consisting of several stages. Here there are groups of subsystems and components, some of which operate concurrently, others sequentially, and all of which must be successful before the mission can be successfully accomplished. Thus the first stage must function satisfactorily to be followed by successful stage separation and then by successful ignition and firing of the second-stage engine, etc. Although these are sequential operations, the idea of a serial system must not be identified uniquely with a time sequence of events. For example, the first stage itself is a serial system. If it consists of several booster engines, then each engine as well as the guidance and control subsystems must function concurrently and successfully for stage success. Generally if the missile is "broken down" into a limited number of major subsystems and these subsystems can be regarded as being independent (in a statistical sense), then the reliability of the missile is the product of the reliabilities of these major components.

9.2.2 LOWER CONFIDENCE LIMITS ON RELIABILITY OF INDEPENDENT SERIAL SYSTEMS

The problem of determining lower confidence limits on the reliability R of an independent serial system is equivalent to that of determining upper confidence limits on probability of failure p (unreliability) as shown in Sec. 8.1. In keeping with the discussion in Appendix 8A, we will present the latter formulation of the problem.

The problem can be expressed as follows: Find a set of numbers $C_{N_1 \cdots N_k}(f_1, \cdots, f_k; \gamma)$ based upon the results of N_j tests and f_j failures on the jth subsystem ($j = 1, 2, \cdots, k$) such that

$$P[p \leq C_{N_1 \cdots N_k}(f_1, \cdots, f_k; \gamma)] \geq \gamma \tag{9.16}$$

where p is the true but unknown probability of failure of the system and γ is the confidence coefficient for the statement that C is an upper confidence limit on p. When the system is *independent serial*, (9.16) becomes

$$P\left[1 - \prod_{j=1}^{k} (1 - p_j) \leq C_{N_1 \cdots N_k}(f_1, \cdots, f_k; \gamma)\right] \geq \gamma \tag{9.17}$$

where p_j is the true but unknown unreliability of the jth subsystem.

One method of constructing a set of numbers $C^*_{N_1 \cdots N_k}(f_1, \cdots, f_k; \gamma)$ such that (9.17) is satisfied is to set

$$C^*_{N_1 \cdots N_k}(f_1, \cdots, f_k; \gamma) = 1 - \prod_{j=1}^{k} [1 - C_{N_j}(f_j; \gamma_j)] \qquad (9.18)$$

where

$$\prod_{j=1}^{k} \gamma_j = \gamma \qquad (9.19)$$

and the number $C_{N_j}(f_j; \gamma_j)$ is an upper confidence limit with at least confidence γ_j on the unreliability of the jth subsystem, e.g., determined by the methods of Appendixes 8A, 8B, or 8C of Chap. 8, or by any other method. In order to show that the numbers C^* defined by Eqs. (9.18) and (9.19) actually do satisfy (9.17), we must evidently show that

$$P\left\{\prod_{j=1}^{k}(1 - p_j) \geq \prod_{j=1}^{k}[1 - C_{N_j}(f_j; \gamma_j)]\right\} \geq \prod_{j=1}^{k} \gamma_j \qquad (9.20)$$

Now, the event $\bigcap_{j} \{1 - p_j \geq 1 - C_{N_j}(f_j; \gamma_j)\}$ is contained in the event

$$\left\{\prod_{j=1}^{k}(1 - p_j) \geq \prod_{j=1}^{k}[(1 - C_{N_j}(f_j; \gamma_j))]\right\}$$

(For $k = 2$, the above statement is the same as:

$$1 - p_1 \geq 1 - C_1 \text{ and } 1 - p_2 \geq 1 - C_2$$

imply

$$(1 - p_1)(1 - p_2) \geq (1 - C_1)(1 - C_2)$$

but not necessarily conversely). Thus, the first event has a probability not greater than the second. Hence:

$$P\left\{\bigcap_{j}[1 - p_j \geq 1 - C_{N_j}(f_j; \gamma_j)]\right\}$$

$$\leq P\left\{\prod_{j=1}^{k}(1 - p_j) \geq \prod_{j=1}^{k}[1 - C_{N_j}(f_j; \gamma_j)]\right\} \qquad (9.21)$$

Now, it is assumed that the experiments on different subsystems give independent outcomes; therefore, the probability of the joint event on the lefthand side of Eq. (9.21) is the product of the individual probabilities, or

$$P\left\{\bigcap_{j}[1 - p_j \geq 1 - C_{N_j}(f_j; \gamma_j)]\right\} = \prod_{j=1}^{k} P[1 - p_j \geq 1 - C_{N_j}(f_j; \gamma_j)]$$

$$(9.22)$$

But the righthand side of (9.22) is not less than

$$\prod_{j=1}^{k} \gamma_j.$$

The desired result (9.20) then follows from (9.21).

EXAMPLE: Let $k = 2, \gamma = 0.90, N_1 = 50, f_1 = 2; N_2 = 100, f_2 = 3$ (N_1 and N_2 fixed in advance of sampling); then

$$C^*_{50,100}(2, 3; 0.90) = 1 - [1 - C_{50}(2; \gamma_1)][1 - C_{100}(3; \gamma_2)]$$

where γ_1 and γ_2 are any positive numbers such that $\gamma_1 \gamma_2 = 0.90$. Thus $\gamma_1 = \gamma_2 = (0.90)^{1/2} = 0.9487$ will satisfy this requirement. From Figs. A.3 and A.4 (pp. 558, 559) (by interpolation in γ) or from Ref. 1, we find

$$C_{50}(2; 0.9487) = 0.1200$$
$$C_{100}(3; 0.9487) = 0.0761$$

Thus $\qquad C^*_{50,100}(2, 3; 0.90) = 0.187$

and a lower confidence limit on reliability of the system, with confidence coefficient 0.90, is therefore 0.813. Note that the choice of $\gamma_1 = \gamma_2$ was somewhat arbitrary. It is possible that a slightly lower value of C^* could be achieved with a different set of γ_1, γ_2 such that $\gamma_1 \gamma_2 = 0.90$. The reader should consider what happens if one chooses $\gamma_1 = 1.0, \gamma_2 = 0.90$, or vice versa.

EXERCISE: An independent serial system consists of two A subsystems and two B subsystems (the A's have the same part number, as do the B's). Subsystem A has been tested in 50 independent tests with two failures, and subsystem B in 100 independent tests with three failures. What is a lower 90 per cent confidence limit on the reliability of the system?

Answer: $\qquad \hat{R}_L = (0.813)^2 = 0.66$

9.2.3 A PREFERRED ALTERNATE METHOD OF COMPUTING SERIAL SYSTEM CONFIDENCE LIMITS

The preceding method is quite general, in the sense that it can be applied in the cases where the numbers of tests per subsystem are different, and the subsystem confidence limits can be based upon any kind of sampling; e.g., binomial sampling with sample size fixed in advance or number of failures fixed in advance (Sec. 6.4). However, the system of lower confidence limits on reliability obtained by this method can be considerably improved by using a different method developed by Buehler (Ref. 2) and Steck (Ref. 3). The detailed proof that the method yields better confidence limits is fairly complicated and is not given here; however, a simple alternative method, suggested by D. L. Lindstrom and J. H. Madden, was found by the authors to yield almost identical confidence

RELIABILITY STRUCTURE MODELS   227

limits to those obtained by the method of Refs. 2 and 3 when the N_j are equal.* The method is as follows:
Compute

$$\hat{R} = \prod_{j=1}^{k} \frac{N - f_j}{N} \qquad (9.23)$$

Thus $\hat{R}$ is a point estimate of the independent serial system reliability. The quantity $N(1 - \hat{R})$ is then considered to represent the number of *system* failures F in N trials of the *system*. Then by Figs. A.1–A.5 (pp. 556-560) determine the lower confidence limit for any chosen confidence coefficient γ. The quantity $N(1 - \hat{R})$ will generally not be an integer; however the desired result can be found by interpolation.

When the N_j are not equal, compute

$$\hat{R} = \prod_{j=1}^{k} \frac{N_j - f_j}{N_j} \qquad (9.24)$$

and the quantity $N_m(1 - \hat{R})$, where N_m is the *minimum* of the N_j. The number $N_m(1 - \hat{R})$ is then considered to be the number of system failures in N_m trials of the system. From this point the method of obtaining a lower confidence limit on system reliability is the same as before.

EXAMPLE: Let $N_1 = N_2 = 100, f_1 = 2, f_2 = 3$. Then $\hat{R} = (0.98)(0.97) = 0.9506$. Thus $F = 4.94$, and we find from Fig. A.3, p. 558, a 90 per cent lower confidence limit (based on $N = 100$, $F = 4.94$) of $1 - 0.090 = 0.910$. If $N_1 = 50$ instead of 100, then $\hat{R} = (0.96)(0.97) = 0.9312$ and we have $F = 50(1 - 0.9312) = 3.44$. Thus, from Fig. A.3 we find a lower 90 per cent confidence limit (based on $N = 50$, $F = 3.44$) of $1 - 0.140 = 0.860$ (compare this value with 0.813 obtained by the previous method).

Intuitively, one can see that the simplified method should yield valid confidence limits on system reliability by the following arguments. We will assume $k = 2$; however, analogous reasoning holds for any value of k. First let $N_1 = N_2 = N$. The two subsystems are tested separately with f_1 and f_2 observed failures. The N test results (f_1 failures and $N - f_1$ successes) for one subsystem can then be paired with the N results for the second subsystem (f_2 failures, $N - f_2$ successes) to form a set of N pairs, each pair representing the outcome of a "system" test. The pairing would be done by "random" selection of each of the subsystem results. Now, with a little thought, it can be seen that the quantity $N(1 - \hat{R})$ is then the *expected* number of pairs in which at least one member of the pair is a

---

* The authors made numerical comparisons for $k = 2, 3$ only, because of the cost and difficulty of calculating confidence limits by the methods of Refs. 2 and 3. However, it is conjectured that the described simple alternate method will give a valid system of confidence limits for higher values of k and unequal N_j. A rather extensive Monte Carlo study would be necessary, however, to gain confidence in the alternate method.

failure. This number (not necessarily an integer) is then treated as if it were the observed number of serial system failures in N system trials.

When $N_1 < N_2$ (say), the procedure states that the smaller number N_1 should be used as the equivalent number of system trials. This is logical in that only N_1 separate system results can be formed by pairing off the subsystem results. Again $N_1(1 - \hat{R})$ is the expected number of system failures.

For example let $k = 2$, $N_1 = 2$, $N_2 = 3$, $f_1 = 1$, $f_2 = 2$; i.e., there are two different subsystems, A, B, of which two trials were made on subsystem A with one failure, and three trials on subsystem B with two failures. Label the two trials of subsystem A as A_1 and A_2 and the three trials of subsystem B as B_1, B_2, B_3. The possible sets of two *system* trials are exhibited below.

$$\begin{pmatrix} A_1B_1 \\ A_2B_2 \end{pmatrix} \quad \begin{pmatrix} A_1B_2 \\ A_2B_1 \end{pmatrix}$$

$$\begin{pmatrix} A_1B_1 \\ A_2B_3 \end{pmatrix} \quad \begin{pmatrix} A_1B_3 \\ A_2B_1 \end{pmatrix}$$

$$\begin{pmatrix} A_1B_2 \\ A_2B_3 \end{pmatrix} \quad \begin{pmatrix} A_1B_3 \\ A_2B_2 \end{pmatrix}$$

Thus

$$\begin{pmatrix} A_1B_1 \\ A_2B_2 \end{pmatrix}$$

represents two system trials in which the first system trial consists of the first trial of A combined with the first trial of B. The second system trial consists of the second trial of A combined with the second trial of B, and similarly for the remaining five possible sets of two system trials.

Let us now agree that A_1 was a failure and B_1 and B_2 were failures. Then the following results could have been obtained:

$$\begin{pmatrix} A_1B_1 \\ A_2B_2 \end{pmatrix} \equiv \begin{pmatrix} FF \\ SF \end{pmatrix} \equiv 2 \text{ system failures} \qquad \begin{pmatrix} A_1B_2 \\ A_2B_1 \end{pmatrix} \equiv \begin{pmatrix} FF \\ SF \end{pmatrix} \equiv 2 \text{ system failures}$$

$$\begin{pmatrix} A_1B_1 \\ A_2B_3 \end{pmatrix} \equiv \begin{pmatrix} FF \\ SS \end{pmatrix} \equiv 1 \text{ system failure} \qquad \begin{pmatrix} A_1B_3 \\ A_2B_1 \end{pmatrix} \equiv \begin{pmatrix} FS \\ SF \end{pmatrix} \equiv 2 \text{ system failures}$$

$$\begin{pmatrix} A_1B_2 \\ A_2B_3 \end{pmatrix} \equiv \begin{pmatrix} FF \\ SS \end{pmatrix} \equiv 1 \text{ system failure} \qquad \begin{pmatrix} A_1B_3 \\ A_2B_2 \end{pmatrix} \equiv \begin{pmatrix} FS \\ SF \end{pmatrix} \equiv 2 \text{ system failures}$$

The expected number of system failures in two system trials is then

$$F = \frac{2 + 1 + 1 + 2 + 2 + 2}{6} = \frac{10}{6}$$

i.e., "one and two-thirds failures" in two system trials. However, this result is directly obtained by finding

$$\hat{R} = \left(\frac{2-1}{2}\right)\left(\frac{3-2}{3}\right) = \left(\frac{1}{2}\right)\left(\frac{1}{3}\right) = \frac{1}{6}$$

then

$$F = N_m(1 - \hat{R}) = 2\left(1 - \frac{1}{6}\right) = 2\left(\frac{5}{6}\right) = \frac{10}{6}$$

EXERCISE: How should N_m be determined when some of the subsystems are repeated within the system? Check your answer by considering the system whose reliability function is $R = R_1^2$.

EXERCISE: Derive a formula, using a normal distribution approximation, for confidence limits on reliability of an independent serial system based on the same type of subsystem sampling results we have been discussing. Part of the answer is as follows:

$$\hat{R} = \prod_{j=1}^{k} \frac{S_j}{N_j}$$

where S_j is the number of successes in N_j trials of the jth subsystem ($S_j = N_j - f_j$, where f_j is the number of failures).

$$\text{Var } \hat{R} = \prod_{j=1}^{k} R_j^2 \left[\prod_{j=1}^{k} \left(1 + \frac{1 - R_j}{N_j R_j}\right) - 1 \right]$$

where

$$R_j = E\left(\frac{S_j}{N_j}\right)$$

Now apply the method of Sec. 7.4.4. Use the formula obtained to find a 90 per cent lower confidence limit on R for $k = 2, N_1 = 50, f_1 = 2, N_2 = 100, f_2 = 3$.

Answer: $\hat{R}_L = 0.891$.

Figures A.6–A.11, pp. 561-566, give lower confidence limits on independent serial system reliability for systems consisting of two or three subsystems; $N_1 = N_2 (= N_3) = N$, where $5 \leq N \leq 40$; confidence coefficient $\gamma = 0.50, 0.90, 0.95$; for various observed failure combinations (F_1, F_2) or (F_1, F_2, F_3). The values shown are based on the methods of Refs. 2 and 3 and were abstracted from Ref. 4.

9.3. THE "WEAKEST LINK" MODEL

This model is a special serial system model in which the reliability of the system is the maximum possible for a serial system, namely, the minimum of the reliabilities of the subsystems (Sec. 9.2). The relation for serial systems can be written as

$$F = F_1 \cup F_2 \cup \cdots \cup F_n \qquad (9.25)$$

which is obtained from Eq. (9.1) by taking the complement (Sec. 5.2.1) of both sides; i.e., system failure occurs when and only when any one or more of the subsystems fail. The additional assumption is now made that the event "system failure" implies that a "distinguished" subsystem fails. Let us label subsystem number one as the distinguished subsystem; thus the assumption is

$$F \subset F_1 \qquad (9.26)$$

However, since $F_1 \subset F$, the two events F, F_1 are equivalent, and therefore their probabilities are the same; i.e.,

$$P(F) = P(F_1) \qquad (9.27)$$

Since any of the F_j imply F, they must also imply F_1, thus

$$F_1 \supset F_j, \qquad j = 2, \cdots, n \qquad (9.28)$$

and therefore

$$P(F_1) \geq P(F_j), \qquad j = 2, \cdots, n \qquad (9.29)$$

The final result is therefore

$$P(F) = P(F_1) = \max_j P(F_j) \qquad (9.30)$$

or

$$R = R_1 = \min_j R_j \qquad (9.31)$$

The result given by (9.30) or (9.31) does not depend on which subsystem is chosen as the "distinguished" subsystem. It is evident also that the system must be a dependent serial system in that the probability that the distinguished subsystem fails, given that failure of any one or more of the other subsystems occurs, is unity.

If we now use the terminology "weakest link" instead of "distinguished subsystem," we see that the model represents a system with the failure properties of a "chain." A chain is made up of "links," one or possibly more than one being the "weakest" link; i.e., its "breaking strength" is the minimum of the strengths possessed by the links. Suppose that a stress is applied to the chain and equally to all the links. If the stress exceeds the strength of any one or more links, the chain fails. The weakest link is always included in the group of links that fail and any one or more links failing mean that the chain fails. Thus the chain represents a more concrete physical realization of the model considered above. Further properties of the chain model will be discussed in Sec. 9.3.1.

Consider an example of the weakest link model. The general cause of failure of a chain, i.e., one of its links breaks under the stress of *external* forces, leads us to look at the circumstances in which the chain would be the appropriate failure model for system failure. For example, the function of a satellite in orbit might be observing, measuring, and transmitting data considered "offensive" by the authorities of a power opposed to the originators of the satellite. One method of destroying the effectivity of the satellite would be to create an explosion, i.e., an instantaneous release of heat radiation in the satellite's approximate vicinity with the result that some of its vital observing, measuring, or transmitting equipment would be destroyed. If the explosion were close, then probably all the equipment would be destroyed; however, if it were sufficiently distant so that most of the "force" were dissipated, only the most susceptible or "weakest" equipment would fail. We see the analogy again of an external force surpassing the minimum of the inherent strengths of the various vital components. The reader will note that the intrinsic failure manifestations, i.e., initial, random, and wearout, discussed in Sec. 6.7 are not involved in this discussion—which of course is conditional that the system has survived until the time that the external force becomes critical. In general, the weakest link model needs to be considered in addition to intrinsic failure models if the system's operation exposes it to severe external forces.

9.3.1 THE CHAIN MODEL

The chain model discussed in this section is a particular case of the weakest link model of Sec. 9.3. Here it is assumed that a "chain" is made up of *identical* "links" in the sense that breaking strengths of all the links in the chain have a *fixed probability distribution*. The chain breaks, i.e., fails, when its weakest link fails. However, the latter occurrence is dependent only upon the variability of the link strengths based on the probability distribution of strengths, i.e., by a calculable chance the weakest link will have a strength of lesser magnitude than the stress applied to it.

Similarly, the stress to be applied to a given link is assumed to have a probability distribution. The probability that the link does not break is then the probability that its strength exceeds the applied stress.

When the links are assembled to form a chain, it is assumed that when a stress is applied to the chain as a whole, it is also applied equally to each of the links (visualize a weightless chain hanging vertically with a weight attached to the bottom of the chain).

The preceding considerations are by no means restricted to chains in an actual or ideal sense. For example, in a computer containing many thousands of electron tubes, one (or more) tubes may fail when a severe shock is applied to the whole system. Although the tubes are designed to be identical with respect to their ability to withstand a given shock, there will nevertheless be an inherent variability in this ability, due to random variations in the materials and manufacturing process.

The following paragraphs describe the model in more detail. First the reliability R_ℓ of a single link is determined. Based on the assumption that the strength of a chain is equal to that of its weakest link, the reliability R_n of an n-link chain is determined. It will then be shown that the chain reliability can range between the lower limit: reliability of a single link raised to the nth power, and the upper limit: reliability of a single link raised to the first power. The achievement of these limits or any reliability in between depends on the variabilities of the strength and stress distributions.

First let us consider a "link," whose important property is its "breaking strength"; i.e., this is a number associated with the link, such that if a "stress" of magnitude equal to or greater than this number is applied to the link, it fails. We assume that the breaking strength, or simply "strength" of a link is known only in terms of a probability distribution of strengths. We say that the population of link strengths has a density function $f(x)$, with associated (assumed continuous) distribution function $F(x)$, such that

$$F(b) - F(a) = \int_a^b f(x)\, dx \qquad (9.32)$$

is the probability that the link strength lies between the magnitudes a and b, where $a < b$. This quantity can be interpreted in the usual sense if we imagine the population of link strengths as actually being an infinite number of links, such that $F(b) - F(a)$ denotes the proportion of links with strengths between a and b. Similarly, we assume stress to be defined by a probability density function $g(y)$ and continuous distribution function $G(y)$, such that

$$G(d) - G(c) = \int_c^d g(y)\, dy \qquad (9.33)$$

is the probability that the stress lies between the magnitudes c and d where $d > c$. It is convenient to define positive random variables $X \equiv$ link strength and $Y \equiv$ applied stress such that

$$P(X \leq x) = F(x)$$

and $\qquad\qquad\qquad\qquad\qquad\qquad\qquad\qquad\qquad\qquad\qquad\qquad (9.34)$

$$P(Y \leq y) = G(y)$$

It is assumed that X and Y are independent random variables defined on the same sample space. The reliability of a link is defined to be the probability that the link does not break; i.e., the probability that the link strength exceeds the applied stress. In symbols.

$$R_\ell = P(X > Y) \qquad (9.35)$$

Now, the quantity $P(X > Y)$ can be written as the integral of the joint density function $f(x)g(y)$ over the shaded region shown in Fig. 9.1. Thus

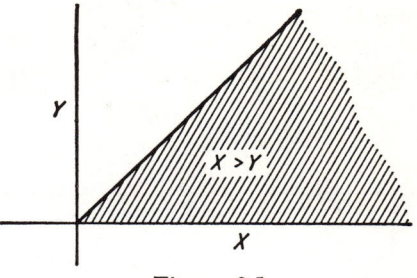

Figure 9.1

$$R_\ell \equiv P(X > Y) = \int_0^\infty \int_y^\infty f(x)g(y)\ dx\ dy$$

or

$$R_\ell = \int_0^\infty g(y)[1 - F(y)]\ dy \qquad (9.36)$$

Evidently, the latter expression can be also written as

$$P(X > Y) = \int_0^\infty \int_0^x f(x)g(y)\ dy\ dx$$

$$= \int_0^\infty f(x)G(x)\ dx \qquad (9.37)$$

9.3.2 CHAIN STRENGTH

Let us now imagine the process of forming a "chain" of n "links" selected from a population whose link strength probability density function is given by $f(x)$. This is the same as selecting a random sample $X_1, \cdots, X_n$ of link strengths from a population with probability density $f(x)$. The chain has strength equal to that of its weakest link; i.e., the strength Y_n of an n-link chain is equal to the minimum of the X_i, $i = 1, \cdots, n$.

The problem is to express the probability distribution of Y_n in terms of $f(x)$. Let $f_n(y)$ denote the probability density function of Y_n; and let $Y_n = X_1$ (arbitrarily picked as the minimum of the X_i; $i = 1, \cdots, n$). Then, visualizing the X_i as points on the x-axis falling into mutually exclusive "pockets" or cells, we can use the multinominal distribution (Sec. 6.3) to express the probability that out of n points, X_1 falls in the interval $(y, y + dy)$, and all other points $X_2, \cdots, X_n$ fall in a region to the right of y (Fig. 9.2). Thus

$$P(y < Y_n < y + dy) = f_n(y)dy = \frac{n!}{1!(n-1)!} f(y)dy[1 - F(y)]^{n-1}$$
(9.38)

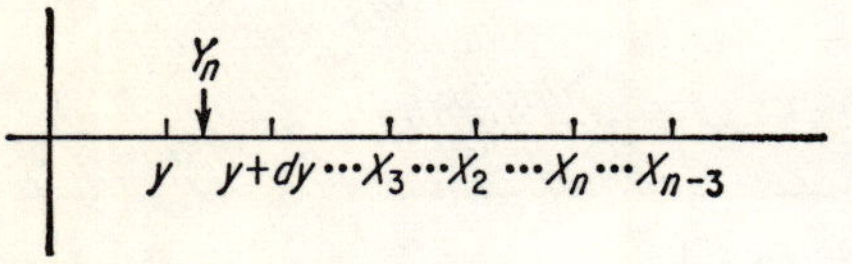

Figure 9.2

and therefore

$$f_n(y) = n[1 - F(y)]^{n-1}f(y)$$
(9.39)

Therefore, the strength distribution function of the n-link chain is given by

$$F_n(y) = \int_0^y f_n(t)\,dt$$
(9.40)

$$= \int_0^y n[1 - F(t)]^{n-1}f(t)\,dt$$

$$= n \int_0^{F(y)} (1 - u)^{n-1}\,du$$
(9.41)

where we have made the transformation

$$u = F(t), \qquad du = F'(t)\,dt = f(t)\,dt$$

Thus

$$F_n(y) = [-(1-u)^n]_0^{F(y)} = 1 - [1 - F(y)]^n$$
(9.42)

Now for any stress Y with probability density $g(y)$, the probability that the chain strength Y_n exceeds the stress Y applied to the chain is, from (9.36),

$$R_n = P(Y_n > Y) = \int_0^\infty g(x)[1 - F_n(x)]\,dx$$
(9.43)

or, using (9.42),

$$R_n = \int_0^\infty g(x)[1 - F(x)]^n\,dx$$
(9.44)

Thus R_n is the reliability of the n-link chain.

9.3.2.1. Limits of R_n. It will now be shown that the reliability of the n-link chain is not less than the reliability of a single link raised to the nth power, R_ℓ^n.

For convenience we denote R_ℓ by R_1, which is correct, since when $n = 1$ in (9.44) we obtain the reliability for a single link as given by (9.36). Thus, we wish to prove

$$R_n \geq R_1^n$$
(9.45)

Consider the expression

$$H(u, v) = \int_0^\infty \{u[1 - F(x)]^{(\nu-1)/2} + v[1 - F(x)]^{(\nu+1)/2}\}^2 g(x)\, dx \quad (9.46)$$

which is evidently non-negative, where u and v are quantities independent of the variable of integration and ν is a positive integer. If we expand the square of the bracketed terms in (9.46) and use (9.44) we obtain

$$H(u, v) = u^2 R_{\nu-1} + 2uv R_\nu + v^2 R_{\nu+1} \geq 0 \quad (9.47)$$

From the theory of quadratic forms we must have

$$R_{\nu-1} R_{\nu+1} \geq R_\nu^2 \quad (9.48)$$

Thus
$$R_0 R_2 \geq R_1^2 \quad (9.49)$$

or, since we would define $R_0 \equiv 1$ (which also follows from (9.44) if it is assumed to hold for $n = 0$),

$$R_2 \geq R_1^2 \quad (9.50)$$

Thus the inequality to be proved holds for $n = 2$. Now *assume* that (9.45) is true for some n and we will now prove that

$$R_{n+1} \geq R_1^{n+1} \quad (9.51)$$

Upon proving (9.51), we will then be able to say that

$$R_2 \geq R_1^2 \Rightarrow R_3 \geq R_1^3 \Rightarrow \cdots \Rightarrow R_m \geq R_1^m \Rightarrow \cdots$$

and thus (9.45) will have been shown correct for any value of n. This is the usual proof by induction on n.

In (9.48), let $\nu = 1, 2, \cdots, n$ and we thereby obtain the series of inequalities:

$$\begin{aligned}
R_1^2 &\leq R_2 & \nu &= 1 \\
R_2^2 &\leq R_1 R_3 & \nu &= 2 \\
R_3^2 &\leq R_2 R_4 & \nu &= 3 \\
&\;\;\vdots & &\;\;\vdots \\
R_n^2 &\leq R_{n-1} R_{n+1} & \nu &= n
\end{aligned} \quad (9.52)$$

If we now multiply all inequalities (9.52) together, we obtain

$$R_1^2 R_2^2 \cdots R_n^2 \leq R_1 R_2^2 R_3^2 R_4^2 \cdots R_{n-1}^2 R_n R_{n+1} \quad (9.53)$$

or

$$R_1 R_n \leq R_{n+1} \quad (9.54)$$

Now, making use of the assumption that (9.45) is true we have

$$R_1(R_1^n) \leq R_1 R_n \leq R_{n+1} \tag{9.55}$$

or
$$R_1^{n+1} \leq R_{n+1} \tag{9.56}$$

By induction, it follows that (9.45) is true for *all* n.

The upper limit to R_n is easily seen to be R_1, since the integrand in (9.36), namely $[1 - F(\cdot)]g(\cdot)$ is always equal to or greater than the integrand in (9.44), namely $[1 - F(\cdot)]^n g(\cdot)$. Thus we have

$$R_1^n \leq R_n \leq R_1 \tag{9.57}$$

9.3.2.2 Upper and Lower Limits of R_n Achieved. The upper and lower limits of R_n as given by (9.57) are achieved for a chain in the following cases:

(1)   Upper limit: $\quad\quad\quad R_n = R_1$

Let the strength distribution of the links be such that the links have identical strengths; i.e.,

$$\begin{aligned} F(x) &= 0 & x < s \\ &= 1 & x \geq s \end{aligned} \tag{9.58}$$

Alternatively, (9.58) is implied by assuming that the standard deviation of the link strength distribution tends to zero, so that all of the probability is concentrated at the point $x = s$; then (9.44) yields

$$R_n = \int_0^s g(x)\, dx = G(s) \tag{9.59}$$

which is the probability that the stress does not exceed the magnitude s. Since the right side of (9.59) is independent of n, the reliability of the n-link chain is evidently equal to the reliability of one of the identical links.

(2)   Lower limit: $\quad\quad\quad R_n = R_1^n$

To demonstrate this limit for R_n, let us integrate (9.44) by parts to obtain

$$R_n = n \int_0^\infty G(x)[1 - F(x)]^{n-1} f(x)\, dx \tag{9.60}$$

Now assume that the stress distribution $G(x)$ concentrates all the probability at the point $x = s$; then from (9.60)

$$R_n = n \int_s^\infty [1 - F(x)]^{n-1} f(x)\, dx \tag{9.61}$$

or
$$R_n = [1 - F(s)]^n \tag{9.62}$$

Since $1 - F(s)$ is the probability that a single link has greater strength than the stress applied to the link when the stress has only one value s, then the chain reliability is the single link reliability raised to the nth power as indicated by (9.62).

9.3.3 STRESS VERSUS STRENGTH ANALYSIS

Let us assume that $f(x)$ and $g(x)$ are both normal probability density functions with means m_1, m_2 and standard deviations σ_1 and σ_2, respectively. As long as m_1 and m_2 are large with respect to σ_1 and σ_2, we can theoretically allow negative values for strengths and stresses, since these will have negligible probability of occurring. Now $\xi = X - Y$ is also normally distributed with mean $m_1 - m_2$, standard deviation $(\sigma_1^2 + \sigma_2^2)^{1/2}$. We wish to find $P(X > Y) = P(\xi > 0)$.

$$R_\ell = \frac{1}{\sqrt{2\pi(\sigma_1^2 + \sigma_2^2)}} \int_0^\infty \exp\left[-\frac{1}{2}\left\{\frac{(t - m_1 + m_2)^2}{(\sigma_1^2 + \sigma_2^2)}\right\}\right] dt \qquad (9.63)$$

By transforming the integration variables, we obtain

$$R_\ell = \Phi\left(\frac{m_1 - m_2}{\sqrt{\sigma_1^2 + \sigma_2^2}}\right) \qquad (9.64)$$

where Φ is the standard normal distribution function (Sec. 6.10).

When m_1 is larger than m_2 and the difference increases, σ_1 and σ_2 remaining fixed, then $\Phi \to 1$, which corresponds with the physical picture that for large mean strength relative to mean stress, the link reliability is large. Figure 9.3 depicts the situation. To calculate the reliability of a 2, 3, $\cdots$-link chain it appears that numerical integration is necessary, when $f(x)$ and $g(x)$ are normal probability density functions.

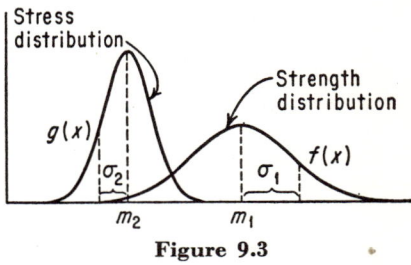

Figure 9.3

The following exercises are related to the problem of calculating link reliability and illustrate some practical aspects of the strength versus stress model.

> EXAMPLE: A total of n test firings of a solid-propellant rocket engine are made, and the maximum value of the chamber pressure is evaluated for each firing. The burst pressures of the rocket chambers are known to be normally distributed with mean m_1, standard deviation, σ_1. Calculate an estimate and a lower confidence limit on the probability that the rocket engine will not blow up. Assume that the chamber pressure maxima are normally distributed.

Let X, Y be the random variables denoting the burst pressure and the maximum chamber pressure, respectively. We wish to find an estimator for R_ℓ, as given by (9.64). Although X has a known normal (m_1, σ_1) distribution, the assumed normal (m_2, σ_2) distribution of Y is unknown and must be estimated. Thus, an estimator for R_ℓ is given by substituting the sample mean, $\bar{y}$, and variance, s^2, of the observed chamber pressure maxima into (9.64) to obtain

$$\hat{R}_\ell = \Phi\left(\frac{m_1 - \bar{y}}{\sqrt{\sigma_1^2 + s^2}}\right)$$

To find a lower confidence limit on R_ℓ, we first find the variance of $\hat{R}_\ell$ by the approximate method of Sec. 8.4 (note that $\bar{y}$ and s^2 are *independent* and therefore *uncorrelated*), which yields

$$\text{Var } \hat{R}_\ell \cong \frac{\phi^2}{n}\left[\frac{1}{1 + \left(\frac{\sigma_1}{\sigma_2}\right)^2} + \frac{\left(\frac{m_1 - m_2}{\sigma^2}\right)^2}{2\left(1 + \left(\frac{\sigma_1}{\sigma_2}\right)^2\right)^3}\right]$$

where $\phi \equiv \phi\left(\dfrac{m_1 - m_2}{\sqrt{\sigma_1^2 + \sigma_2^2}}\right)$ is the standard normal p.d.f.

A *numerical* estimate for Var $\hat{R}_\ell$, denoted by $\langle \text{Var } \hat{R}_\ell \rangle$ or V, may then be obtained by replacing m_2, σ_2, wherever they appear, by their estimators $\bar{y}$, s, respectively. Then by the method of Sec. 7.4.4, an approximate lower confidence limit, with confidence coefficient γ, on R_ℓ is given by

$$\hat{R}_L = \hat{R}_\ell - K_{1-\gamma} V^{1/2}$$

For example, if $m_1 = 800$ psia, $\sigma_1 = 100$ psia, $n = 20$, $\bar{y} = 450$ psia, $s = 25$ psia, and $\gamma = 0.90$, then

$$\hat{R}_\ell = \Phi(3.3955) \cong 0.999658$$

$$V \equiv \langle \text{Var } R_l \rangle = \frac{(0.0012512)^2}{20}\left[\frac{1}{1 + \left(\frac{100}{25}\right)^2} + \frac{\left(\frac{800 - 450}{25}\right)^2}{2\left(1 + \left(\frac{100}{25}\right)^2\right)^3}\right]$$

$$= 6.166 \times 10^{-8}$$

Thus $V^{1/2} = 0.0002483$

Hence $\hat{R}_L = 0.999658 - 1.282\,(0.0002483)$

$$= 0.99934$$

EXERCISE: Do the preceding problem when, instead, the burst pressures of the chambers are measured from a sample of size m.

9.4. PARALLEL SYSTEMS AND REDUNDANCY

A parallel system is defined as a system consisting of n subsystems such that system failure occurs when and only when *all* subsystems fail. Equivalently, the system is successful when *at least one* of its subsystems is successful. It is easily shown that all of the properties given in Sec. 9.2 for serial systems can be dualized to give the corresponding properties for parallel systems by simply replacing any event by its complementary event; i.e., change S_j to F_j, S to F or vice versa; however, note that a formula which calculates *reliability* for a serial system dualizes to a formula which calculates *unreliability* for a parallel system.

EXERCISE: Formulate the corresponding reliability properties for parallel systems by dualizing the formulas of Sec. 9.2.

EXERCISE: Convince yourself that a lower confidence limit on reliability of an independent parallel system can be found by computing

$$\hat{R} = 1 - \prod_{j=1}^{k} \frac{f_j}{N_j} \qquad (9.65)$$

then determining the "equivalent number of system failures" by

$$F = N_m(1 - \hat{R}) \qquad (9.66)$$

where N_m is the minimum value of the N_j, and the confidence limit is obtained from Figs. A.1–A.5, pp. 556-560 (or directly by the method of Appendix 8A).

EXERCISE: Show that

$$1 - \prod_{j=1}^{k} C_{N_j}(f_j, \gamma_j)$$

is a lower confidence limit on the reliability of an independent parallel system, with confidence coefficient

$$\gamma = \prod_{j=1}^{k} \gamma_j$$

where $C_N(f_j, \gamma_j)$ is an upper confidence limit, with confidence coefficient γ_j on the unreliability of the jth subsystem for which f_j failures in N_j trials are observed.

The property of parallel systems is frequently described by the engineering term *redundancy*. This relates to the concept that there are alternate components, existing within the system, to help the system operate successfully in case of failure of one or more of the original components. A further example of redundancy relates to the situation when all the components are operating but less than all of them are necessary for the system to be successful. A measure of redundancy could be defined as the percentage ratio of the total number of components available for a specific task to the number actually necessary. Sections 9.4.1, 9.4.2, and 9.4.3 are examples of the variety of applications that redundancy may have.

9.4.1 EXAMPLE OF A PARTIALLY PARALLEL SYSTEM

A rocket engine igniter system is designed to have n separate igniters connected in a parallel circuit. This design is intended to decrease the probability of failure by allowing ignition even if several of the separate igniter circuits are "open." However, owing to a restriction on over-all dimensions of the system, i.e., each separate igniter is required to be "small," the energy per single igniter may or may not be sufficient by itself to successfully ignite the propellant. In fact there is a probability P_k equal to or less than one that any group of k igniters working together will result in successful ignition.

Let q be the probability that an individual igniter will work, i.e., its circuit is not open, then the probability Q_k that exactly k igniters out of a total of n will work is

$$Q_k = \binom{n}{k} q^k p^{n-k} \tag{9.67}$$

where $p \equiv 1 - q$. Thus the probability R that the igniter system will successfully ignite the propellant is

$$R = \sum_{k=1}^{n} P_k Q_k \tag{9.68}$$

where Q_k is given by (9.67).

The probabilities P_k may be assumed to be increasing in k; thus, physically, the true probability that $k + 1$ working igniters successfully cause ignition of the propellant should be greater than the probability of the same event for k working igniters. A special assumption could be made, namely that j or more working igniters are certain to ignite the propellant successfully, while fewer than j working igniters will never ignite the propellant. In this special case

$$P_k = \begin{cases} 0, & 0 \leq k < j \\ 1, & j \leq k \leq n \end{cases} \tag{9.69}$$

Equation (9.68) then becomes

$$R = \sum_{k=j}^{n} \binom{n}{k} q^k p^{n-k} \tag{9.70}$$

Thus successful operation of the system is equivalent to the event that at least j subsystems are successful. If $j = 1$, then the system is *completely parallel*; otherwise we say it is *partially parallel*.

Figure 9.4 presents values of R from Eq. (9.70) when $n = 8$, for values of $j = 3, 4, 5, 6, 7$, and 8, where q ranges from 0.50 to 0.999. For example, if it is known that $j = 4$, then if the reliability of a single igniter is $q = 0.88$, the system reliability will be 0.999.

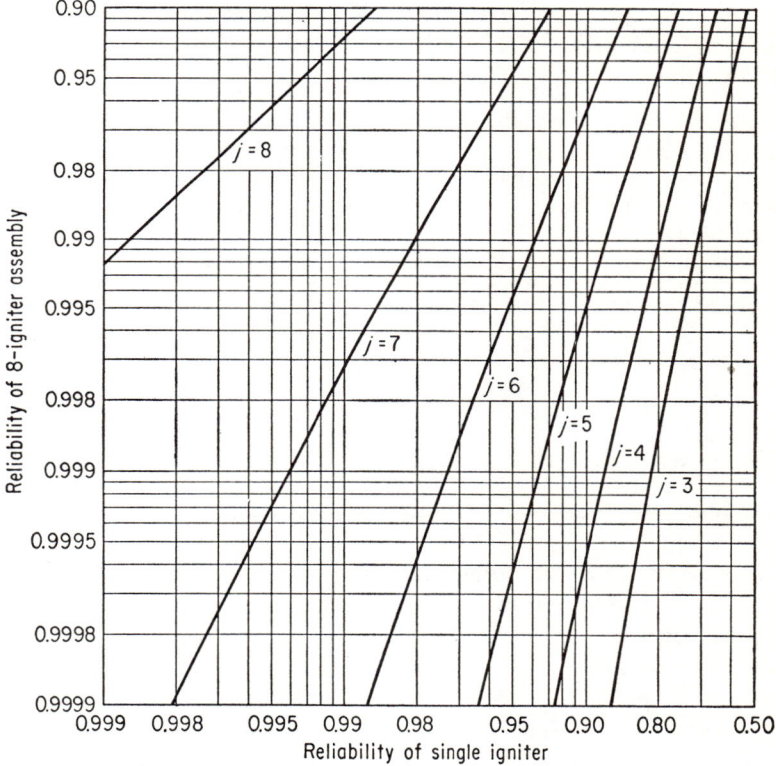

Fig. 9.4 Reliability of 8-igniter assembly *vs.* reliability of single igniter when j or more working igniters are certain to ignite propellant, while less than j working igniters cannot ignite propellant.

For the general case given by Eq. (9.68), the values of P_k would be determined from test data. A detailed discussion of the most efficient design of such an experiment is outside the scope of this section; however, one can imagine a series of groups of tests of the n-igniter system in which (say) j "live" igniters ($j \leq n$) are used in each test of a given group. The proportion of successful ignitions then gives an estimate of P_j. Another group of igniter systems is then tested in which a different value of j is used; and so on. The up-and-down method (Sec. 13.3) might also be used. An important point here, also, is that we have assumed that the event that "one igniter works" is equivalent to the event that its individual circuit is not open. However, it is physically possible that if a sufficient number of igniters are working, a "nonworking" igniter, owing to its proximity to the others, may actually operate to add its energy to that of the rest. Here is an example of the possibility of "enhancing interaction" (Sec. 4.6). In this case the value of Q_k would be larger than that given by Eq. (9.67). Furthermore, considering the type of experiments just described, the interaction effect could be measured by deliberately inserting "live" but open circuit igniters into the system.

> EXERCISE: Consider a system composed of two subsystems. Let τ be the time-to-failure of the system, and τ_1, τ_2 be the respective times-to-failure of the subsystems. (a) Show that the system is a serial system if and only if $\tau = \min(\tau_1, \tau_2)$. (b) Show that the system is a parallel system if and only if $\tau = \max(\tau_1, \tau_2)$.
>
> *Answer*: (a) The event $\tau > t$ means success of the system (where t is some required time of operation). Since the events $[\tau > t]$, $[\min(\tau_1, \tau_2) > t]$ and $[(\tau_1 > t) \cap (\tau_2 > t)]$ are all equivalent (the first two by definition, the last two by a simple algebraic argument), then
>
> $$P(S) \equiv P(\tau > t) = P(\tau_1 > t, \tau_2 > t) \equiv P(S_1 \cap S_2)$$
>
> The answer to part (b) is obtained by similar reasoning.

9.4.1.1. Example: Stand-by Redundancy. Consider a system composed of two subsystems. The system becomes operative by turning on subsystem 1 with subsystem 2 "standing by." If subsystem 1 should fail before time t (the required time of operation), then subsystem 2 is immediately turned on; otherwise subsystem 2 is not needed for system success. System failure occurs if and only if subsystem 2 fails before time t (where time t is measured from the time subsystem 1 is turned on). It is further assumed that the times-to-failure τ_1, τ_2 of subsystems 1 and 2 are independent random variables; i.e., the probability that subsystem 2 fails prior to any given time is not affected by the knowledge that subsystem 1 has failed at some particular time. To determine the system reliability we see that the event of system failure occurs when both of the events $(u < \tau_1 < u + du)$ and $(\tau_2 < t - u)$ occur, where u is any fixed number between 0 and t. The probability of the first event is $f_1(u)\,du$, the probability density element of the random variable τ_1. The probability of the second event is $F_2(t - u)$, where $F_2(\cdot)$ is the distribution function of τ_2. Since τ_1 and τ_2 are independent, the probability of joint occurrence of the two events is the product; i.e., $f_1(u)\,du \cdot F_2(t - u)$. Now, failure occurs for each value of u between 0 and t. For different values of u, the events $(u < \tau_1 < u + du) \cap (\tau_2 < t - u)$ are mutually exclusive; hence the probability of system failure is obtained by summing over all possible values of u. This is done by integrating the previous probability with respect to u from 0 to t; i.e.,

$$P(\text{system failure}) = \int_0^t F_2(t - u) f_1(u) \, du$$

The last expression is simply the probability of the event $\tau_1 + \tau_2 < t$, which is, in a sense, the "obvious" equivalent event to system failure. Thus if τ denotes time-to-failure of the system, then

$$1 - R = P(\tau < t) = P(\tau_1 + \tau_2 < t)$$

Note that the reliability of such a system is never less than the reliability of a simple *parallel* system, since for a parallel system failure we must have $\tau_1 < t$ and $\tau_2 < t$. However, the event $\tau_1 + \tau_2 < t$ implies the event $[(\tau_1 < t) \cap (\tau_2 < t)]$, but not necessarily conversely. Hence the probability of the former event must be equal to or less than the probability of the latter event; i.e., the probability of failure of the stand-by system is not greater than the probability of failure of the parallel system.

> EXERCISE: In a stand-by redundant system consisting of two components, assume that the times-to-failure τ_1, τ_2 have exponential probability density functions with failure rates λ_1, λ_2. Calculate the probability that the system will operate for at least time t (see Ref. 10).

9.4.2 EXAMPLE OF A MIXED PARALLEL-SERIAL SYSTEM

The following example considers the design of optimum circuitry and combination of pressure switches used in an abort sensing system. When a pressure switch detects a significant drop in pressure in the fuel manifold of a liquid-propellant rocket engine it responds by opening the portion of the circuit containing it. The pressure switch may fail to operate properly in two ways:

1. by not opening the circuit when a significant pressure drop occurs (an abort condition exists, but is not detected)
2. by opening the circuit when a significant pressure drop has not occurred (the abort condition does not exist, but is falsely detected).

The abort sensing system, as a whole, fails if it does not sense an abort condition when it exists, or if it erroneously senses an abort condition when it in fact does not exist. When an abort condition is sensed, whether false or true—i.e., the entire circuit is open—the system gives the signal to abort the mission. Thus, an individual pressure switch can fail by not opening when a true abort condition exists, and also by opening, or being open, when an abort condition does not exist. However, provided that other switches are in the circuit, individual failure of a pressure switch may not cause a false abort nor be the cause of the system's not sensing the true abort condition.

To illustrate the last statement, Table 9.1 presents all distinct circuits utilizing as many as three pressure switches, together with the probabilities of failure for the corresponding circuit. The general n-switch serial circuit and the m-switch parallel circuit are also shown in Table 9.1. To calculate the respective probabilities of failure for each of the circuit configurations, let α be the probability of an abort condition, q_1 the probability that a single pressure switch fails to open when an abort condition exists, and q_2 the probability that a single pressure switch opens when an abort condition does not exist.

For example, in configuration 3c, we note first that failure occurs in two mutually exclusive ways: (1) If an abort condition exists (probability α), then the circuit will remain closed (unsensed abort) if 1, 2, 3 fail to open; both 1 and 2 fail to open and 3 opens; both 1 and 3 fail to open and 2 opens. These events are mutually exclusive and their probabilities are, respectively: q_1^3, $q_1^2(1-q_1)$, $q_1^2(1-q_1)$. Thus, the probability of unsensed abort is $\alpha[q_1^3 + 2q_1^2(1-q_1)]$. (2) If an abort condition does not exist, the circuit will be open (false abort) when any one of these five mutually exclusive events occurs: 1, 2, 3 open; 1, 2 open and 3 closed; 1, 3 open and 2 closed; 2, 3 open and 1 closed; 1 open and 2, 3 closed. The respective probabilities are

$$q_2^3, q_2^2(1-q_2), q_2^2(1-q_2), q_2^2(1-q_2), q_2(1-q_2)^2$$

Thus the probability of false abort is

$$(1-\alpha)[q_2^3 + 3q_2^2(1-q_2) + q_2(1-q_2)^2]$$

Therefore, since unsensed abort and false abort are mutually exclusive events:

$$P_{3c} = \alpha[q_1^3 + 2q_1^2(1-q_1)] + (1-\alpha)[q_2^3 + 3q_2^2(1-q_2) + q_2(1-q_2)^2]$$

or simplifying,

$$P_{3c} = \alpha q_1^2(2-q_1) + (1-\alpha)q_2[1 + q_2 - q_2^2]$$

as given in Table 9.1. The other probabilities are derived in a similar manner. Note that circuit 3e is equivalent to a circuit with two switches in parallel (2b). Hence, circuit 3e would not be preferred to the latter, since circuit 2b utilizes fewer switches.

Table 9.1. PROBABILITY OF FAILURE OF ABORT SENSING SYSTEM

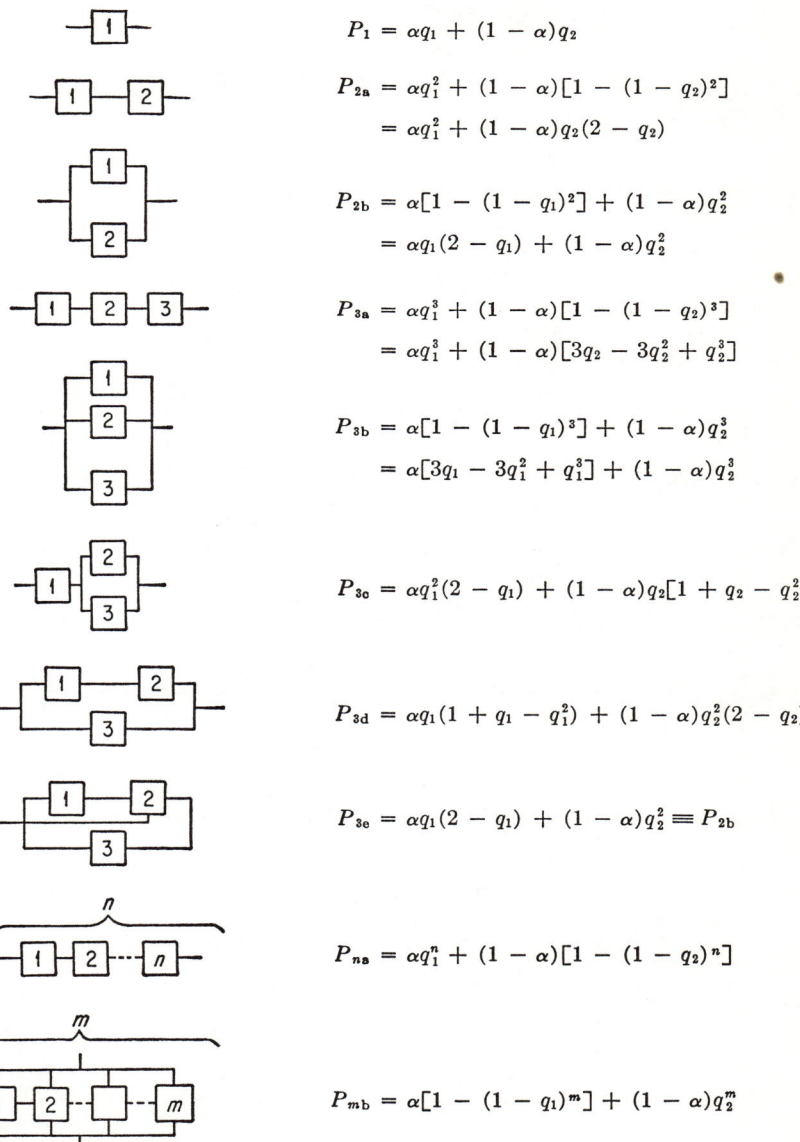

$P_1 = \alpha q_1 + (1 - \alpha) q_2$

$P_{2a} = \alpha q_1^2 + (1 - \alpha)[1 - (1 - q_2)^2]$
$= \alpha q_1^2 + (1 - \alpha) q_2 (2 - q_2)$

$P_{2b} = \alpha[1 - (1 - q_1)^2] + (1 - \alpha) q_2^2$
$= \alpha q_1 (2 - q_1) + (1 - \alpha) q_2^2$

$P_{3a} = \alpha q_1^3 + (1 - \alpha)[1 - (1 - q_2)^3]$
$= \alpha q_1^3 + (1 - \alpha)[3q_2 - 3q_2^2 + q_2^3]$

$P_{3b} = \alpha[1 - (1 - q_1)^3] + (1 - \alpha) q_2^3$
$= \alpha[3q_1 - 3q_1^2 + q_1^3] + (1 - \alpha) q_2^3$

$P_{3c} = \alpha q_1^2 (2 - q_1) + (1 - \alpha) q_2 [1 + q_2 - q_2^2]$

$P_{3d} = \alpha q_1 (1 + q_1 - q_1^2) + (1 - \alpha) q_2^2 (2 - q_2)$

$P_{3e} = \alpha q_1 (2 - q_1) + (1 - \alpha) q_2^2 \equiv P_{2b}$

$P_{na} = \alpha q_1^n + (1 - \alpha)[1 - (1 - q_2)^n]$

$P_{mb} = \alpha[1 - (1 - q_1)^m] + (1 - \alpha) q_2^m$

If one compares only the circuits 1, 2a, 2b, the condition that

$$P_{2a} \leq P_1 \leq P_{2b}$$

is easily seen to be equivalent to

$$q_2(1 - q_2) \leq \frac{\alpha}{1 - \alpha} q_1(1 - q_1) \tag{9.71}$$

or

$$q_2 \geq 0.5 + \left[0.25 - \frac{\alpha}{1 - \alpha} q_1(1 - q_1)\right]^{1/2}$$

$$q_2 \leq 0.5 - \left[0.25 - \frac{\alpha}{1 - \alpha} q_1(1 - q_1)\right]^{1/2} \tag{9.72}$$

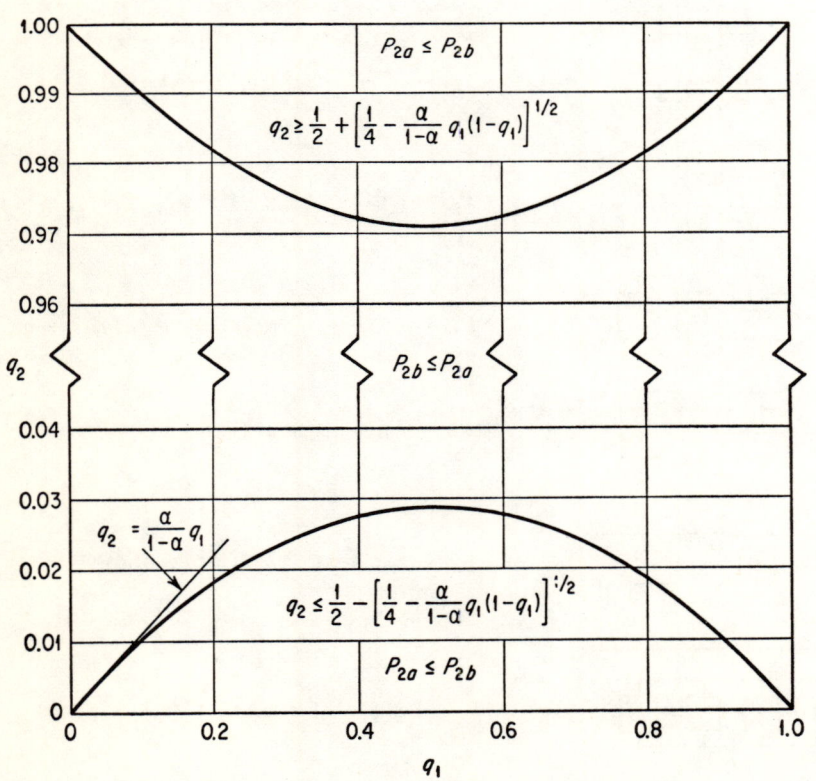

Fig. 9.5 Optimum regions for two-switch circuits: series (2a) and parallel (2b).

RELIABILITY STRUCTURE MODELS

If we take $\alpha = 0.10$ (this value will be assumed in all subsequent calculations), Fig. 9.5 illustrates the regions for which circuit 2a is optimum and for which 2b is optimum. The single switch is nowhere optimum among the three circuits 1, 2a, 2b, its probability of failure being equal to or greater than one of 2a or 2b throughout the region $0 \leq q_1 \leq 1, 0 \leq q_2 \leq 1$.

While Fig. 9.5 shows the entirety of the regions where either circuit 2a or 2b has minimum probability of failure, the region where $q_1 \ll 1$, $q_2 \ll 1$ is the one of real interest, since q_1, q_2 are expected to be small in practice. Thus, by taking only the asymptotic expressions (as $q_1, q_2 \to 0$) in (9.71), we obtain

$$q_2 \leq \frac{\alpha}{1-\alpha} q_1 \tag{9.73}$$

as the condition that $P_{2a} \leq P_{2b}$. The reverse of inequality (9.73) gives the condition that $P_{2b} \leq P_{2a}$. Figure 9.5 also shows the "straight-line" approximation (9.73) to the exact regions of optimality.

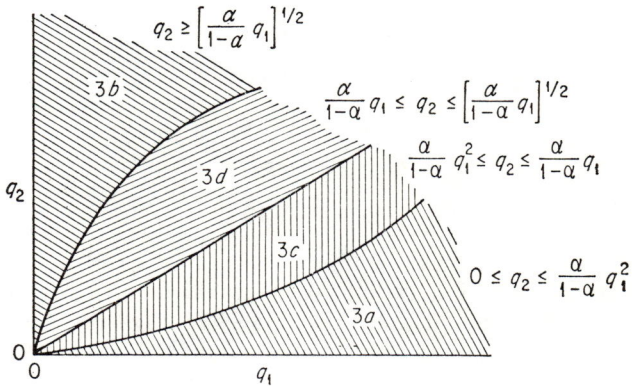

Figure 9.6 Optimum regions for three-switch circuits (from Table 9.1).

For circuits with as many as three switches, it is necessary to compare circuits 2a, 2b, 3a, 3b, 3c, and also 3d. This can be done as above, by setting, e.g., P_{3a} equal to or less than each of the other probabilities P_{2a}, P_{2b}, P_{3b}, P_{3c}, P_{3d}, and in turn, discarding in each case all but the lowest powers of both q_1 and q_2. When this is done for each of the six circuits, one obtains the optimum regions as illustrated in Fig. 9.6 (not to scale). Note from the figure that of all the circuits considered, only those with three switches are optimum in mutually exclusive and exhaustive sectors.*

---

* This and the preceding results would imply that the more switches in the circuit, the more optimum are the choices for the circuit. The authors did not check this, however. If true, additional restrictions, e.g., a space limitation, would be necessary in order to determine the truly optimum circuit in the noted context.

A numerical example is as follows: Let $\alpha = 0.10$, $q_1 = 0.01$, $q_2 = 0.001$; then by inspection

$$\frac{0.10}{1 - 0.10}(0.01)^2 \leq q_2 \leq \frac{0.10}{1 - 0.10}(0.01)$$

or
$$0.0000111 \leq q_2 \leq 0.00111$$

so that (q_1, q_2) falls in the region where circuit 3c is optimum. From Table 9.1,

$$P_{3c} = 0.10(0.0001)(1.99) + 0.90(0.001)(1 + 0.001 - 0.000001)$$

$$= 0.0000199 + 0.0009009$$

$$= 0.0009208$$

The values of the probabilities for the other circuits are:

$$P_1 = 0.0019$$
$$P_{2a} = 0.0018$$
$$P_{2b} = 0.0020$$
$$P_{3a} = 0.0027$$
$$P_{3b} = 0.0030$$
$$P_{3d} = 0.0010$$

which are all larger than P_{3c}.

In view of the fact that circuit 3e is equivalent to a circuit with two switches in parallel, one suspects that no advantage will be gained (in circuits with four or more switches) when additional circuit links are put in from the main terminals to any of the switches.

Since the optimum circuitry depends on the values of q_1 and q_2 a problem may arise if these parameters are not determined accurately. The circuit chosen may not be truly optimum if the estimated (q_1, q_2) falls in, e.g., region 3c, whereas the true value of (q_1, q_2) is in region 3d of Fig. 9.6. This problem is not quite as serious as one might at first suppose, since the minimum probability of failure is continuous across the boundaries of the regions. Thus, the error committed would be a continuous function of the "distance" between the true (q_1, q_2) and the estimated (q_1, q_2), for fixed α. This problem therefore can be considered merely one of determining a sufficient sample size and number of tests of pressure switches in order to ascertain the accuracy of estimates of the failure probabilities within a given limit of error.

> EXERCISE: Check the statement that the minimum probability of failure is continuous across the boundaries of the regions in Fig. 9.6.

EXERCISE: Study the effect of this additional restriction: "Ratio of false abort to unsensed abort should be greater than ten-to-one."

9.4.3 AN EXAMPLE OF RELIABILITY DESIGN ANALYSIS

The example which is presented in this section concerns an electronic system; however, the concepts apply to all types of equipment. Electronic equipment was chosen for illustration because the type of prediction described is standard practice and also because it allows the topic of "part application" to be introduced. Although with other equipment one should not neglect part application as to its effect on reliability, generally it is only for electronic and electromechanical equipment that sufficient information has been generated and documented to be of extensive analytical use. The reason is that most components of non-electronic types of equipment are generally designed specifically for particular system configurations and requirements. This is true, for example, of turbopumps, thrust chambers, gas generators, etc. in liquid rocket engines, and cases, insulation, nozzles, etc. in solid rocket engines. In these and similar examples the component failure information is generally limited and specific to the particular system. Thus, it does not readily apply to other systems of the same type. On the other hand, electronic parts are frequently standard off-the-shelf items, and they are often repeated many times within the system configuration.

The second topic, redundancy, which is discussed further at some length, is in no way unique to electronic equipment and should be utilized when appropriate in all kinds of equipment reliability design analyses.

9.4.3.1. Part Application. In most cases electronic systems and assemblies consist of an aggregation of electronic and electromechanical parts with known failure rates. The factors which determine the effect of the combination of these parts into a system are essentially twofold: (1) the reliability structure of the equipment, and (2) the application of the parts.

A great amount of usage experience and failure information exists which has been organized and tabulated in the reference documents (Refs. 5 and 6). During the course of this work it was observed that the failure rate of the part at a "standard electrical stress" and in a "standard environment," i.e., the laboratory, could be correlated with the failure rate in different environments and at different electrical stresses. These correlations are in the form of factors, sometimes called *application factors*, K_A, which multiply the failure rate of a given part to "correct it" according to its intended electrical and environmental application. The K_A factors, which can be larger or smaller than unity, have also been tabulated extensively in the previously mentioned references. Therefore, before use can be made of any part's numerical failure rate, it is first necessary to determine its application. This can be done after the mock-up equipment has been packaged according to the best design technique. The local environments, that is, the internal environments of the system package—particularly with respect to temperature and vibration, the two major degrading environmental factors—are measured for each part. The physical makeup or packaging of the components therefore plays an important role in creating the internal environment. The level of the second factor; i.e., the electrical stress, is of course known from circuit considerations and is part of the basic design exercise.

The external environment must also be considered. This is established by system usage requirements. By this is meant that the particular equipment in question will be operating in certain environments such as stationary ground equipment, missile equipment (static test), missile equipment (in-flight), satellite equipment, etc. The factors which relate these latter environments to that of the "standard environment" are called *operating mode factors*, K_{op}. Values of K_{op} for various operating environments are given in Ref. 5. As an example, it is considered that the general effect of environment on an equipment mounted on a missile in flight is 1000 times more severe, i.e., $K_{op} = 1000$, than for similar equipment placed in a typical laboratory computer. If it were placed in a particular spot in a missile interstage compartment, such as next to the rocket nozzles, then K_{op} would be even greater.

Thus to establish the expected failure rate for a particular part as it will be used in the given system we proceed with the following steps:

1. List the part by name, number.
2. Obtain its "standard failure rate."
3. Obtain its K_A factor from the appropriate reference according to its expected electrical and environmental stress level.
4. Select the appropriate K_{op} factor.
5. Multiply the numbers in 2, 3, and 4 together to give the part's "application failure rate." This is the expected failure rate of a particular kind of part for the given system's design and use.

This process is repeated for all types of parts in the system. When a given part type has several different applications in the system, then the above process should be performed for each application.

Frequently in electronic equipment when an exponential distribution of time-to-failure is assumed and no redundancy exists within the assembly, then the reliability of the assembly can be estimated almost immediately. This is done simply by adding all the parts' application failure rates to obtain the system's failure rate, which can be readily transposed into a reliability estimate (see Sec. 8.1). An illustration of this technique is contained in the example (Table 9.2) where the reliability of an individual component (a solar panel control circuit) is evaluated from the failure rates of its parts. However, it is not possible to apply such a simple tech-

RELIABILITY STRUCTURE MODELS

nique when the previously mentioned assumptions do not exist, and it must be considered as a "shortcut" method only. The "general" method for system reliability estimation is to combine the parts' reliabilities according to the system's reliability structure, taking into account the effects of redundancy, which we discuss further in the following section. If it is found, when the assessment is completed, that the over-all reliability is inadequate, then there are several recourses. Higher-quality parts may be used with higher stress ratings (lower application factors), the packaging may be redesigned to improve the local environments, or redundancy in the system might be increased.

9.4.3.2. Redundancy. The concept of redundancy is easily understood by diagrammatic representation such as in Table 9.1 and Figs. 9.8, 9.9, and 9.10 and its advantages appreciated by means of the corresponding reliability equations. It is important to note that in both the discussion and example in this section all the reliability structures are assumed to consist of *independent* serial and parallel systems. Figures 9.7 and 9.8 are structures discussed in Secs. 9.2 and 9.4. Figures 9.9 and 9.10 are direct extensions of the first two models in Figs. 9.7 and 9.8. The reader is urged to write down the equations for Figs. 9.9 and 9.10 directly from the equations associated with Figs. 9.7 and 9.8.

Alternately the reliability equations may be obtained directly from the reliability structure diagrams by use of the intersection and union operators ∩ and ∪ (Sec. 5.2.1).

Thus in Fig. 9.7 we have

$$\text{system success} = [C_1 \text{ success}] \cap [C_2 \text{ success}] \cap \cdots \cap [C_n \text{ success}]$$

$$R_s = P\{[C_1 \text{ success}] \cap [C_2 \text{ success}] \cap \cdots \cap [C_n \text{ success}]\}$$

$$= P[C_1 \text{ success}]P[C_2 \text{ success}] \cdots P[C_n \text{ success}]$$

or $$R_s = R_1 R_2 \cdots R_n$$

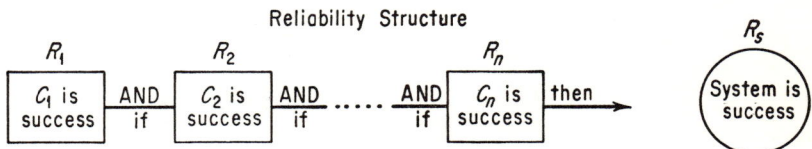

Fig. 9.7 Simple serial system, i.e., no redundancy.

Reliability Equation

$$R_s = R_1 R_2 \cdots R_n$$

or $$R_s = R^n \quad \text{when all } R_i \text{ equal } R$$

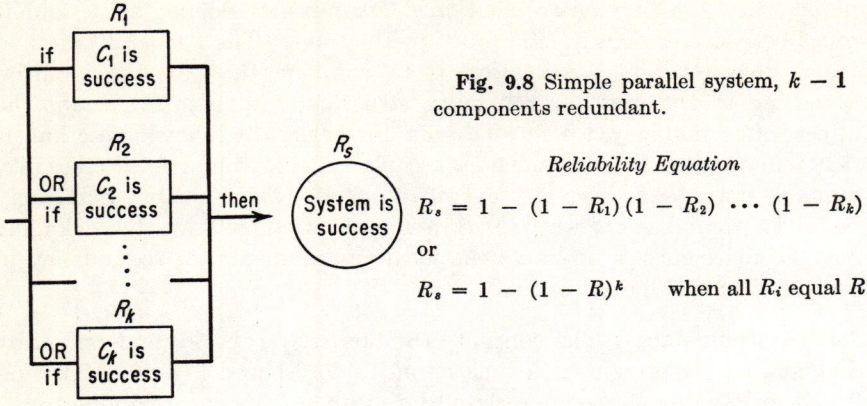

Fig. 9.8 Simple parallel system, $k-1$ components redundant.

Reliability Equation

$$R_s = 1 - (1-R_1)(1-R_2)\cdots(1-R_k)$$

or

$$R_s = 1 - (1-R)^k \qquad \text{when all } R_i \text{ equal } R$$

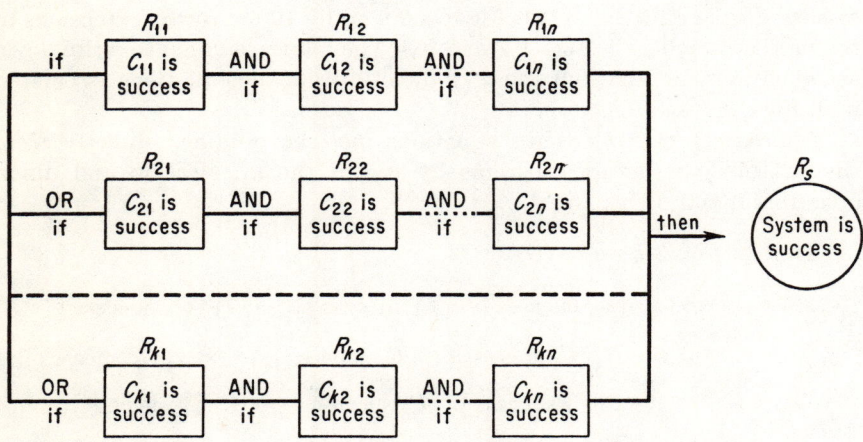

Fig. 9.9 Serial-parallel system, i.e., redundancy of systems in parallel.

Reliability Equation

$$R_s = [1 - (1 - R_{11}R_{12}\cdots R_{1n})(1 - R_{21}R_{22}\cdots R_{2n})\cdots(1 - R_{k1}R_{k2}\cdots R_{kn})]$$

or

$$R_s = 1 - (1 - R_1 R_2 \cdots R_n)^k \qquad \text{when } R_{ij} = R_j$$

or

$$R_s = 1 - (1 - R^n)^k \qquad \text{when all } R_{ij} \text{ are equal to } R$$

The reader may visualize Fig. 9.9 as Fig. 9.8 but with each component $C_1, C_2, \cdots, C_k$ replaced with a serial system of components (Fig. 9.7), thus allowing the reliability equations to be written down immediately.

RELIABILITY STRUCTURE MODELS

In Fig. 9.8

system success = $[C_1 \text{ success}] \cup [C_2 \text{ success}] \cup \cdots \cup [C_k \text{ success}]$

$$R_s = P\{\overline{[\overline{C_1 \text{ success}}] \cap [\overline{C_2 \text{ success}}] \cap \cdots \cap [\overline{C_k \text{ success}}]}\}$$

$$= 1 - P\{[\overline{C_1 \text{ success}}] \cap [\overline{C_2 \text{ success}}] \cap \cdots \cap [\overline{C_k \text{ success}}]\}$$

$$= 1 - P[\overline{C_1 \text{ success}}]P[\overline{C_2 \text{ success}}] \cdots P[\overline{C_k \text{ success}}]$$

or $\qquad R_s = 1 - (1 - R_1)(1 - R_2) \cdots (1 - R_k)$

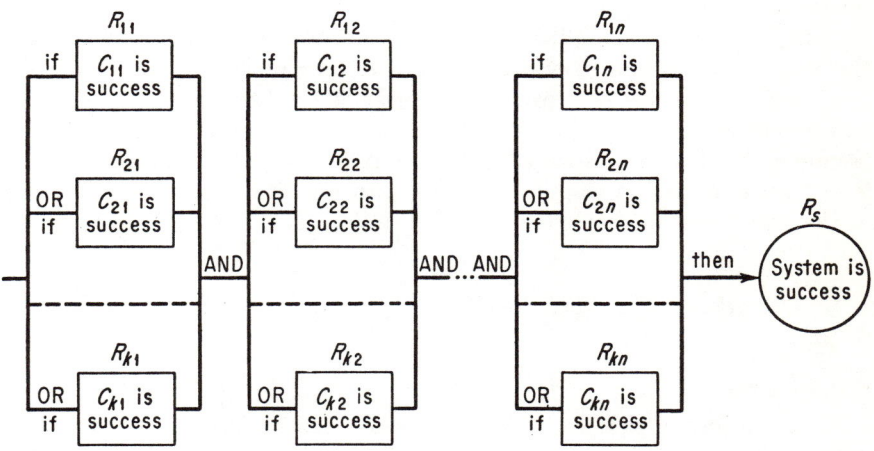

Fig. 9.10 Parallel-serial systems, i.e., redundancy of components in parallel.

Reliability Equation

$$R_s = [1 - (1 - R_{11})(1 - R_{21}) \cdots (1 - R_{k1})]$$
$$\times [1 - (1 - R_{12})(1 - R_{22}) \cdots (1 - R_{k2})] \cdots$$
$$[1 - (1 - R_{1n})(1 - R_{2n}) \cdots (1 - R_{kn})]$$

or

$$R_s = [1 - (1 - R_1)^k][1 - (1 - R_2)^k] \cdots [1 - (1 - R_n)^k] \qquad \text{when } R_{ij} = R_j$$

or

$$R_s = [1 - (1 - R)^k]^n \qquad \text{when all } R_{ij} \text{ are equal to } R$$

For Fig. 9.10 the reader may visualize Fig. 9.7 but with each component $C_1, C_2, \cdots, C_n$ replaced by a parallel system of components (Fig. 9.8), again allowing the reliability equations to be deduced immediately.

We see from the variety of reliability equations in Figs. 9.7 through 9.10 that the magnitude of R_s can theoretically be changed by the number of components and the way in which they are arranged. It would at first appear that R_s could be made to approach unity as closely as desired simply by making k as large as necessary. However, there are generally practical reasons which prevent this. The weight or space limitation is an immediate example of restrictions imposed on aerospace equipment. Another problem is created by the sheer physical complexity of the configuration or circuitry resulting from the additional redundant components. Also it is sometimes necessary to introduce further components which "sense" failures and divert or switch the operation to the next or stand-by component if this component is not already operating. The switching component also will have a reliability factor which must be taken into consideration when the reliability equations are being derived.

The four basic reliability structures which we have just reviewed are the most simple found in practice. We have seen in Sec. 9.4.1 an example of a partially parallel system, and it is easy to see how this form of structure might be incorporated into the examples of Figs. 9.9 and 9.10. For instance, if we were to hypothesize in Fig. 9.9 that for system success at least r series out of the k parallel series must operate, then the probability of the occurrence of this event from Eq. (9.70) is

$$R_s = \sum_{i=r}^{k} \binom{k}{i} [R_{11}R_{12}\cdots R_{1n}]^i [1 - R_{11}R_{12}\cdots R_{1n}]^{k-i} \qquad (9.74)$$

assuming $R_{1j} = R_{2j} = \cdots = R_{kj}$

The purpose of this discussion is to show that, in general, equations of more sophisticated redundancy relationships can be built up from the simple models described so far in this chapter. Thus when the reliability structure equation of a system is being obtained, we may proceed in the following manner. Write out the sequence or set of operations that must occur for the success of the over-all system using the type of diagrammatic format illustrated in Figs. 9.7–9.10. In general, most systems will consist of major subsystems in series, with the subsystems sometimes containing redundancy of components or parts. This first step having been performed, the structures of each of the lesser assemblies within the subsystems can be established in a similar manner. We proceed in this way until just before that level of system breakdown is reached at which there is no longer a discrete success-failure relationship* between one component and its neighbor. The reliability value known or estimated (either observationally or analytically) is then associated with each part or component,

---

* In this situation there frequently will be one of the variables interactions of the kinds described in Secs. 8.7, 9.3.3, 9.5, *et seq.*

and the reliability of the system is assessed by the methods and equations similar to those just described and to be illustrated in the following example.

9.4.3.3. Example of Reliability Design Analysis of a Space Vehicle's Temperature Control Subsystem.*

The equipment used in this example is the temperature control subsystem which is part of a space vehicle designed for scientific lunar probes. The vehicle consists of a propulsion system, airframe structure, power supply, flight control system, communication and telemetry system, and a temperature control subsystem, together forming a serial system.

The purpose of the temperature control subsystem is to protect the internal instrumentation from the extremes of temperature which would be experienced during the flight and, after landing on the moon, to provide a favorable operating environment of between 40° and 75°F.

The method of controlling the temperature is by use of radiation equilibrium during flight and lunar daytime, and by use of a chemical furnace for the lunar night. The control of emission of radiation depends on louvers which open and move according to instructions from a solar panel control. The power supply for this operation is obtained from a solar cell array via the solar panel control, which is therefore involved in two operations. The furnace is a simple design which is activated when the temperature of the vehicle falls to a certain value owing to heat dissipation in the lunar night.

The configuration of the temperature control subsystem consists of three solar cell arrays, S, (each consisting of 22 solar cell strings), three solar panel control units, C, three louvers, L, and the chemical furnace, which contain redundancy characteristics as illustrated in Fig. 9.11. For success of this mission with respect to the temperature control subsystem, it is necessary to have success both during the flight and lunar daytime operations and during the lunar night. We shall consider these two operations separately for our analysis.

Reliability during flight and lunar day. For an adequate power supply at least half of the total of 66 solar cell strings must operate. This in turn means that no fewer than two solar panel controls, C, each of which is in series with a solar cell array, S, must operate. In addition the proper operation of the louver, L, depends on its associated solar panel control circuit's working successfully. Thus,

if R_C = reliability of one solar panel control circuit, C.

R_L = reliability of one set of temperature control louvers, L.

R_{S_1} = reliability of solar cell arrays, S, given zero panel control failures

R_{S_2} = reliability of solar cell arrays, S, given exactly one panel control failure

* Based on unpublished material by A. C. Reed, of Aerospace Corporation, Los Angeles.

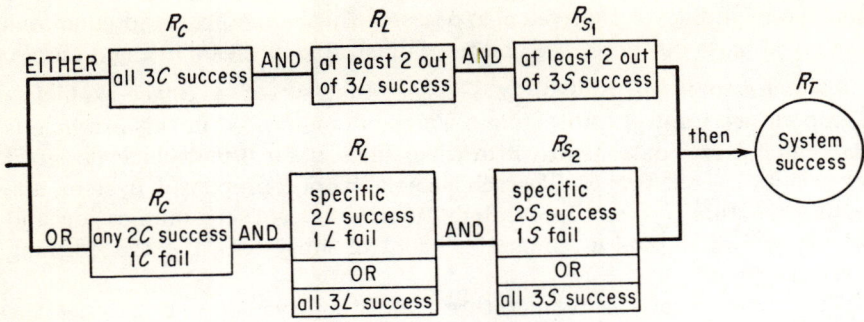

Fig. 9.11 Reliability structure diagram for temperature control subsystem during flight and lunar day.

then the complete temperature control system reliability structural diagram or model for flight and lunar day is as shown in Fig. 9.11. This figure allows us to write down the reliability equation (9.75) immediately, bearing in mind that when only two of the solar panel control circuits operate, their associated solar cell arrays and louvers must be successful also for system success. That is to say,

$$R_T = R_C^3 \left[\binom{3}{2} R_L^2 (1 - R_L) + R_L^3 \right] \left[\binom{3}{2} R_{S_1}^2 (1 - R_{S_1}) + R_{S_1}^3 \right]$$

$$+ 3 R_C^2 (1 - R_C) [R_L^2 (1 - R_L) + R_L^3][R_{S_2}^2 (1 - R_{S_2}) + R_{S_2}^3] \qquad (9.75)$$

which reduces to

$$R_T = R_C^3 [3 R_L^2 - 2 R_L^3][3 R_{S_1}^2 - 2 R_{S_1}^3] + 3 R_C^2 (1 - R_C) R_L^2 R_{S_2}^2 \qquad (9.76)$$

where R_{S_1} and R_{S_2} are computed as follows:

If q is the probability of success for a single solar cell string, then the probability of success of at least r out of k solar cell strings is

$$P = \sum_{i=r}^{k} \binom{k}{i} q^i (1 - q)^{k-i} \qquad (9.77)$$

for R_{S_1} we compute the probability of at least 33 solar cell strings out of 66 being successful, i.e.,

$$R_{S_1} = \sum_{i=33}^{66} \binom{66}{i} q^i (1 - q)^{66-i} \qquad (9.78)$$

for R_{S_2} we compute the probability of at least 33 solar cell strings out of 44 being successful, i.e.,

$$R_{S_2} = \sum_{i=33}^{44} \binom{44}{i} q^i (1 - q)^{44-i} \qquad (9.79)$$

RELIABILITY STRUCTURE MODELS 257

The probability q was known to be approximately equal to 0.97, resulting in both R_{S_1} and R_{S_2} tending to unity from Eqs. (9.78) and (9.79). Thus Eq. (9.76) can be approximated by

$$R_T = 3R_C^2 R_L^2 - 2R_C^3 R_L^3 = R_C^2 R_L^2 (3 - 2R_C R_L) \qquad (9.76)'$$

Values of R_C, R_L for the 90-hour flight. R_C for a 90-hour flight was obtained as follows:

The parts list, failure rates, and the reliability computations, assuming an exponential distribution of time-to-failure for the solar panel controls, are shown in Table 9.2. (Reliability of structural parts associated with

Table 9.2. ONE SOLAR PANEL CONTROL CIRCUIT

Part	Part application failure rate (%/1000 hours)	Total	Total part failure rate (%/1000 hours)
Transistors	0.018	20	0.360
Diodes	0.0036	14	0.050
Resistors:			
Carbon Composition	0.0015	22	0.033
Carbon Film	0.002	23	0.046
Wire Wound	0.015	1	0.015
Capacitors	0.001	8	0.008
Tantalum Capacitors:			
Solid	0.001	1	0.001
Foil	0.005	1	0.005
Transformer and Chokes	0.020	2	0.040
Motors	0.050	1	0.050
Harmonic Drive	<0.001	1	0
Switches	0.020	6	0.120
Relays	0.030	3	0.090
Total			0.818

the solar panels is taken to be near unity for the duration of the flight, landing, and operations on the lunar surface.)

For a 90-hour flight, the reliability potential of a solar panel control circuit is, therefore, from Table 9.2:

$$R_C(90 \text{ hours}) = \exp[-(90)(0.818 \times 10^{-5})] = e^{-0.000736}$$

$$= 0.999264 \qquad (9.80)$$

Based on past experience, a conservative estimate of the reliability potential of one set of temperature control louvers and its sensor-transducer actuator device exceeds 0.995 for 90 hours. This will be taken as a conservative estimate; i.e.,

$$R_L(90 \text{ hours}) = 0.995 \tag{9.81}$$

Substituting for R_C, R_L, in Equation (9.76′) we have for the 90-hour transit

$$R_T(\text{flight}) = (0.999264)^2(0.995)^2[3 - 2(0.999264)(0.995)]$$

$$\underline{R_T(\text{flight}) = 0.999902} \tag{9.82}$$

Values of R_C, R_L for the lunar day. The reliability model for the temperature control louvers and for the solar panel controls on the lunar surface during daytime will remain the same as during the flight. The reliability potential of a single set of louvers for 15 earth days (360 hours), on the same basis as for the 90-hour flight, may be taken very conservatively as R_L (360 hours) = 0.98. Using the failure rate previously computed in Table 9.2, we have

$$R_C = \exp[-(360)(0.818 \times 10^{-5})]$$

$$= e^{-0.00295}$$

$$= 0.99706$$

The reliability of the solar cell strings can be shown to be essentially unity for 410 hours, so that the approximate model still holds, and from (9.76′)

$$R_T(\text{day}) = (0.99706)^2(0.98)^2[3 - 2(0.99706)(0.98)]$$

or

$$R_T(\text{day}) = 0.99845 \tag{9.83}$$

Reliability during lunar night. A different reliability model must be used for nighttime operations. At night the solar panel or its associated temperature control louvers must close to prevent heat loss. The design is such that these two components operate in parallel, i.e., so that the louver is redundant to its solar panel. Also, a solar panel and louver which fail to close will fail their associated solar cell array because of exposure to the lunar night temperatures. Thus at least two of the solar panels must close during the nighttime operation; otherwise there will be a power-supply failure upon subsequent return to daytime operation. When the operation is completed, the chemical furnace is activated by a thermostat when the temperature inside the vehicle falls to a preset level.

The reliability structure diagram for lunar night operations without the furnace is shown in Fig. 9.12. That is,

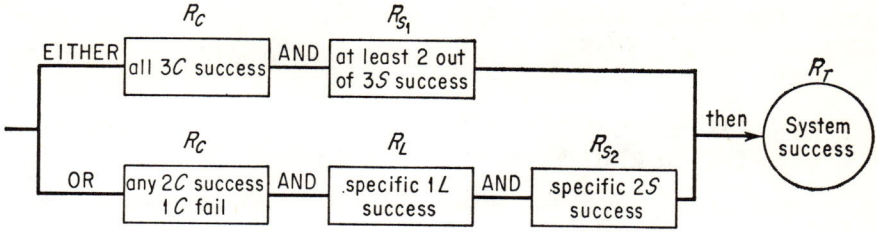

Fig. 9.12 Reliability structure diagram for temperature control subsystem during lunar night.

$$R_T(\text{night}) = R_C^3\left[\binom{3}{2}R_{S_1}^2(1 - R_{S_1}) + R_{S_1}^3\right] + \binom{3}{2}R_C^2(1 - R_C)R_L R_{S_2}^2$$

$$= R_C^3[3R_{S_1}^2 - 2R_{S_1}^3] + 3R_C^2(1 - R_C)R_L R_{S_2}^2$$

$$= R_C^3 + 3R_C^2(1 - R_C)R_L, \quad \text{when } R_{S_1}, R_{S_2} \approx 1$$

For 360 hours of lunar night, taking $R_C = 0.99706$ and $R_L = 0.98$ as before, we have

$$R_T(\text{night}) = 0.99354 \tag{9.84}$$

The total reliability potential of solar panels and louvers for lunar operations is the product of the day and night survival probabilities.

$$R_T(\text{lunar day and night}) = (0.99845)(0.99354)$$

$$\underline{R_T(\text{lunar day and night}) = 0.992} \tag{9.85}$$

Launch and landing shock. Past experience has indicated no unreliability of solar panels or temperature control louvers during launch and powered flight so that $\underline{0.9999}$ may be taken as a conservative estimate of reliability potential during this period. For landing shock, a reliability potential estimate of $\underline{0.999}$ will be assumed.

Chemical furnace. The chemical furnace is a simple, straightforward design and utilizes redundancy to improve survival probability. On this basis, the chemical furnace is conservatively estimated to have a reliability potential of at least $\underline{0.995}$, allowing for possible damage during the flight and landing.

Reliability estimate for the over-all temperature control system operation. The reliability estimate for the complete operation consists of the product of the separate estimated reliabilities for the flight, day, and nighttime operation of the solar panel controls, louvers, and solar cell arrays together with a successful operation of the chemical furnace. Thus

R_T (for flight, lunar day, and night)
$$= (0.999902)\,(0.992)\,(0.9999)\,(0.999)\,(0.995) = 0.986$$

9.5. SYSTEM PERFORMANCE VARIABILITY AS A FUNCTION OF SUBSYSTEM PERFORMANCE VARIABILITY

A problem of frequent importance is to specify the probability distribution of one, or possibly more than one, output performances of a system which depend on the input variables and the "transfer functions" of the sub-systems. For example, a rocket engine nozzle produces steady-state thrust, F, as a function of essentially independent* variables: chamber pressure, p_c, and nozzle throat area, A_t, by the relation

$$F = C_F p_c A_t \qquad (9.86)$$

where C_F is the thrust coefficient.

A second example is in the resultant velocity increment given to a payload by an upper stage solid-propellant rocket engine on a voyage to the moon. Assuming a vacuum and a gravitation-free trajectory, we have

$$V = gI_{sp} \log\left(1 + \frac{W_p}{W_M}\right) \qquad (9.87)$$

where g equals 32.174 ft sec^{-2} lb$_m$ lb$_f^{-1}$.

The quantity I_{sp} is the *specific impulse* of the propellant-nozzle combination (lb$_f$ sec lb$_m^{-1}$); W_p is the propellant weight and W_M is the weight of the rocket engine metal parts plus the payload weight. It is important that the velocity increment be sufficient so that the payload, when separated from the rocket engine stage, will arrive in the vicinity of the moon. On the other hand, the velocity increment should not be so large that the small payload rockets will not be able to inject the payload into a lunar orbit. Thus the upper stage should produce a fairly precise velocity increment to the payload. Variation or uncertainty in specific impulse I_{sp} and the weights W_p and W_M will produce variation in the velocity V. One can ask for the magnitude of the variation of V as a function of known variations in the variables I_{sp}, W_p, and W_M, or conversely, the required limits on variation of the latter variables, given allowable limits on the velocity V.

In general, the required or allowable limits on one or more parameters can be expressed as those limits which will be exceeded with a certain probability, when it is assumed that the parameters have, individually, normal distributions. This assumption, together with a method of calcu-

---

* Strictly speaking, C_F is a function of p_c and A_t and other variables, but is relatively insensitive to changes in these variables under optimum expansion conditions. Also, p_c is a strong function of A_t in the sense that steady-state chamber pressure level is determined by the throat area; however, variation in chamber pressure (from rocket engine to rocket engine) is due mainly to variation in propellant flow rates or propellant burning rate.

lating expected values and variances of functions of several variables, constitutes a very useful and practical approach to the problem at hand. We will come back to these examples after we have considered the general methods involved.

9.5.1 METHOD OF CALCULATING MEANS AND VARIANCES OF A FUNCTION OF SEVERAL VARIABLES

In what follows, we have Y as a single performance output, which is a function of several random variables $X_1, X_2, \cdots, X_n$. Thus

$$Y = H(X_1, X_2, \cdots, X_n) \qquad (9.88)$$

The means and standard deviations of the X_j are denoted by $\bar{X}_j$ and σ_j, respectively. We shall also consider that the X_j might be correlated, so that $\rho(X_i, X_j) \equiv \rho_{ij}$, although in most practical cases it will suffice to assume that the correlation coefficients are all zero. The latter assumption is also implied by the assumption of *pairwise independence* of the X_j.

The expected or mean value of the output Y is given approximately by

$$\bar{Y} = H(\bar{X}_1, \bar{X}_2, \cdots, \bar{X}_n) \qquad (9.89)$$

This formula is exact when H is a linear function of the X_j; i.e.,

$$H = \sum_{j=1}^{n} a_j X_j \qquad (9.90)$$

where the a_j are constants. In general, the greater the deviation of H from linearity (a hyperplane) over the ranges of the random variables X_j, the greater the error in Eq. (9.89). However, if the function H has first partial derivatives in a neighborhood of the mean $(\bar{X}_1, \bar{X}_2, \cdots, \bar{X}_n)$ it is essentially linear in that neighborhood; and if the standard deviations and therefore the ranges of the variables X_j are small enough to guarantee with high probability that the X_j are within the neighborhood, then Eq. (9.89) is very nearly correct. It is possible also to place a crude bound on the error of Eq. (9.89) if it is assumed that the second partial derivatives exist and are bounded over the ranges of the X_j, although this will not be given here.

The variance of Y can be derived by expanding H in a Taylor's series about the point $(\bar{X}_1, \bar{X}_2, \cdots, \bar{X}_n)$, neglecting all terms involving second or higher order derivatives, or products of more than two first derivatives, squaring the difference $Y - \bar{Y}$ and taking expected values, noting that $E(X_j - \bar{X}_j) = 0$. The result is

$$\text{Var } Y = \sum_{j=1}^{n} \left(\frac{\partial H}{\partial X_j}\right)^2 \sigma_j^2 + 2 \sum_{i<j} \left(\frac{\partial H}{\partial X_i}\right)\left(\frac{\partial H}{\partial X_j}\right) \rho_{ij} \sigma_i \sigma_j \qquad (9.91)$$

The derivatives in (9.91) are all evaluated at the mean $(\bar{X}_1, \bar{X}_2, \cdots, \bar{X}_n)$ and are treated as constants when taking expected values to derive the result.

Note the similarity of Eqs. (9.89) and (9.91) with Eqs. (8.15) and (8.16). In the latter case, however, the random variables X_j were considered as sample statistics based on a random sample of size n from a fixed population. It was, therefore, meaningful in that case to allow n to become large, so that asymptotically as $n \to \infty$, the expressions corresponding to Eqs. (9.89) and (9.91) are exact. In the present problem n is fixed, so that the "correctness" of Eq. (9.91) depends in some way upon the "smoothness" of the function H and the size of the standard deviations σ_j, and possibly the higher moments of the X_j.

EXERCISE 1: Derive Eqs. (9.89) and (9.91) by the method outlined.

EXERCISE 2: Use Eq. (9.91) to show that if $Y = X_1^a X_2^b$, then

$$\frac{\text{Var } Y}{\bar{Y}^2} = a^2 \frac{\sigma_1^2}{\bar{X}_1^2} + b^2 \frac{\sigma_2^2}{\bar{X}_2^2} + 2ab \frac{\text{Cov }(X_1, X_2)}{\bar{X}_1 \bar{X}_2}$$

EXERCISE 3: Let $Y = X^2$. Find the exact expected value of Y and the percentage error of the value given by Eq. (9.89) from the exact value.

EXAMPLE 1: Let $Y_1 = H_1(X_1, X_2)$, $Y_2 = H_2(X_1, X_3)$. What is the covariance of Y_1 and Y_2, assuming that X_1, X_2, X_3 are independent? By definition

$$\text{Cov }(Y_1, Y_2) = E[(Y_1 - \bar{Y}_1)(Y_2 - \bar{Y}_2)]$$

We have

$$Y_1 - \bar{Y}_1 = \left(\frac{\partial H_1}{\partial X_1}\right)(X_1 - \bar{X}_1) + \left(\frac{\partial H_1}{\partial X_2}\right)(X_2 - \bar{X}_2) + \cdots$$

$$Y_2 - \bar{Y}_2 = \left(\frac{\partial H_2}{\partial X_1}\right)(X_1 - \bar{X}_1) + \left(\frac{\partial H_2}{\partial X_3}\right)(X_3 - \bar{X}_3) + \cdots$$

Hence

$$(Y_1 - \bar{Y}_1)(Y_2 - \bar{Y}_2) = \left(\frac{\partial H_1}{\partial X_1}\right)\left(\frac{\partial H_2}{\partial X_1}\right)(X_1 - \bar{X}_1)^2$$

$$+ \left(\frac{\partial H_1}{\partial X_2}\right)\left(\frac{\partial H_2}{\partial X_1}\right)(X_2 - \bar{X}_2)(X_1 - \bar{X}_1)$$

$$+ \left(\frac{\partial H_1}{\partial X_1}\right)\left(\frac{\partial H_2}{\partial X_3}\right)(X_1 - \bar{X}_1)(X_3 - \bar{X}_3)$$

$$+ \left(\frac{\partial H_1}{\partial X_2}\right)\left(\frac{\partial H_2}{\partial X_3}\right)(X_2 - \bar{X}_2)(X_3 - \bar{X}_3) + \cdots$$

RELIABILITY STRUCTURE MODELS 263

The expected values of the last three terms vanish, so that

$$\text{Cov}(Y_1, Y_2) = \left(\frac{\partial H_1}{\partial X_1}\right)\left(\frac{\partial H_2}{\partial X_1}\right)\sigma_1^2 + \cdots \quad (9.92)$$

EXERCISE 4: Suppose, in the preceding example, that H_1 and H_2 are functions of X_1 only. What is the correlation coefficient of Y_1, Y_2?

Answer:

$$\rho(Y_1, Y_2) = \frac{\left(\frac{dH_1}{dX_1}\right)\left(\frac{dH_2}{dX_1}\right)\sigma_1^2}{\left|\left(\frac{dH_1}{dX_1}\right)\right|\left|\left(\frac{dH_2}{dX_1}\right)\right|\sigma_1^2}$$

$$= \begin{cases} +1 & \text{if } \left(\frac{dH_1}{dX_1}\right), \left(\frac{dH_2}{dX_1}\right) \text{ have the same sign} \\ -1 & \text{if } \left(\frac{dH_1}{dX_1}\right), \left(\frac{dH_2}{dX_1}\right) \text{ have opposite signs} \end{cases}$$

9.5.2 EVALUATION OF TOLERANCES

Suppose now that each of the "independent" variables and the performance output variable are to have tolerances placed on their nominal or mean values. We will assume that all variables are normally distributed; and by "tolerances" on the variable X is meant the values $\bar{X} - a$, $\bar{X} + a$ such that

$$P(\bar{X} - a \leq X \leq \bar{X} + a) = 1 - \alpha$$

where α is small, say 0.10, 0.05, 0.01, etc. The probability α is assumed the same for all variables in a given problem; so that if we specify the tolerances of the variables $X_1, X_2, \cdots, X_n$ and Y as $X_1 \pm a_1, X_2 \pm a_2, \cdots, X_n \pm a_n$, and $Y \pm b$, it is meant that

$$P(X_j \geq \bar{X}_j + a_j) = \frac{\alpha}{2} \quad (9.93)$$

for all X_j (and similarly for Y). From (9.93) we obtain

$$P\left(\frac{X_j - \bar{X}_j}{\sigma_j} \geq \frac{a_j}{\sigma_j}\right) = \frac{\alpha}{2} \quad (9.94)$$

and since $(X_j - \bar{X}_j)/\sigma_j$ has a standard normal distribution, then

$$\frac{a_j}{\sigma_j} = K_{\alpha/2} = \frac{b}{\sigma_Y} \tag{9.95}$$

where $K_{\alpha/2}$ is the standard normal deviate exceeded with probability $\alpha/2$. For $\alpha = 0.10, 0.05, 0.01, 0.005$, and 0.001, the values of $K_{\alpha/2}$ are 1.645, 1.960, 2.576, 2.807, and 3.291, respectively, correct to three decimal places.

Equation (9.95) allows one to state the tolerance in terms of the standard deviation, or conversely, the standard deviation in terms of the tolerance.

EXAMPLE: Consider the expression for the thrust given by Eq. (9.86). Typical 99 per cent specifications ($\alpha = 0.01$) are

$$p_c = 500 \text{ psia} \pm 3\%$$

$$A_t = 138.0 \text{ in.}^2 \pm 2\%$$

and $\quad C_F = 1.45$, assumed constant
(zero standard deviation)

The above data imply (see (9.94))

$$\frac{\sigma_{p_c}}{\bar{p}_c} = \frac{0.03}{2.576} = 0.0163$$

$$\frac{\sigma_{A_t}}{\bar{A}_t} = \frac{0.02}{2.576} = 0.00776$$

From Eq. (9.89) and Exercise 2 of Sec. 9.5.1, the value of $\sigma_F/\bar{F}$ is found to be

$$\frac{\sigma_F}{\bar{F}} = [0.000266 + 0.000060]^{1/2} = 0.0181$$

Thus the tolerance on thrust is $(2.576)(0.0181)(100\%) = 4.66\%$.

EXERCISE: Given only the tolerances on pressure p_c and thrust F, in the above example, what is the maximum allowable tolerance in inches on the diameter of the nozzle throat? If four nozzles with the same throat diameters were used instead of a single nozzle, what would be the maximum allowable tolerance on each throat diameter?

EXERCISE: Consider the example of the upper stage rocket given at the beginning of this section. For given tolerances on the velocity increment, specific impulse, and propellant weight, compute the maximum allowable tolerance on the rocket metal parts plus payload weight in pounds. Typical values might be

$$I_{sp} = 290 \pm 1\%$$

$$W_p = 200 \pm 0.5\%$$

$$V = 3140 \pm 1.5\%$$

Assume 99% tolerances.

REFERENCES

1. *Tables of the Cumulative Binomial Probability Distribution*, Harvard University Press, Cambridge, Mass., 1953.
2. R. J. Buehler, "Confidence Intervals for the Product of Two Binomial Parameters," *J. Am. Stat. Assoc.*, **52**, 482–493 (1953).
3. G. P. Steck, "Upper Confidence Limits for the Failure Probability of Complex Networks," Sandia Corporation Research Report, December 1957 (obtainable from OTS, Department of Commerce, Washington 25, D.C.).
4. Space Technology Laboratories, Inc., Report No. TR–59–0000–00756, "Tables of Upper Confidence Limits on Failure Probability of 1, 2 and 3 Component Serial Systems," July 1959.
5. J. H. Bollman, "Instructions and Data for Failure and Prediction," Bell Telephone Laboratories Report, October 11, 1957.
6. RCA Technical Report 59–46–1, "Reliability Stress Analysis for Electronic Equipment," January 15, 1959.
7. A. Albert, "A Measure of the Effort Required to Increase Reliability," Technical Report No. 43, November 5, 1958, Applied Mathematics and Statistics Laboratory, Stanford University, Contract No. N6onr-25140 (NR 342–022).
8. H. Cramér, *Mathematical Methods of Statistics*, Princeton University Press, Princeton, N.J., 1946.
9. H. S. Carslaw and J. C. Jaeger, *Operational Methods in Applied Mathematics*, 2d ed., Oxford University Press, 1943.
10. L. A. Aroian and R. H. Myers, "Redundancy Considerations in Space and Satellite Systems," *Proc. 7th National Symposium on Reliability and Quality Control, Philadelphia, Pa., January 9–11, 1961*.

ADDITIONAL READING

Balaban, H. S., "Some Effects of Redundancy on System Reliability," *I.R.E., EIA, ASQC, and AIEE, Proc. 6th Natl. Symp. on Reliability and Quality Control in Electronics, January 11–13, 1960*, 388–402.

Bellman, R., and Dreyfus, S., "Dynamic Programming and the Reliability of Multicomponent Devices," *Operations Research*, **6:2**, 200–206 (April 1958).

Brown, H. B., "The Role of Specifications in Predicting Equipment Performance," *Proc. 2d National Symposium on Quality Control and Reliability, Washington, D.C., January 9–10, 1956*, 133–48.

Cohen, G. D., "Predicting Performance Failures," *Machine Design*, **29:20**, 106–11 (October 3, 1957).

Cox, D. R., and Smith, W. L., "On the Superposition of Renewal Processes," *Biometrika*, Vol. 41, 1954, 91–99.

Creveling, C. J., "Increasing the Reliability of Electronic Equipment by the Use of Redundant Circuits," *Proc. I.R.E.*, **44**, 509–515 (April 1956).

DiToro, M. J., "Reliability Criterion for Constrained Systems," *I.R.E. Trans. Reliability and Quality Control*, **PGRQC-8,** 1–6 (September 1956).

Drenick, R. F., "The Failure Law of Complex Equipment," *J. Soc. Indust. Appl. Math.*, Vol. 8, No. 4, December 1960, 680–90.

Firstman, Sidney I., "The Use of Reliability Estimates in the Design of Missile Prelaunch Checkout Equipment," *Proc. 6th Joint Military-Industry Guided Missile Reliability Symposium, 15–17 February 1960*, 2–147.

Gordon, R., "Optimum Component Redundancy for Maximum System Reliability," *Operations Research*, **5**:2, 229–43 (April 1957).

Howard, W. J., "Chain Reliability: A Simple Failure Model for Complex Mechanisms," RAND Report No. RM-1058 (27 March 1953).

Kahn, L. B., "A Statistical Model for Evaluating the Reliability of Safety Systems for Plants Manufacturing Hazardous Products," *Technometrics*, **1**:3, 293–307 (August 1959).

Krohn, Charles A., "Improve Circuit Reliability," *Electronic Design*, **7**:7, 20–25 (April 1, 1959).

McLean, J. P., and Moskowitz, F., "Some Reliability Aspects of Systems Design," *I.R.E. Trans. Reliability and Quality Control*, **PGRQC-8,** 7–35 (September 1956).

Meltzer, Sanford A., "Designing for Reliability," *I.R.E. Trans. Reliability and Quality Control*, **PGRQC-8,** 36–43 (September 1956).

Moan, O. B., "Reliability of Series-Parallel Combinations," *Control Eng.*, **4,** 99–100 (April 1957).

Morrison, S. James, "The Study of Variability in Engineering Design," *Applied Statistics*, **6**:2, 133–38 (June 1957).

Wagner, D. H., "Numerical Reliability Requirements (AGREE)," *Proc. 3rd Natl. Symp. Reliability and Quality Control in Electronics, I.R.E., January 1957*, 154–58.

Wehrfritz, Frank W., "Uses of the Propagation of Errors," *National Convention Trans., ASQC, 1956*, 673–76.

Welker, E. L., and Bradley, C. E., "A General Method for Determining Logistic Requirements for a Satellite System," *Proc. 6th Joint Military-Industry Guided Missile Reliability Symposium, 15–17 February 1960*, 2–351 to 2–378.

Whiteman, I. R., "Reliability Starts With the Design," *I.R.E. Proc. 5th Natl. Symp. on Reliability and Quality Control in Electronics, January 12–14, 1959*, 98–102.

APPENDIX 9A

A TECHNIQUE FOR RELIABILITY APPORTIONMENT

The importance of reliability apportionment was discussed in Sec. 2.8. In this section, we will indicate a technique that could be followed during the course of a development program in which system reliability is to be increased to a given requirement. The technique is based on a theorem of A. Albert (Ref. 7) which is stated in the following paragraphs.

At any stage of a design and development program, it is assumed that the system can be partitioned into a number of subsystems and the initial reliability estimated for each subsystem, based, if possible, on tests of the given subsystem. Let $R_1, R_2, \cdots, R_n$ denote the subsystem reliabilities; and assuming that the "product-rule" holds (Eq. 9.7) the system reliability R would be given by

$$R = R_1 R_2 \cdots R_n \qquad (9A.1)$$

Let $\bar{R}$ be the required reliability of the system, where $\bar{R} > R$. It is then required to increase *at least* one of the values of the R_i, to the point that the required reliability $\bar{R}$ will be met, in accordance with Eq. (9A.1). To accomplish such an increase takes a certain *effort*, which is to be allotted in some way among the subsystems. The amount of effort would be some function of numbers of tests, amount of manpower applied to the task, and so forth. Under fairly general assumptions on the properties of such an *effort function* it was shown in Ref. 7 that the technique of increasing R to $\bar{R}$ with *minimum effort* is as follows:

(A) Order the known reliabilities $R_1, R_2, \cdots, R_n$ in nondecreasing order (we assume now that such an ordering is implicit in the notation) so that

$$R_1 \leq R_2 \leq \cdots \leq R_n \qquad (9A.2)$$

(B) Increase each of the reliabilities $R_1, R_2, \cdots, R_{K_0}$ to the *same* value $\bar{R}_0$; but do not attempt to increase the reliabilities $R_{K_0+1}, \cdots, R_n$.

The number K_0 is determined as

$$K_0 = \text{maximum value of } j \text{ such that}$$

$$R_j < \left[\frac{\bar{R}}{\prod_{i=j+1}^{n+1} R_i}\right]^{1/j} = r_j \text{ (say)} \tag{9A.3}$$

where $R_{n+1} = 1$ by definition.

The number $\bar{R}_0$ is determined as

$$\bar{R}_0 = \left[\frac{\bar{R}}{\prod_{j=K_0+1}^{n+1} R_j}\right]^{1/K_0} \tag{9A.4}$$

(C) It is evident that the system reliability will then be $\bar{R}$, since

$$\text{new reliability} = \bar{R}_0^{K_0} R_{K_0+1} \cdots R_n = \bar{R}_0^{K_0} \prod_{j=K_0+1}^{n+1} R_j \tag{9A.5}$$

and by using (9A.4) we immediately obtain

$$\text{new reliability} = \bar{R} \tag{9A.6}$$

EXAMPLE: Let $R_1 = 0.70$, $R_2 = 0.80$, $R_3 = 0.90$; then $R = 0.504$. The required value of system reliability is $\bar{R} = 0.65$. Suppose one did not consider the selection of K_0 by (9A.3) but arbitrarily decided to set $K_0 = 1$ and use Eq. (9A.4). One would then obtain

$$\bar{R}_0 = \left[\frac{0.65}{(0.80)(0.90)(1.00)}\right]^{1/1} = 0.903 \text{ approximately} \tag{9A.7}$$

and he would have

$$\bar{R} = 0.65 = (0.903)(0.80)(0.90)$$

as desired. However, the theorem tells us that *effort* to increase reliability has not been allotted in an optimum manner; i.e., more effort has been used than is necessary. Rather, one should determine K_0 by using (9A.3). To do this we calculate the three quantities

$$r_1 = \left(\frac{\bar{R}}{R_2 R_3 (1.00)}\right)^1, \quad r_2 = \left(\frac{\bar{R}}{R_3(1.00)}\right)^{1/2}, \quad r_3 = \left(\frac{\bar{R}}{(1.00)}\right)^{1/3} \tag{9A.8}$$

We have

$$r_1 = 0.903$$
$$r_2 = (0.722)^{1/2} = 0.850$$
$$r_3 = (0.65)^{1/3} = 0.866$$

Since $R_1 < r_1$, $R_2 < r_2$, but $R_3 > r_3$ then $K_0 = 2$, since 2 is the largest subscript j such that $R_j < r_j$. Thus, from (9A.4) $\bar{R}_0 = (0.65/0.90)^{1/2} = (0.722)^{1/2} = 0.850$, which means that the effort is to be allotted so that subsystem no. 1 increases in reliability from 0.70 to 0.850, and subsystem no. 2 increases in reliability from 0.80 to 0.850; whereas subsystem no. 3 is left alone with reliability 0.90. The resulting reliability of the entire system is, as required, $0.65 = (0.850)^2(0.90)$.

When the apportionment procedure (A)–(C), illustrated by the example, is followed, the results of the corresponding effort allotment can be re-examined at a later point of time. Perhaps, for example, six months from the time of the first forecast, which is (say) 18 months prior to the time at which the reliability requirement is to be met, we might re-estimate R_1 to be 0.80, $R_2 = 0.87$, $R_3 = 0.90$ (using the previous example). In fact, for various reasons effort might have been actually put on subsystem no. 3 (which increased its reliability to (say) 0.93), even though the original planning did not call for this effort. Also, reliability estimates might have greater precision at the later date. In any case, the apportionment procedure could then be repeated, yielding different subsystem goals and corresponding amounts of effort to achieve them. At subsequent periodic intervals, the results of the program would be examined, and a new apportionment made as necessary.

So far we have not specified the properties of the effort function used in Ref. 7 to derive the procedure. This is done as follows.

First, each subsystem has associated with it the *same* effort function $G(R_i, \bar{R}_i)$ which measures the amount of effort needed to increase the reliability of the ith subsystem from R_i to $\bar{R}_i$. The assumption that the effort function for each subsystem is the same has some drawbacks, but it represents a good first approximation to the problem. The remaining assumptions are:

1. $G(x, y) \geq 0$
2. $G(x, y)$ nondecreasing in y for fixed x and nonincreasing in x for fixed y; e.g.,

$$G(0.35, 0.65) \leq G(0.35, 0.75)$$

and $$G(0.25, 0.65) \geq G(0.35, 0.65)$$

3. If $x \leq y \leq z$, $G(x, y) + G(y, z) = G(x, z)$, which states that the amount of effort to increase the reliability from x to z is equal to the sum of the efforts to increase the reliability from x to y, then from y to z
4. $G(0, x)$ has a derivative $h(x)$ such that $xh(x)$ is strictly increasing in $(0 < x < 1)$.

With these assumptions, it was shown in Ref. 7 that the unique solution to the problem is given by the procedure A–C, when

$$\sum_{i=1}^{n} G(R_i, \bar{R}_i) \tag{9A.9}$$

is to be minimized subject to the condition

$$\prod_{i=1}^{n} \bar{R}_i = \bar{R} \tag{9A.10}$$

that is, the unique solution is

$$\bar{R}_i = \begin{cases} \bar{R}_0, & i \leq K_0 \\ R_i, & i > K_0 \end{cases} \tag{9A.11}$$

where K_0 is determined by (9A.3), $\bar{R}_0$ by (9A.4), and the ordering specified by (9A.2) is used.

EXERCISE: Show that

$$G(x, y) = \begin{cases} a \log \dfrac{y + b}{x + b}, & 0 \leq x \leq y \leq 1 \\ & a, b > 0 \\ 0, & x \geq y \end{cases}$$

satisfies the previously given assumptions on $G(x, y)$. Suppose we make the additional assumption that $\lim_{y \to 1} G(x, y) = +\infty$, where $0 \leq x < 1$. Is this assumption reasonable? Show that

$$G(x, y) = \begin{cases} a \log \dfrac{1 - x}{1 - y}, & 0 \leq x \leq y \leq 1 \\ & a > 0 \\ 0, & x \geq y \end{cases}$$

satisfies the additional assumption.

EXERCISE: Show that $G(x, y)$ must be of the form $H(y) - H(x)$.

Hint: Show that

$$\frac{\partial G(x, y)}{\partial y} = \frac{dG(0, y)}{dy}$$

which is only a function of y. Then verify the property $G(x, x) = 0$ from the original assumptions and apply it to obtain the desired result.

EXERCISE: Verify that the effort required to change (0.70, 0.80, 0.90) to (0.903, 0.80, 0.90) is greater than the effort required to change to (0.85, 0.85, 0.90). Use as the effort function $G(x, y) = y^{1/2} - x^{1/2}$.

APPENDIX 9B
RELIABILITY OF A SYSTEM WITH COMPONENT REPLACEMENT

In this appendix, we wish to show that the reliability of a complex system tends, in a certain sense, to be described by the exponential distribution of time-to-failure.

Assume that the system is serial with a large number of components, each of which fails independently of the others (see Sec. 9.2). Each component is assumed to have the same time-to-failure distribution function $F(t)$ and probability density function $f(t)$. As the system is operated, if one of the components fails, it is immediately replaced by a new component. At that time, however, the other components have age equal to the given time (assuming that every component is new at the beginning of operation), so that the expected remaining life of any one of them may be less than that of a newly installed component. As the system is operated for a longer period of time other components can fail, which are also immediately replaced, so that the system is "always operating," but the components at any time are of different ages.

At the start of operation of the system with all new components (all age zero) we would define the reliability (survival probability) as a function of time t to be

$$R(t) = [1 - F(t)]^N \tag{9B.1}$$

where N is the number of components. However, when some components have positive ages at the point when we start counting time, Eq. (9B.1) must be modified to take the initial ages of the components into account. If at any particular time t_0 the ages are $x_1, \cdots, x_N$, then the (conditional) probability of surviving an additional time t would be (see Sec. 6.8)

$$R(t; t_0 \mid x_1, \cdots, x_N) = \prod_{j=1}^{N} \frac{1 - F(x_j + t)}{1 - F(x_j)} \tag{9B.2}$$

The x_j are random variables depending on the time selected, t_0, as well as the time-to-failure distribution F. Thus, if we knew the probability density functions of the ages x_j, (say) $g(x_j; t_0)$, then the absolute probability of surviving a time t, when we start counting time at t_0 would be

$$R(t; t_0) = \overbrace{\int_0^\infty \cdots \int_0^\infty}^{N} R(t; t_0 \mid x_1, \cdots, x_N) \prod_{j=1}^{N} g(x_j; t_0) \, dx_1 \cdots dx_N \tag{9B.3}$$

or, using Eq. (9B.2) in Eq. (9B.3)

$$R(t; t_0) = \left[\int_0^\infty \frac{1 - F(x + t)}{1 - F(x)} g(x; t_0) \, dx \right]^N \tag{9B.4}$$

The probability density function $g(x; t_0)$ is given later by Eq. (9B.22). At this point we only consider the limiting distribution of ages when $t_0 \to \infty$. From Eq. (9B.22) and the remark following Eq. (9B.29), we have

$$\lim_{t_0 \to \infty} g(x; t_0) = \frac{1 - F(x)}{\mu} \tag{9B.5}$$

where μ is the mean of the time-to-failure distribution F.

In other words, if we look at one of the replacements of any particular one of the components after a long time, the probability that its age lies in the interval $(x, x + dx)$ is approximately $(1 - F(x)) \, dx/\mu$. (The right-hand side of Eq. (9B.5) is evidently a probability density function, since its integral from zero to infinity is one.) Thus if we incorporate (9B.5) into (9B.4), then

$$R(t; \infty) = \frac{1}{\mu^N} \left[\int_0^\infty [1 - F(x + t)] \, dx \right]^N \tag{9B.6}$$

Now, let us consider the expression

$$[R(t; \infty)]^{1/N} \equiv \tilde{R} = \int_0^\infty \frac{1 - F(x + t)}{\mu} \, dx \tag{9B.7}$$

Let $x + t = y$, then $dx = dy$ and (9B.7) becomes

$$\tilde{R} = \int_t^\infty \frac{1 - F(y)}{\mu} \, dy \tag{9B.8}$$

or

$$\tilde{R} = \int_0^\infty \frac{1 - F(y)}{\mu} \, dy - \int_0^t \frac{1 - F(y)}{\mu} \, dy \tag{9B.9}$$

$$= 1 - \frac{t}{\mu} + \int_0^t \frac{F(y)}{\mu} \, dy \tag{9B.10}$$

RELIABILITY STRUCTURE MODELS

We will now assume that the variance σ^2 of the time-to-failure distribution F exists. It can then be shown (Ref. 8, p. 256) that for $y < \mu$

$$F(y) \leq \frac{\sigma^2}{\sigma^2 + (\mu - y)^2} \tag{9B.11}$$

(The inequality is true for any distribution function F with mean μ and variance σ^2.)

Thus

$$\int_0^t \frac{F(y)}{\mu} \, dy \leq \frac{1}{\mu} \int_0^t \frac{\sigma^2 \, dy}{\sigma^2 + (\mu - y)^2} \tag{9B.12}$$

The integral on the right side of (9B.12) can be transformed by letting $(\mu - y)/\sigma = x$, so that it becomes

$$\frac{\sigma}{\mu} \int_{(\mu-t)/\sigma}^{\mu/\sigma} \frac{dx}{1 + x^2} = \frac{\sigma}{\mu} \left[\arctan\left(\frac{\mu}{\sigma}\right) - \arctan\left(\frac{\mu - t}{\sigma}\right) \right] \tag{9B.13}$$

Now since

$$\tan(A - B) = \frac{\tan A - \tan B}{1 + \tan A \tan B}$$

the previous expression becomes

$$\frac{\sigma}{\mu} \arctan \left[\frac{\dfrac{\mu}{\sigma} - \dfrac{\mu - t}{\sigma}}{1 + \left(\dfrac{\mu}{\sigma}\right)\left(\dfrac{\mu - t}{\sigma}\right)} \right]$$

$$= \frac{\sigma}{\mu} \arctan \left[\frac{\dfrac{\sigma t}{\mu^2}}{\left[\left(\dfrac{\sigma}{\mu}\right)^2 + 1 - \dfrac{t}{\mu}\right]} \right] \tag{9B.14}$$

$$\leq \frac{\sigma}{\mu} \arctan \left[\frac{\dfrac{\sigma t}{\mu^2}}{1 - \dfrac{t}{\mu}} \right] \tag{9B.15}$$

Since arctan $x \leq x$, then the last expression is not more than

$$\frac{\left(\dfrac{\sigma}{\mu}\right)^2}{\dfrac{\mu}{t} - 1}$$

Thus
$$\tilde{R} = 1 - \frac{t}{\mu} + H \qquad (9\text{B}.16)$$

where
$$H \leq \frac{\left(\dfrac{\sigma}{\mu}\right)^2}{\dfrac{\mu}{t} - 1} \qquad (9\text{B}.17)$$

Now, for fixed t, $\mu = \alpha N$ and $\sigma = o(N)$, we see that H is bounded by a function which is small order of N^{-1}. Since

$$\left[1 - \frac{t}{\alpha N} + o\left(\frac{1}{N}\right)\right]^N \to e^{-t/\alpha} \qquad (9\text{B}.18)$$

as $N \to \infty$, then
$$R(t; \infty) \to e^{-t/\alpha} \qquad (9\text{B}.19)$$

Thus for large, but fixed N
$$R(t; \infty) \approx e^{-tN/\mu} \qquad (9\text{B}.20)$$

which is the desired result.

If the system consists of N_1 components with time-to-failure distribution function F_1, N_2 components with time-to-failure distribution function F_2, and so on, it is then plausible that

$$R(t; \infty) \simeq \exp\left[-t\left(\frac{N_1}{\mu_1} + \frac{N_2}{\mu_2} + \cdots\right)\right] \qquad (9\text{B}.21)$$

where $\mu_1, \mu_2, \cdots$ are the respective means of the distributions $F_1, F_2, \cdots$.

It was stated previously that the limiting age probability density function was given by Eq. (9B.5). The age distribution at any time t_0, assuming that a component is new at time 0, can be shown to be

$$g(x; t_0) = \begin{cases} u(t_0 - x)(1 - F(x)), & x < t_0 \\ \delta(x - t_0)(1 - F(x)), & x = t_0 \\ 0, & x > t_0 \end{cases} \qquad (9\text{B}.22)$$

where $\delta(x - t_0)$ is the Dirac delta function, such that

$$\int \delta(x - t_0)(1 - F(x))\, dx = 1 - F(t_0)$$

and $u(t)$ is the probability of replacement at time t ($u(t)$ is not a probability density function), satisfying the (renewal) equation:

$$u(t) = f(t) + \int_0^t u(t - y)f(y)\, dy \qquad (9\text{B}.23)$$

Equation (9B.23) is interpreted as follows: $u(t)\, dt$ is the probability of any replacement in the interval $(t, t + dt)$, no matter how many previous replacements have been made. On the other hand $f(t)\, dt$ is the probability that a replacement is required in the time interval $(t, t + dt)$ for the first time. The integrand in Eq. (9B.23), when multiplied by dt, represents the probability that a replacement is required for the first time in the interval $(y, y + dy)$, followed by replacement at $t - y$ time units later in the time interval $(t - y, t - y + dt)$. This quantity, when integrated with respect to y, gives the total of the probabilities for all possible ways of replacement in the time interval $(t, t + dt)$, except for the case when replacement occurs for the *first* time in the interval $(t, t + dt)$. The latter expression is given by the first term of the righthand side of Eq. (9B.23) multiplied by dt. Thus, dividing through by dt, we obtain Eq. (9B.23).

Equation (9B.22) can be interpreted as follows: the probability that the age of a component falls in the interval $(x, x + dx)$ at time t_0 is equal to the probability that the component had been replaced at time $t_0 - x$ prior to time t_0 and has survived the time x. This is correct for $x < t_0$. Since we are assuming the component was new at time 0, the probability density function $g(x; t_0)$ must be zero for $x > t_0$ since the component could not be older than age t_0. For $x = t_0$, the expression used gives the probability that the component has survived exactly time t_0, in probability density notation.

The general theory tells us that the limit of $u(t)$, as $t \to \infty$ is $1/\mu$; thus Eq. (9B.5) follows from Eq. (9B.22) where $t_0 \to \infty$. To see that $\lim_{t \to \infty} u(t)$ is $1/\mu$, we take the Laplace transform of Eq. (9B.23), obtaining

$$\bar{u}(s) = \bar{f}(s) + \bar{u}(s)\bar{f}(s) \qquad (9\text{B}.24)$$

where

$$\bar{u}(s) = \int_0^\infty e^{-st} u(t)\, dt, \text{ the Laplace transform of } u(t) \qquad (9\text{B}.25)$$

and $\bar{f}(s)$ is defined similarly. Equation (9B.24) can be rewritten as

$$s\bar{u}(s) = \frac{s\bar{f}(s)}{1 - \bar{f}(s)} \qquad (9\text{B}.26)$$

Now, we use a known theorem on Laplace transforms (Ref. 9, p. 256):

$$\lim_{s \to 0} s\bar{h}(s) = \lim_{t \to \infty} h(t) \tag{9B.27}$$

where $\bar{h}(s)$ is the Laplace transform of $h(t)$.

Since $f(t) \to 0$ as $t \to \infty$, then $s\bar{f}(s) \to 0$ as $s \to 0$ by (9B.27). Also, the denominator of the right side of (9B.26) approaches 0 as $s \to 0$; i.e.,

$$\lim_{s \to 0} \int_0^\infty e^{-st} f(t)\, dt = \int_0^\infty f(t)\, dt = 1 \tag{9B.28}$$

Hence to obtain the limit of $s\bar{f}(s)/1 - \bar{f}(s)$ we differentiate numerator and denominator separately, obtaining

$$\lim_{s \to 0} s\bar{u}(s) = \lim_{s \to 0} -s - \frac{\bar{f}(s)}{\bar{f}'(s)}$$

$$= -\lim_{s \to 0} \frac{\bar{f}(s)}{\bar{f}'(s)}$$

$$= \frac{1}{\int_0^\infty t f(t)\, dt} = \frac{1}{\mu} \tag{9B.29}$$

Thus by (9B.27) $u(t) \to 1/\mu$ as $t \to \infty$.

Consider the following example: Let $f(t) = (1/\mu)e^{-t/\mu}$; then $\bar{f}(s) = 1/(\mu s + 1)$. Thus $\bar{u}(s) = 1/\mu s$ and therefore $u(t) = 1/\mu$. From Eq. (9B.22)

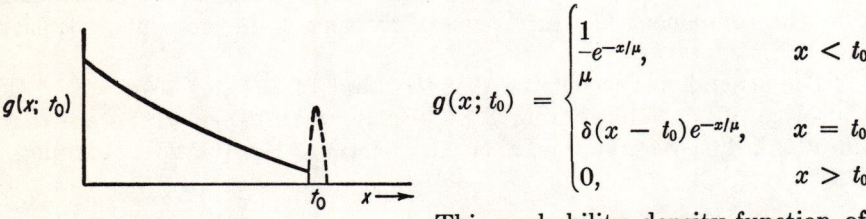

$$g(x; t_0) = \begin{cases} \dfrac{1}{\mu} e^{-x/\mu}, & x < t_0 \\ \delta(x - t_0) e^{-x/\mu}, & x = t_0 \\ 0, & x > t_0 \end{cases}$$

Figure 9B.1

This probability density function of age is as shown in Fig. 9B.1. The "spike" in Fig. 9B.1 contains probability $e^{-t_0/\mu}$, so that

$$\int_0^\infty g(x; t_0)\, dx = \int_0^{t_0} \frac{1}{\mu} e^{-x/\mu}\, dx + e^{-t_0/\mu} + 0$$

$$= 1 - e^{-t_0/\mu} + e^{-t_0/\mu} = 1$$

RELIABILITY STRUCTURE MODELS

As another example, consider

$$f(t) = \frac{4t}{\mu^2} e^{-2t/\mu} \tag{9B.30}$$

We have

$$\bar{f}(s) = \frac{4}{(\mu s + 2)^2}$$

Thus

$$\bar{u}(s) = \frac{4}{\mu s(\mu s + 4)} = \frac{1}{\mu s} - \frac{1}{\mu s + 4}$$

and therefore

$$u(t) = \frac{1}{\mu}(1 - e^{-4t/\mu}) \tag{9B.31}$$

The age probability density function is, for $x < t_0$,

$$g(x; t_0) = \frac{1}{\mu}(1 - e^{-4(t_0-x)/\mu}) e^{-2x/\mu}\left(1 + \frac{2x}{\mu}\right) \tag{9B.32}$$

Using (9B.32) in (9B.4), (or using (9B.5) directly) we find that for $t_0 \to \infty$,

$$\tilde{R} = \left(1 + \frac{t}{\mu}\right) e^{-2t/\mu} \tag{9B.33}$$

For t/μ small

$$\tilde{R} = e^{-t/\mu} + o\left(\frac{t}{\mu}\right) \tag{9B.34}$$

Note that $e^{-t/\mu}$ approximates $\tilde{R}$ better than does $e^{-2t/\mu}$ since the error in the latter case is $O(t/\mu)$ as $t/\mu \to 0$. Table 9B.1 gives the relative error in per cent for each of the two approximations to $\tilde{R}$.

The preceding results justify, in a certain sense, the use of the exponential time-to-failure distribution as a description of the failure characteristic of a system with many components which are of a variety of ages due to previous failures and replacements.

Table 9B.1

$\dfrac{t}{\mu}$	$\tilde{R} = \left(1 + \dfrac{t}{\mu}\right) e^{-2t/\mu}$	$e^{-2t/\mu}$	Per cent error	$e^{-t/\mu}$	Per cent error
0.5	0.552	0.368	−33.	0.607	+10.
0.2	0.804	0.670	−17.	0.818	+1.7
0.1	0.900	0.818	−9.1	0.904	+0.44
0.05	0.950	0.905	−4.7	0.9512	+0.13

The problem of determining a *replacement policy*, that is, a rule for making replacements with the purpose of preventing failures or eliminating marginally operating components before actual failure, can be based on some of the considerations of this appendix. One would also consider the effect of maintenance time involved in replacing or checking the operation of components. However, the act of replacing a component would have the same effect as failure of that component if the equipment were required to operate continuously. In the latter instance it would be desirable to have two or more complete equipments (such as radar sets) so that one would always be operating. Even so, all equipments could be in a failed state at the same time; but in general the probability of this occurrence would be smaller, the larger the number of equipments.

The above considerations and problems cannot be treated here in any detail. However, some of the additional reading references given at the end of this chapter cover this aspect of reliability.

CHAPTER TEN

RELIABILITY DEMONSTRATION AND DECISIONS

10.1. INTRODUCTION

In this chapter we shall be concerned with *sampling plan* procedures to determine whether or not reliability requirements are met. There is a close connection between these procedures and the results of Chaps. 7 and 8 on estimation and confidence intervals. The types of sampling that were discussed involved random sampling from attribute populations (success or failure) and from continuous populations (e.g., time-to-failure distributions). In either case the problem was to estimate a function of the parameters of the underlying population which could be defined as the reliability R; i.e., probability of success, or probability that time-to-failure exceeds T. The precision of the estimate was expressed by calculating a confidence limit, generally a lower one-sided limit $\hat{R}_L$ on the function R, such that the result of the sampling was the statement: "$\hat{R}_L < R$." A priori, the probability that the statement was to be true was fixed at γ (or at least γ), the confidence coefficient.* Thus, in the circumstance that a long series of such statements were to be made, i.e., each after a sample had been observed, one would expect that 100γ per cent (or at least 100γ per cent) of the statements would be true.

On the other hand, the quantity $100(1 - \gamma)$ represents the risk in per cent, when making any such statement, of being in error. If it were agreed in advance of making observations that a certain action A would be taken when the results of the sampling led one to state "$\hat{R}_L \leq R$," when in reality $R < \hat{R}_L$, evidently action A would be in error, perhaps seriously or even disastrously so.

---

* We are anticipating the notation used in Sec. 10.6, *et seq*. However, in Sec. 10.3 the notation $R_2 \equiv 1 - p_2$ (as is customary in binomial sampling) is used to denote the reliability to be demonstrated. Also, in Sec. 10.3 the quantity β is used to denote the risk of error in making the statement "$R_2 < R$." Thus (R_2, β), $(\hat{R}_L, 1 - \gamma)$ are equivalent notations in this chapter.

To be more precise we now wish to specify the quantity $\hat{R}_L$ together with the maximum risk of being in error in making the statement "$\hat{R}_L < R$." Then the problem will be to select a sampling plan, together with a *criterion*, which, depending upon the result of the sampling, tells us to make the statement "$\hat{R}_L < R$," or not. If the statement is in fact false, the probability of stating "$\hat{R}_L < R$" will be at the most equal to the specified maximum risk, owing to the properties of the chosen sampling plan.

Another equivalent way of stating the specification is as follows: *A reliability $\hat{R}_L$ is to be demonstrated with confidence γ.* The quantity $1 - \gamma$ is identified with the maximum risk just mentioned. It will be found that the numerical specification of the quantities $\hat{R}_L$ and γ is not sufficient to specify a *unique* sampling plan. This is shown particularly in Sec. 10.3 and in subsequent sections of this chapter.

In Sec. 10.2, binomial sampling plans are discussed from a very general point of view, extending the material of Sec. 6.6.

Section 10.4 presents the well-known Wald sequential binomial sampling plans. In particular, the optimum feature of the Wald plans (minimum average sample size needed for testing) is shown to be essentially a direct implication of the considerations of Sec. 10.2.

In Secs. 10.5–10.9, sampling plans based on time-to-failure distributions are presented; those based on the exponential distribution are emphasized. Benjamin Epstein and Milton Sobel (see the references at the end of Chap. 10) are responsible for a good part of the theory and applications of sampling plans based on the exponential distribution.

In addition, sampling plans based on the more general time-to-failure distribution given in Sec. 6.8.3.1 are also discussed briefly in Sec. 10.6.

Sampling plans based on the normal distribution are not presented here, since they are adequately covered in the literature, notably in Refs. 1, 2, and 3.

10.1.1 NONSTATISTICAL FACTORS IN SELECTION OF SAMPLING PLANS

Generally, the greater the information used in making a decision or taking an action, the less our chances of making an error. To increase the information and thereby decrease this risk, we can increase the sample size or the number of tests used to reach a decision. However, suppose the risk could be expressed in terms of dollar loss. If testing is also costly, in terms of dollars, the attempt to decrease the risk of making a wrong decision to a negligible value, and thereby prevent incurring a large dollar loss, would be nullified by the expense of testing.

Cost is not the only "nonstatistical" factor that influences the choice of a sampling plan. Before any one plan can be formulated, the practical circumstances of the situation must be thoroughly examined. In fact, considerations of these circumstances will invariably result in a narrowing down of the variety of plans from which to choose. For instance, the type of equipment being tested and the method of testing will determine whether we observe attribute or variables data. Of course there may be some freedom of choice in the type of data obtained. For example, if the test items are continuously operating, perhaps it would be desirable to measure times-to-failure. However, since it might be costly or impractical to install automatic equipment or a full-time observer to take these measurements, we may have to be satisfied with counting numbers of failures at the end of discrete periods (e.g., each morning). This would therefore change the sampling model. Again, the failure characteristics of the equipment itself and/or its complexity of operation might not indicate any amenable time-to-failure distribution, so that an attribute test result of failure or success would probably be the most realistic type of observation to make.

Further considerations, which would strongly influence the choice of a sampling plan, would result from the manner in which the items were made available for testing as well as from the urgency of generating the test results. For example, it might be more efficient to use a sequential plan whether or not all items were available at one time or were being produced sequentially. On the other hand, perhaps the schedules and high reliability goals would require more data than could be generated with the testing of single items, one at a time.

The method of truncating the testing might depend on such factors as the necessity to stop testing by a given date (as for a scheduled milepost requirement); or the fact of having a limited number of items available for a reliability demonstration program; or that of having a limited amount of test facility time available.

The restricting considerations given above, though typical of those found in industry, are by no means all-inclusive, and the reader can probably add many of his own from the particular field with which he is most familiar.

It is considered, therefore, that the nature of the problem of establishing a reliability demonstration sampling plan is to plan the testing and sampling so that, within the framework of the above practical considerations, the risks of making wrong decisions are acceptably small in some sense.

10.2. RELIABILITY DECISIONS BASED ON BINOMIAL SAMPLING

We will consider in this section methods of making one of two decisions based on binomial sampling. One decision is to "accept" an item as sufficiently reliable for its intended purpose; the other is the complementary decision, i.e., to "reject" the item as too unreliable for its intended purpose.

Binomial sampling, or a *binomial sampling plan* as we shall call it, consists of testing a set of the items one by one, i.e., making trials, such that either success or failure is observed on each trial. The probability of failure on any one trial is constantly equal to p, and the result of any trial is statistically independent of the results of any other trials. Furthermore, the sampling process terminates in a specified manner, at which point one of the two aforementioned decisions is made. In keeping with standard terminology on the subject, the criteria used in setting up such plans will be in terms of the parameter $p \equiv 1 - R$, the unknown unreliability of the item under consideration.

To make a decision to accept or reject the item, we have to decide first of all what range of values of p should be considered as "acceptable" and what range of values as unacceptable. There is therefore some value of unreliability p_c such that if $p \leq p_c$, it is desirable that the sampling plan lead to a decision to accept the item as sufficiently reliable. If, however, $p > p_c$, it is desirable that the sampling plan lead to a decision to reject the item as not sufficiently reliable for its intended use. We already know, however, that no sampling plan can be infallible in this respect in that there is always some uncertainty associated with sampling. Thus the sampling plan can lead to a decision to reject the item when in fact $p \leq p_c$; or it can lead to a decision to accept the item when in fact $p > p_c$. If either of these two events occurs, the decision will be in error, in the context of the previous discussion.

The seriousness of the error, it is evident, should depend upon the "distance" of p from p_c. Thus if p were actually very small compared to p_c and the sampling plan led to a decision to reject the item, the error would be more serious than if p were just slightly less than p_c. Also, if p were much larger than p_c and the sampling plan led to a decision to accept the item, this would be regarded as a more serious error than if p were just slightly larger than p_c.

For a sampling plan to be effective, therefore, it must almost certainly not allow serious errors to be made. Thus the probability of rejecting when $p \ll p_c$ should be sufficiently small; and the probability of accepting when $p \gg p_c$ should be sufficiently small.

10.2.1 THE OPERATING CHARACTERISTIC FUNCTION

The type of sampling plans to be considered have the characteristic that as $p < p_c$ and $p \rightarrow 0$, the probability of rejecting the item approaches zero. Also if $p > p_c$ and $p \rightarrow 1$, the probability of accepting the item as suitable approaches zero. This behavior is to be exhibited by the *operating characteristic* (*OC*) *function* $L(p)$ of the plan, whose general features are shown in Fig. 10.1. The function $L(p)$ is defined as the probability that the sampling plan leads to a decision to accept the item. The OC function is a measure of the ability of a plan to distinguish between reliable and unreliable items. Ideally, a plan with the OC function shown in Fig. 10.1(a) would be best, since for all values of $p \leq p_c$ the plan would lead to a decision of acceptance, and for all values of $p > p_c$ the plan would lead to a decision of rejection. However, it can be shown that to achieve this type of "error-free" decision rule, the number of trials would have to be infinite. In general, then, when only a given finite number of trials are to be used to reach a decision, the *steeper* the OC function, the better the ability of the plan to reach a decision with minimum error. It is necessary, however, to consider other criteria related to the allowable or maximum errors of incorrect decisions before a unique plan can be specified; we will discuss these in Sec. 10.3.

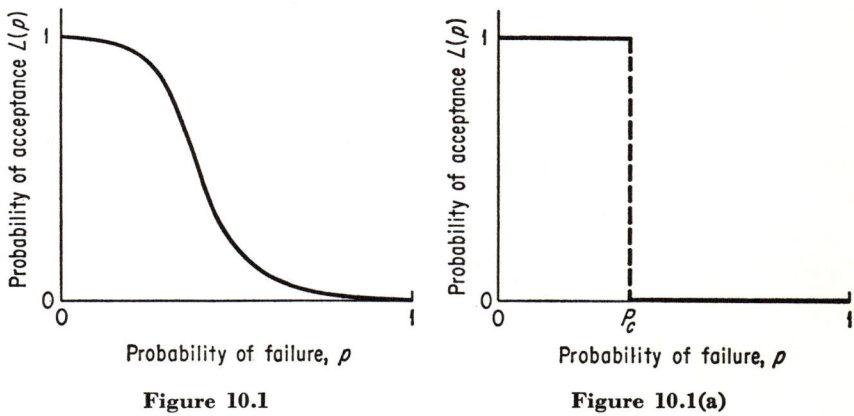

Figure 10.1   Figure 10.1(a)

10.2.2 GENERAL PROPERTIES OF A BINOMIAL SAMPLING PLAN

Before we consider some of the criteria that might be used to specify a binomial sampling plan, let us now examine the general properties of such a plan. In Sec. 6.6 the most general binomial sampling scheme was described. The *sample space* or set of all possible outcomes of sampling was specified by *boundary points* in the failures-vs.-trials plane. To specify a binomial sampling plan that leads to one of two decisions, the set of boundary points must be divided into two groups. If the result of some trial is a boundary point belonging to the first group, then the decision is to reject the item. Similarly, if the result of some trial is a boundary point belonging to the second group, the decision is to accept the item. The first group of boundary points is often called the *critical region*. Figure 10.2 illustrates a binomial sampling plan, where the boundary points constituting the critical region are indicated by X's.

The probability $L(p)$ that the result of the binomial sampling plan of Fig. 10.2 leads to a decision to accept can be found by summing the proba-

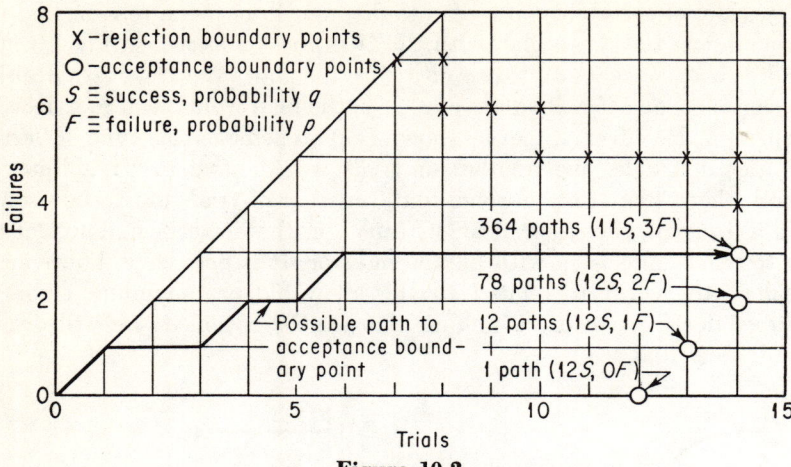

Figure 10.2

bilities attached to each of the four acceptance boundary points. These probabilities can be found by the method of Sec. 6.6. Thus the OC function is

$$L(p) = q^{12} + 12q^{12}p + 78q^{12}p^2 + 364q^{11}p^3 \qquad (10.1)$$

Figure 10.3 presents a graph of $L(p)$ given by Eq. (10.1).

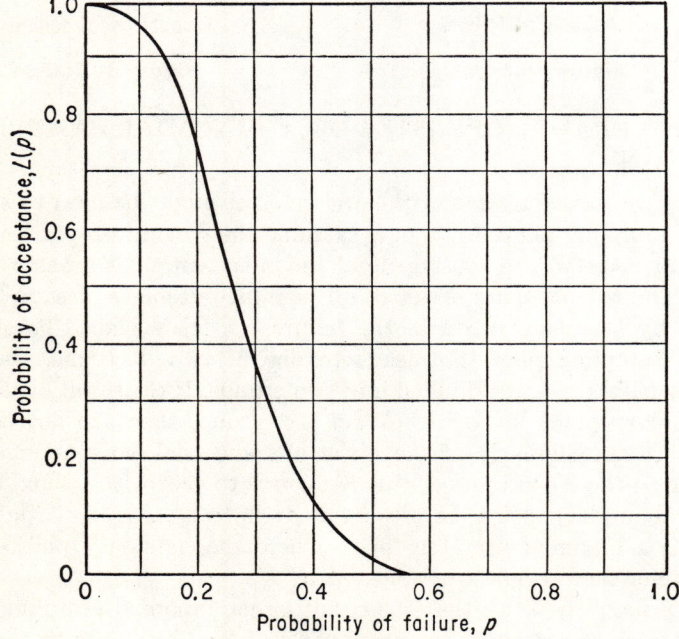

Figure 10.3

10.2.3 THE AVERAGE SAMPLE NUMBER FUNCTION

Suppose that the rejection boundary points of the plan shown in Fig. 10.2 were moved in any manner (if necessary adding to the number of rejection points to keep the boundary closed) such that none of the paths to acceptance points shown were interrupted [see, for example, Figs. 10.4(a) and (b)]. The function $L(p)$ would still be given by Eq. (10.1)

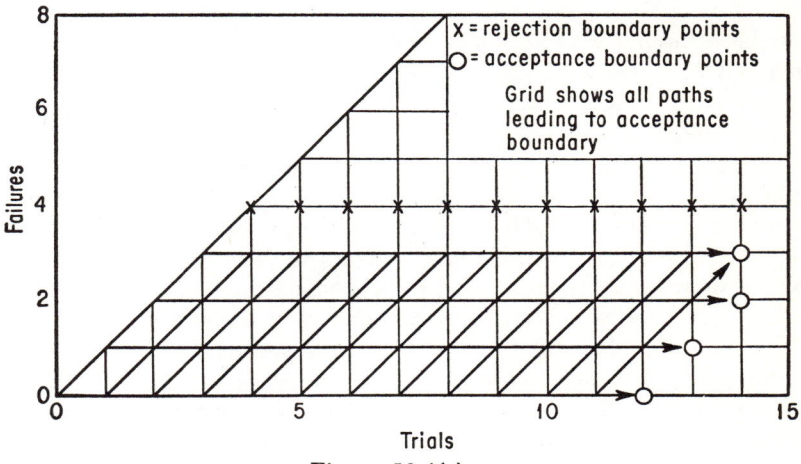

Figure 10.4(a)

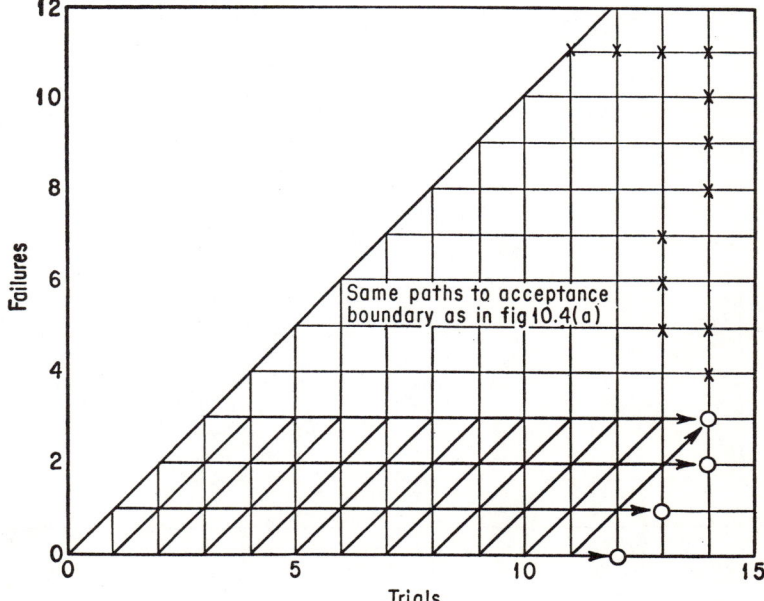

Figure 10.4(b)

since the probabilities of reaching each of the acceptance points are unchanged. What, then, is the difference between the binomial sampling plans of Figs. 10.4(a), 10.4(b), and 10.2? The answer is: the number of trials to reach one of the two decisions. In each of the three plans the maximum possible number of trials is 14. However, for any given value of p, there is a probability distribution of the number of trials; for example, the range of this distribution would be between 4 and 14 trials for the plan of Fig. 10.4(a). A simple way to characterize this aspect of a given binomial sampling plan is to plot the *expected* number of trials needed to reach a decision as a function of p. This is commonly called the *average sample number* function, abbreviated ASN (see Fig. 10.8).

It is evident that of the three plans shown in Figs. 10.2, 10.4(a), 10.4(b), viz. the one shown in Fig. 10.4(a), will have the smallest ASN; i.e., its graph will generally lie below those of Figs. 10.2 and 10.4(b). The savings in number of trials to reach a decision will occur mainly for higher values of p, since if the item is unreliable, the plot of failures vs. trials will most probably move quickly up to a rejection boundary point.

Now, it is also possible to shift the acceptance points (12, 0), (13, 1), (14, 2) to (11, 0), (12, 1), and (13, 2) respectively, in Fig. 10.4(a), leaving the acceptance point (14, 3) unchanged, without changing the function $L(p)$ given by Eq. (10.1). This is evident because the arrival in the course of sampling at the points (11, 0), (12, 1), (13, 2) *guarantees* a decision of acceptance. These changes would tend to decrease the ASN function further, leading to a greater savings in numbers of trials to reach

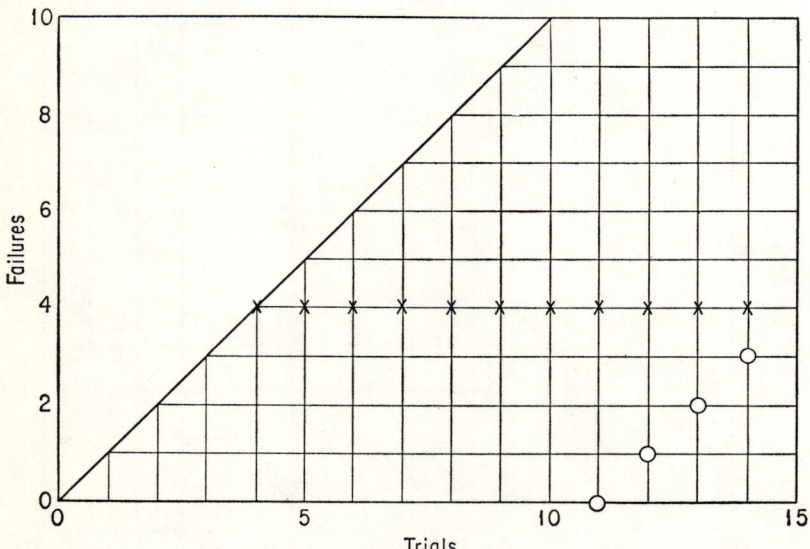

Fig. 10.5 Curtailed sampling plan.

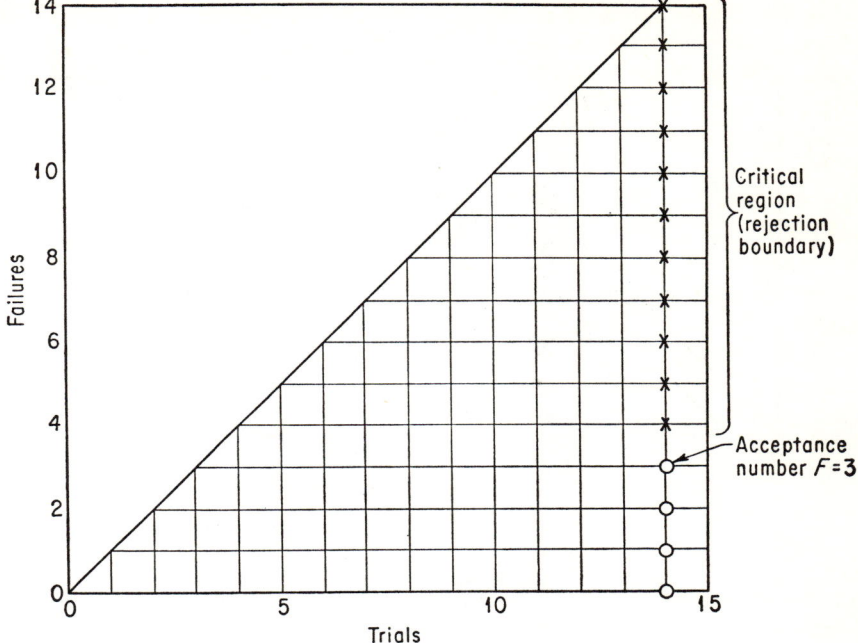

Fig. 10.6 Fixed sample size sampling plan.

a decision, compared to the plan of Fig. 10.4(a). Figure 10.5 shows the resulting plan. This is known as a *curtailed* sampling plan and commonly replaces the *fixed sample size* sampling plan shown in Fig. 10.6. The curtailed sampling plan of Fig. 10.5 evidently will have the "lowest" ASN function of all of the plans given in Figs. 10.2, 10.4, and 10.6.

The ASN function for the plan of Fig. 10.5 can be obtained analytically in terms of the binomial distribution function. For the plans of Figs. 10.2 or 10.4, it would be easier to compute the ASN directly by summing, over all boundary points, the products of the probability attached to each boundary point times the number of trials to reach the given boundary point.

10.2.3.1. Formula for the ASN Function for a Curtailed Binomial Sampling Plan. We will now compute the ASN function for the general *curtailed* sampling plan which limits the maximum number of trials to N, and the maximum number of failures for acceptance to F. (In Fig. 10.5, $N = 14$, $F = 3$). Figure 10.7 depicts the general curtailed sampling plan. First we deal with the rejection points, whose coordinates are $(j, F + 1)$, where j runs from $F + 1$ to N. The probability associated with the point $(j, F + 1)$ is given by $p^{F+1}q^{j-F-1}$ times the number of paths to the point $(j, F + 1)$. This number of paths is the same as the number of paths to the point

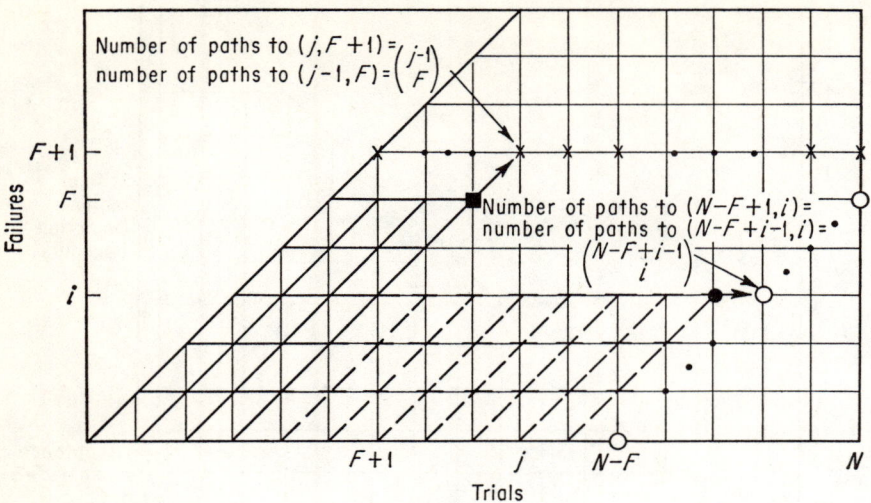

Figure 10.7

$(j - 1, F)$ since the point $(j, F + 1)$ is accessible only from the point $(j - 1, F)$. Thus the probability for $(j, F + 1)$ is

$$P(j, F + 1) = \binom{j - 1}{F} p^{F+1} q^{j-F-1} \tag{10.2}$$

Since the number of trials to reach the point $(j, F + 1)$ is j, then that part of the average number of trials corresponding to rejection points only is

$$E_r = \sum_{j=F+1}^{N} j \binom{j - 1}{F} p^{F+1} q^{j-F-1} \tag{10.3}$$

For the acceptance points we find similarly, referring to Fig. 10.7,

$$E_a = \sum_{i=0}^{F} (N - F + i) \binom{N - F + i - 1}{i} p^i q^{N-F} \tag{10.4}$$

and the ASN $= E_r + E_a$.

The expected values E_r and E_a will now be expressed in terms of the binomial distribution function (Sec. 6.2), since the result will enable us to compute the ASN function from Ref. 1. We use the notation

$$B(n, p; X) = \sum_{j=X}^{n} \binom{n}{j} p^j q^{n-j} \tag{10.5}$$

which represents "the probability of X or more failures in n binomial trials"; i.e., the "tail" of a binomial distribution function. To determine

RELIABILITY DEMONSTRATION AND DECISIONS 289

E_r, we make use of the identity given by Eq. (6.56), calling the left side of Eq. (10.6) H:

$$H \equiv \sum_{j=F+1}^{N} \binom{j-1}{F} p^{F+1} q^{j-F-1} \equiv B(N, p; F+1) \qquad (10.6)$$

Now,

$$q \frac{dH}{dq} = \sum_{j=F+1}^{N} (j - F - 1)\binom{j-1}{F} p^{F+1} q^{j-F-1} \qquad (10.7)$$

$$= E_r - (F+1)B(N, p; F+1) \qquad (10.8)$$

using (10.6).

Next, we have to find $q(dH/dq)$ directly. From (10.7), since the summand is zero when $j = F + 1$, we have, with some slight manipulation of the binomial coefficient,

$$q \frac{dH}{dq} = \frac{q(F+1)}{p} \sum_{j=F+2}^{N} \binom{j-1}{F+1} p^{F+2} q^{j-F-2} \qquad (10.9)$$

$$= \frac{q(F+1)}{p} B(N, p; F+2) \qquad (10.10)$$

using (10.6) again. Thus, from (10.8) and (10.10),

$$E_r = \frac{q(F+1)}{p} B(N, p; F+2) + (F+1)B(N, p; F+1) \qquad (10.11)$$

To find E_a, we transform the index of summation in (10.4) to $j = N - F + i$ and define $f + 1 \equiv N - F$, thereby obtaining

$$E_a = \sum_{j=f+1}^{N} j\binom{j-1}{f} q^{f+1} p^{j-f-1} \qquad (10.12)$$

The righthand side of (10.12) matches the expression (10.3) for E_r except that f replaces F and q replaces p; hence, using (10.11)

$$E_a = \frac{p(f+1)}{q} B(N, q; f+2) + (f+1)B(N, q; f+1) \qquad (10.13)$$

Now, from (10.5) it is easy to see that

$$B(n, q; X) \equiv 1 - B(n, p; n - X + 1) \qquad (10.14)$$

Hence from (10.13), using (10.14) and the definition of f, we obtain

$$E_a = \frac{N-F}{q} - \frac{(N-F)p}{q} B(N, p; F) - (N-F)B(N, p; F+1) \qquad (10.15)$$

and

$$\text{ASN} = E_r + E_a = \frac{N-F}{q} + \frac{q(F+1)}{p} B(N, p; F+2)$$

$$- (N - 2F - 1)B(N, p; F+1) - \frac{(N-F)p}{q} B(N, p; F) \quad (10.16)$$

The expression for the ASN given by (10.16) can be further simplified to

$$\text{ASN} = \frac{F+1}{p} B(N+1, p; F+2)$$

$$+ \frac{N-F}{q}[1 - B(N+1, p; F+1)] \quad (10.17)$$

by making use of the identity

$$\binom{N+1}{j} \equiv \binom{N}{j} + \binom{N}{j-1} \quad (10.18)$$

EXERCISE: Note that the identity (10.18) is a simple consequence of the fact that the number of unrestricted paths to the point $(N+1, j)$ is the sum of the number of unrestricted paths to the points (N, j) and $(N, j-1)$. Use (10.18) and (10.5) to derive (10.17) from (10.16).

The ASN function for the plan shown in Fig. 10.5 is given by (10.17) where $N = 14$, $F = 3$, and is shown plotted as the solid curve in Fig. 10.8. Note that the *expected* savings in number of trials to reach a decision is

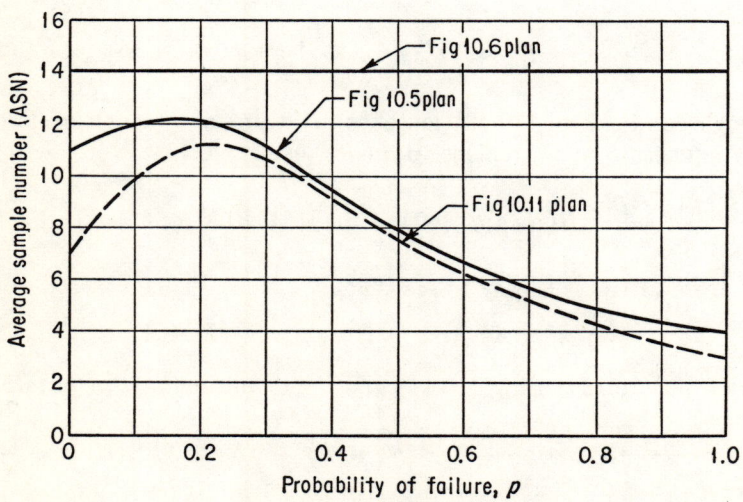

Figure 10.8

considerable for larger values of p, as compared to the fixed-sample-size plan of Fig. 10.6, for which 14 trials are to be made irrespective of any sampling results.

10.2.4 MODIFICATIONS TO OPTIMIZE THE CURTAILED SAMPLING PLAN

We have considered, so far, a set of sampling plans with the same OC function, but with different ASN functions. Of this set, the curtailed sampling plan presents the "smallest" ASN function and therefore would be considered the best plan of the set. *It is highly remarkable that there exist binomial sampling plans which can further decrease the ASN function by significant amounts and still leave the OC function essentially unchanged.* If in our attempt to reduce the ASN from that of a curtailed plan we require that the maximum possible number of trials remain fixed (in the example we have been using for illustration, $N = 14$ is this maximum), the OC function will change only for the worse; i.e., it will become less steep. This means that the values of $L(p)$, for values of p less than but near to p_c, will tend to decrease and, for values of p greater than but near to p_c, will tend to increase. If, however, we allow the maximum number of trials to increase somewhat, or even allow an *unlimited* number of trials to reach a decision, the OC function can, as stated, be made to remain essentially unchanged or even improved, yet the ASN function can be decreased.*
The points just stated will be illustrated in the following paragraphs. We shall first discuss the curtailed sampling plan shown in Fig. 10.5 and then proceed to modify it in order to decrease the ASN.

The sampling plan of Fig. 10.5 is shown again as Fig. 10.9, with numbers of paths to each point noted. The first modification is to change the

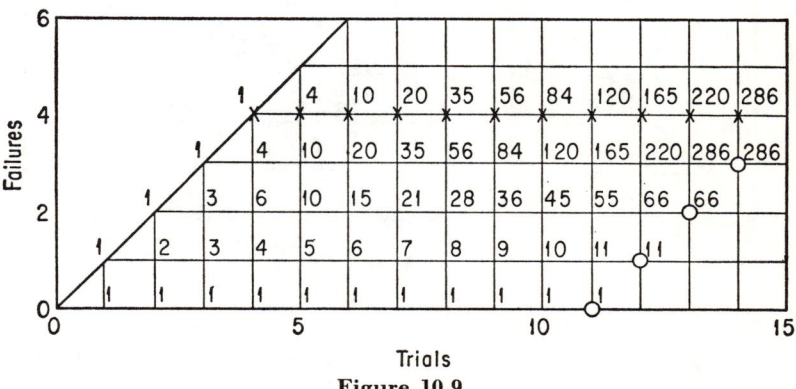

Figure 10.9

* With unlimited trials allowed, the variability of the number of trials to reach a decision becomes important. See Ref. 5.

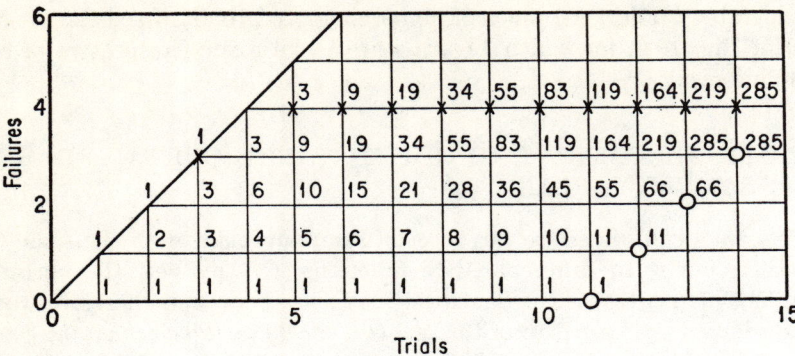

Figure 10.10

rejection boundary point (4, 4) to (3, 3). Figure 10.10 shows the resulting plan and numbers of paths to each point. The only effect on the OC function is indicated by the reduction in number of paths to the acceptance point (14, 3) from 286 to 285. Thus if $L(p)$ is the OC function for the plan of Fig. 10.9,* then $L_1(p) = L(p) - q^{11}p^3$ is the OC function for the plan of Fig. 10.10. Thus $L_1(p) < L(p)$ for all values of p, and the maximum reduction is approximately 0.00069 and occurs at $p = \frac{3}{14} = 0.214$. The ASN function also decreases for all values of p by the amount $p^2(1 - q^{11})$, which amounts to one full trial reduction when $p = 1$ (as is obvious) but is much less for small values of p, in fact zero when $p = 0$.

EXERCISE: Show that the decrease in the ASN function is given by $p^2(1 - q^{11})$.

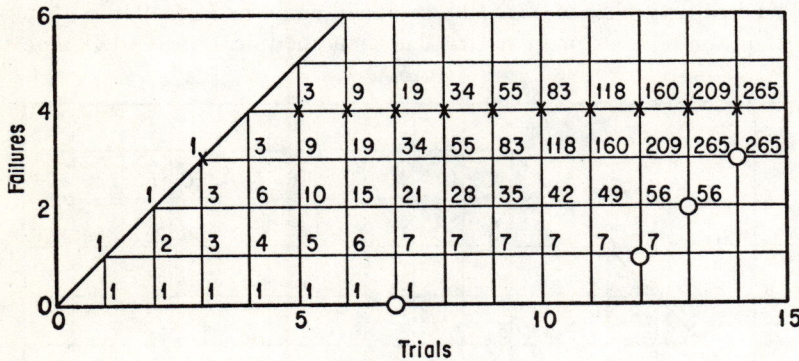

Figure 10.11

* $L(p)$ was given by Eq. (10.1). Note that $L(p)$ from Figs. 10.5 or 10.9 would be obtained directly as

$$L(p) = q^{11} + 11q^{11}p + 66q^{11}p^2 + 286q^{11}p^3 \qquad (10.1a)$$

(Show that the two expressions (10.1) and (10.1a) are the same.)

In order to decrease the ASN function for small values of p, the next modification is to move the *acceptance* point (11, 0) for the plan of Fig. 10.10 closer to the origin. For reasons which will be noted later, the new acceptance point is chosen as (7, 0). Figure 10.11 depicts the second modification of the curtailed sampling plan of Fig. 10.9.

The OC function $L_2(p)$ for the plan shown in Fig. 10.11 is given by

$$L_2(p) = q^7 + 7q^{11}p + 56q^{11}p^2 + 265q^{11}p^3 \qquad (10.19)$$

Thus

$$L(p) - L_2(p) = q^{11} - q^7 + 4q^{11}p + 10q^{11}p^2 + 21q^{11}p^3 \qquad (10.20)$$

$$= q^{11}(1 + 4p + 10p^2 + 21p^3) - q^7 \qquad (10.21)$$

By numerical computation it can be determined that $L_2(p)$ differs from $L(p)$ by, at most, approximately 0.01 (when $p \simeq 0.3$). The ASN function for the plan shown in Fig. 10.11 is plotted in Fig. 10.8 as the dashed curve. For values of $p \leq 0.2$, the expected saving in sample size over the curtailed plan of Fig. 10.5 (or 10.9) is at least one and as much as four.

The procedure of modifying the curtailed sampling plan can be carried further; i.e., by "pulling in" toward the origin both the rejection boundary points corresponding to low numbers of successes and the acceptance boundary points corresponding to low numbers of failures. Figure 10.12 illustrates a resulting nearly "optimum" plan whose OC function differs by only a small amount from that of the curtailed sampling plan of Fig. 10.5 or 10.9, but whose ASN function is essentially as small as possible. Figures 10.13 and 10.14 illustrate the OC and ASN functions for the plan shown in Fig. 10.12. We will see later in Sec. 10.4 that the plan of Fig. 10.12 is essentially equivalent to a truncated Wald sequential probability ratio test for binomial proportion. The latter (untruncated) test is known to be optimum, or nearly optimum, under certain conditions, where "opti-

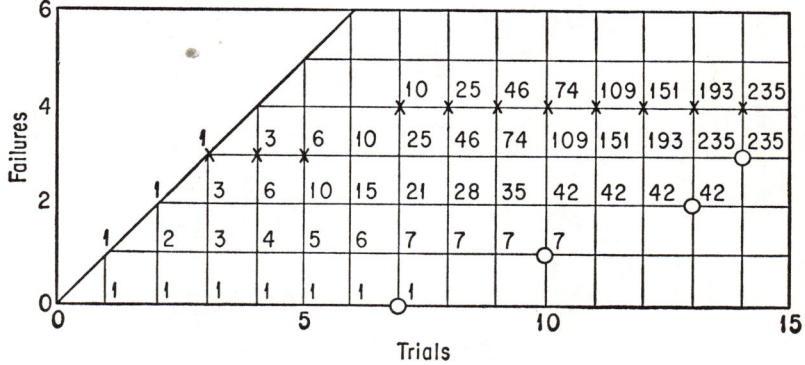

Figure 10.12

294 RELIABILITY DEMONSTRATION AND DECISIONS

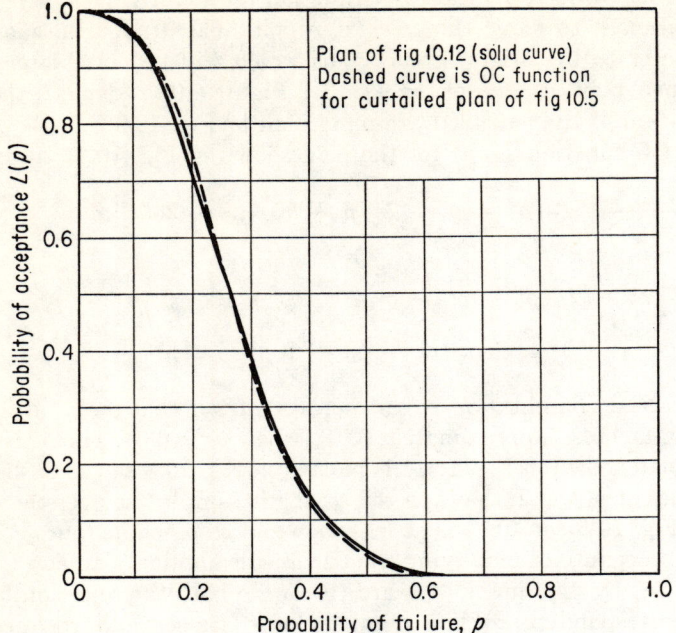

Figure 10.13

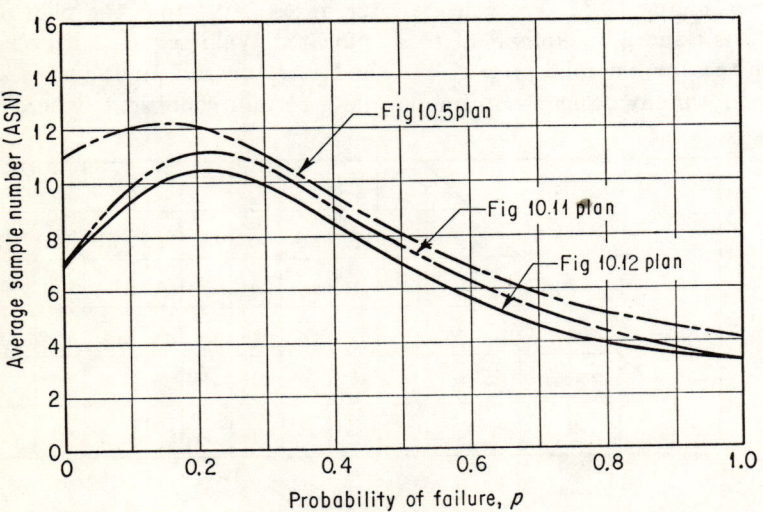

Figure 10.14

mum" means essentially that for a given OC function, the ASN function is as small as possible.

10.3. CRITERIA FOR SPECIFICATION OF A BINOMIAL SAMPLING PLAN (FIXED SAMPLE SIZE OR CURTAILED)

In the opening paragraphs of Sec. 10.2 it was stated in particular that an effective sampling plan should have a small probability of leading to a decision to reject when $p \ll p_c$; and correspondingly, a small probability of leading to a decision to accept when $p \gg p_c$. This requirement can be used to formulate the following criteria for specifying one of a class of binomial sampling plans:

Select a value of $p_1 < p_c$ and a value $p_2 > p_c$ such that the probability is to be at most α that the sampling plan leads to a decision of rejection when $p = p_1$, and such that the probability is at most β that the sampling plan leads to a decision of acceptance when $p = p_2$. The quantities α and β are some specified, suitably small, numbers; e.g., $\alpha = 0.05$, $\beta = 0.10$. Evidently, this is a specification of the values of the operating characteristic function at two points of its domain, p_1 and p_2. Thus we have

$$L(p_1) = 1 - \alpha$$
$$L(p_2) = \beta$$
(10.22)

Since it is required that the OC function be a decreasing function of p,* we see that for values of $p \leq p_1$, $L(p) \geq 1 - \alpha$, and for values of $p \geq p_2$, $L(p) \leq \beta$.

We now apply the foregoing criteria to the class of *fixed-sample-size* binomial plans (see Fig. 10.6). Two parameters can be used to specify uniquely a plan of this class, namely N, the number of trials to be made, and F, the maximum number of failures allowable in the N trials in order to reach a decision of acceptance. Thus the parameters N and F can be determined by expression (10.22) in the form

$$L(p_1) = \sum_{j=0}^{F} \binom{N}{j} p_1^j q_1^{N-j} \geq 1 - \alpha$$

$$L(p_2) = \sum_{j=0}^{F} \binom{N}{j} p_2^j q_2^{N-j} \leq \beta$$
(10.23)

The reason for the inequality signs in (10.23) is that exact integer values of N and F cannot in general be found to satisfy Eqs. (10.22); however,

* A sufficient condition for this is that the ratio f/n for all acceptance boundary points (n, f) of the plan be less than the same ratio for all rejection boundary points.

it is required that the risks of error for $p \leq p_1$, $p \geq p_2$ be *at most* α and β, respectively. For example, values of N_1, F_1 might be found such that $L(p_1)$ is somewhat larger than $1 - \alpha$ and $L(p_2)$ slightly larger than β; then possibly the values $N_1 + 1$, F_1 will *just* satisfy inequalities (10.23), since $L(p)$ is a decreasing function of N for any fixed p between zero and unity. When N and F turn out to be small (say $N \leq 20$, $F \leq 4$), it will generally be difficult to do this without making $L(p_1)$, $L(p_2)$ *too* small. For example, if $\alpha = \beta = 0.10$; $p_1 = 0.13$, $p_2 = 0.41$, it is found that $N = 14$, $F = 3$ yield $L(0.13) = 0.9021$, $L(0.41) = 0.1095$; whereas $N = 15$, $F = 3$ yield $L(0.13) = 0.8796$, $L(0.41) = 0.0785$. The former values closely satisfy Eqs. (10.22) but violate the second inequality of (10.23).

EXERCISE: Use the Ref. 4 tables to find parameters N and F for a binomial sampling plan when $p_1 = 0.126$, $p_2 = 0.242$; $\alpha = \beta = 0.10$.

Answer: $N \simeq 70$, $F = 12$. What is the expected number of trials if the plan is curtailed and the true reliability is 0.874 ($= 1 - p_1$)?

We already know from Sec. 10.2.3 that the identical protection against making either kind of error (i.e., to accept when reliability is low or reject when reliability is high) can be obtained even when we curtail the fixed-sample-size sampling plan. Thus when one uses the criteria $(p_1, \alpha; p_2, \beta)$ the operating characteristic function of the curtailed plan may be most easily obtained by assuming the plan is a fixed-sample-size plan and using the tables of Ref. 4. The average sample number function is given by Eq. (10.17), and its values can be obtained also by using the tables of Ref. 4.

It may happen that only the values of p_2 and β are specified. This can be the case when a customer specifies a "minimum acceptable" reliability $R_2 = 1 - p_2$, which is to be "demonstrated" with "confidence" $\gamma = 1 - \beta$ by the producer of the item. For example, if $R_2 = 0.90$, $\gamma = 0.90$, Table A.1 or Fig. A.3 shows that a sample size $N = 22$, with $F = 0$, gives a test procedure such that if no failures occur in 22 trials the reliability requirements will have been met. However, Fig. A.3 shows that a variety of test procedures are available which, if carried out, will also allow the requirements to be demonstrated; e.g., $(N, F) = (22, 0), (38, 1), (52, 2), (65, 3), (78, 4)$, and so on. If the producer were paying for the cost of the N trials or N items to be tested, it would appear that the smallest N, namely 22, would be most suitable. However, the producer must consider the possibility of not meeting the demonstration requirements and thereby losing immediate or future business from his customers.

In the example in which $N = 22$, $F = 0$, the actual reliability will have to be 0.9952 or higher in order that the probability of meeting the demonstration requirements be at least 90 per cent. If the producer believes that his product has a reliability of only 95 per cent when used in the customer's application, he will conclude that he has only a 32.4 per cent chance of meeting the demonstration requirement.

Evidently, the producer in this situation must use a plan with a larger number of trials, if he wishes a high probability, say 90 per cent, of meeting the requirement. In this case, we see that the specification for the demonstration plan should be $p_1 = 0.05$, $\alpha = 0.10$; $p_2 = 0.10$, $\beta = 0.10$; and using the previously given techniques (Eqs. (10.22)), $N \simeq 190$, $F = 13$ would also satisfy the additional specifications.

Now, if the consequences (as many as 190 tests) of the producer's specifications ($\alpha = 0.10$) are considered by him to be too expensive, then only one choice is left: the actual reliability must be increased.

EXERCISE: Show for $p_1 = 0.04$, $\alpha = 0.10$; $p_2 = 0.10$, $\beta = 0.10$, that $N \simeq 117$, $F = 7$ satisfy these specifications. What are suitable values of N and F if $p_1 = 0.03$, $\alpha = 0.10$; $p_2 = 0.10$, $\beta = 0.10$? Calculate in each case the expected number of trials for curtailed plans, when $R = 1 - p_1$.

When the customer (say, the government) is going to pay the cost of the demonstration, the producer will frequently find that not only p_2 and β but also an upper limit to the number of trials are specified. The producer must then either have a product of sufficiently high reliability or be willing to pay the cost of additional tests.

10.4. APPLICATION OF WALD SEQUENTIAL BINOMIAL PLANS TO RELIABILITY DEMONSTRATION

In Sec. 10.3.1 some examples of the consequences of specified reliability demonstration requirements were given for the class of fixed sample size or curtailed sampling plans. In this section we shall present the application of the Wald sequential binomial plans to this problem.

The specifications are as given in Sec. 10.3, namely $(p_1, \alpha; p_2, \beta)$. Wald showed (Ref. 6, Chap. 5),* using an optimum method of choosing a rejection boundary (critical region) for a *fixed-sample-size* binomial sampling plan, that the acceptance and rejection boundaries for a more general sampling plan could be described *approximately* by two parallel straight lines with slope $s < 1$ in the failures-vs.-trials plane.† An example of a Wald sequential plan was shown in Fig. 6.4.

---

* It should be noted here that we are using the notation (p_1, p_2) instead of (p_0, p_1) as used in the reference.

† Wald sequential sampling plans have quite general application; i.e., to situations other than binomial sampling. See Ref. 6, Chap. 3, and the subsequent applications in Chaps. 7–9.

The formulas for the location of the acceptance and rejection boundary points (n, a_n) and (n, r_n), respectively, are

$$a_n = \left[\frac{\log \dfrac{\beta}{1-\alpha}}{\log \dfrac{p_2}{p_1} - \log \dfrac{1-p_2}{1-p_1}} + n \frac{\log \dfrac{1-p_1}{1-p_2}}{\log \dfrac{p_2}{p_1} - \log \dfrac{1-p_2}{1-p_1}} \right] \quad (10.24)\dagger$$

$$-r_n = \left[\frac{\log \dfrac{\alpha}{1-\beta}}{\log \dfrac{p_2}{p_1} - \log \dfrac{1-p_2}{1-p_1}} + n \frac{\log \dfrac{1-p_2}{1-p_1}}{\log \dfrac{p_2}{p_1} - \log \dfrac{1-p_2}{1-p_1}} \right] \quad (10.25)\dagger$$

As described in Sec. 10.2, sampling stops with the decision to "accept" when the coordinates of the cumulative plot of a sequence of failures and/or successes (n, f) coincide for the first time with *any* one of the points (n, a_n) in the failures-vs.-trials plane. Similarly, when $(n, f) \equiv (n, r_n)$ for the first time, the sampling stops, and a decision is made to "reject." The numbers a_n, r_n depend only upon the specifications $(p_1, \alpha; p_2, \beta)$ and can be computed for as many values of n as necessary before the sampling plan is commenced.

Table 10.1

n	a_n	r_n	n	a_n	r_n
1	..	..	14	2	6
2	..	..	15	2*	6
3	..	3	16	2*	6
4	..	3	17	3	7*
5	..	3	18	3*	7
6	0	4*	19	3*	7
7	0*	4	20	3*	7
8	0*	4	21	4	8*
9	0*	4	22	4*	8
10	1	5*	23	4*	8
11	1*	5	24	4*	9*
12	1*	5	25	5	9
13	1*	6*			

* Inaccessible points.

Table 10.1 gives the acceptance and rejection boundary points for number of trials equal to or less than 25 for a plan with $(p_1, \alpha; p_2, \beta) = (0.15, 0.10; 0.42, 0.10)$, computed from Eqs. (10.24) and (10.25).

The appearance of the accessible boundary points given in Table 10.1

†The brackets denote the greatest integer equal to or less than the quantity in the brackets. Thus in (10.24), if the expression in brackets is, e.g., 3.25, then $a_n = 3$. In (10.25), if the expression in brackets is, e.g., -4.35, then $-r_n = -5$ or $r_n = 5$.

RELIABILITY DEMONSTRATION AND DECISIONS 299

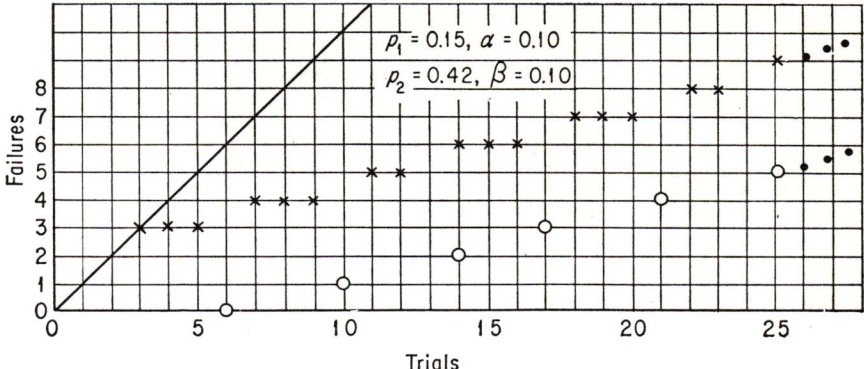

Figure 10.15 Wald sequential binomial plan.

is shown in Fig. 10.15. The operating characteristic function for the plan shown in Fig. 10.15 may be obtained from the pair of equations:

$$L(p) = \frac{1 - \left(\frac{\alpha}{1 - \beta}\right)^h}{1 - \left(\frac{\alpha}{1 - \beta}\right)^h \left(\frac{\beta}{1 - \alpha}\right)^h} \qquad (10.26)*$$

and

$$p = \frac{\left(\frac{1 - p_1}{1 - p_2}\right)^h - 1}{\left(\frac{p_2}{p_1}\right)^h \left(\frac{1 - p_1}{1 - p_2}\right)^h - 1} \qquad (10.27)*$$

To solve for $L(p)$ as a function of p, a fixed value of h between $-\infty$ and $+\infty$ is selected, and corresponding values of $L(p)$ and p are then calculated from (10.26) and (10.27) respectively.

EXERCISE: Show that $h = +\infty, 1, 0, -1, -\infty$ result in values of

$$L(0) = 1$$
$$L(p_1) = 1 - \alpha$$
$$L(s) = \frac{\log \dfrac{\alpha}{1 - \beta}}{\left(\log \dfrac{\beta}{1 - \alpha} + \log \dfrac{\alpha}{1 - \beta}\right)}$$
$$L(p_2) = \beta$$
$$L(1) = 0$$

* These remarkable formulas for the OC function are a special case (for sequential binomial sampling plans) of formulas for more general Wald sequential sampling plans derived by A. Wald in 1944.

respectively, where s is the common "slope" of the acceptance and rejection boundaries given by the coefficient of n in Eq. (10.24). Note that one must find the limits of (10.26) and (10.27) as $h \to 0$ by first differentiating numerator and denominator separately with respect to h.

Once values of $L(p)$ and p have been found for some $h > 0$, the values of $L(p)$ and p for $-h$ can be found more simply than by using (10.26) and (10.27), by the formulas:

$$p' = \left(\frac{p_2}{p_1}\right)^h p \qquad (10.28)$$

and

$$L(p') = \left(\frac{\beta}{1-\alpha}\right)^h L(p) \qquad (10.29)$$

where p', $L(p')$ are the coordinates of the OC function corresponding to $-h$.

The values for the average sample number function can be found, once the OC function is known, by the following formula:

$$\text{ASN} = \frac{L(p) \log \dfrac{\beta}{1-\alpha} - [1 - L(p)] \log \dfrac{\alpha}{1-\beta}}{p \log \dfrac{p_2}{p_1} - (1-p) \log \dfrac{1-p_1}{1-p_2}} \qquad (10.30)$$

In general, but not always, the maximum value of the ASN occurs near $p = s$, and in this case

$$\text{ASN}(s) = \frac{\left(\log \dfrac{\beta}{1-\alpha}\right)\left(\log \dfrac{\alpha}{1-\beta}\right)}{\log \dfrac{p_2}{p_1} \log \dfrac{1-p_1}{1-p_2}} \qquad (10.31)$$

Note that it is conceivable that sampling could proceed indefinitely without reaching a decision. As remarked in Sec. 6.6, however, the probability of this event is zero. Consequently, the number of trials ν to reach a decision has a proper probability distribution; i.e., there is no positive probability that $\nu = \infty$. It has also been shown that all moments (mean, variance, etc.) of this distribution exist and are finite.* The formula for $E(\nu) \equiv \text{ASN}$ is given by Eq. (10.30). Graphs showing the standard deviation (square root of the variance) of ν have been given for selected examples of sequential binomial sampling plans (Ref. 5). The referred-to

* C. Stein, "A Note on Cumulative Sums," *Ann. Math. Stat.*, Vol. 17 (1946), pp. 498–99.

graphs show that the variability of the number of trials can be considerable, which means that the actual number of trials to reach a decision by use of a Wald plan may, with fair probability, exceed the corresponding number of trials for a fixed-sample-size sampling plan (Sec. 10.3) with the same specifications p_1, α; p_2, β. However, if one compares the ASN for the Wald plan with the fixed number of trials for the corresponding fixed-sample-size plan (or even with the ASN for the curtailed plan), he will nearly always find that the Wald plan has the smallest *expected* number of trials to reach a decision.

EXERCISE: Compute the OC and ASN functions for the Wald plan whose specifications are $(p_1, \alpha; p_2, \beta) = (0.15, 0.10; 0.42, 0.10)$. Plot these functions vs. true failure probability p. Compare the fixed-sample-size plan with the same specifications.

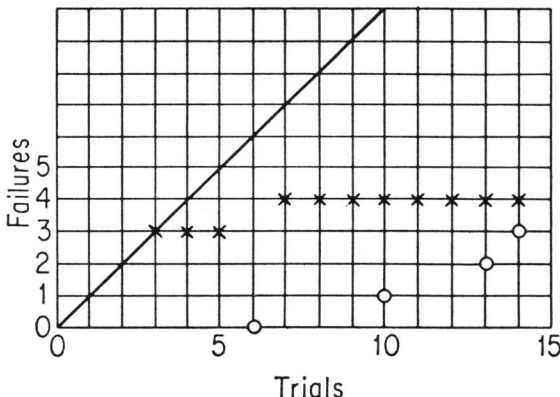

Figure 10.16

EXERCISE: Truncate the plan shown in Fig. 10.15 at 14 trials maximum (first make the point (14, 3) an acceptance boundary point). Show that the truncated plan then becomes as shown in Fig. 10.16. Note that this plan differs from the plan of Fig. 10.12 only at the acceptance boundary point nearest to the origin (6, 0) instead of (7, 0). Compute the OC and ASN functions for the truncated Wald plan by the methods of Sec. 10.2.

The previous exercise illustrates the remark made at the end of Sec. 10.2.4. It further indicates that the effort to determine an "optimum" truncated binomial sampling plan may be eased by starting directly with a Wald sequential plan with specifications $(p_1, \alpha; p_2, \beta)$, then truncating and proceeding as in the exercise. One should first consider using values of α and β slightly smaller than the actual specification values, as in general the effect of truncating the ordinary Wald sequential plan is to slightly increase these risks.

10.5. RELIABILITY DEMONSTRATION BASED ON TIME-TO-FAILURE

The methods of Secs. 10.2–10.4 can apply without change to the case when a time-to-failure model is appropriate to describe the failure characteristic of a device. In this case there exists a distribution function $P(\tau \leq t) = F(t)$, which may be of known or unknown form, with one or more parameters which also may be known or unknown. If the required time of operation of the device is T; if each of a sample of such devices is operated for exactly time T, with only success or failure noted (the information given by the observed times-to-failure is not used); then it is evident that in this case binomial sampling methods apply to reliability demonstration.

It is emphasized at this point that so far in this chapter we have not made use of any information regarding the "shape" of the time-to-failure distribution. If this can be done, i.e., if we know or can make valid assumptions concerning the failure characteristics, then we can expect to further improve the sampling techniques. That is, we can increase our confidence and demonstrated reliability and/or reduce the sample sizes necessary for demonstration of a specified reliability number.

This raises two related problems when the form of the time-to-failure distribution is known.

1. Can reliability demonstration be accomplished when it is not feasible or desirable to test each device for time T, but possibly for some other time T'.
2. How can information on actual times-to-failure be utilized, since we know (intuitively at this point) that additional information will result in even further savings in sample size and still allow the same reliability objectives to be demonstrated?

The solution to both these problems evidently depends on the knowledge of the mathematical form of the time-to-failure distribution as well as knowledge of some of its parameters.

In the following sections, several sampling plans for reliability demonstration based on the one-parameter exponential distribution (Sec. 6.8.1) will be presented.

In Sec. 10.6 we will extend the methods of binomial sampling to apply to the experimental situation where it is desired to test each *item* for time t, which may be *different* than the required time of operation T. It will also be shown that the same methods can be applied when the underlying time-to-failure distribution is of the general form mentioned in Sec. 6.8.3.1. Subsequently, Secs. 10.7–10.9 present sampling plans for reliability demonstration in which the test time is at our disposal, and/or the information given by the actually observed times of failure is utilized. However, in these cases the methods as presented cannot in general be applied when the underlying time-to-failure distribution is of the general form of Sec. 6.8.3.1, i.e., other than the simple exponential. This is unfortunate, for it has tended to place an undue emphasis on the exponential distribution, especially in the electronics field, as the underlying distribution for reliability estimation and sampling plans. Although the failure characteristics of many components do follow an exponential distribution, some do not, and careful examination of the validity of the assumption should be made. Epstein* describes tests to check precisely this assumption. The consequences of assuming the underlying distribution to be exponential when it is actually Weibull are discussed by Zelen and Dannemiller.† However, if the exponential distribution can be assumed to describe the time-to-failure characteristics of a component (as we have already seen it in Appendix 9B to be theoretically justified under certain conditions in the case of a complex system) then a variety of sampling procedures can be generated from this basis.

10.6. RELIABILITY DEMONSTRATION WHEN TEST TIME IS UNEQUAL TO REQUIRED TIME

The basic requirement is that a device operates for time T; the probability that it does so is defined as its reliability R. When the time-to-failure τ has the exponential distribution, then

$$R = e^{-T/\theta} \qquad (10.32)$$

where θ is the mean time-to-failure or mean "life." In accordance with the principle stated in Sec. 10.1, we must specify a lower confidence limit $\hat{R}_L$ with confidence coefficient γ; i.e., such that

$$P[\hat{R}_L < R] \geq \gamma \qquad (10.33)$$

If the result of a sampling plan allows us to state with confidence γ that $\hat{R}_L < R$, we say that the reliability $\hat{R}_L$ has been demonstrated with confidence γ.

The sampling procedure is specified as follows: A random sample of N items are *simultaneously*‡ put on test for time t, where t is not necessarily

---

* B. Epstein, "Tests for the Validity of the Assumption that the Underlying Distribution of Life is Exponential," Part I, *Technometrics*, **2**:1, 83 (February 1960); Part II, *Technometrics*, **2**:2, 167 (May 1960).

† M. Zelen and M. C. Dannemiller, "The Robustness of Life Testing Procedures Derived from the Exponential Distribution," *Technometrics*, **3**:1, 29 (February 1961).

‡ This is to save time in reaching a decision. The reader will quickly note that items could be tested one by one, each for time t (a lengthy procedure), and the methods of Secs. 10.2–10.4 would apply directly to this problem. A *replacement* procedure, wherein failed items are immediately replaced with new items, and which would further reduce expected waiting time to complete the test, is not considered here (*cf.* Sec. 10.7.2 and Ref. 9).

equal to T. At the end of time t, the number of failures is noted. If the number of failures does not exceed a certain specified limit F, then a specified numerical reliability $\hat{R}_L$ has been demonstrated with a specified confidence γ; otherwise, one states that the specified reliability has not been demonstrated with the specified confidence γ.

To see how the test parameters (N, F) are to be calculated, first let $\hat{R}_L$ be the specified reliability to be demonstrated with specified confidence γ, where T is the required time of operation. Then a lower confidence limit $\hat{\theta}_L$ on the mean life θ, with confidence coefficient γ, can be determined by solving (see Sec. 8.1)

$$e^{-T/\hat{\theta}_L} = \hat{R}_L \qquad (10.34)$$

for $\hat{\theta}_L$, and obtaining

$$\hat{\theta}_L = \frac{T}{\log \dfrac{1}{\hat{R}_L}} \qquad (10.35)$$

Now, if the *test* time is specified as t, the probability of failure of a single device during its test is $p_\theta = 1 - e^{-t/\theta}$. Therefore an upper confidence limit on p_θ with confidence coefficient γ is $\hat{p}_U = 1 - e^{-t/\hat{\theta}_L}$. We now see that we can apply binomial sampling methods to the problem, since all that need be done is to determine values of N and F so that if no more than F failures occur, then by the methods of Appendix 8A the (now specified) quantity $\hat{p}_U$ will be an upper confidence limit on p_θ.

Notice that there is an important difference between the binomial sampling procedure described here and that described in Sec. 10.2. The difference is in putting all N items on test simultaneously. The sample space still consists of the set of observations at time t: 0, 1, 2, $\cdots$, N, which denote all possible numbers of failed items (failing at some time during their tests); however, there are no *paths* to observe, i.e., sequences of successes and failures. Nevertheless, one should keep in mind the pictorial aspect of binomial sampling in the failures-vs.-trials plane, because as we shall show later, it is possible to *partially curtail* the sampling plan now being described.

> EXAMPLE: A certain type of electron tube is required to operate for 1000 hours with no drift in performance outside of certain specification limits in its intended application. The customer requires that a reliability of 0.90 with 90 per cent confidence be demonstrated. Three weeks (500 hours) is the allotted test time, and 75 tubes are available for testing. Assuming that the time-to-failure has an exponential distribution, where failure occurs the instant that specification limits are exceeded, what is the maximum number of failures F which allows demonstration of the required reliability, with the required confidence?

Solution: We have $\hat{R}_L = 0.90$, $T = 1000$, hence, from Eq. (10.35)

$$\hat{\theta}_L = \frac{1000}{0.1053} = 9500 \text{ hours}$$

Since $t = 500$, therefore

$$\hat{p}_U = 1 - e^{-500/9500} = 1 - e^{-0.5263} = 0.0513$$

From Fig. A.3 ($\gamma = 0.90$), we find $F = 1$. Note that a variety of values of N and F are at our disposal when neither is specified. We will take up this point again.

For convenience, Tables A.2–A.6 supply sample sizes N corresponding to values of F for given values of $t/\hat{\theta}_L$ and γ.* As a supplement, Figs. A.1–A.5 may be used (as was Fig. A.3 in the previous example) for ranges of F and N beyond those of the tables.

We now show how the sampling plan may be partially curtailed. Assume that it is possible to observe the testing continuously and therefore to know at any time in the interval $(0, t)$ the total number of failures up to the given time. Then at the instant the number of failures becomes $F + 1$, we may simply shut down the whole test and draw the conclusion that the required demonstration has not been met. This is evidently the same as moving all of the *rejection points* (N, j), where $F + 1 \leq j \leq N$, in the failures-vs.-trials plane to the corresponding positions $(N - j + F + 1, F + 1)$, as was shown in Sec. 10.2.3. Note that we *cannot* curtail any *acceptance* points since no matter how short the time remaining until every trial is completed (at time t), any number up to N failures could occur.

The described curtailing is interesting from the following points of view:

1. The $N - F - 1$ items which do not fail are just as good as new, even though they were operated for a certain amount of time $< t$, since the time-to-failure distribution is exponential (the basic assumption). Thus by our basic assumption the testing is nondestructive on nonfailing items.

2. The curtailing process introduces a random variable τ_1, the time required to complete the test, whose range is the interval $(0, t)$. The *expected* test time was shown by Epstein† to be

$$E_\theta(\tau_1) = \sum_{k=1}^{F+1} \binom{N}{k} p_\theta^k (1 - p_\theta)^{N-k} E_\theta(X_{k,N}) \qquad (10.36)$$

---

* Tables A.2–A.6 are reprinted with the kind permission of the Editorial Department, Institute of Radio Engineers, Inc., from Reference 7.

† B. Epstein, "Truncated Life Tests in the Exponential Case," *Ann. Math. Stat.*, **25** (1954), 555–564, Eqs. (20), (1), and (10); B. Epstein and M. Sobel, "Sequential Life Tests in the Exponential Case,"*Ann. Math. Stat.*, **26** (1955), pp. 82–93, see p. 86.

where
$$p_\theta = 1 - e^{-t/\theta}$$
$$E_\theta(X_{k,N}) = \theta \sum_{j=1}^{k} \frac{1}{N-j+1} \simeq \theta \log \frac{N+\frac{1}{2}}{N-k+\frac{1}{2}}$$

The expected test time is, of course, a function of the true mean time-to-failure θ of the item, as is shown by Eq. (10.36). For very small values of θ, one would expect quick termination of the test; when θ is large, the test would tend to take the full time t. For example, when $F = 0$ we have only one term in (10.36):

$$E_\theta(\tau_1) = \theta p_\theta (1 - p_\theta)^{N-1} \qquad (10.36')$$

When $\theta \to 0$, $E_\theta(\tau_1) \to 0$; and when $\theta \to \infty$, $E_\theta(\tau_1) \to t$, since $\lim_{\theta \to \infty} \theta(1 - e^{-t/\theta}) = t$.

The previous discussion has not considered the probability of the test procedure's demonstrating the required reliability when the true mean life θ is much greater than $\hat{\theta}_L$. Table 10.2* presents the ratio of θ to $\hat{\theta}_L$ such that when the true mean life is θ and the allowable number of failures F is used with the requirement $\hat{\theta}_L$ for $\gamma = 0.90$ (Table A.5, Fig. A.3), then the probability of demonstration is 0.95. It turns out that the ratio $\theta/\hat{\theta}_L$ is practically independent of test time t, so that the parameter t is absent from Table 10.2.

Table 10.2

F	0	1	2	3	4	5	6	7	8	9	10
$\theta/\hat{\theta}_L$	46	11	6.7	5.0	4.1	3.6	3.3	3.1	2.9	2.7	2.5

Table 10.2 can be used to uniquely specify a plan from Table A.5 ($\gamma = 0.90$) when further information in the form of some knowledge of true mean life θ is available.

EXAMPLE: Let $\hat{\theta}_L = 9500$ hours, $t = 500$ hours, $\tau = 1000$ hours, $\gamma = 0.90$. Suppose further that it is known (or believed) that the true θ is approximately 47,500 hours. Then from Table 10.2 we see that $\theta/\hat{\theta}_L = 5.0$ gives a value $F = 3$. Hence from Fig. A.3, we would test a sample of $N \simeq 130$ items for 500 hours and consider that 0.90 reliability (or 9500 hours mean life) had been demonstrated with 90 per cent confidence if no more than three failures occurred. The probability of passing the demonstration would be 95 per cent. Of course $F = 0$, $N = 44$, $t = 500$ would be an equivalent demonstration plan, except that the true mean life would have to be $46(9500) = 437,000$ hours in order to have a 95 per cent probability of passing the demonstration. Evi-

---

* Table 10.2 reprinted with the kind permission of the Editorial Department, Institute of Radio Engineers, Inc., from Ref. 7.

dently when true mean life is not too much larger than the minimum acceptable value $\hat{\theta}_L$, a large number of items must be put on test in order to have a reasonably high probability of passing the demonstration. In this situation, the cost of testing may become of major importance, and some consideration must be given to balancing the loss caused by testing a large number of items and the loss resulting from not passing the demonstration requirements.

10.6.1 GENERALIZATION TO PARTIALLY KNOWN HAZARD FUNCTIONS

The methods of the previous sections can be immediately applied to life-test situations where the underlying distribution function of time-to-failure is given by

$$G(t;\theta) = 1 - \exp\left[-\frac{1}{\theta} H(t; \beta_1, \beta_2, \cdots)\right] \qquad (10.37)$$

where $\beta_1, \beta_2, \cdots$, are known (or assumed known) parameters, and θ is the only unknown parameter. This generalization was discussed in Sec. 6.8.3.1.

EXAMPLE: Assume that the underlying probability distribution function of time-to-failure is

$$P(\tau \leq t) \equiv p_\theta = 1 - e^{-t^2/\theta}, \quad t > 0 \quad \text{(a Weibull d.f.)}$$

where $\theta > 0$ is an unknown parameter. (Show that the mean life μ is $\mu = \frac{1}{2}\sqrt{\pi\theta}$.) To demonstrate a required reliability of 0.90 ($T = 1000$ hours) with confidence $\gamma = 0.90$ using a test time $t = 500$ hours, compute

$$\hat{\theta}_L = \frac{T^2}{\log{(1/\hat{R}_L)}} = 9.5 \times 10^6$$

Then

$$\hat{p}_U = 1 - e^{-t^2/\hat{\theta}_L} = 0.0260$$

Then if $N = 150$, we see from Fig. A.3 that $F = 1$, which is noted to be a more stringent requirement than when the underlying distribution is the simple exponential (cf. the example on pp. 304–5).

EXERCISE: Consider the two-parameter Weibull d.f. (Sec. 6.8.2), $P(\tau \leq t) \equiv p_\theta = 1 - e^{-t^\beta/\theta}$, $t > 0$, $\theta > 0$, where it is known that $\beta = \frac{1}{2}$, using the data of the previous example. (Reference 8 presents an extensive set of sampling plans for the Weibull distribution of time-to-failure).*

* The Ref. 8 plans are stated in terms of demonstrating a specified mean life. To convert the requirement into reliability demonstration, it is necessary to specify the required time of operation T and relate the reliability function

$$R(T) = \exp{(-T^\beta/\theta)}$$

to mean life by Eq. (6.103) where $\alpha \equiv \beta$ and $\lambda \equiv 1/\theta$.

EXERCISE: Apply the methods of this section to the extreme value distribution (Sec. 6.8.3).

10.7. RELIABILITY DEMONSTRATION UTILIZING TIME-TO-FAILURE INFORMATION—EXPONENTIAL DISTRIBUTION

10.7.1 TESTING UNTIL ALL ITEMS FAIL

In this case, a sample of N items are put on test. The test is concluded when all items have failed, and the times-to-failure $t_1, \cdots, t_N$ (each measured from the time the item is "turned on") are recorded. The required life of the item in its intended application is T. It is desired to demonstrate reliability $\hat{R}_L$ with confidence γ. What sample size N and what criteria should be used to determine whether the item has demonstrated its reliability requirements?

The method for doing this is based on Eq. (8.14). In the present notation, we must then have

$$\hat{R}_L \leq \exp\left[-\frac{T\chi^2_{2N;1-\gamma}}{2N\hat{\theta}}\right] \tag{10.38}$$

where

$$\hat{\theta} = \sum_{j=1}^{N} \frac{t_j}{N}$$

or

$$\frac{\hat{\theta}}{T} \geq \frac{\chi^2_{2N;1-\gamma}}{2N \log (1/\hat{R}_L)} \tag{10.39}$$

EXAMPLE: Let $\hat{R}_L = 0.90$, $\gamma = 0.90$, $T = 1000$ hours, and $N = 75$. What criterion is used to decide whether the required reliability is demonstrated? From (10.39) and Table A.8,

$$\frac{\chi^2_{2N;1-\gamma}}{2N} = \frac{\chi^2_{150;0.10}}{150} = 1.155, \qquad \log \frac{1}{\hat{R}_L} = 0.10535$$

hence,

$$\hat{\theta} \geq \frac{(1.155)(1000)}{0.10535} = 10{,}960 \text{ hours}$$

Therefore, if the sample mean of the times-to-failure exceeds 10,960 hours, then the required reliability of 0.90 will have been demonstrated, with 90 per cent confidence.

The entries in Table A.7 (p. 551) give lower limits to the statistic $\hat{\theta}/T$; for $\gamma = 0.80, 0.90, 0.95$; $\hat{R}_L = 0.90, 0.95, 0.99, 0.995$; and $N = 5, 10, 15, 25, 50, 100, 250,$ and 500. If the appropriate entry in Table A.7 is exceeded

RELIABILITY DEMONSTRATION AND DECISIONS 309

by $\hat{\theta}/T$, then the corresponding reliability $\hat{R}_L$ has been demonstrated with confidence γ.

EXAMPLE: Suppose the sample size N is not specified in the previous example, but it is desired that if the true reliability $R = e^{-T/\theta}$ is 0.95, then the probability of passing the demonstration plan should be 0.95. What are the sample size N and criterion to be used to make this decision?

Let $L(\theta)$ be the probability of passing the demonstration, i.e., the probability of acceptance, for any true value of θ. Then (*cf.* Eq. (8.12)),

$$P\left(\frac{2N\hat{\theta}}{\theta} \leq \chi^2_{2N;\,L(\theta)}\right) = 1 - L(\theta)$$

Hence if $R = e^{-T/\theta}$, and $L(\theta)$ is specified, we easily obtain

$$\frac{\hat{\theta}}{T} \leq \frac{\chi^2_{2N;\,L(\theta)}}{2N \log(1/R)} \qquad (10.40)$$

In this example $L(\theta) = 0.95$, $R = 0.95$; therefore both the following inequalities must hold:

$$\frac{\hat{\theta}}{T} \leq \frac{\chi^2_{2N;\,0.95}}{2N \log(1/0.95)} \qquad (10.41\text{a})$$

$$\frac{\hat{\theta}}{T} \geq \frac{\chi^2_{2N;\,0.10}}{2N \log(1/0.90)} \qquad (10.41\text{b})$$

The righthand side of (10.41a) is an increasing function of N and the righthand side of (10.41b) is a decreasing function of N (*cf.* Table A.8); there is therefore a unique minimum value of N which satisfies both requirements. In this example, the joint inequality (10.41a), (10.41b) becomes

$$\frac{\chi^2_{2N;\,0.10}}{2N} \leq 2.054 \, \frac{\chi^2_{2N;\,0.95}}{2N} \qquad (10.42)$$

since

$$\frac{\log(1/0.90)}{\log(1/0.95)} = \frac{0.10535}{0.05219} = 2.054.$$

From Table A.8, we find $2N = 35$ or $N \simeq 18$ is the required sample size. Hence the demonstration criterion is

$$\hat{\theta} \geq T \frac{\chi^2_{35;\,0.10}}{35} \frac{1}{0.10535}$$

$$= 1000 \, \frac{1.32}{0.10535} = 12{,}530 \text{ hours}$$

One should note in comparison with the previous example that although fewer items are tested (18 instead of 75), the *observed* mean life must be larger in order to demonstrate the reliability requirement.

10.7.2 TESTING UNTIL r ITEMS FAIL ($r \leq N$)

In this case, a sample of N items are put on test. The test is concluded when $r \leq N$ items have failed, and the times-to-failure $t_1, \cdots, t_r$ (each measured from the time the item is "turned on") are recorded. We see that this sampling procedure includes that of Sec. 10.7.1 as a special case. There is in addition another consideration, which also applies in Sec. 10.6; this is the idea of immediate *replacement* of failed items.* In this case the number of items on test is always equal to N, whereas in the *nonreplacement* case, the number of items on test eventually drops to $N - r + 1$. The usefulness of the type of testing described in this section is that reliability decisions may be reached in *less time* than if we wait for all items on test to fail, *provided* that the cost of putting additional items on test is worth the time saved in reaching a decision. The "additional items" are not only those which replace failed items; but also the required increase in N compared to the sample size one would use if all items were required to fail in order to conclude testing, under the same reliability demonstration requirement.

It can be shown that in either the replacement or nonreplacement case (Ref. 10) $2r\hat{\theta}_{r,N}/\theta$ has a χ^2 distribution with $2r$ "degrees of freedom," where $r\hat{\theta}_{r,N}$ is given by the total test time (for all items both failed and nonfailed) as of termination of the test. This means that the reliability demonstration criteria of Sec. 10.7.1 can be directly applied to the present experimental situation. All one has to do is replace $\hat{\theta}$ by $\hat{\theta}_{r,N}$ and N by r in Table A.7 (p. 551).

EXAMPLE: Let $\hat{R}_L = 0.90$, $\gamma = 0.90$, $T = 1000$ hours, $r = 75$, and $N \geq 75$. The reliability requirements data are the same as those of the first example of Sec. 10.7.1. Hence if $\hat{\theta}_{75,N} \geq 10{,}960$ hours, the required reliability of 0.90 would be demonstrated, with 90 per cent confidence.

Let us now examine the significance of N in the previous example. It was shown in Ref. 11 that the *expected waiting time* for the rth failure to occur (waiting time τ_1 to reach a decision) when N items are on test in a *nonreplacement* procedure is

$$E_{NR}(\tau_1) = \theta \sum_{j=1}^{r} \frac{1}{N - j + 1} \tag{10.43}$$

Correspondingly, for the *replacement* procedure it was shown in Ref. 9 that

$$E_R(\tau_1) = \frac{\theta r}{N} \tag{10.44}$$

---

* Reference 9 presents the same test procedure as given in Sec. 10.6 by using a different approach, and it also considers the replacement case, which was not discussed in Sec. 10.6.

RELIABILITY DEMONSTRATION AND DECISIONS 311

A. In the *nonreplacement* procedure we may therefore determine the expected saving in time to reach a decision for any sample size $N \geq r$ as compared to the case when the sample size is $N = r$, by computing the ratio

$$\lambda_{NR} = \frac{\theta \sum_{j=1}^{r} \frac{1}{N-j+1}}{\theta \sum_{j=1}^{r} \frac{1}{r-j+1}} = \frac{\sum_{j=1}^{r} \frac{1}{N-j+1}}{\sum_{j=1}^{r} \frac{1}{j}} \qquad (10.45)$$

It can be shown that a good approximation for λ_{NR} is

$$\lambda_{NR} \simeq \frac{\log \frac{N + \frac{1}{2}}{N - r + \frac{1}{2}}}{\log (r + \frac{1}{2}) + E} \qquad (10.46)*$$

EXERCISE: In the previous example, what should the sample size N be in order to reduce the expected waiting time to reach a decision by 50 per cent as compared to the case when $N = r = 75$?

Answer: $N = 81.6 \simeq 82$.

B. In the *replacement* procedure, the ratio of expected waiting times is

$$\lambda_R = \frac{\theta r/N}{\theta r/r} = \frac{r}{N} \qquad (10.47)$$

EXERCISE: For the sampling plan to demonstrate the requirements $\hat{R}_L = 0.90$, $\gamma = 0.90$, $T = 1000$, where $r = 75$, compute the expected waiting times to reach a decision when $\theta = \hat{\theta}_L = 9492$ hours for both replacement and nonreplacement procedures, for $N = 75, 82, 100, 200,$ and 500.

Answer:

Expected waiting time (hours)

N	Replacement	Nonreplacement
75	9,492	4.9014(9492) = 46,524
82	8,682	2.3979(9492) = 22,761
100	7,119	1.3715(9492) = 13,018
200	3,560	0.4685(9492) = 4,447
500	1,424	0.16234(9492) = 1,541

* Cf. Sec. 11.3.2, where

$$C_1 = \sum_{k=1}^{N} \frac{1}{k} \simeq \log \left(N + \frac{1}{2} \right) + E$$

where E is Euler's constant $= 0.577215665\cdots$. This expression is easily made use of in (10.45) to obtain (10.46).

Note in the previous exercise that the expected waiting times may not be strictly comparable for the same value of N, since in the *replacement* case a total of $N + r - 1$ items are used in testing, whereas only N items are used in the *nonreplacement* case. If, however, one assumes that the nonfailing items are "as good as new" (as is theoretically true) and that there is no cost, either in increased test facilities or in tying up additional items for test (rather than selling them at a profit, say), then one would prefer the replacement procedure from the time standpoint, since in either case r items are "used up." In the following example, a cost model, due to Epstein (Ref. 12), is presented which shows that: (1) for each procedure (replacement or nonreplacement), values of N can be chosen which minimize the cost of the demonstration, and (2) the least expensive *nonreplacement* procedure does not have higher expected cost than the least expensive *replacement* procedure.

10.7.3 COST MODEL FOR RELIABILITY DEMONSTRATION, AND EXAMPLE

Suppose that certain reliability requirements are to be demonstrated by either a replacement or a nonreplacement procedure. From the given requirements, in either procedure, the same value of r is chosen as the number of failing items at which the test is terminated. Let N_R be the number of items to be put on test initially and maintained by replacing failing items with new ones in the replacement procedure; and let N_{NR} be the number of items initially placed on test in the nonreplacement procedure. Further, let C_1 be the cost per unit time of waiting until the test is completed, and let C_2 be the cost per item of placing an item on test. Then the *expected total cost* in the replacement procedure is, from (10.44):

$$C_R = C_1 \theta \frac{r}{N_R} + C_2(N_R + r - 1) \tag{10.48}$$

and in the nonreplacement procedure, from (10.43), is

$$C_{NR} = C_1 \theta \sum_{j=1}^{r} \frac{1}{N_{NR} - j + 1} + C_2 N_{NR} \tag{10.49}$$

or

$$C_{NR} \simeq C_1 \theta \log \frac{N_{NR} + \frac{1}{2}}{N_{NR} - r + \frac{1}{2}} + C_2 N_{NR} \tag{10.49a}$$

(using the approximation indicated in (10.46)), where θ in each case is the actual mean life.

In each procedure, separately, we can find the value of N_R or N_{NR} which minimizes the expected total cost by the usual methods for finding minimum values of a function. We then obtain for the optimum replacement procedure (using the one of two closest integer values that minimizes C_R)

$$\bar{N}_R \simeq \sqrt{\frac{C_1 \theta r}{C_2}} \qquad (10.50)$$

and for the optimum nonreplacement procedure (using (10.49a), and the one of two closest integers that minimizes C_{NR}):

$$\bar{N}_{NR} \simeq \frac{r - 1 + \sqrt{4(C_1 \theta r / C_2) + r^2}}{2} \qquad (10.51)$$

EXAMPLE: If $C_1 = \$1/\text{hr}$, $C_2 = \$100/\text{item}$, $r = 10$, $\theta = 1000$ hours, then $\bar{N}_R = 10$ and $\bar{N}_{NR} = 16$. The minimized total expected cost in each case would be, from (10.48),

$$\bar{C}_R = \$1000 + \$1900 = \$2900 \qquad (10.48')$$

and, from (10.49a),

$$\bar{C}_{NR} = \$932 + \$1600 = \$2532 \qquad (10.49a')$$

Hence, we see in this example that the *nonreplacement* procedure would be less expensive as a reliability demonstration test than the *replacement* procedure. It is now proved generally that $\bar{C}_{NR}$ is *never* more than $\bar{C}_R$.

*Proof:**
For convenience let

$$\frac{C_1 \theta r}{C_2} \equiv \alpha$$

and denote optimum values for N_R, C_R; N_{NR}, C_{NR} by $\bar{N}_R$, $\bar{C}_R$; $\bar{N}_{NR}$, $\bar{C}_{NR}$, respectively. From (10.48)

$$\frac{\bar{C}_R}{C_2} = \frac{\alpha}{\bar{N}_R} + \bar{N}_R + r - 1 \qquad (10.52)$$

Now, *define* $\tilde{N}_{NR} \equiv r + \bar{N}_R - 1 \geq r$. The integer $\tilde{N}_{NR}$ is *not necessarily* the value of N_{NR} which minimizes C_{NR}, hence

$$\frac{\bar{C}_{NR}}{C_2} \leq \frac{\tilde{C}_{NR}}{C_2} \qquad (10.53)$$

where $\tilde{C}_{NR}$ is the expected total cost for the nonreplacement procedure when $N_{NR} = \tilde{N}_{NR}$. Using the *exact* formula (10.49) for C_{NR} and the definition of $\tilde{N}_{NR}$

$$\frac{\tilde{C}_{NR}}{C_2} = \frac{\alpha}{r}\left[\frac{1}{r + \bar{N}_R - 1} + \cdots + \frac{1}{\bar{N}_R}\right] + r + \bar{N}_R - 1$$

$$\leq \frac{\alpha}{r}\left[r\left(\frac{1}{\bar{N}_R}\right)\right] + r + \bar{N}_R - 1 \qquad (10.54)$$

---

* This proof is due to J. D. Riley, of Space Technology Laboratories, Inc.

(when $r > 1$, the inequality is strictly "less than"; when $r = 1$, strict equality holds). Using (10.52), we obtain

$$\frac{\bar{C}_{NR}}{C_2} \leq \frac{\alpha}{\bar{N}_R} + r + \bar{N}_R - 1 = \frac{\bar{C}_R}{C_2} \qquad (10.55)$$

and combining (10.53) and (10.55)

$$\bar{C}_{NR} \leq \bar{C}_R \qquad (10.56)$$

which concludes the proof.

Remark 1: When $r = 1$, it is evident that $\bar{C}_{NR} = \bar{C}_R$.

Remark 2: When $C_1 = 0$, $C_2 > 0$, we must have $\bar{N}_R = 1$, $\bar{N}_{NR} = r$, and $\bar{C}_R = C_2 r = \bar{C}_{NR}$.

Remark 3: When $C_1 > 0$, $C_2 = 0$; i.e., there is no cost associated with placing an item on test; the optimum values $\bar{N}_R$ and $\bar{N}_{NR}$ are infinite, which in turn makes both $\bar{C}_R$ and $\bar{C}_{NR}$ equal to zero.

Remark 4: In all other cases $\bar{C}_{NR} < \bar{C}_R$; i.e., strict inequality holds.

10.8. A TRUNCATED AND CENSORED LIFE TEST PROCEDURE

In this case each item is put on test for at most time t. If the (say) jth item fails at time t_j where $t_j \leq t$, this time is recorded. If the item does not fail on or before time t, the result as a *success* is recorded; in either case a new item is immediately selected and put on test. The procedure continues until a total of r items fail, yielding the results $t_1, \cdots, t_r$ together with an observed number of items which did not fail. The number r is specified in advance; hence, the total number of items put on test is unknown in advance and is therefore a random variable. The time t is called the *truncation* time; the test is said to be *censored*, because one observes only the *number* of items that *would have failed after* time t and not their actual times-to-failure.*

When the underlying distribution of time-to-failure is exponential with mean time-to-failure θ, it can be shown (Ref. 13) that $2\hat{S}/\theta$ has a chi-square distribution with $2r$ degrees of freedom, where $\hat{S}$ is the total

---

* See Ref. 14 for an alternative to the procedure of this section whereby the number of items N to be tested is fixed in advance; all N tests can thereby be run simultaneously; and each item is "turned off" at time t if it did not fail. The number r of observed failures at times $t_1, \cdots, t_r$ is therefore a random variable. An example where this procedure has application is given in Sec. 12.6.7 as the basis for a statistically designed experiment. However, one drawback to this procedure is that exact statistical procedures are not available.

operating time of all items, both failed and nonfailed; i.e.,

$$\hat{S} = \sum_{j=1}^{r} t_j + (N - r)t \tag{10.57}$$

and where N is the actual number of items put on test.

When a reliability requirement $\hat{R}_L$ is to be demonstrated with confidence γ, and T is the required operating time for the item, we may proceed as in Sec. 10.7.1 to obtain the demonstration criterion. Thus from Eq. (10.39) with $\hat{\theta}$ replaced by $\hat{S}/r$ and N replaced by r, the criterion is

$$\frac{\hat{S}/r}{T} \geq \frac{\dfrac{\chi^2_{2r;1-\gamma}}{2r}}{\log\left(\dfrac{1}{\hat{R}_L}\right)} \tag{10.58}$$

EXAMPLE: Let $\hat{R}_L = 0.90$, $\gamma = 0.90$, $T = 1000$ sec, and $r = 75$. What criterion is used to decide whether the required reliability is demonstrated? From (10.58) and Table A.8 (by interpolation)

$$\frac{\chi^2_{2r;1-\gamma}}{2r} = \frac{\chi^2_{150;0.10}}{150} = 1.155$$

$$\log \frac{1}{\hat{R}_L} = 0.10535$$

hence

$$\hat{S} \geq 75 \frac{(1.155)(1000)}{0.10535} = 75(10{,}960) = 822{,}000 \text{ sec}$$

Therefore, if the total operating life (waiting time to completion of test) of all items exceeds 822,000 sec, then the required reliability of 0.90 would have been demonstrated, with 90 per cent confidence.

As in Secs. 10.7.1 and 10.7.2, the entries in Table A.7 give lower limits to the statistic $\hat{S}/rT$; for $\gamma = 0.80, 0.90, 0.95$; $\hat{R}_L = 0.90, 0.95, 0.99, 0.995$, and $r = 5, 10, 15, 25, 50, 100, 250,$ and 500. If the appropriate entry in Table A.7 is exceeded by $\hat{S}/rT$, then the corresponding reliability $\hat{R}_L$ has been demonstrated with confidence γ.

The test procedure as stated may require an excessive amount of time to reach a decision. In this case the expected waiting time is $r\theta$, which can be determined from the fact that the expected value of a random variable with a chi-square distribution with $2r$ degrees of freedom is just $2r$ (prove this by making use of Eq. (7.13) and the fact that $2\hat{S}/\theta$ has such a distribution). Note that the individual test time t does not enter into either the demonstration criterion or the waiting time; the choice of t is, so far, arbitrary; however, we will see shortly that its value affects the expected number of items placed on test.

A way of reducing the expected waiting time is to initially put r items on test simultaneously. If all items fail on or before time t, the test is over. Otherwise, if fewer than r, say f_1 items fail on or before time t, place $r - f_1$ items on test for another time period t. Repeat this as necessary, each time placing on test the number of items equal to the difference between r and the accumulated number of failures. When the total of r failures occur, the testing is completed. It can readily be seen that the expected waiting time to completion is reduced, and that the test criterion remains unchanged.

Another way to reduce the expected waiting time is to stop testing as soon as the demonstration criterion (10.58) is met, if indeed this event happens, even though r failures have not yet occurred. Additional test time can only increase the value of $\hat{S}/rT$ further above its required limit. Thus, the test can be *truncated* with the decision of acceptance (i.e., the reliability requirement is met) but not with that of rejection, which is opposite to the situation in the test procedure of Sec. 10.6.

The expected number of items tested, $E(N)$, may be easily computed by noting that independent trials with constant probability of failure $p = 1 - e^{-t/\theta}$ are being made until the rth failure occurs (provided that the test is not truncated). Thus, from Eq. (6.37), the probability distribution of N is

$$P(N = k) = \binom{k-1}{r-1}(1 - e^{-t/\theta})^r (e^{-t/\theta})^{k-r} \qquad (10.59)$$

and therefore from the remark following Eq. (6.53) the expected value of N is

$$E(N) = \frac{r}{1 - e^{-t/\theta}} \qquad (10.60)$$

Here we see that the choice of the test time t governs the number of items to be placed on test. Large values of t make $E(N)$ decrease to r (obviously the lower limit), whereas for smaller and smaller values of t approaching zero, Eq. (10.60) shows that the expected number of items to be placed on test approaches infinity.

Since the test time t does not affect the waiting time, it is evident that the most economical testing of the type described in this section would require the use of the largest value of t possible. In fact, if $t = \infty$, one simply waits until each item fails (which it must do at some finite time, with probability one); then when r items have been put on test and have failed, the test is completed. However, it is then worthwhile to put r items on test simultaneously until all fail. The last two procedures would have equivalent demonstration criterion; but the second procedure, which is the same as given in Sec. 10.7.1, would evidently require far less test time. To carry this one step further, we also know from Sec. 10.7.2 that by placing *more* than r items on test and waiting for the first r failures, further decrease in waiting time can be achieved in order to demonstrate a given reliability requirement.

It is evident that the procedure described in this section is comparatively inefficient when the test time t is at our disposal.* It should only be used when t is fixed by the physical conditions of the test and simultaneous testing is difficult or costly. The following two examples illustrate cases in which use of the procedure of this section would be feasible. The second example, interestingly enough, shows that the methods we have been discussing apply to situations other than life testing.

EXAMPLE: A piece of electronic equipment whose time-to-failure distribution is known to be exponential is required to demonstrate a certain reliability for a deep space probe. Rather than use the standard sources of electrical power available for the vibration and low-temperature test chamber, it was decided to test the equipment in the chamber in conjunction with the type of battery to be used in the space vehicle. This would also allow evaluation of the battery's behavior characteristics as well as system compatibility of the two equipments. However, the duration of the battery's operation was specifically designed to be short, since it would be expected to burn up upon the space probe's reentry into the atmosphere (approximately within a week); and there was a limited production of this type of battery. Consequently, each test of the electronic equipment had to be limited to a maximum duration of t hours.

EXAMPLE: A series of guided missiles are launched to a target for the purpose of demonstration of missile reliability, which is defined as the probability that the impact point is on the target. The target is circular with radius a. The radial distances from the center of the target of the impact points inside a circle of radius b are measured. Only the *number* of impacts outside the circle of radius b are recorded.†

The following assumptions are made:

1. The true center of impact is the center of the target (no impact bias).
2. Both range and deflection errors are independently and normally distributed with the same standard deviation σ.

Under these circumstances, the *square* of the radial distance from the target center has an *exponential* distribution with mean $2\sigma^2$ (see Ref. 15, p. 236, or the second exercise following).

If it is further specified that the decision as to whether a given reliability $\hat{R}_L$ is demonstrated with confidence γ is made as soon as r impacts within the obser-

* The extent to which testing in groups affects the waiting time, as described in this section, is not known to the authors.

† Alternatively, assume that a camera with a ground-level circular field of view (radius b) is stationed at a high altitude directly above the center of the target. Impacts within the field of view would be photographed and could therefore be measured.

vation circle (radius b) are achieved, then we have the truncated and censored test situation described in this section.

EXERCISE: Suppose $\hat{R}_L = 0.90$, $\gamma = 0.90$, $a = 3$ miles, $b = 2$ miles, and $r = 75$ impacts within the observation circle of radius b. What is the demonstration criterion?

Answer: Exactly analogous to Eq. (10.57), we have

$$\hat{S} = \sum_{j=1}^{r} d_j^2 + b^2(N - r)$$

where d_j is the radial distance of the jth impact point and N is the number of missiles launched until r impacts within the observation circle are achieved. From Eq. (10.58) the criterion to decide that the missile demonstrates its reliability requirements is therefore

$$\hat{S} \leq \frac{-ra^2}{\log(1-\hat{R}_L)} \times \frac{\chi^2_{2r;\gamma}}{2r} = \frac{-75(3)^2\, 0.85517}{-2.3026} = 250.69 \text{ (miles)}^2$$

It is interesting to note that the expected number of missiles launched is, from (10.60), replacing θ by $2\sigma^2$ and t by b^2

$$E(N) = \frac{r}{1 - e^{-b^2/2\sigma^2}} \qquad (10.60')$$

What is the expected number of missiles to be launched when $\sigma =$ one mile?

Answer: $E(N) \simeq 87$.

EXERCISE: Show that the sum of squares of two independent normally distributed random variables each with mean zero, standard deviation σ, has an exponential distribution with mean $2\sigma^2$.

Solution: First we find the distribution of ζ^2 where ζ is normal with mean zero, standard deviation σ. We have for the d.f. of ζ

$$P(\zeta \leq X) = \frac{1}{\sigma\sqrt{2\pi}} \int_{-\infty}^{X} e^{-t^2/2\sigma^2}\, dt$$

Let $\eta = \zeta^2$, then $\eta \geq 0$, and the event $\eta \leq y$, where $y \geq 0$, is equivalent to the event $(-\sqrt{y} \leq \zeta \leq \sqrt{y})$; hence $P(\eta \leq y) = P(-\sqrt{y} \leq \zeta \leq \sqrt{y})$. Therefore

$$P(\eta \leq y) = \frac{1}{\sigma\sqrt{2\pi}} \int_{-\sqrt{y}}^{\sqrt{y}} e^{-t^2/2\sigma^2}\, dt$$

$$= \frac{2}{\sigma\sqrt{2\pi}} \int_{0}^{\sqrt{y}} e^{-t^2/2\sigma^2}\, dt$$

The p.d.f. of η is obtained by differentiating $P(\eta \leq y)$ with respect to y:

$$f(y) = \frac{2}{\sigma\sqrt{2\pi}} \frac{1}{2\sqrt{y}} e^{-y/2\sigma^2}$$

or

$$f(y) = \begin{cases} \dfrac{1}{\sigma\sqrt{2\pi y}} e^{-y/2\sigma^2}, & y \geq 0 \\ 0, & y < 0 \end{cases}$$

The next step is to find the m.g.f. (Sec. 5.8) of the random variable η. We have

$$\psi(t) = E(e^{\eta t}) = \int_{-\infty}^{\infty} e^{yt} f(y)\, dy$$

$$= \frac{1}{\sigma\sqrt{2\pi}} \int_0^{\infty} y^{-1/2} \exp\left[-y\left(\frac{1}{2\sigma^2} - t\right)\right] dy$$

$$= \frac{1}{\sigma\sqrt{2\pi}} \frac{\Gamma\left(\dfrac{1}{2}\right)}{\left(\dfrac{1}{2\sigma^2} - t\right)^{1/2}}$$

$$= \frac{1}{\sigma\sqrt{2}} \frac{1}{\left(\dfrac{1}{2\sigma^2} - t\right)^{1/2}}$$

As noted in Sec. 5.8, the m.g.f. for the sum of two such independent random variables is $(\psi(t))^2$, or

$$(\psi(t))^2 = \frac{1}{2\sigma^2} \frac{1}{\left(\dfrac{1}{2\sigma^2} - t\right)} = \frac{1}{1 - 2\sigma^2 t}$$

The last expression, however, is the m.g.f. for a random variable which has an exponential distribution with mean $2\sigma^2$.

10.9. APPLICATION OF WALD SEQUENTIAL LIFE TEST PLANS TO RELIABILITY DEMONSTRATION

It was stated in Sec. 10.4 that Wald sequential sampling plans have quite general application, i.e., to situations other than binomial sampling. In Ref. 16 an application is made to the life test sampling procedure in which the underlying distribution of time-to-failure is exponential. A random sample (Sec. 5.10) of N items is obtained from a population whose reliability function is

$$R(t) = e^{-t/\theta} \qquad (10.61)$$

where θ is the true mean life. The N items are put on test, and observations on total amount of life and number of failures are observed continuously in real time. As soon as certain criteria, which we shall describe shortly, are satisfied, the test terminates with a decision that the required reliability is demonstrated, or it is not.

Similarly to Sec. 10.7.2, both replacement and nonreplacement methods can be used. In the replacement procedure, a total of N items is continuously kept on test by immediate replacement of failed items; and in the nonreplacement procedure, failed items are not replaced. Generally, when the replacement procedure is used, the expected waiting time to completion of the test is shorter than if the nonreplacement procedure is used.

If, however, one considers the cost model of Sec. 10.7.3 as applied to the sequential sampling plan of this section, it can be inferred that the same situation holds here as in Sec. 10.7.3—that the "optimum" nonreplacement procedure would not have greater expected total cost than the corresponding "optimum" replacement procedure (see Sec. 10.9.5). It is difficult to justify this inference theoretically, however, since exact formulas for the expected waiting times are not easily available.

An added complication occurs when the nonreplacement procedure is used for the Wald sequential sampling plan of this section. It may happen that all N items fail before the criteria of the plan allow a decision to be reached. In this case, one must either put more items on test or specify (beforehand) a rule of *truncation*, i.e., a separate criterion for reaching one of the two aforementioned decisions. We will not treat the truncation problem here, however (see Ref. 16).

10.9.1 SPECIFICATIONS AND CRITERIA OF THE PLAN

The specifications are as given in Sec. 10.3, namely $(p_1, \alpha; p_2, \beta)$. Recall that the correspondence to reliability demonstration requirements $\hat{R}_L$, γ is $p_2 \equiv 1 - \hat{R}_L$ and $\beta \equiv 1 - \gamma$. The quantities $p_1 \equiv 1 - R_1$ and α are to be specified in order to obtain a *unique* sampling plan.

It can be shown, based upon the general Wald sequential probability ratio test (Ref. 6) or directly from Ref. 16, that a good approximation to the decision boundaries is given by two parallel straight lines in the *total life vs. failures* plane, with the following equations:

$$V(t) = h_1 + fs, \qquad \text{the ``acceptance'' boundary}$$
$$V(t) = -h_2 + fs, \qquad \text{the ``rejection'' boundary} \qquad (10.62)$$

RELIABILITY DEMONSTRATION AND DECISIONS 321

where $V(t)$ is the total observed life at real time t given by Eqs. (10.66) and (10.67) for the replacement and nonreplacement cases, respectively;

$$h_1 = \frac{T \log \frac{1-\alpha}{\beta}}{\log \frac{1-p_1}{1-p_2}} \tag{10.63}$$

$$h_2 = \frac{T \log \frac{1-\beta}{\alpha}}{\log \frac{1-p_1}{1-p_2}} \tag{10.64}$$

$$s = \frac{T \log \left[\frac{\log (1-p_2)}{\log (1-p_1)} \right]}{\log \frac{1-p_1}{1-p_2}} \tag{10.65}$$

Here, T is the *required life* of the item.

The total observed life $V(t)$ at any time t is given by

$$V(t) = Nt \tag{10.66}$$

for the replacement case, and

$$V(t) = \sum_{j=0}^{f} t_j + (N-f)t \tag{10.67}$$

for the nonreplacement case. The times t_j are the observed times-to-failure reckoned from the time the N items are first put on life test, and $t_0 \equiv 0$.

The quantities h_1, h_2, and s are functions of the specifications only, hence the sampling plan boundaries can be determined and plotted as a graph on the $V(t)$-vs.-f plane before testing starts, for as large a range of values of $V(t)$ and f as necessary. The appearance of the graph is as given in Fig. 10.17. As soon as a plot of the total life vs. failures touches or crosses, for the first time, one of the two boundaries in Fig. 10.17, the appropriate decision is made.*

10.9.2 PROPERTIES OF THE PLAN

The operating characteristic (OC) function for the Wald sequential life test plan can be found from the following expressions, which are similar

* See Sec. 10.9.4, where an example with reference to Fig. 10.17 is presented.

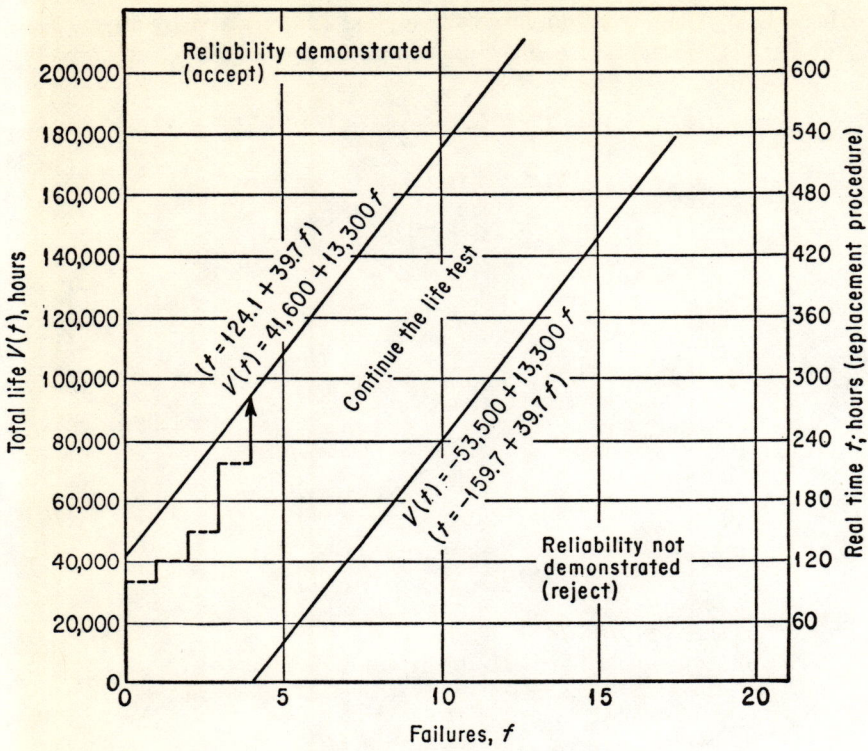

Fig. 10.17 Wald sequential life test plan.

to those given in Sec. 10.4 for the Wald sequential binomial plans:

$$L(p) = \frac{1 - \left(\dfrac{\alpha}{1-\beta}\right)^h}{1 - \left(\dfrac{\alpha}{1-\beta}\right)^h \left(\dfrac{\beta}{1-\alpha}\right)^h} \qquad (10.68)$$

$$\log(1-p) = \frac{h \log \dfrac{1-p_2}{1-p_1}}{\left[\dfrac{\log(1-p_2)}{\log(1-p_1)}\right]^h - 1} \qquad (10.69)$$

In expressions (10.68) and (10.69) the parameter h runs through all real values from $-\infty$ to $+\infty$, as in Sec. 10.4. The reader will easily verify that values of $h = +\infty$, $+1$, 0, -1, and $-\infty$ correspond to values of $p = 0$, p_1, $p_s \equiv 1 - \exp(-T/s)$, p_2, and 1, respectively.

RELIABILITY DEMONSTRATION AND DECISIONS

EXERCISE: Show that the values of $L(p)$ at the five points $p = 0, p_1, p_s, p_2$, and 1 are, respectively,

$$L(p) = 1,\ 1 - \alpha,\ \frac{\log \frac{1-\beta}{\alpha}}{\log \frac{(1-\beta)(1-\alpha)}{\alpha\beta}},\ \beta,\ \text{and } 0$$

10.9.3 EXPECTED NUMBER OF FAILURES AND EXPECTED WAITING TIME TO REACH A DECISION

It can be shown (Ref. 16) that the expected total life is related to the expected number of failures to reach a decision, $E(\mathbf{f})$, by the following equation:*

$$E[V(\tau_1)] = \frac{TE(\mathbf{f})}{\log \frac{1}{1-p}} \tag{10.70}$$

where τ_1 is the waiting time to reach a decision, $\mathbf{f}$ is the number of failures occurring when a decision is reached, and p is the true probability of failure (probability that time-to-failure of the item is less than T).

It can also be shown (Ref. 16) that

$$E(\mathbf{f}) \simeq \begin{cases} \dfrac{\dfrac{h_2}{T} - L(p)\dfrac{h_1 + h_2}{T}}{\dfrac{s}{T} + \dfrac{1}{\log(1-p)}}, & p \neq 1 - \exp\left(-\dfrac{T}{s}\right) \\[2ex] \dfrac{h_1 h_2}{s^2}, & p = p_s = 1 - \exp\left(-\dfrac{T}{s}\right) \end{cases} \tag{10.71}$$

To obtain an expression in the *replacement* case for the expected waiting time to reach a decision $E_R(\tau_1)$ for any value of p, we first take expected values in Eq. (10.66), (replacing t by τ_1)

$$E[V(\tau_1)] = NE_R(\tau_1) \tag{10.72}$$

Hence for the replacement procedure, $E_R(\tau_1)$ may be obtained from Eqs. (10.70)–(10.72).

---

* The boldface $\mathbf{f}$ denotes the random variable "number of failures that occur when a decision is reached", whereas f denotes merely the number of failures occurring at any time; i.e., the abscissa of the $V(t)$-vs.-f plot.

In the *nonreplacement* case an approximate formula (Ref. 16) for the expected waiting time is

$$E_{NR}(\tau_1) \simeq \frac{T}{\log \frac{1}{1-p}} \log \left[\frac{N + \frac{1}{2}}{N - E(\mathbf{f}) + \frac{1}{2}} \right] \qquad (10.73)$$

It is not too difficult to show, using Eq. (10.73), that $E_{NR}(\tau_1)$ exceeds $E_R(\tau_1)$ by approximately Δ, where

$$\Delta = \frac{T(E(\mathbf{f}))^2}{2\left(N + \frac{1}{2}\right)^2 \log \frac{1}{1-p}} \qquad (10.74)$$

Note that the expected number of failures to reach a decision, $E(\mathbf{f})$, is independent of whether a replacement or nonreplacement procedure is used, as has been indicated by the notation.

Table 10.3 presents, for various sets of specifications $(p_1, \alpha; p_2, \beta)$, the Wald sequential test plan parameters h_1/T, h_2/T, s/T; the OC function $L(p)$ for $p = p_s = 1 - \exp(-T/s)$; the expected number of failures $E(\mathbf{f})$ for $p = p_1, p_s, p_2, 1$ (when $p = 0$, $E(\mathbf{f}) = 0$; and also, of course, $L(p) = 1, 1 - \alpha, \beta, 0$ when $p = 0, p_1, p_2, 1$, respectively); and the expected waiting times for the replacement case, $NE_R(\tau_1)/T$; i.e., normalized by the required life, T, and by the sample size N, for $p = 0, p_1, p_s, p_2$ (when $p = 1, E_R(\tau_1) = 0$).*

Estimates of expected waiting time (normalized) for the nonreplacement case may be obtained by adding the quantity $N\Delta/T$ to the corresponding value of $NE_R(\tau_1)/T$, where Δ is given by Eq. (10.74).

10.9.4 EXAMPLE OF WALD SEQUENTIAL LIFE TEST FOR RELIABILITY DEMONSTRATION

Let $p_1 = 0.05$, $\alpha = 0.05$; $p_2 = 1 - \hat{R}_L = 0.10$, $\beta \equiv 1 - \gamma = 0.10$; and $T = 1000$ hours. From Table 10.3 we obtain

$$h_1 = 41.6 \times 1000 = 41,600 \text{ hr}$$

$$h_2 = 53.5 \times 1000 = 53,500 \text{ hr}$$

$$s = 13.3 \times 1000 = 13,300 \text{ hr}$$

---
* Note that $p = 1$ implies not just that the item *must* fail at some time in the interval $0 < t < T$, but that it must fail *immediately*, owing to the nature of the exponential distribution in having positive probability in any interval $t < \tau < \infty$.

Suppose that about three weeks (500 hr), more or less, are allowable for completion of the demonstration life test, and we decide to use a replacement procedure. We note in Table 10.3 that the expected waiting time (roughly maximum when $p = p_s$) would be

$$E_R(\tau_1) \simeq \frac{167.2 \times 1000}{N}$$

and this is set equal to 500 hours. Thus to have reasonable assurance of reaching a decision within the allowable time, the sample size to be put on test and maintained is

$$N \simeq \frac{167.2 \times 1000}{500} \simeq 335$$

The expected number of failures is given in Table 10.3 as $E(\mathbf{f}) = 12.6$ for $p = p_s = 1 - \exp(-T/s) = 0.1104$. Hence, we should have a reasonable assurance of not having to place more than about 350 items on test.

Figure 10.17 shows the plan in graphical form. Note that real time, t, may be shown on a proportional scale to total life when the replacement plan is used by simply dividing the total life by the sample size N, in this case equal to 335. For a nonreplacement plan, the real time t is obtained by subtracting the total of the times-to-failure from the total life, then dividing by the number of items remaining on test, as indicated by Eq. (10.67).

Suppose the results of the life test were as follows: $t_1 = 100$ hr, $t_2 = 120$ hr, $t_3 = 150$ hr, $t_4 = 220$ hr, and no further failures on or before $t = 283$ hr. Then as Fig. 10.17 shows, or as can be determined algebraically, a decision would be reached at 283 hr real time that the required reliability of 0.90 with 90 per cent confidence had been demonstrated. If the nonreplacement procedure were used, a decision of acceptance would be made under the same conditions at 284.6 hr.

> EXERCISE: Verify the preceding statements. Also, show in general that when a given number of failures have occurred, the nonreplacement procedure then requires a subsequently longer failure-free interval than does the replacement procedure, in order to reach a decision of acceptance. *Hint*: Let t_R and t_{NR} be the corresponding real times for the replacement and nonreplacement procedures to reach a decision of acceptance. The total lives at that moment, in either procedure, must be equal; i.e.,
>
> $$\sum_{j=0}^{f} t_j + (N - f)t_{NR} = Nt_R$$
>
> Now show that $t_R/t_{NR} \leq 1$.

Table 10.3 Parameters for Wald Sequential Life Test Plans

p_1	$1 - \hat{R}_L$ $= p_2$	α	$1 - \gamma = \beta$	$\dfrac{h_1}{T}$	$\dfrac{h_2}{T}$	$\dfrac{s}{T}$	$L(p)^*$ $p = p_s$	$E(\mathbf{f})$† $p = p_1$	$p = p_s$	$p = p_2$	$p = 1$	$NE_R(\tau_1)/T$‡ $p = p_1$	$p = p_s$	$p = p_2$
0.01	0.10	0.05	0.05	30.9	30.9	24.7	0.500	0.37	1.57	1.84	1.25	36.9	38.7	17.5
		0.05	0.10	23.6	30.3		0.562 0.0397	0.28	1.18	1.66	1.23	27.9	29.0	15.8
		0.10	0.05	30.3	23.6		0.438	0.33	1.18	1.40	0.96	33.1	29.0	13.3
		0.10	0.10	23.1	23.1		0.500	0.25	0.87	1.25	0.94	24.5	21.6	11.8
0.01	0.05	0.05	0.05	71.4	71.4	39.5	0.500	1.07	3.26	3.23	1.81	106.9	129.0	62.9
		0.05	0.10	54.6	70.1		0.562 0.0250	0.81	2.45	2.91	1.77	80.2	96.8	56.7
		0.10	0.05	70.1	54.6		0.438	0.96	2.45	2.45	1.38	95.6	96.8	47.7
		0.10	0.10	53.3	53.3		0.500	0.71	1.82	2.18	1.35	70.7	71.8	42.5
0.05	0.20	0.05	0.05	17.1	17.1	8.56	0.500	1.41	4.01	3.81	2.00	27.5	34.3	17.0
		0.05	0.10	13.1	16.8		0.562 0.1104	1.06	3.01	3.43	1.97	20.7	25.8	15.4
		0.10	0.05	16.8	13.1		0.438	1.26	3.01	2.89	1.53	24.6	25.8	12.9
		0.10	0.10	12.8	12.8		0.500	0.94	2.23	2.57	1.49	18.2	19.1	11.5
0.05	0.10	0.05	0.05	54.5	54.5	13.3	0.500	7.9	16.7	12.9	4.09	154.5	222.8	122.4
		0.05	0.10	41.6	53.5		0.562 0.0723	6.0	12.6	11.6	4.02	116.3	167.2	110.3
		0.10	0.05	53.5	41.6		0.438	7.1	12.6	9.8	3.13	138.6	167.2	92.8
		0.10	0.10	40.6	40.6		0.500	5.3	9.3	8.7	3.05	102.5	124.0	82.8

1. To determine h_1, h_2, and s, multiply the corresponding values in the table by T, the required life of the item.
2. In a plot of total life (ordinate) vs. failures (abscissa), h_1 and $-h_2$ are the respective intercepts of the acceptance and rejection boundaries on the total life axis.
3. To determine $E_R(\tau_1)$, multiply the corresponding values in the table by T/N, where N is the size of the sample maintained on life test.‡

* $L(0) = 1$, $L(p_1) = 1 - \alpha$, $L(p_2) = \beta$, $L(1) = 0$.
† $E(\mathbf{f}) = 0$ when $p = 0$.
‡ $NE_R(\tau_1)/T = h_1/T$ when $p = 0$.

10.9.5 COST MODEL

As in Sec. 10.7.3, the total expected cost for each of the replacement and nonreplacement cases may be determined, bearing in mind that the approximations of the Wald sequential plans are being used. We will make a numerical comparison, for $p = p_s$ only, using the plan specified in Sec. 10.9.4; the optimum sample sizes $\bar{N}_R$ and $\bar{N}_{NR}$, and corresponding minimum total expected costs are then to be found.

Let $C_1 = \$1.00/\text{hr}$, $C_2 = \$100/\text{item}$. We have, for the total expected cost in the replacement and nonreplacement cases,

$$C_R = C_1 E_R(\tau_1) + C_2(N_R + E(\mathbf{f}) - 1) \tag{10.75}$$

$$C_{NR} = C_1 E_{NR}(\tau_1) + C_2 N_{NR} \tag{10.76}$$

The quantities to be inserted in Eqs. (10.75) and (10.76) may be taken directly from Table 10.3, and we have

$$C_R = 1 \cdot \frac{167.2(1000)}{N_R} + 100 \cdot (N_R + 12.6 - 1) \tag{10.75'}$$

$$C_{NR} = 1 \cdot \frac{1000}{0.0751} \log\left(\frac{N_{NR} + \frac{1}{2}}{N_{NR} + \frac{1}{2} - 12.6}\right) + 100 N_{NR} \tag{10.76'}$$

where the value of $E_{NR}(\tau_1)$ is obtained from Eq. (10.73).

The reader may verify that the optimum sample sizes are $\bar{N}_R = 41$ and $\bar{N}_{NR} = 47$, and the minimum total expected costs are $\bar{C}_R = \$9338$ and $\bar{C}_{NR} = \$8804$. These results confirm the statement made at the beginning of Sec. 10.9.

EXERCISE: Calculate the values of $\bar{N}_R$, $\bar{N}_{NR}$, $\bar{C}_R$, and $\bar{C}_{NR}$ when $C_1 = \$10/\text{hr}$ and $C_2 = \$10/\text{item}$, all other data being the same as in the example given.

EXERCISE: Compare the consequences of using a Wald sequential replacement plan with those of using the plan of Sec. 10.7.2 (replacement case). Use as reliability specifications $p_1 = 0.05$, $\alpha = 0.05$; $p_2 = 0.10$, $\beta = 0.10$. Also compare costs. In addition consider the nonreplacement cases.

EXERCISE: The sampling plan criteria of this section may be expressed in terms of specifications on mean time-to-failure; i.e., θ_1, α; θ_2, β. Convert Eqs. (10.63)–(10.65), (10.68)–(10.71), (10.73), and (10.74) to expressions in terms of θ_1, θ_2 (see Ref. 16). Note that the required life T does not enter into this formulation of the life test sampling plan criteria. Note also that the same conversion can be applied to all of the life test sampling plans of this chapter.

REFERENCES

1. A. H. Bowker and G. J. Lieberman, *Engineering Statistics*, Prentice-Hall, Inc., Englewood Cliffs, N. J., 1959, pp. 467–535.

2. G. J. Lieberman and G. J. Resnikoff, "Sampling Plans for Inspection by Variables," *J. Am. Stat. Assoc.*, **50** (June 1955), 457–516.
3. G. J. Resnikoff and G. J. Lieberman, *Tables of the Non-Central t-Distribution*, Stanford University Press, Stanford, Calif., 1957.
4. *Tables of the Cumulative Binomial Distribution*, Harvard University Press, Cambridge, Mass., 1955.
5. C. Eisenhart, M. W. Hastay, and W. A. Wallis, *Techniques of Statistical Analysis*, McGraw-Hill Book Company, Inc., New York, 1947, Chap. 6.
6. A. Wald, *Sequential Analysis*, John Wiley & Sons, Inc., New York, 1947.
7. M. Sobel and J. A. Tischendorf, "Acceptance Sampling with New Life Test Objectives," *Proc. 5th National Symposium on Reliability and Quality Control in Electronics, Philadelphia, January 12–14, 1959*, pp. 108–118.
8. H. P. Goode and J. H. K. Kao, "Sampling Plans Based on the Weibull Distribution," *Proc. 5th National Symposium on Reliability and Quality Control in Electronics, Philadelphia, January 9–11, 1961*, pp. 24–40.
9. B. Epstein, "Truncated Life Tests in the Exponential Case," *Ann. Math. Stat.*, **25** (1954), 555–64.
10. B. Epstein, "Testing of Hypotheses," *Wayne State Univ. Tech. Report No. 3* (October 1, 1958). (Prepared under Contract Nonr-2163(00), NR-042-018, for Office of Naval Research.) (This report is also included in Ref. 12).
11. B. Epstein and M. Sobel, "Life Testing," *J. Am. Stat. Assoc.*, **48** (1953), 486–502.
12. *Statistical Techniques in Life Testing*, PB171580, U. S. Department of Commerce, Office of Technical Services, Washington 25, D.C., (July 1961).
13. J. Nadler, "Inverse Binomial Sampling Plans When an Exponential Distribution is Sampled with Censoring," *Ann. Math. Stat.*, **31** (December 1960), 1201–204.
14. D. Mendenhall and E. H. Lehman, Jr., "An Approximation to the Negative Moments of the Positive Binomial Useful In Life Testing," *Technometrics*, **2** (May 1960), 227–42.
15. H. Cramér, *Mathematical Methods of Statistics*, Princeton University Press, Princeton, N. J., 1946.
16. B. Epstein and M. Sobel, "Sequential Life Tests In The Exponential Case," *Ann. Math. Stat.*, **26** (1955), 82–93.

ADDITIONAL READING

Albert, G. E., "Accurate Sequential Tests on the Mean of an Exponential Distribution," *Ann. Math. Stat.*, **27,** 460–70 (June 1956).

Bechofer, R., "A Note on the Limiting Relative Efficiency of the Wald Sequential Probability Ratio Test," *J. Am. Stat. Assoc.*, **55,** 660–663 (1960).

Breakwell, J. V., "Economically Optimum Acceptance Tests," *J. Am. Stat. Assoc.*, **51,** 243–256 (June 1956).

Burr, I. W., "Average Sample Number under Curtailed or Truncated Sampling," *Ind. Qual. Control*, **13**:8, 5–7 (February 1957).

Chernoff, H., and Lieberman, G. J., "Sampling Inspection by Variables with No Calculations," *Ind. Qual. Control*, **13**:7, 5–7 (January 1957).

Epstein, B., "Statistical Developments in Life Testing," *Proc. 3rd National Symposium on Reliability and Quality Control, Washington, D.C., January 14–16, 1957*, pp. 106–112.

Harter, H. L., "Circular Error Probabilities," *J. Am. Stat. Assoc.* **55**, 723–731 (1960).

Lindley, D. V., "Binomial Sampling Schemes and the Concept of Information," Parts 1 and 2, *Biometrika*, **44**, 179–86 (June 1957).

Moranda, P. B., "Effect of Bias on Estimates of the Circular Probable Error," *J. Am. Stat. Assoc.*, **55**, 732–735 (1960).

Moriguti, Sigeiti, "Efficiency of a Sampling Inspection Plan," *Reports of Statistical Applications Research, Union of Japanese Scientists and Engineering*, **4**:3, 1–7 (July 1956).

Sobel, M., "Statistical Techniques for Reducing the Experiment Time in Reliability Studies," *Bell System Tech. J.*, **35**, 179–202 (January 1956).

Stevens, C. F., "A Sequential Test for Comparing Component Reliabilities," *I.R.E. Trans. Reliability and Quality Control*, **RQC–12**, 37–47 (November 1957).

Walsh, J. E., "Estimating Future from Past in Life Testing," *Ann. Math. Stat.*, **28**, 432–41 (June 1957).

CHAPTER ELEVEN
RELIABILITY GROWTH MODELS

11.1. INTRODUCTION

The periodic recording of reliability estimates generates a time sequence of data from which trends can be observed. We are interested in determining and studying reliability growth curves for the following reasons.

First, we are concerned with the current reliability estimate and its rate of growth up to the present time and whether it is possible to extrapolate or extend the growth curve into the future. If this can be done, then the ultimate level of reliability may be estimated, permitting us to determine whether or not any future reliability requirements will be satisfied. Second, a reliability growth study should be more than a periodic recording of data. It should seek the reasons for the growth patterns, so that decisions regarding the future of the program can be made on a more objective basis. Thus we might determine how growth of a system is related to such factors as the numbers of tests, cost, design reviews, manpower application, efficiency of failure analysis, numbers and types of design changes, and so on. As a consequence, we should be able to apply our money, manpower, and time more effectively for future programs.

It has been emphasized that it is of the highest importance to design reliability into the system (Sec. 12.5). This becomes particularly obvious when the growth curves of certain types of equipment are studied. For example, in the case of liquid rocket engines, if the reliability is plotted against numbers of tests, it will generally be found that the reliability curve begins to "level off." Of course, this is expected if the value of reliability is approaching unity; however, this "phenomenon" can occur even when reliability is significantly less than unity. In other words, we find that continued testing, failure reporting, analysis, and minor engineering corrective action frequently reach, after a time, "a point of diminishing returns." To all intents and purposes the ultimate inherent reliability of the design has been reached. If this value is not sufficient, then a major design change or a major system concept change might have to be introduced into the program. It is usually found that the new design will start at a lower reliability value than its predecessor, which it will rapidly overtake, only to start leveling off at a higher reliability number. The level of this higher number is dependent on both our intrinsic design capability and our learning ability. Thus it becomes of prime importance with respect to both time and cost to be able to detect at what time and what reliability level of the system's development the "point of diminishing returns" will occur.

Generally, we are dealing with a complex system which is continuously undergoing a variety of experiences, and as a consequence the plot of reliability points will be expected to show a certain amount of fluctuation, owing to the effect of these factors as well as to statistical random variation of the test results. This fluctuation tends to mask the true reliability growth curve and therefore lessen our ability to detect the probable level of the ultimate and inherent reliability of the system.

This chapter therefore illustrates, by means of simple models, methods for determining the reliability growth curves and assigning confidence limits to estimates or predictions derived from them. It also considers the type of mathematical model which might generate a growth function.

11.2. A SIMPLE MODEL OF RELIABILITY GROWTH

Suppose a device is being developed for a given application. The application is such that the device, when put into operation, either succeeds or fails to accomplish what it is designed to do. Suppose further that the device is so simple that it fails in only one possible way, and that the whole purpose of the development effort on the device is to discover what the cause of failure is and then attempt to redesign or fix the device so that it won't fail at all. Let us further assume that any redesign effort either is completely successful or else does not change the inherent probability of failure p. In other words, the probability of failure is either p or 0 at any time.

The development effort then consists of repeated attempts at operation (trials) of the device. If the device operates successfully on any given trial, the designer or development engineer decides to take no redesign action prior to the next trial on the chance that he has already fixed the device or that its probability of failure is zero. If it fails on any given trial, the engineer goes to work on it and has a probability α of fixing the device permanently prior to the next trial.*

---

* This model is not as trivial as might at first be thought. Occasionally a situation is experienced in a development program wherein a particular component repeatedly fails. While every effort is made to analyze the failure, the solution to the problem may not be readily obtainable. This is perhaps because we are working at the limits of our knowledge and/or because system requirements (with respect to type of materials, weight, dimensions, etc.) limit our ability to "engineer away" completely the problem for this particular component without significantly compromising the system's configuration or performance. An example of this type occurred in the early development work of movable nozzles for solid propellant rocket engines. Throat inserts of various materials and configurations failed sufficiently frequently to indicate that a variety of designs were marginal until a significant breakthrough was made.

To put these assumptions in mathematical form, let $P_n(1)$ denote the probability that the device is still unreliable just before trial n is to be made; i.e., it is in state number 1. Correspondingly let $P_n(0)$ denote the probability that it is completely reliable (state zero) just before trial n. Before the first trial is made $(n = 1)$, the device is known to be in state 1; i.e., $P_1(1) = 1$. Now, if we ask ourselves how the device could be in state 1 just before trial n, we see that this result could have come about in two mutually exclusive ways: (1) The device was in state 1 at trial $n - 1$ and the device was successful in trial $n - 1$. This probability is

$$P_{n-1}(1) \cdot (1 - p)$$

(2) The device was in state 1 at trial $n - 1$; it failed in trial $n - 1$, and an unsuccessful attempt was made to fix it prior to trial n. This probability is

$$P_{n-1}(1) \cdot p \cdot (1 - \alpha)$$

Since the probability of being in state 1 at trial n is the sum of the probabilities of the mutually exclusive ways it can occur,

$$P_n(1) = P_{n-1}(1) \cdot (1 - p) + P_{n-1}(1) \cdot p \cdot (1 - \alpha) \tag{11.1}$$

or

$$P_n(1) = P_{n-1}(1)[1 - p\alpha] \tag{11.2}$$

Using the initial condition $P_1(1) = 1$, we easily see by induction that

$$P_n(1) = (1 - p\alpha)^{n-1} \tag{11.3}$$

To obtain $P_n(0)$, we note that the device could be in state zero at trial n if: (1) it was in state 0 at trial $n - 1$; (2) it was in state 1 at trial $n - 1$, failed, and was fixed. Hence

$$P_n(0) = P_{n-1}(0) + P_{n-1}(1) \cdot p\alpha \tag{11.4}$$

or, using (11.3), with n replaced by $n - 1$

$$P_n(0) = P_{n-1}(0) + p\alpha(1 - p\alpha)^{n-2} \tag{11.5}$$

Since $P_1(1) = 1$, then $P_1(0) = 0$; because the device cannot be in state 0 at trial 1 if it is in state 1 at trial 1.

Successively, letting $n = 2, 3, \cdots$ we see that

$$P_n(0) = p\alpha[1 + (1 - p\alpha) + \cdots + (1 - p\alpha)^{n-2}] \tag{11.6}$$

$$= \frac{p\alpha[1 - (1 - p\alpha)^{n-1}]}{1 - (1 - p\alpha)} \tag{11.7}$$

or

$$P_n(0) = 1 - (1 - p\alpha)^{n-1} \tag{11.8}$$

We could have obtained this result directly by noting that the device is in either state 1 or state 0 at any trial. Since the probability that it is

RELIABILITY GROWTH MODELS 333

in state 1 at trial n was given by (11.3), then the probability that it is in state 0 at trial n must be this value subtracted from unity, which is the result given by (11.8).

Now we define the reliability R_n of the device at trial n as the probability, as of the time just before the first trial is made, that the device will not fail in trial n. The event: no failure in trial n occurs in two mutually exclusive ways:

1. It is in state 0 at trial n (probability $1 - (1 - p\alpha)^{n-1}$)
2. It is in state 1 at trial n and does not fail (probability $(1 - p\alpha)^{n-1}(1 - p)$).

Thus
$$R_n = 1 - (1 - p\alpha)^{n-1} + (1 - p\alpha)^{n-1}(1 - p) \quad (11.9)$$

or
$$R_n = 1 - (1 - p\alpha)^{n-1}p \quad (11.10)$$

The reader can also easily verify that if it is assumed that the probability is β that the device is in state 1 at trial 1, i.e., $P_1(1) = \beta$, then

$$R_n = 1 - (1 - p\alpha)^{n-1}\beta p \quad (11.11)$$

Note that when $\beta < 1$, the reliability at the nth trial is larger than when it is certain that the device was in state 1 at trial 1, which is as it should be.

The type of reliability growth exhibited by (11.11) is exponential. This can be more directly illustrated by writing (11.11) as:

$$R_n = 1 - Ae^{-C(n-1)} \quad (11.12)$$

where
$$A \equiv \beta p \quad (11.13)$$

$$C \equiv \log \frac{1}{1 - p\alpha} > 0 \quad (11.14)$$

Methods for estimating parameters of reliability growth models will be presented in Sec. 11.3.

11.2.1 FURTHER PROPERTIES OF THE MODEL

The question arises for the previous model: Is the result of any trial *dependent* upon the results of other trials? Intuitively the answer is yes; and we will show that this answer is correct. This result might mean more generally, that if, *a priori*, we postulate a reliability growth curve; e.g., that given by Eq. (11.12), and attempt to fit the curve by success-failure data, we should have to be cautious about assuming independence of the trials. The "caution" actually would be taken only when attempting to make predictions or set confidence limits, since estimators for the parameters of the growth curve can be found without making any assumptions relating to dependence or independence.

The question of dependence of trials is now analyzed. Let T_n be a random variable which has the value one if a success occurs in the nth trial, and zero if a failure occurs in the nth trial. Let V_n be another random variable such that $V_n = 1$ if the device is in state 1 just prior to trial n; $V_n = 0$ if the device is in state 0 just prior to trial n. From Eqs. (11.3), (11.8), and (11.10) we have

$$\begin{aligned} P(V_n = 0) &= 1 - \Delta^{n-1} \\ P(V_n = 1) &= \Delta^{n-1} \\ P(T_n = 0) &= p\Delta^{n-1} \\ P(T_n = 1) &= 1 - p\Delta^{n-1} \end{aligned} \quad (11.15)$$

where $\Delta \equiv 1 - p\alpha$.

Our aim is now to find the joint distribution of T_m and T_n. To do this it is convenient to define an auxiliary random variable X_n as follows:

$X_n = 0$, if device is in state 1 and fails in trial n: $P(X_n = 0) = p\Delta^{n-1}$

$X_n = 1$, if device is in state 1 and is successful in trial n: $P(X_n = 1) = (1 - p)\Delta^{n-1}$

$X_n = 2$, if device is in state 0 (and therefore must be successful in trial n): $P(X_n = 2) = 1 - \Delta^{n-1}$

If we denote conditional probabilities as follows:

$$P(X_{n+1} = j \mid X_n = i) = p_{ij} \quad (11.16)$$

where $i, j = 0, 1, 2$, then the conditional probabilities can be put in the form of a matrix

$$\mathbf{P} \equiv (p_{ij}) = \begin{bmatrix} (1-\alpha)p & (1-\alpha)(1-p) & \alpha \\ p & 1-p & 0 \\ 0 & 0 & 1 \end{bmatrix} \quad (11.17)$$

For example, the value of p_{01} is given above by $p_{01} = (1 - \alpha)(1 - p)$. To obtain this expression, we note that if the device is in state 1 and fails in trial n ($X_n = 0$) it can be in state 1 and successful in trial $n + 1$ ($X_{n+1} = 1$) only by not being "fixed" (probability $1 - \alpha$) then by not failing (probability $1 - p$). Thus the required conditional probability is given by $(1 - \alpha)(1 - p)$. The other probabilities are found by similar reasoning.

EXERCISE: Check the values in the matrix $\mathbf{P}$ given by Eq. (11.17).

The values of p_{ij} in the matrix $\mathbf{P}$ are known as the "one-step" transition probabilities of a Markoff chain with constant transition probabilities

RELIABILITY GROWTH MODELS

(Ref. 1, p. 340). They give the probability of going in a single trial from a given "state i" to another "state j." These are not the same states which we used in defining the random variable V_n; so we will use quotes in the following discussion when we mention "state," for X_n.

From the theory of Markoff chains, it turns out that we can find the "k-step" transition probabilities $p_{ij}^{(k)}$ by raising the matrix $\mathbf{P}$ to the kth power (Ref. 1, pp. 347–348). The quantities $p_{ij}^{(k)}$ denote the conditional probability that our "system" is in "state j" when k trials are made after the system is in "state i." Thus $p_{ij}^{(k)}$ is the conditional probability that $X_{n+k} = j$ given that $X_n = i$.

EXERCISE: Show that

$$\mathbf{P}^k = \begin{bmatrix} (1-\alpha)p\,\Delta^{k-1} & (1-\alpha)(1-p)\Delta^{k-1} & 1-(1-\alpha)\Delta^{k-1} \\ p\,\Delta^{k-1} & (1-p)\Delta^{k-1} & 1-\Delta^{k-1} \\ 0 & 0 & 1 \end{bmatrix} \quad (11.18)$$

Hint: Use the fact that the sum of probabilities for any row must be unity, since, for example, in the second row, if the system is in "state 1" it must either stay in "state 1" or go to "state 0" or "state 2."

Now, consider the four events:

$$\begin{aligned} &\text{(a)} \quad T_m = 0, T_n = 0 \\ &\text{(b)} \quad T_m = 0, T_n = 1 \\ &\text{(c)} \quad T_m = 1, T_n = 0 \\ &\text{(d)} \quad T_m = 1, T_n = 1 \end{aligned} \quad (11.19)$$

where $n \leq m - 1$

Event (a) occurs when the device is in state 1 and fails on trial n, i.e., "state 0" in trial n, and is in state 1 and fails in the mth trial, i.e., "state 0" after $m - n$ trials. Thus the probability of event (a) is the product of the following probabilities:

$$(a_1) \quad P(X_n = 0) = p\Delta^{n-1}$$
$$(a_2) \quad p_{00}^{(m-n)} = (1-\alpha)p\Delta^{m-n-1}$$

or
$$P(T_m = 0, T_n = 0) = p^2(1-\alpha)\Delta^{m-2} \quad (11.20)$$

The probability of event (b) is the product of the probabilities:

$$(b_1) \quad P(X_n = 1) = (1-p)\Delta^{n-1}$$
$$(b_2) \quad p_{10}^{(m-n)} = p\Delta^{m-n-1}$$

Thus
$$P(T_m = 0, T_n = 1) = p(1-p)\Delta^{m-2} \quad (11.21)$$

The probability of event (c) is the product of the probabilities:

(c_1) $P(X_n = 0) = p\Delta^{n-1}$

(c_2) $p_{01}^{(m-n)} + p_{02}^{(m-n)} = (1-\alpha)(1-p)\Delta^{m-n-1} + 1 - (1-\alpha)\Delta^{m-n-1}$
$$= 1 - p(1-\alpha)\Delta^{m-n-1}$$

Thus
$$P(T_m = 1, T_n = 0) = p[\Delta^{n-1} - p(1-\alpha)\Delta^{m-2}] \quad (11.22)$$

The probability of event (d) is the product of the probabilities:

(d_1) $P(X_n = 1) = (1-p)\Delta^{n-1}$

(d_2) $p_{11}^{(m-n)} + p_{12}^{(m-n)} = (1-p)\Delta^{m-n-1} + 1 - \Delta^{m-n-1}$
$$= 1 - p\Delta^{m-n-1}$$

plus the probability

(d_3) $P(X_n = 2) = 1 - \Delta^{n-1}$

Thus
$$P(T_m = 1, T_n = 1) = 1 - p[\Delta^{n-1} + (1-p)\Delta^{m-2}] \quad (11.23)$$

Summarizing, the joint probability distribution of (T_m, T_n) is:

$$\begin{aligned} P(T_m = 0, T_n = 0) &= p^2(1-\alpha)\Delta^{m-2} \\ P(T_m = 0, T_n = 1) &= p(1-p)\Delta^{m-2} \\ P(T_m = 1, T_n = 0) &= p[\Delta^{n-1} - p(1-\alpha)\Delta^{m-2}] \\ P(T_m = 1, T_n = 1) &= 1 - p[\Delta^{n-1} + (1-p)\Delta^{m-2}] \end{aligned} \quad (11.24)$$

where $n \leq m - 1$.

EXERCISE: Write down the probabilities $P(T_m = 0, T_n = 0)$, etc., where $n \geq m + 1$.

EXERCISE: Check the joint distribution given by Eqs. (11.24) for consistency by showing that the probabilities add up to one, and also by finding the marginal distributions for T_m (or T_n).

Now, we wish to find the covariance of T_m and T_n and, incidentally, the correlation coefficient as well. We have, for $n \leq m - 1$

$$\begin{aligned} \text{Cov}(T_m, T_n) &= E(T_m T_n) - E(T_m)E(T_n) \\ &= p^2\Delta^{m-2}(1 - \alpha - \Delta^n) \end{aligned} \quad (11.25)$$

$$\text{Var } T_m \cdot \text{Var } T_n = p^2\Delta^{m+n-2}(1 - p\Delta^{m-1})(1 - p\Delta^{n-1}) \quad (11.26)$$

Thus, the correlation coefficient is given by

$$\rho(T_m, T_n) = \frac{p\Delta^{[(m-n)/2]-1}(1 - \alpha - \Delta^n)}{[(1 - p\Delta^{m-1})(1 - p\Delta^{n-1})]^{1/2}} \quad (11.27)$$

RELIABILITY GROWTH MODELS 337

EXERCISE: Work out the details leading to Eqs. (11.25), (11.26), and (11.27). Use the fact that $T_m T_n = 1$ if and only if $T_m = 1$ and $T_n = 1$. Consider also the case when $n \geq m + 1$.

If we inspect Eq. (11.27), we see that the correlation coefficient will be positive or negative as the quantity $1 - \alpha - \Delta^n$ is positive or negative. Since $\Delta \equiv 1 - p\alpha$ is greater than $1 - \alpha$, but $\Delta < 1$, then for sufficiently large n the correlation coefficient becomes positive for all values of $m \geq n + 1$. However, for "small" values of n, and for m just slightly larger than n, the correlation coefficient can be negative.

Before we summarize the results of this section and Sec. 11.2, let us reconsider the model of Sec. 11.2. We still assume that only one mode of failure of the device exists; however, let us consider the following modification of "trials": Let the nth trial consist of a group of N tests, and assume that only if *all* N tests are successful does the trial result in a success, but if *at least* one test of the group is a failure, then the trial results in failure, and there is, as before, probability α of "fixing" the device prior to the next group of N tests on the $(n + 1)$st trial. Thus if the probability of failure for any test of the nth trial is p', then the probability that the trial is a failure is given by $p = 1 - (1 - p')^N$. This would mean that all we have to do is replace the quantity p in all of the preceding equations by $1 - (1 - p')^N$.

Evidently, to set up a reliability growth model in which the device has more than one mode of failure might be a difficult task. However such models have been considered (Ref. 2).

This and the preceding section can be considered as preliminary, in a sense, to the problem of fitting reliability growth curves to chronological groups of success-failure data. The statement made at the beginning of this section is re-emphasized—that the results of testing may very likely be correlated or dependent, as illustrated with the very simple reliability growth model of Sec. 11.2. This should, if possible, be taken into account in reliability prediction and determination of confidence limits.

In the next section, we shall consider the problem of fitting a reliability growth curve to success-failure data, first by the maximum likelihood method, then by a "least-squares" method. In order to use the maximum likelihood method on *dependent* observations, we should, properly, set up a model allowing correlation between the groups of observations of successes and failures. To make any use of the results of this section as to how such a model would "look," we would have had to find the joint distribution of the whole group of the random variables $T_1, \cdots, T_n, \cdots$. However, only the joint distribution of any *pair* T_m, T_n was found. The latter result is applicable when we use the least-squares method of estimation; the reason is that variances of the estimators can be obtained by finding expected values of products of pairs of observations only; i.e., we only have to consider a problem of simple correlation rather than complete dependence. Unfortunately we cannot go into this problem in further detail.

11.3 FITTING A CURVE TO A RELIABILITY GROWTH MODEL

The model considered here is based on the following assumptions and data. A "test-program" is conducted in N stages; each stage consists of a certain number of tests or trials of the item under test, and the only data recorded is whether the item was successful or failing in each test. All tests in a given stage of testing are conducted on items with a fixed reliability. The results of each stage of testing are used to improve the item for further testing in the next stage.

When the Nth stage of the test program has been completed, it is desired to fit a growth curve to the N groups of success-failure data. For the kth group of data, taken in chronological order, there are n_k tests with S_k observed successes. The growth function chosen for illustration is

$$R_k = R_\infty - \frac{\alpha}{k} \qquad (11.28)$$

where R_k is the actual reliability during the kth stage of testing, R_∞ is the ultimate value of reliability which would be attained if $k \to \infty$, and $\alpha > 0$ modifies the rate of growth, which is "hyperbolic" in this case.

11.3.1 MAXIMUM LIKELIHOOD ESTIMATORS

We will use the method of maximum likelihood to determine estimators for the parameters R_∞ and α. We have for the kth stage alone

$$L_k = \text{const.}\ R_k^{S_k}(1 - R_k)^{n_k - S_k} \qquad (11.29)$$

and assuming that the results of each stage of testing are statistically independent

$$L = \prod_{k=1}^{N} L_k = \text{const.} \prod_{k=1}^{N} R_k^{S_k}(1 - R_k)^{n_k - S_k} \qquad (11.30)$$

Thus

$$\mathcal{L} \equiv \log L$$
$$= \log \text{const.} + \sum S_k \log\left(R_\infty - \frac{\alpha}{k}\right)$$
$$+ \sum (n_k - S_k) \log\left(1 - R_\infty + \frac{\alpha}{k}\right) \qquad (11.31)*$$

* All summations are from 1 to N.

RELIABILITY GROWTH MODELS

The likelihood equations are

$$\frac{\partial \mathcal{L}}{\partial R_\infty} = \sum \frac{S_k}{R_\infty - \frac{\alpha}{k}} - \sum \frac{n_k - S_k}{1 - R_\infty + \frac{\alpha}{k}} = 0 \qquad (11.32)$$

$$\frac{\partial \mathcal{L}}{\partial \alpha} = -\sum \frac{\frac{S_k}{k}}{R_\infty - \frac{\alpha}{k}} + \sum \frac{\frac{n_k - S_k}{k}}{1 - R_\infty + \frac{\alpha}{k}} = 0 \qquad (11.33)$$

Equations (11.32) and (11.33) can be rewritten as:

$$\frac{\partial \mathcal{L}}{\partial R_\infty} = \sum \frac{\frac{S_k}{n_k} - \left(R_\infty - \frac{\alpha}{k}\right)}{\frac{1}{n_k}\left(R_\infty - \frac{\alpha}{k}\right)\left(1 - R_\infty + \frac{\alpha}{k}\right)} = 0 \qquad (11.32')$$

$$\frac{\partial \mathcal{L}}{\partial \alpha} = -\sum \frac{\frac{1}{k}\frac{S_k}{n_k} - \left(R_\infty - \frac{\alpha}{k}\right)\frac{1}{k}}{\frac{1}{n_k}\left(R_\infty - \frac{\alpha}{k}\right)\left(1 - R_\infty + \frac{\alpha}{k}\right)} = 0 \qquad (11.33')$$

Equations (11.32′) and (11.33′) can only be solved by trial and error. To obtain initial values of α and R_∞ assume that $1 - R_\infty \ll \alpha/k \ll 1$. Also replace n_k by $\bar{n} = (1/N) \sum n_k$. With these assumptions the term retained in the denominators of the above sums is α/k, since $R_\infty - \alpha/k \simeq R_\infty$ and $1 - R_\infty + \alpha/k \simeq \alpha/k$; the remaining constants are multiplied out, and we obtain

$$\frac{1}{\bar{n}} \sum k S_k = \frac{N(N+1)}{2} R_\infty - \alpha N \qquad (11.34)*$$

$$\frac{1}{\bar{n}} \sum S_k = N R_\infty - \alpha \sum \frac{1}{k} \qquad (11.35)*$$

---

* In Sec. 11.3.2 "least square" estimators are found, which could be used as initial values for maximum likelihood estimators, instead of the ones obtained from Eqs. (11.34) and (11.35).

Table 11.1

k	n_k	S_k	kS_k	S_k/n_k
1	10	5	5	0.500
2	8	5	10	0.600
3	9	6	18	0.667
4	9	7	28	0.778
5	10	6	30	0.600
6	10	7	42	0.700
7	10	8	56	0.800
8	10	7	56	0.700
9	10	6	54	0.600
10	11	7	70	0.636
11	10	9	99	0.900
12	11	10	120	0.909
13	12	9	117	0.750
14	10	8	112	0.800
15	10	7	105	0.700
16	10	8	128	0.800
17	10	9	153	0.900
18	10	9	162	0.900
19	10	10	190	1.000
20	10	9	180	0.900

$N = 20$ $\bar{n} = 10$ $\Sigma S_k = 152$ $\Sigma k S_k = 1735$

Solving Eqs. (11.34) and (11.35)

$$\hat{\alpha} = \frac{\frac{1}{\bar{n}}\left(\sum kS_k - \frac{N+1}{2}\sum S_k\right)}{\frac{N+1}{2}C_1 - N} \tag{11.36}$$

$$\hat{R}_\infty = \frac{\frac{1}{\bar{n}}\left(\frac{C_1}{N}\sum kS_k - \sum S_k\right)}{\frac{N+1}{2}C_1 - N} \tag{11.37}$$

where
$$C_1 \equiv \sum_{k=1}^{N}\frac{1}{k} \simeq \log\left(N + \frac{1}{2}\right) + E \tag{11.38}*$$

EXAMPLE: Estimate α and R_∞ by Eqs. (11.36) and (11.37) using the data in Table 11.1.

From (11.38) we obtain $C_1 \simeq 3.598$; and from Eqs. (11.36) and (11.37) $\hat{\alpha} = 0.782, \hat{R}_\infty = 0.901$.

* $E \equiv$ Euler's constant $= 0.577215665\cdots$.

RELIABILITY GROWTH MODELS

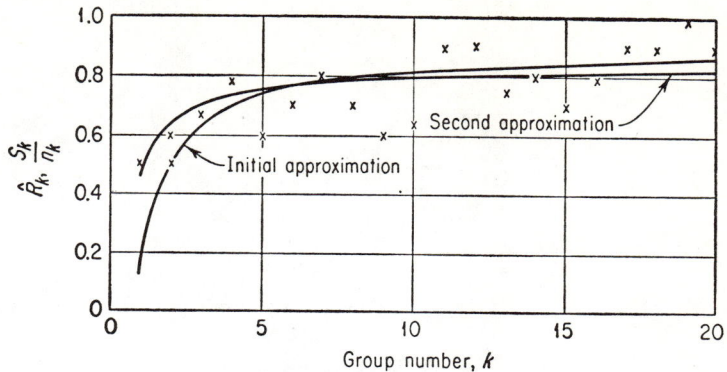

Figure 11.1

Figure 11.1 shows a plot of the curve $\hat{R}_k = 0.901 - 0.782/k$ superimposed on S_k/n_k.

Improved estimators for α and R_∞ can be obtained by evaluating the denominators of Eqs. (11.32′) and (11.33′) using the initial values of α and R_∞ obtained from Eqs. (11.34) and (11.35). The resulting estimating equations are

$$\sum \frac{kS_k}{D_k n_k} = R_\infty \sum \frac{k}{D_k} - \alpha \sum \frac{1}{D_k} \qquad (11.32'')$$

$$\sum \frac{S_k}{n_k D_k} = R_\infty \sum \frac{1}{D_k} - \alpha \sum \frac{1}{kD_k} \qquad (11.33'')$$

where

$$D_k = \frac{k}{n_k}\left(\hat{R}_\infty - \frac{\hat{\alpha}}{k}\right)\left(1 - \hat{R}_\infty + \frac{\hat{\alpha}}{k}\right)$$

$$= \frac{k}{n_k}\left(0.901 - \frac{0.782}{k}\right)\left(0.099 + \frac{0.782}{k}\right)$$

The values of the various coefficients appearing in Eqs. (11.32″) and (11.33″) are, from the data in Table 11.1,

$$\sum \frac{kS_k}{D_k n_k} = 1067$$

$$\sum \frac{k}{D_k} = 1389$$

$$\sum \frac{1}{D_k} = 240.2$$

$$\sum \frac{S_k}{n_k D_k} = 154.6$$

$$\sum \frac{1}{k D_k} = 121.6$$

Hence Eqs. (11.32″) and (11.33″) become

$$1067 = 1389 R_\infty - 240.2 \alpha$$

$$154.6 = 240.2 R_\infty - 121.6 \alpha$$

Solving the above equations, we obtain for the second approximation

$$\hat{\alpha} = 0.374, \qquad \hat{R}_\infty = 0.833$$

The third approximation using the above values of $\hat{\alpha}$ and $\hat{R}_\infty$ yields

$$\hat{\alpha} = 0.407, \qquad \hat{R}_\infty = 0.831$$

which indicates that the process is converging fairly well.

11.3.1.1. Variances and Covariances. The second partial derivatives of $\mathcal{L}$ with respect to R_∞ and α are, from Eqs. (11.32) and (11.33),

$$\frac{\partial^2 \mathcal{L}}{\partial R_\infty^2} = -\sum \frac{S_k}{\left(R_\infty - \dfrac{\alpha}{k}\right)^2} - \sum \frac{n_k - S_k}{\left(1 - R_\infty + \dfrac{\alpha}{k}\right)^2} \qquad (11.39)$$

$$\frac{\partial^2 \mathcal{L}}{\partial \alpha^2} = -\sum \frac{\dfrac{S_k}{k^2}}{\left(R_\infty - \dfrac{\alpha}{k}\right)^2} - \sum \frac{\dfrac{n_k - S_k}{k^2}}{\left(1 - R_\infty + \dfrac{\alpha}{k}\right)^2} \qquad (11.40)$$

$$\frac{\partial^2 \mathcal{L}}{\partial R_\infty \, \partial \alpha} = \frac{\partial^2 \mathcal{L}}{\partial \alpha \, \partial R_\infty} = \sum \frac{\dfrac{S_k}{k}}{\left(R_\infty - \dfrac{\alpha}{k}\right)^2} + \sum \frac{\dfrac{n_k - S_k}{k}}{\left(1 - R_\infty + \dfrac{\alpha}{k}\right)^2} \qquad (11.41)$$

Thus, estimated values of the second partial derivatives can be calculated by substituting in (11.39), (11.40), and (11.41) the previously

RELIABILITY GROWTH MODELS 343

estimated values of α and R_∞, and using the data in Table 11.1. We obtain, using $\alpha = 0.39$, $R_\infty = 0.83$:

$$\left\langle \frac{\partial^2 \mathcal{L}}{\partial R_\infty^2} \right\rangle \simeq -1155.3$$

$$\left\langle \frac{\partial^2 \mathcal{L}}{\partial \alpha^2} \right\rangle \simeq -69.95$$

$$\left\langle \frac{\partial^2 \mathcal{L}}{\partial \alpha \, \partial R_\infty} \right\rangle \simeq 184.40$$

Thus, taking the parameters in the order R_∞, α:

$$\mathbf{A} = \begin{bmatrix} -1155.3 & 184.40 \\ 184.40 & -69.95 \end{bmatrix}$$

$$Det\ \mathbf{A} = 46810$$

Therefore

$$-\mathbf{A}^{-1} = \begin{bmatrix} 0.001494 & 0.003939 \\ 0.003939 & 0.02468 \end{bmatrix} = \begin{bmatrix} \langle \text{Var } \hat{R}_\infty \rangle & \langle \text{Cov } (\hat{R}_\infty, \hat{\alpha}) \rangle \\ \langle \text{Cov } (\hat{R}_\infty, \hat{\alpha}) \rangle & \langle \text{Var } \hat{\alpha} \rangle \end{bmatrix}$$

(The brackets "$\langle\ \rangle$" denote estimated values.) To obtain the variance of $\hat{R}_k \equiv \hat{R}_\infty - \hat{\alpha}/k$, we have

$$\langle \text{Var } \hat{R}_k \rangle = \langle \text{Var } \hat{R}_\infty \rangle + \frac{1}{k^2} \langle \text{Var } \hat{\alpha} \rangle - \frac{2}{k} \langle \text{Cov } (\hat{R}_\infty, \hat{\alpha}) \rangle$$

or
$$\langle \text{Var } \hat{R}_k \rangle = 0.001494 + \frac{0.02468}{k^2} - \frac{0.007878}{k} \qquad (11.42)$$

A lower confidence limit with confidence coefficient γ can now be computed as (see Sec. 7.4.4)

$$\hat{R}_{L,k} = \hat{R}_k - K_{1-\gamma} \sqrt{\langle \text{Var } \hat{R}_k \rangle} \qquad (11.43)$$

Figure 11.2 shows the maximum likelihood estimator, $\hat{R}_k$, and also the lower confidence limit, $\hat{R}_{L,k}$, as a function of k for $\gamma = 0.90$; $\hat{\alpha} = 0.39$, $\hat{R}_\infty = 0.83$. A valid inference can then be made for reliability as of some future group number k, i.e., based on the data and using Eq. (11.43). One would state that "the true reliability of (say) group $k = 40$ is contained in the interval $(\hat{R}_{L,40}, 1)$," with confidence γ. However, the inference would probably not be valid if (1) the model was incorrect, or (2) the model was correct, but the future data were from a population with different values of the constants R_∞ and α. Also it would only be proper to make one such inference; in fact, the group number k at which an inference is to be made should be chosen in advance of obtaining the data, or at least independently

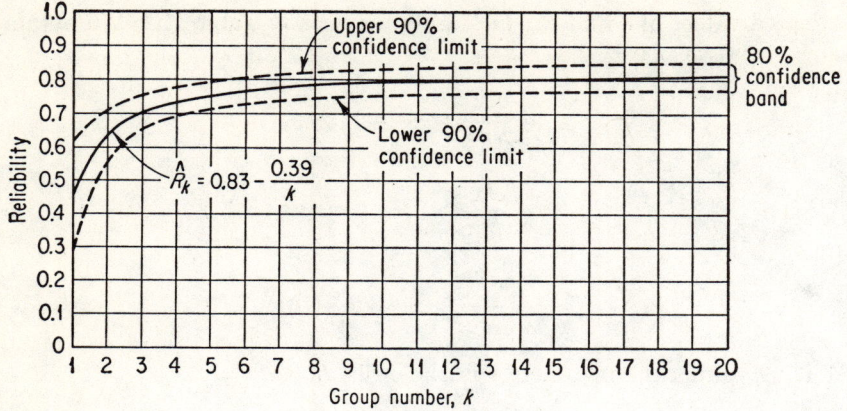

Figure 11.2

of any results of testing. No provision is given here for testing the correctness of the model chosen, although such a test is possible, in principle.*

11.3.2 LEAST-SQUARE ESTIMATORS

Let us now attempt to obtain estimators for the parameters R_∞ and α by a "least-squares" method. To do this we minimize the sum of squares Q of the deviations of the observed success-ratio S_k/n_k from its expected value $R_\infty - \alpha/k$ with respect to the parameters R_∞ and α. Thus

$$Q = \sum_{k=1}^{N} \left(\frac{S_k}{n_k} - R_\infty + \frac{\alpha}{k} \right)^2 \tag{11.44}$$

$$\frac{\partial Q}{\partial R_\infty} = -2 \sum \left(\frac{S_k}{n_k} - R_\infty + \frac{\alpha}{k} \right) = 0 \tag{11.45}†$$

$$\frac{\partial Q}{\partial \alpha} = 2 \sum \left(\frac{S_k}{n_k} - R_\infty + \frac{\alpha}{k} \right) \frac{1}{k} = 0 \tag{11.46}†$$

We obtain two equations, linear in the two unknowns, R_∞ and α:

$$\sum \frac{S_k}{n_k} = NR_\infty - \alpha C_1 \tag{11.47}$$

$$\sum \frac{S_k}{kn_k} = R_\infty C_1 - \alpha C_2 \tag{11.48}$$

where $\qquad C_1 = \sum \frac{1}{k} \qquad C_2 = \sum \frac{1}{k^2}$

---

* See Ref. 3, pp. 321–322, in which a test of *linearity* is given for a linear regression problem in which the x_i (our R_k) is distributed normally about the regression line.

† Summations from 1 to N.

RELIABILITY GROWTH MODELS

Solving Eqs. (11.47) and (11.48) for R_∞ and α, denoting the solutions by $\hat{R}_\infty$, $\hat{\alpha}$, we obtain the least-square estimators:

$$\hat{R}_\infty = \frac{C_2 \sum (S_k/n_k) - C_1 \sum (S_k/kn_k)}{NC_2 - C_1^2} \tag{11.49}$$

$$\hat{\alpha} = \frac{C_1 \sum (S_k/n_k) - N \sum (S_k/kn_k)}{NC_2 - C_1^2} \tag{11.50}$$

The above estimators are easily shown to be unbiased (Sec. 7.2.1) by taking expected values of both sides of Eqs. (11.49) and (11.50), using the fact that

$$E\left(\frac{S_k}{n_k}\right) = R_\infty - \frac{\alpha}{k}$$

Numerical values for C_1 and C_2 are most easily obtained by

$$C_1 = \sum_{k=1}^{N} \frac{1}{k} \simeq \log\left(N + \frac{1}{2}\right) + E$$

$$C_2 = \sum_{k=1}^{N} \frac{1}{k^2} \simeq \frac{\pi^2}{6} - \frac{1}{N + \frac{1}{2}}$$

where $E = 0.577215665\cdots$ is Euler's constant and $\pi^2/6 = 1.64493407\cdots$.

The variances and covariances of the least-square estimators $\hat{R}_\infty$ and $\hat{\alpha}$ may be found exactly, directly from Eqs. (11.49) and (11.50). However, the calculations for the general formulas are fairly complicated. *The least-square estimators can be used nonetheless to obtain good initial values for the maximum likelihood estimators.* Then the iteration procedure given in Sec. 11.3.1 can be used to obtain the nearly exact values. The computation of the estimates of the variances and covariances as given in Sec. 11.3.1 is tedious but straightforward; and it may be simpler, in general, than using the exact formulas referred to above. For example, using the data from Sec. 11.3.1, the least-square estimators turn out to be

$$\hat{R}_\infty = 0.802, \quad \hat{\alpha} = 0.248$$

These values are much closer to the values obtained in Sec. 11.3.1 in the second iteration, indicating that perhaps only one iteration would be sufficient starting with the above values.

Other rate-of-growth functions might be suitable in the model. Thus a more general model would be

$$R_k = R_\infty - \alpha f(k) \tag{11.51}$$

where $f(k)$ is a positive, decreasing function of k. In order to obtain least-square estimators of R_∞ and α it is necessary to evaluate sums of the form

$$C_1 = \sum_{k=1}^{N} f(k) \quad \text{and} \quad C_2 = \sum_{k=1}^{N} \overline{f(k)}^2$$

For convenience, Table 11.2 gives approximation formulas for a few $f(k)$ that might be considered.

Table 11.2 Approximation Formulas for $C_1 = \sum_{k=1}^{N} f(k)$ and $C_2 = \sum_{k=1}^{N} \overline{f(k)}^2$

$f(k)$	C_1	C_2
$k^{-1/2}$	$2\left(N+\dfrac{1}{2}\right)^{1/2} - 1.460$	$\log\left(N+\dfrac{1}{2}\right) + E$
k^{-1}	$\log\left(N+\dfrac{1}{2}\right) + E$	$\dfrac{\pi^2}{6} - \left(N+\dfrac{1}{2}\right)^{-1}$
k^{-2}	$\dfrac{\pi^2}{6} - \left(N+\dfrac{1}{2}\right)^{-1}$	$\dfrac{\pi^4}{90} - \dfrac{1}{3}\left(N+\dfrac{1}{2}\right)^{-3}$
k^{-3}	$1.2020569 - \dfrac{1}{2}\left(N+\dfrac{1}{2}\right)^{-2}$	$\dfrac{\pi^6}{945} - \dfrac{1}{5}\left(N+\dfrac{1}{2}\right)^{-5}$

$$E = 0.577215665, \quad \frac{\pi^2}{6} = 1.64493407$$

$$\frac{\pi^4}{90} = 1.08232323, \quad \frac{\pi^6}{945} = 1.01734306$$

11.3.3 SUMMARY OF ESTIMATION PROCEDURE

Probably the simplest estimation procedure is to use the least-squares method to obtain estimators of the growth curve parameters, since the equations are usually easier to solve than those obtained by the maximum likelihood method. To obtain estimates of variances and covariances, one can insert the least-squares estimators directly into the second partial derivatives of the log-likelihood function [e.g., Eqs. (11.39)–(11.41)]. However, it is recommended that the least-squares estimators be used as initial approximations to the maximum likelihood estimators, which would then be obtained by successive iteration (*cf.* Sec. 11.3.1). In those cases where the maximum likelihood equations are easier to solve than those obtained by the least squares method, the maximum likelihood procedure would be followed throughout. The recommended procedure is in accordance with the principle that maximum likelihood estimators tend to have smallest variances (*cf.* Sec. 7.5.3).

11.4 CONCLUDING REMARKS

It is worthwhile to comment on the contents of this chapter as well as to add a few points of generality, since the methods illustrated do not represent a universal approach to the problem of reliability growth.

11.4.1 PROBLEMS OF NON-INDEPENDENCE

In Sec. 11.2 we considered a very simple model of reliability growth, which, as was indicated, could be applicable to the final stages of a development program in which one component or subsystem constitutes the major problem area. The model was further investigated in order to describe mathematically how the test results would be correlated, and it was concluded that in order to make confidence statements on predicted reliabilities, correlation of test results should, if possible, be taken into account. However, in Sec. 11.3, it was assumed that test results were independent and therefore uncorrelated, in order to simplify the discussion and keep algebraic computation to a minimum.

One method of introducing correlation into the least squares estimation process of Sec. 11.3.2 is to assume that a variable correlation coefficient exists, one that is negative for closely adjacent trials and positive, tending to zero, for widely spaced trials [this is in keeping with the remarks on Eq. (11.27)]. Then, when variances and covariances of the growth curve parameter estimators are to be computed, the assumed correlation function enters into the computation of expected values of quantities such as $\Sigma\, S_k/n_k$ or $(\Sigma\, S_k/n_k)^2$ (*cf.* Sec. 5.9.2). It appears to be a formidable problem to set up a model to *estimate* such a correlation function, and probably the most feasible procedure is to assign values arbitrarily to the correlation coefficients, modifying them somehow as experience dictates.

11.4.2 OTHER POSSIBLE TYPES OF MODELS

In Sec. 11.3, we considered as a measure of reliability the ratio of successes to trials; i.e., the binomial success-failure model. Equally, the individual tests or trials could have consisted of measurements of (say) time-to-failure. In this case, provided that the exponential distribution of time-to-failure applies, a suitable model for reliability growth could be given by

$$R_k = e^{-\alpha_k T} \tag{11.52}$$

where α_k is some specified decreasing function of k with one or more unspecified parameters involved, and T denotes, as usual, the required life. One could then apply the maximum likelihood or least-squares methods to obtain estimators of the unknown parameters and, consequently, an estimator for R_k in the manner illustrated in Sec. 11.3.1 or Sec. 11.3.2.

Other, entirely different approaches could be used to predict reliability. We shall give just one example, as follows.

Let $\hat{R}_k$ be the estimator for the reliability as of the kth trial (the kth trial could represent the tests on items during the kth month after the start of a test program, for example). Let $\hat{X}_k$ represent an estimator for the reliability of those items tested *during* the kth trial. Then $\hat{R}_k$ is defined as

$$\hat{R}_k = \alpha \hat{X}_k + (1 - \alpha) \hat{R}_{k-1} \tag{11.53}$$

where α is a suitable constant between zero and one. It is evident that $\hat{R}_k$ is a weighted estimate of the (current) result of the kth trial and the results of all of the previous trials. When α is small, say $\alpha = 0.1$, the result is to weight the previous data heavily relative to the current result. Conversely, a large value of α, say 0.9, would result in assigning more importance to the current estimate than the results of previous testing. The quantity α could be dependent upon the complexity of the device being developed, the frequency of design changes, and other factors.

REFERENCES

1. W. Feller, *An Introduction to Probability Theory and Its Applications*, 2d ed., John Wiley & Sons, Inc., New York, 1957, vol. I.
2. H. K. Weiss, "Estimation of Reliability Growth in A Complex System with a Poisson-type Failure," *Operations Research*, **4**:5, 532–45 (October 1956).
3. A. M. Mood, *Introduction to the Theory of Statistics*, McGraw-Hill Book Company, Inc., New York, 1950.

ADDITIONAL READING

Ayer, M., Brunk, H. D., Ewing, G. M., Reid, W. T., Silverman, E., "An Empirical Distribution Function for Sampling with Incomplete Information," *Ann. Math. Stat.*, **26**, 641–47 (1955).

Gabriel, K. R., "The Distribution of the Number of Successes in a Sequence of Dependent Trials," *Biometrika*, **46**, Parts 3 and 4, 454–60 (December 1959).

Mandel, J., "Fitting a Straight Line to Certain Types of Cumulative Data," *J. Am. Stat. Assoc.* **52**, 552–66 (December 1957).

Noether, G. E., "Two Sequential Tests Against Trend," *J. Am. Stat. Assoc.*, **51**, 440–50 (September 1956).

Wheeler, R. E., "A Variable Probability Distribution Function," *Ann. Math. Stat.*, **27**, 196–99 (March 1956).

CHAPTER TWELVE
EXPERIMENTATION AND TESTING

12.1. THE REASONS FOR TESTING

A test has been defined as "a subjection to conditions that show the real character of the thing." However, testing is no unique operation performed at a specific point in the development of a complex system but is a continuing operation to provide information throughout the complete evolution of the system. The type of information derived at any particular stage of development of a program changes, and therefore it must be expected that many distinct testing techniques will exist.

The purposes of testing are numerous. In the initial stages a test may be performed to see whether a certain configuration or item is feasible. Later in the development we may wish to determine which of several configurations is the optimum with respect to performance, reliability, cost, modes of behavior under varying conditions, etc. As the product evolves to a more established configuration, we might wish to make more sensitive comparisons to further improve economy, maintenance, use of standard parts, and so on. Still later in the program there are various qualifying tests to demonstrate whether the item is adequate to meet the requirements of performance and reliability. Finally, there should be a thorough investigation of the latent capabilities of the item under severer or more diverse conditions than those immediately anticipated. Such testing seeks to prove that the inherent design is good enough to withstand severer conditions and thus allows less costly methods to be used in handling and transporting, or allows more diversified application of the item.

12.2. PHILOSOPHIES OF TESTING

In the field of reliability there are several philosophies of testing. Each has its merits, and where there is conflict of opinion it is usually a result of a lack of appreciation of the basic operation of the item being tested, its state of development, and the real objectives of the test.

Let us consider these aspects further. By the phrase "basic operation" we mean both the functional and physical interrelationship of the components with respect to each other and their system. For the purpose of this immediate discussion, we shall refer to only two levels of equipment complexity: *component* and *system*. These are relative terms and we use them in the sense that a system is a major functional unit which is composed of a complex of interacting components. The degree of interrelationship is one of the factors which affect the test philosophy. To take two extremes, we can consider that the propellant of a solid rocket motor is much more an *integral* component of the engine system than is a transistor in an electronic computer. Both are necessary for the operation of the system, but the test of a transistor by itself has a more definable meaning in regard to its end effect on the system than has the test of a propellant sample, which might produce interaction effects with the case, liner, or nozzles, for example. We must, therefore, consider the relationship of the information generated by a test on equipment against its end use, which will in turn affect our approach to testing.

12.3. SYSTEM AND COMPONENT TESTING

The level of development also has implications for test philosophy. One system may consist of an aggregation of components, most of which are standard off-the-shelf items, whereas a second system may consist of components which have had to be developed from almost first principles specifically for the system concerned. In the first case, if the behavior of the components is well established, provided that the system is not subjecting the components to functional or physical experiences beyond their known limits of operation, then most of the testing should consist of system tests. However, when the components are undeveloped themselves, they must be subjected to a certain amount of separate testing. Component testing forms an important part of development; it is at this stage that various components should be tested over a wide range of conditions to insure that the best of several alternates will be chosen, and that it will give a satisfactory performance at other than nominal conditions when integrated into a larger system. However, one of the problems of component testing is that of realistically simulating system environments, including parametric input and variation to the component. Also, the extremely high reliability required of a single component necessitates large numbers of tests in order to demonstrate its reliability. *Thus, component testing is better suited to improving reliability by optimum selection than to determining the absolute value of reliability.*

When testing is on a system level there is obviously no need to simulate internal environments. The system has a lower reliability requirement relative to its components, which makes it easier to demonstrate an absolute reliability number (dependent of course on the cost and/or number of units available for testing). However, one of the disadvantages of entering into system testing too early is that many failures will occur in components not sufficiently proven out, thus making failure tracking difficult. Another disadvantage is that if too many component failures occur, then the remainder will be subjected to too many start operations, which are perhaps severer than steady-state operation; consequently, a false impression of the failure distributions will occur, compared with those expected in operation.

Testing systems for their reaction to external environmental conditions can present quite a problem. For instance, while a large rocket engine can be temperature conditioned and cycled, it is impossible to simulate all of the acceleration forces experienced in flight. The only way to establish the effect of this factor (for example) would be to compare flight tests with ground tests, but here again effects other than acceleration can confound the difference. If this is the case, then component tests should attempt to cover adequately those of the external-to-the-system environmental conditions which the system cannot receive in test. In this way, for example, acceleration-sensitive components may be strengthened or modified to eliminate this weakness. System testing focuses on the question, "Is the component reliable within the system?" rather than, "Does the component meet the specifications?" Too much emphasis should not be placed on component testing-to-specifications. The specifications are only the means to the end of having a reliable and workable system. System testing does not eliminate component testing but helps to pinpoint the faulty components, so that they may be replaced or modified by superior products. System testing is a way of realistically evaluating reliability as well as guiding component improvement by systematically discovering component problems and weaknesses, thereby helping to establish an apportionment of reliability criticality.

The specific objectives of tests vary with time, but the over-all object is to derive information as efficiently as possible so that the development program will be concluded successfully and rapidly. This should be accomplished in an optimum manner. Consequently, any specific test should be planned to be coordinated with the total program so that the derived information has the maximum possible value for continuing application throughout later stages of the program.

12.4 TECHNIQUES OF TESTING

In addition to the foregoing factors we must consider the methods of testing effectively so that true behavior of the equipment is established.

Let us return to our original definition: "Subjection to conditions which show the real character of the thing." Although there is no question that the end operation is the real test, it is also obvious that if we have not anticipated the end operation by suitable tests earlier in the development program, we may find at too late a date that "the real character of the thing" is not adequate. In order to prevent this situation from occurring, the equipment should be subjected to several or all of the tests discussed briefly in the following paragraphs. These tests will enable us to determine the equipment's weaknesses, behavior characteristics, and modes of failure.

12.4.1 TIME TESTING

Time or life testing, which is discussed extensively in Chaps. 8 and 10 from the viewpoint of estimating and demonstrating numerical reliability, is extremely important for other considerations. We are interested not only in *when* an item fails, but also in *which part* in a component or *which component* in a system it is that fails. Further, we seek to determine the *mode* or *modes of failure*; that is, the types of failure, as exemplified by performance drift, erratic performance, catastrophic failure, etc., and also the *mechanism of failure*; that is, the reasons for failure caused by poor design, part misapplication, etc: in other words, the *how* and *why* of failure. Time-to-failure testing by actually generating failure, together with the subsequent failure analysis, helps to find answers to these questions when *time* is the critical parameter of the item. Much electronic, electromechanical, and hydraulic equipment falls into this category when it is continuously operating or experiences a large number of cycles wherein the transient conditions of starting and stopping are not more severe than the accumulation of time. Whether the underlying failure distribution is exponential or some other simple distribution, life testing will indicate how much more (or less) life the equipment has than is required for operational use. This in turn allows priorities of criticality for reliability improvement to be established. When time is not a critical parameter, then one of the following methods is frequently more appropriate for "precipitating" the failure and determining the mode and mechanism of failure.

12.4.2 EVENT TESTING

In this case samples of the equipment are tested repeatedly through their cycle of operation until failure. This testing, which is analogous to time-to-failure testing, becomes the more meaningful test when starting and stopping operations are more destructive than the mere accumulation of time. The important parameter in this form of testing is $\bar{T}$ = mean number of cycles to failure. In this case the point estimate of reliability is $\hat{R} = 1 - (1/\bar{T})$. This equation would be used only if the length of the test cycle were equal to the operational time cycle or if the amount of time between starting and stopping had a negligible effect upon the failure rate.

12.4.3 PERIPHERAL TESTING

Peripheral or overstress testing has an important place in reliability assurance, but great care must be taken in its application. For instance, in many cases we can establish a great deal of confidence in the ability of the component or system to withstand much greater stresses than would be expected in operational use. However, if this evidence turns out to be nonconclusive, i.e., the device fails at those high levels, then the test engineer will not know at which point the stress level of the factor produces a critical stress in the item. Thus, provisions, such as those described in Sec. 12.6.7.3, should be made in the test plan for the procedures to be followed if this occurs.

12.4.4 ENVIRONMENTAL TESTING

As its name implies, this form of testing represents a survey of the reaction of the item to the various environments. It is usually required in *qualification* tests* and is frequently introduced in the development stage, usually at less numerous or less severe environmental levels. Thus by investigating a broad spectrum of the environmental space, we will have much greater confidence in the equipment than had we subjected the same number of equipments to only ambient conditions. However, as indicated previously, unusually extreme or unrealistic environmental levels should be avoided because of the difficulty of their interpretation.

Testing is not only concerned with the conditions to which the device is subjected but also with the *sequence* or *pattern of testing*. It is the purpose of the remainder of this chapter and Chapter 13 to discuss the various ways in which testing should be undertaken to derive real information as efficiently as possible.

12.5 THE SCIENTIFIC METHOD

The *scientific method* is the analytical and philosophical foundation of statistical experimentation. The rationale of its methodology is as follows:

1. An hypothesis is established which is based on theory or empiricism.
2. Tests are then formulated and observations taken.
3. The observations are compared with the expected behavior, i.e., the original hypothesis.
4. The essential information is abstracted and conclusions are reached.
5. An improved hypothesis is developed.

* See, for example, MIL-R-25534A (USAF) Military Specification, Rocket Motors, Aeronautical, Qualification Test for.

6. Steps 2 through 5 are repeated until an hypothesis is satisfactorily evolved and demonstrated.

In a typical industrial example, the sequence would start with the formulation of a design of a system required to perform a stated function: this would represent the original hypothesis. The design would then be tested, modified, retested, and so on until it performed as required and with demonstrated reliability. The rapidity with which the problem of meeting the end requirements is solved would depend *first* upon step 1, which depends on the inherent capability of the designer, the materials available for his utilization, and the ultimate requirements of the design, and *second*, upon the efficiency of steps 2, 3, 4, and 5. It is evident that the major task lies with the designer; the closer he can come to satisfying the end requirement, the fewer steps will be needed between the conception of the system and its realization. However, as discussed in earlier sections of the book, there are so many contingencies and interactions in today's complex equipment that it is generally impossible to anticipate all of them in the initial design and original configuration. As most of these contingencies will manifest themselves in operation, they should be sought out by efficient and adequate testing before the system or component has stabilized its configuration. Redevelopment after this phase is expensive and leads to the loss of valuable time.

12.6 STATISTICAL EXPERIMENTATION

The design and analysis of experiments—as it is frequently referred to in statistical literature—utilizes the philosophy of the scientific method with one major addition. It associates a degree of assurance with any statements or hypotheses that are established from sample observations. This assurance is measured in terms of numerical probability so that objective comparisons can be made. Thus decisions are made on a mathematical basis rather than on a subjective basis open to various interpretations.

The design of the test is no less important than its analysis, since correct formulation will allow valid comparisons to be made. Later, a detailed example is given (Sec. 12.6.5) which illustrates how wrong decisions can be made when experiments are not properly planned.

12.6.1 EXPERIMENTAL CONSIDERATIONS

When an experiment or test is being planned, there are many considerations to take into account. Some of these are enumerated below:

1. What are the objectives of the test?
2. How will (should) the results be interpreted?

3. How can the information gathered from this test be integrated with earlier/later tests on identical or similar equipment?
4. How many items are available and how are they available (in batches, sequentially, etc.)?
5. How many factors might influence the item's behavior?
6. How many factors are the items to be subjected to in the test?
   (a) Number of types of factors.
   (b) Number of levels of each factor.
7. Which of these factors interact, and which have negligible interaction?
8. What type of data are to be measured?
9. Is the sample homogeneous?
10. Is the sample representative of the population?
11. Are the tests destructive?
12. Are the items tested expensive?
13. How much control over the testing will we have?
14. How complicated may the test plan be considering possibility of
    (a) human error,
    (b) change of schedule of delivery of item being tested,
    (c) effect on schedule of complicated test procedures,
    (d) loss of data?

12.6.2 THE ELEMENTS OF VALID COMPARISON

The check points of Section 12.6.1 must be reviewed before the experimental design can be chosen. A variety of statistical plans is then available, which by specifying the sequence of factor-level application, allows both efficient and valid comparisons of factor effects to be made. This is the prime intent of statistical experimentation. A further "bonus" is that once a statistical design has been chosen, its analysis is also to a large extent established, and therefore there is less chance of misinterpretation.

Statistical experimentation differs from the *classical* or *traditional* method of testing by varying more than one factor at a time while determining its effect. The traditional method for investigating several factors is to keep all but one constant, allowing this single factor to vary over its several levels. This procedure is repeated for each factor in turn. It can be shown that this method not only is inefficient but also can produce erroneous results; generally, the optimum response, if sought, cannot be found by this method. Further, if this traditional method is used no interaction of factors can be discovered. The fundamental basis of statistical experimentation is to apply the factors in such a manner that their effects can be mathematically factored out of the design and the variations caused by the factors' effects can be measured and compared with each other and against random variation. We can get a clearer understanding by considering some simple examples, which have been constructed for this purpose.

12.6.3 A SIMPLE EXAMPLE*

Suppose we are investigating the effect of two factors, A and B, on a piece of equipment whose measurable response is possibly affected by these factors. Factor A has three levels A_1, A_2, A_3; factor B has two levels B_1, B_2. There are six possible factor combinations (conditions) which the equipment can experience. Tables 12.1 represent three sets of hypothetical

Tables 12.1

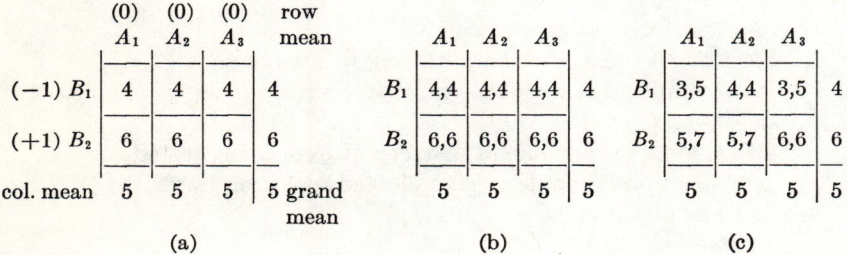

	(0) A_1	(0) A_2	(0) A_3	row mean		A_1	A_2	A_3			A_1	A_2	A_3	
$(-1)\ B_1$	4	4	4	4	B_1	4,4	4,4	4,4	4	B_1	3,5	4,4	3,5	4
$(+1)\ B_2$	6	6	6	6	B_2	6,6	6,6	6,6	6	B_2	5,7	5,7	6,6	6
col. mean	5	5	5	5 grand mean		5	5	5	5		5	5	5	5
	(a)					(b)					(c)			

data which might be obtained by testing a set of items under the six conditions. The magnitude of the *main*† effects are shown by the parenthesized numbers.

12.6.3.1. Table 12.1(a). Here the various levels of A have apparently no effect on the result while factor B does have an effect. B_1 has an average effect of (-1) on the grand mean and B_2 an effect of $(+1)$ on the grand mean. These numbers are shown in parentheses at the left of Table 12.1(a).

12.6.3.2. Table 12.1(b). Here the same conclusions can be drawn as in Table 12.1(a), but by repeating the experiment completely once, we have tended to verify the results. Also, there is apparently no measurement error (sometimes called random error, sampling error, error variation) since each result is exactly repeated.

12.6.3.3. Table 12.1(c). This table shows that B_1 and B_2 have the same average effect as in the previous two examples, as does A; however, there

---

* Reproduced by kind permission of Prof. Frank J. Massey from a series of lectures given by him, U.C.L.A., 1958.

† The main effect of a factor is its influence on the response and is due only to that factor's presence. Also, the main effect excludes any additional influence produced by that factor's *interaction* with other factors.

is large sampling variation. In such an instance, had only one experiment been performed, it probably would not have led to the conclusions that A has no effect and that B has an average of (-1) to $(+1)$ effect on the grand mean.

Tables 12.2 illustrate the variation in response when both factors A and B produce main effects.

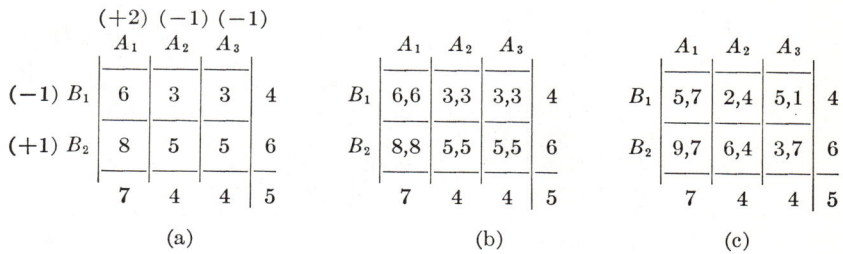

Tables 12.2

	$(+2)$ A_1	(-1) A_2	(-1) A_3			A_1	A_2	A_3			A_1	A_2	A_3	
$(-1)\,B_1$	6	3	3	4	B_1	6,6	3,3	3,3	4	B_1	5,7	2,4	5,1	4
$(+1)\,B_2$	8	5	5	6	B_2	8,8	5,5	5,5	6	B_2	9,7	6,4	3,7	6
	7	4	4	5		7	4	4	5		7	4	4	5
	(a)					(b)					(c)			

12.6.3.4. Table 12.2(a). Under the influence of A_1 the observation is increased by $(+2)$ units over the grand mean, for A_2 the effect is (-1) unit, and for A_3 the effect is (-1) unit. As in Tables 12.1, B_1 changes the measurement by (-1) and B_2 by $(+1)$.

12.6.3.5. Table 12.2(b). Here the same influences are present with replication tending to verify the situation. The error variation is zero.

12.6.3.6. Table 12.2(c). The levels of factors A and B still have the same average effect but, in this example, sampling variation is present as can be seen by the variation within the cells.

Tables 12.3 introduce a further effect. In addition to the main (separate) effects of A and B as in Tables 12.2, we now have the *interaction* effect AB.

12.6.3.7. Table 12.3(a). The difference observed between Table 12.3(a) and Table 12.2(a) is due to interaction effects. In fact, if Table 12.2(a) is subtracted from Table 12.3(a), we obtain Table 12.3(d), i.e., the magnitude of the interaction effects.

12.6.3.8. Table 12.3(b). Again we have no sampling variation as in 12.2(b) and 12.1(b).

12.6.3.9. Table 12.3(c). In this case, the average effects of A, B, and AB (denoting interaction) are the same as in 12.3(a) and 12.3(b), but sampling variation has been introduced.

12.6.3.10. Table 12.3(d). Here we see both effects of A and B separately; i.e., main effects, and the interaction effects of AB. The magnitudes

Tables 12.3

(a)

	A_1	A_2	A_3	
B_1	5	2	5	4
B_2	9	6	3	6
	7	4	4	5

(b)

	A_1	A_2	A_3	
B_1	5,5	2,2	5,5	4
B_2	9,9	6,6	3,3	6
	7	4	4	5

(c)

	A_1	A_2	A_3	
B_1	4,6	2,2	4,6	5
B_2	8,10	4,8	1,5	6
	7	4	4	4

(d)

	(+2) A_1	(−1) A_2	(−1) A_3	0
(−1) B_1	(−1)	(−1)	(+2)	0
(+1) B_2	(+1)	(+1)	(−2)	0
	0	0	0	

of the interaction effects are shown by the parenthesized numbers in the cells. The implication of interaction is that the total effect of A is different for different levels of B and/or vice versa.

It is interesting to note the agreement between 12.3(a) and the first set of observations in 12.2(c). Had 12.2(c) not been replicated, it would have been impossible to differentiate between sampling variation and factor interaction.

12.6.4. COMPONENTS OF VARIATION

The foregoing examples suggest how the factors and their interactions affect the observed data. It will be noted that the number in any cell is a composite of various factors and that the following relationship exists: The observed response = grand mean + main effect due to A + main effect due to B + interaction effect due to AB + random variation—or, in symbols,

$$y_{ij} = \mu + a_i + b_j + (ab)_{ij} + \epsilon_{ij} \tag{12.1}$$

where $i = 1, 2, 3; j = 1, 2$, and where

$$\sum a_i = 0, \ \sum b_j = 0, \ \sum_i (ab)_{ij} = \sum_j (ab)_{ij} = 0, \ \sum \epsilon_{ij} = 0 \tag{12.2}$$

We can rewrite Eq. (12.1) as

$$(y_{ij} - \mu) = a_i + b_j + (ab)_{ij} + \epsilon_{ij}$$

Squaring both sides and summing over i j we have

$$\sum_i \sum_j (y_{ij} - \mu)^2 = \sum_i \sum_j a_i^2 + \sum_i \sum_j b_j^2 + \sum_i \sum_j (ab)_{ij}^2 + \sum_i \sum_j \epsilon_{ij}^2$$

(12.3)

Note that the cross-product terms have all vanished by condition (12.2).

Equation (12.3) states that

$$\begin{pmatrix} \text{Total} \\ \text{variation} \end{pmatrix} = \begin{pmatrix} \text{Variation due to} \\ \text{factor } A \end{pmatrix} + \begin{pmatrix} \text{Variation due to} \\ \text{factor } B \end{pmatrix}$$
$$+ \begin{pmatrix} \text{Variation due to} \\ \text{interaction } AB \end{pmatrix} + \begin{pmatrix} \text{Variation due to} \\ \text{sampling error} \end{pmatrix} \quad (12.4)$$

In practice each of these variations can be computed by convenient formulae. For example, see Table 12.7. The importance of Eq. (12.3), or its equivalent (12.4), is that we can subdivide the total variation of measured observations into variations assignable to separate effects. Statistical sampling distributions and analysis of variance theorems now permit us to determine the significance between the variation due to main effects and interactions compared with the variation due to sampling error. Thus, we can determine whether there are definite indications of interaction or whether the AB interaction is not distinguishable from the sampling variation. Similarly, we can compare the variation due to A (or B) with the random variation to see whether it is statistically significant.*

The foregoing brief discussion holds true for any number of factors and any number of levels. However, it can be seen that as the number of factors and levels increases, so does the number of experimental combinations. There are many techniques† for reducing the number of combinations that must be *observed*, but it is not within the scope of this chapter to discuss in detail the various types of experimental designs. The literature (Refs. 1, 2, 3) contains many designs together with their analyses.

12.6.5. AN EXAMPLE OF A NONSTATISTICAL EXPERIMENT

We have seen the basis for statistically designed experiments and the implications of their analysis; let us now consider, by means of an example, the consequences of a test which has not received statistical test planning. The test which is described received a reasonable amount of planning, but no consideration was given to the statistical concepts.

The example is a prequalification test program for a solid-propellant rocket (Ref. 4). The program as it was originally presented is shown in Table 12.4. It was proposed that 24 units be tested under 18 different

---
* See footnote p. 367.
† See footnote p. 375.

Table 12.4. PREQUALIFICATION TEST PROGRAM

Unit no.	Temperature cycle	Vibrate	Drop	Static fire
1	−75°/175°			−65°
2	175°/−75°			150°
3		−75°		150°
4		175°		−65°
5		−75°		150°
6		175°		−65°
7			−65°	150°
8			165°	−65°
9				150°
10	−75°/175°	−75°		150°
11	175°/−75°	175°		−65°
12		−75°	−65°	150°
13		175°	165°	−65°
14	−75°/175°	−75°	−65°	150°
15	175°/−75°	175°	165°	−65°
16		−75°		150°
17		175°		−65°
18			−65°	−65°
19			165°	150°
20		−75°	−65°	−65°
21		175°	165°	150°
22		−75°	−65°	150°
23		175°	165°	−65°
24				−65°

environmental conditions. These conditions were selected combinations of four key environmental factors: temperature cycling, vibration, shock by dropping, and firing temperature. Some units, for example, were temperature-cycled beginning at $-75°F$, some were temperature-cycled beginning at $+175°F$, while others were not temperature-cycled at all. For additional details, the reader is referred to Table 12.4. The environmental factors and specific test levels used were selected on the basis of engineering and operational usage considerations. All units were fired statically (captive test).

One of the merits of this program is that the unit is tested under several different conditions. However, it has a major weakness, in that it is a series of tests which do not form a unified whole. It is difficult to appreciate this feature by visual inspection of Table 12.4 alone. Therefore, this same information is recast into the systematic array shown in Table 12.5. The array is a simple device for pictorializing all of the possible combinations of the environmental factors and test levels considered. The lack of balance and lack of an efficiently organized plan now become apparent. The numbers in the cells of Table 12.5 refer to the unit number given in Table 12.4.

Another major weakness is that the test program as it stands does not permit the data to be analyzed without a great deal of arbitrary interpretation. The effects of the environmental factors, their variations, and interactions are intermixed or confounded. A pre-qualification test program generally represents the first significant exposure of the engine to environments that it will experience during the qualification test itself and during usage operations. Thus, whereas the limits of the environments might not be so extreme as those for the later test programs, there should be an adequate coverage of the factors so that there will be an indication of the type of behavior that can be expected. If the program is completely successful, a certain amount of confidence will result in the engine's ability to perform without failure under environmental conditions and we can progress to the qualification test program, expecting success and being able to plan for limited production. However, if the exposure of the engine to the environmental factors is not successful, it is most important to be able to determine the particular environmental conditions to which it is sensitive. The test program as it has been presented does not permit this. If we can determine the environmental factors which cause unsatisfactory performance or lack of performance reproducibility, we can take the appropriate steps to make the engine "insensitive" to that environment. This is a fundamental reason behind the establishment and need for statistically designed test programs. Also, the results are valuable in developing similar engines and in gaining a greater understanding of the potentialities of application of the present engine.

We shall now show analytically that the environmental effects cannot be assessed from the specified program.

Table 12.5. Prequalification Test Program

(same information as in Table 12.4 but rearranged to show lack of balance)

		Temperature Cycle									
		None			−75°/175°			+175°/−75°			
		Vibrate			Vibrate			Vibrate			
		None	−75°	+175°	None	−75°	+175°	None	−75°	+175°	No. of Tests
No Drop	Fire −65	24		17 4 6	1					11	6
											12
	Fire +150°	9	16 3 5			10		2			6
Drop −65°	Fire −65°	18	20								2
											6
	Fire +150°	7	12 22			14					4
Drop +165°	Fire −65	8		13 23						15	4
											6
	Fire +150°	19		21							2
No. of Tests		6	6	6	1	2	0	1	0	2	24
			18			3			3		

In the example, suppose we wish to compare the effect of firing temperature upon unit performance under no-drop conditions. In Table 12.5 we might compare the average response of tests in Row 1 with the average response of tests in Row 2. To do this, consider the mathematical model (without interaction):

$$y = \mu + t + v + d + f + \epsilon$$

where y = net response (the value of some parameter, eg., thrust, specific impulse which describes the unit's disposition to success or failure)

μ = mean response (that value which would obtain if environmental influences were excluded)

t = effect of temperature cycling upon unit behavior
v = effect of vibration upon unit behavior
d = effect of dropping upon unit behavior
f = effect of firing temperature upon unit behavior
ϵ = the contribution of sampling variation to the value of y (i.e., random error).

Using this model, the sums of observed values in Row 1 and Row 2 are expressed as follows:

Row 1:

$$6\bar{y}_{-65°F} = 6\mu + 4t_0 + t_1 + t_2 + 2v_0 + 4v_2 + 6d_0 + 6f_0 + \epsilon_1 + \cdots + \epsilon_6$$

Row 2:

$$6\bar{y}_{+150°F} = 6\mu + 4t_0 + t_1 + t_2 + 2v_0 + 4v_1 + 6d_0 + 6f_2 + \epsilon_7 + \cdots + \epsilon_{12}$$

where the subscripts 0, 1, and 2 (except those in the error terms) denote lower, middle, and upper test levels of the factors concerned. To compare the effects of the two temperatures, we take the difference between Row 1 and Row 2 averages:

$$\bar{y}_{+150°F} - \bar{y}_{-65°F}$$
$$= (f_2 - f_0) + \tfrac{2}{3}(v_1 - v_2) + \tfrac{1}{6}[(\epsilon_7 + \cdots + \epsilon_{12}) - (\epsilon_1 + \cdots + \epsilon_6)]$$

The last term on the righthand side of the equation, i.e., the error term, tends to zero as the sample size increases; the first term is a measure of the difference in performance due to firing temperatures (the estimate at which we are trying to arrive), while the second term is a measure of the difference due to vibration. It is obvious that if vibration effects are real, that is, the term $(v_1 - v_2)$ is not zero, then vibration contributes to the difference between row means and therefore influences the estimate of difference due to firing temperature. This feature, which is not apparent without the mathematical model, is known in statistics as *confounding*. Its practical implications are that, if row differences were unacceptably large, these would be deduced as being due to the effect of the firing temperature. Thus corrective action through design changes to the rocket engine would be taken to reduce the influence of temperature, whereas the real trouble factor might be vibration. This point illustrates the value of systematic experimentation in preventing erroneous conclusions and misdirected action.

Other comparisons can be made to estimate environmental effects, but they too will be confounded in nearly all instances unless contrasts are made using observations within two (single) cells. However, the disadvantages in doing this are (1) that the sampling error is large because sample sizes are small, and (2) that the comparison applies to specific rather than general conditions.

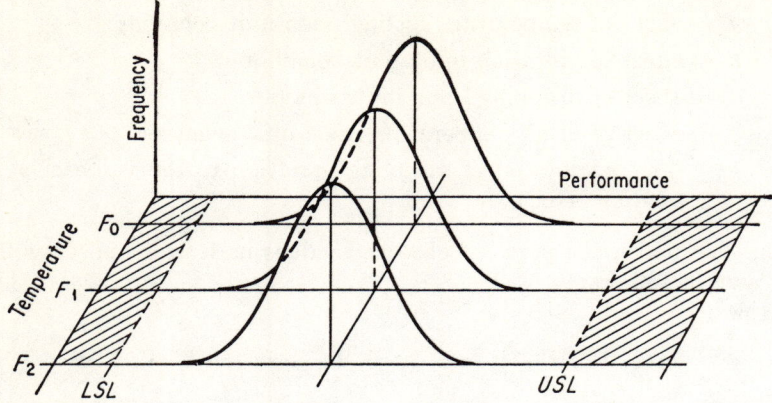

Fig. 12.1 When firing temperature has no perceptible influence upon unit performance.

Another weakness of the program is that it furnishes insufficient information to give a good measure of variation for each of the test levels and factors considered. This is a very important limitation, because precise estimates of performance limits relative to requirements can be given only if precise estimates of the mean and variability of each factor's influence on the engine's performance are available. If measures of the mean and variability can be derived, the importance of the need for action can be made on a sound mathematical basis using the methods of Sec. 8.7. For example, see Figs. 12.1 and 12.2. In both figures, the mean responses lie within the specification limits (USL and LSL), but, because of random variation about these means, a portion of the responses could possibly lie outside the limits, as in Fig. 12.2, and thus result in unreliability. In Fig. 12.1, there would be no failures; however, if the spread of the distribution

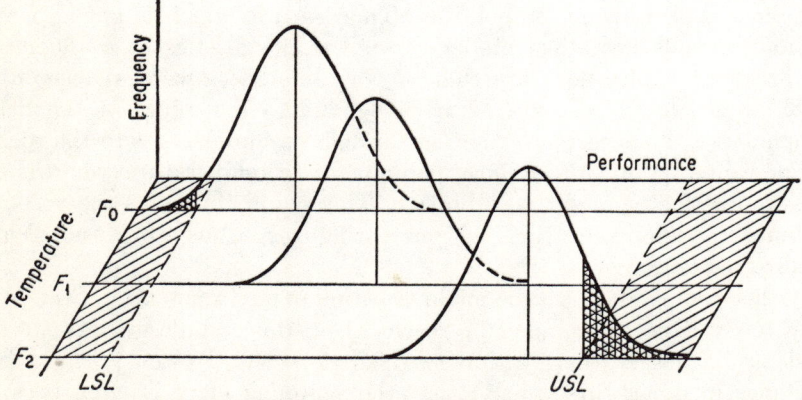

Fig. 12.2 When firing temperature does influence unit performance. (Probability of failure is indicated by the cross-hatch regions under the curves)

EXPERIMENTATION AND TESTING

were larger, some failures might be experienced. This kind of information is not forthcoming from the program we have just discussed.

12.6.6. A STATISTICALLY DESIGNED PROGRAM FOR SOLID ROCKET ENGINES

An alternate program was proposed, taking into account the test planning considerations in Sec. 12.6.1. Twenty-four units were available for the test; however, it will be seen that the information desired on the four environments could be obtained in 18 firings when there is no statistical interaction between the environments. Thrust, burning rate, impulse, and

Table 12.6. Modified Prequalification Test Program (Suggested Design For Test Program without Significant Factor Interaction)

	Temperature Cycle				
	None t_0	$-75°/175°$ t_1	$175°/-75°$ t_2	Number of Tests	Sum of Observations
No Drop d_0	$v_0 f_1$ 2	$v_1 f_2$ 2	$v_2 f_0$ 2	6	D_0
Drop $-65°$ d_1	$v_1 f_0$ 2	$v_2 f_1$ 2	$v_0 f_2$ 2	6	D_1
Drop $165°$ d_2	$v_2 f_2$ 2	$v_0 f_0$ 2	$v_1 f_1$ 2	6	D_2
Number of Tests	6	6	6	18	
Sum of of Observations	T_0	T_1	T_2		Grand Total G

Legend

v_0 = No vibration
v_1 = Vibration $-75°$
v_2 = Vibration $175°$

f_0 = Fire $-65°$
f_1 = Ambient fire $(65°)$
f_2 = Fire $150°$

V_0 = Sum of observations vibrated at v_0
V_1 = Sum of observations vibrated at v_1
V_2 = Sum of observations vibrated at v_2

F_0 = Sum of observations fired at $-65°$
F_1 = Sum of observations fired at $65°$
F_2 = Sum of observations fired at $150°$

Note: All engines are fired statically.

chamber pressure were some of the data measured. The sample was regarded as homogeneous and representative of normal production items. The tests were destructive, since these were solid propellant rockets, sufficiently expensive to warrant a limit of 20 to 25 units. The hazardous nature of the item meant that the statistician was located quite far from the test-area operations. Consequently, the design proposed in place of the example of Sec. 12.6.5 was chosen so that it would be one of the simple statistical designs and easy to administer (see also Sec. 12.6.8). The test plan shown in Table 12.6 was evolved as an alternative to the example of Table 12.5.

The design is a 3 × 3 Graeco-Latin Square, involving nine different environmental conditions, spread evenly over the environmental space. It is a balanced design and contains one more level of temperature, +65°F, than the original test plan. The design is so constructed that it is possible to estimate the effects of each factor upon unit performance, assuming, as was done in the model, that interactions either do not exist or else are small compared to main effects. This would not be assumed unless developmental environmental tests had demonstrated no interaction. An important feature is that the variation about estimated mean performance values can be calculated. Therefore, on the basis of mathematical probability, estimates can be made, with specified levels of assurance, as to how frequently the performance, with respect to temperature influences or vibration influences, etc., can be expected to lie within certain required limits. When the design is analyzed in this manner, it becomes obvious which of the factors, if any, critically influence performance and what their respective safety margins are. Consequently, appropriate redesign measures can be taken if this step is necessary.

It will be noticed that the revised program requires only nine engines for a single *replication*. However, to increase confidence in the information derived from the test program, it is desirable to make a second full replication. Such a procedure utilizes a total of 18 units, six less than the initial program, yet it provides the following number of observations for comparison:

	Test Levels		
Factor	0	1	2
Temperature cycling	6	6	6
Vibration	6	6	6
Drop	6	6	6
Firing temperature	6	6	6

Assuming negligible interactions, the effects of all four factors are determinable from the experiment with the same accuracy and precision as though the entire experiment had been devoted to the study of a single factor alone. In fact, comparisons made in the proposed design are more realistic than comparisons made in single-factor experiments since they

Table 12.7. Analysis of Variance for the Modified Test Program

Source of variation	Degrees of freedom	Sum of squares (SS)	Mean square	F-ratio
Dropping............	2	$(D_0^2 + D_1^2 + D_2^2)/3 - G^2/18$	$SS_D/2$	$9SS_D/2SS_E$
Temperature cycling.	2	$(T_0^2 + T_1^2 + T_2^2)/3 - G^2/18$	$SS_T/2$	$9SS_T/2SS_E$
Vibration............	2	$(V_0^2 + V_1^2 + V_2^2)/3 - G^2/18$	$SS_V/2$	$9SS_V/2SS_E$
Firing temperature...	2	$(F_0^2 + F_1^2 + F_2^2)/3 - G^2/18$	$SS_F/2$	$9SS_F/2SS_E$
Error...............	9	By subtraction of above four SS from total.	$SS_E/9$	
Total	17	Σ (each obs.)$^2 - G^2/18$		

are based on average responses over widely different but general conditions. In addition, 9 degrees of freedom are available to estimate testing error. The formulas* for the *analysis of variance* are shown in Table 12.7.

* For those readers unfamiliar with methods of analysis of variance, the authors again suggest References 1, 2, 3 for an extensive presentation of this subject. Here we shall only indicate the utility of tables such as Table 12.7. The number in the second column of the table entitled "degrees of freedom" and associated with a particular source of variation is equal to the number of individual terms that can be assigned arbitrarily within the constraints of certain identities with comprise the total sum of squares. The mean square, being the sum of squares divided by the degrees of freedom, is an estimate of a certain linear combination of variances of the factors. The ratio of two variances estimated from two independent samples from a Normal population has an F-distribution which depends only on the numbers of degrees of freedom associated with each sample. Thus in the last column of Table 12.7, we compute the F-ratio by dividing the mean square of any given factor by the mean square of the error (random or sampling) variation. If this ratio lies outside the tabulated F value at a prechosen confidence level (Ref. 1, 2, 3), the initial assumption that a factor has no significant variation is considered invalid; i.e., the factor variation is not of the same magnitude as would be expected if it were due only to sampling variation. This implies that the factor does actually affect the behavior of the item. Each source of variation can thus be examined on a quantitative basis to determine whether interaction effects and main effects are significant at the chosen confidence level. In addition, but not shown in Table 12.7, the estimates of the average effect and variation for each factor and interaction can be obtained by simple arithmetic by equating each number in the "mean square" column to the linear combination of variances it is estimating; i.e., its expected mean square. This very valuable aspect of analysis of variance is not stressed in References 1, 2, 3 but is discussed in statistical literature usually under the title of "components of variance." Specific discussion may be found in Part II of a series of three articles on "Fundamentals of Analysis of Variance" by Professor Charles R. Hicks in *Industrial Quality Control* **13,** September 1956. Henry Scheffé's *The Analysis of Variance*, published by John Wiley and Sons, 1st Edition, 1959, discusses point and interval estimation of variance components, p. 228, *et seq*; and F. Lemus gives an example of estimating variance components in "A Mixed Model Factorial Experiment in Testing Electrical Connectors," *Industrial Quality Control*, **17,** December 1960.

Thus, once such a program is completed successfully, a great deal of assurance would prevail that the engine would function under all environmental conditions in operational use. In the case of zero failures in 18 trials, the reliability demonstrated would be 85 per cent with 95 per cent confidence (*cf.* Fig. A.4).

12.6.7. AN EXAMPLE OF STATISTICAL EXPERIMENTATION WITH ELECTRONIC EQUIPMENT

This example is taken from a development program (Ref. 5). It illustrates how the design and analysis of the experiment might differ because of the type of equipment being tested, the amount of information already known, its level of development, and the type of information desired. Here the circumstances are such that it is important to discover the effects of factor interaction. The mathematical form of the failure distribution of the equipment is known. The failure times were expected to vary according to the environmental factors. In addition, test time was limited and this fact had to be incorporated into the design of the experiment.

A piece of electronic equipment was found to be critical within a system during environmental testing. It underwent a certain amount of redesign and was then to be investigated with respect to its life behavior characteristics under several environmental conditions.

The conditions, the effects of which were being estimated, were vibration, humidity, and temperature cycling. The time-to-failure distribution of the equipment being tested was known to be exponential. However, testing time was limited so that the tests had to be terminated at time T. The following environmental levels were regarded as representative of environmental stress:

Vibration...	a_0	a_1	a_2
Humidity...	b_0	b_1	
Temperature cycling.............................	c_0	c_1	

There are $3 \times 2^2 = 12$ combinations of environmental factors and levels, and as the interaction of the factors as well as the main effects were under study, it was decided that a simple factorial design replicated N times would be the most suitable test plan.

The experiment was carried out by subjecting 12 groups of N equipments each to a particular combination of environmental levels, operating each equipment until failure or test truncation at time T, whichever occurred first. The mean failure rate (the reciprocal of mean-time-to-failure) for each combination of environments was estimated by Eq. (12.5) as follows:

For each environmental combination $(a_i, b_j, c_k) \equiv (i, j, k)$, $(i = 0, 1, 2;$

EXPERIMENTATION AND TESTING

$j = 0, 1; k = 0, 1$), we can compute (obtained from the likelihood equation, Ref. 5):

$$\hat{\mu}_{ijk} = \frac{n_{ijk}}{t_{ijk} + (N - n_{ijk})T} \tag{12.5}$$

where n_{ijk} = number of failures occurring before time T in environmental combination (i, j, k)

t_{ijk} = total time experienced by those items failing in (i, j, k)

The mean failure rate is simply the number of failures experienced in (i, j, k) divided by the total time on all items in (i, j, k), whether failing or not. Therefore, each combination of environments will provide a separate estimate of the mean time-to-failure. The resulting statistical array is shown in Table 12.8.

Table 12.8. Mean Life Estimates under Applied Environmental Combinations

		Vibration					
		a_0		a_1		a_2	
		Humidity		Humidity		Humidity	
		b_0	b_1	b_0	b_1	b_0	b_1
Temperature Cycle	c_0	$\hat{\mu}_{000}$	$\hat{\mu}_{010}$	$\hat{\mu}_{100}$	$\hat{\mu}_{110}$	$\hat{\mu}_{200}$	$\hat{\mu}_{210}$
	c_1	$\hat{\mu}_{001}$	$\hat{\mu}_{011}$	$\hat{\mu}_{101}$	$\hat{\mu}_{111}$	$\hat{\mu}_{201}$	$\hat{\mu}_{211}$

12.6.7.1. Establishing the Relative Importance of the Effects of the Environmental Factors. Table 12.8 suggests the use of analysis of variance techniques; however, it can be shown (see Ref. 5) that Var $\hat{\mu} = \mu^2/N[1 - \exp(-\mu T)]$; therefore it is necessary to make a transformation (Ref. 3, Sec. 8.5) of the data to vitiate the relationship between the mean and the variance. If μT is large, then a logarithmic transformation is appropriate; if μT is small, then a square-root transformation can be used. Table 12.8 is therefore rewritten with the data transformed and the standard analysis of variance techniques can be applied. The significance of the effects of the different environments and their levels on the lives of the components is measurable, and we now have a quantitative basis for directing any necessary corrective action. Again, interactions between two or more environments which cannot be detected using the classical method of testing will be indicated by the analysis of variance.

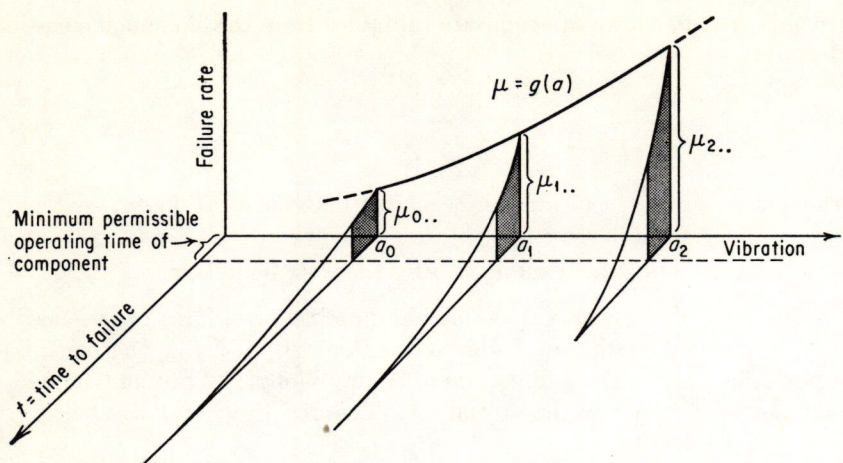

Fig. 12.3 Vibration frequency response surface (shaded region represents expected failure region).

12.6.7.2. The Estimate of Reliability. Since the time-to-failure distribution is known to be exponential, knowledge of μ_{ijk}, the mean failure rate, completely determines the life distribution under the environment (i, j, k). Therefore from Table 12.8 we can generate environmental time-to-failure response surfaces similar to the one shown in Fig. 12.3 for vibration. This figure is constructed as follows: take the mean of the μ's for each level of vibration; this gives the estimate of $\mu_0..$, the expected value of the mean failure rate for a_0, as

$$\mu_0.. = \frac{\sum_{j,k} \mu_{0jk}}{4}$$

Similarly, for a_1 and a_2 the estimates are

$$\mu_1.. = \frac{\sum_{j,k} \mu_{1jk}}{4} \quad \text{and} \quad \mu_2.. = \frac{\sum_{j,k} \mu_{2jk}}{4}$$

Now, these estimated values of $\mu_i..$ for each level of vibration are averaged values over the remainder of the environmental space. Consequently, for each life curve we have a more realistic sample of the behavior characteristics of the components. This differs from the standard method of life testing where only the environmental factor being studied has its levels changed. The construction of a frequency environmental response surface can be carried out for each factor, graphically indicating whether there is any correlation between failure rate and environmental level.

EXPERIMENTATION AND TESTING 371

If t_0 is the minimum required operating time for the component, the probability of its operating longer than t_0, the component having experienced a vibrational level of a_0, is given by

$$P(a_0, t_0) = \int_{t_0}^{\infty} \hat{\mu}_{0..} \exp\left[(-\hat{\mu}_{0..})t\right] dt$$

$$= \exp\left[(-\hat{\mu}_{0..})t_0\right]$$

The surface generated and shown in Fig. 12.3 is a "template frequency surface" because the levels of the factors are discrete. However, if sufficient levels are studied and a regression equation between the failure rate and environmental factor can be determined, $\mu = g(a)$, (say), where a denotes vibration level and μ is the failure rate, the probability that the component will operate for a time greater than t_0, having received a vibration intensity of level a_*, is given by

$$P(a_*, t_0) = \exp\left[-g(a_*)t_0\right]$$

12.6.7.3. A Practical Consideration When Applying the Technique.* So far we have been concerned with the investigation of the item's behavior and its reliability over a range of environmental conditions. However, it may not be necessary to investigate the behavior characteristics over the complete environmental range, if during the course of the experimentation it becomes apparent that the item does have a high reliability. Generally, we are interested in finding the areas of failure and therefore we *want* some of the items to fail; consequently, it is suggested that the most extreme of environmental combinations (i.e., $a_2 b_1 c_1$ in the example) be applied first. The reason is that if no units fail, or very few fail, thereby indicating such a low failure rate under the most extreme conditions that we may extrapolate† back to the nonextreme or benign conditions and conclude that the smallness of the failure rate under the extreme conditions is sufficient guarantee. For instance, if we could state with a high degree of confidence that 99 per cent of the items could be expected to exceed the minimum operating time under the extreme condition, then this might be a sufficient assurance that the item was satisfactory for all conditions without further testing. When the item or testing is expensive, it would not be prudent to spend more time or money than is necessary to assure ourselves of the item's competence when operating separately from its higher levels of system assembly, since the final proof of the equipment's reliability is its ability to perform satisfactorily within the complete system.

* This concept is expanded in Sec. 13.2.
† If appropriate, the "application factors" K_A (*cf.* Sec. 9.4.3.1) might be used for the purpose of extrapolation. Otherwise, engineering judgment based on the mode of failure occurring and previous relatable experience might be acceptable.

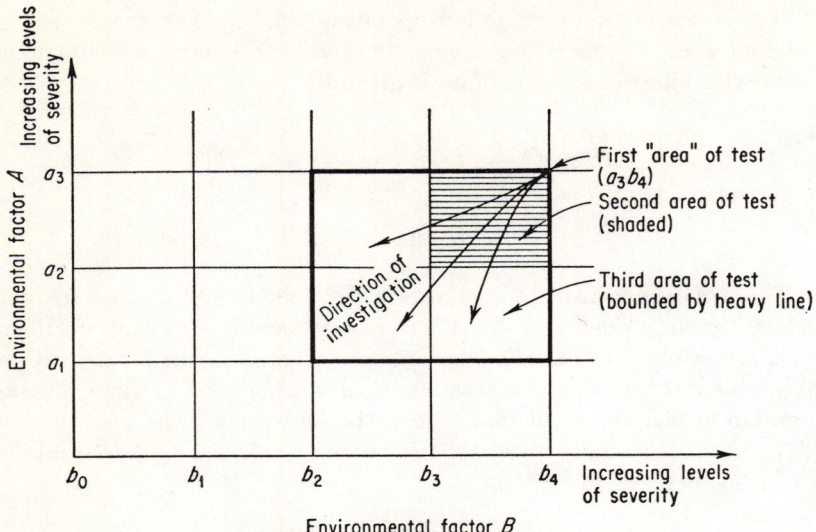

Fig. 12.4 Progression of testing.

However, if no firm decision can be deduced from the testing under extreme conditions, the "demonstration" necessarily becomes an investigation of the item's behavior characteristics, because the theory of peripheral testing can no longer be applied. Consequently, sampling should then progress from the extreme conditions in such a manner that the area of the environmental space under investigation is "self-sufficient" for statistical analysis. Consider Fig. 12.4 and a two-environmental example for the purposes of illustration. Here there are two environments being investigated, A and B. Not only are the main effects of A and B of interest, but also their interaction. As our test sample we have a modified component which has been theoretically designed to work under environmental conditions which produced failure in an earlier design. We might proceed as follows: Suppose the conditions which produced failures in the first design were a_2b_3 (see Fig. 12.4). We test this new design at a_3b_4 so that if successful we would have a great deal of confidence in the ability of the design to perform successfully at level a_2b_3, since this is less severe than a_3b_4. If, however, there are reliability and confidence goals to demonstrate, then the sample size is increased until the demonstration has been achieved. For instance, Table A.6 shows that to demonstrate with 95 per cent confidence that the item will have a mean life of $\hat{\theta}_L$ when tested for a period of $\hat{\theta}_L/20$, sixty items must be tested with zero failures. Another sampling plan would allow one or less failures in 97 tests for acceptance or demonstration.* However, if the demonstration under a_3b_4 is not satisfactory, then

* See also Sec. 10.6.

we extend the area of *investigation* to a_3b_3; a_2b_3; a_2b_4, so that together with a_3b_4 a simple 2×2 factorial design is performed, enabling main effects, interactions, and variabilities to be estimated. This process is extended until the experimentation has provided sufficient information on the failure-producing environments or has demonstrated that the item will perform under the conditions required of it.

12.6.8. THE ROLE OF STATISTICAL EXPERIMENTATION IN RELIABILITY

It is perhaps worthwhile to conclude this chapter on a note of caution. The field of statistical experimentation is highly developed and, at times, a very sophisticated subject, but it should be employed only as a beneficial tool, available to increase the growth of reliability. The statistician must not try to make the behavior of the equipment fit the statistical model but rather the reverse. A great deal of judgment is involved before a design is established. Generally the statistician will not have as much detailed understanding of the equipment as the design engineer, and therefore both must work together so that the statistician can provide the most suitable design from his experience.*

It is advisable, therefore, to keep the designs as simple as is consistent with the considerations listed previously. Theoretically, some information may be lost by doing this but unless the statistician is able to control the testing and circumstances surrounding it, frequently contingencies may occur that make it quite difficult to analyze the design as first planned. For example, an environmental test stand may be damaged or destroyed in the early part of a test program, or not as many items may be delivered or available as first expected once testing has commenced. Experimental design textbooks discuss analyses with missing data, but such analysis obviously has its limitations. However, if the design is relatively simple and a catastrophy occurs, usually much more information can be salvaged from it than from a highly sophisticated design.

REFERENCES

1. O. L. Davies, *Design and Analysis of Industrial Experiments*, Oliver and Boyd, London and Edinburgh, 1954.
2. W. G. Cochran, and G. M. Cox, *Experimental Designs*, 2d ed., John Wiley & Sons, New York, 1957.
3. O. Kempthorne, *The Design and Analysis of Experiments*, John Wiley & Sons, New York, 1952.

* See F. R. Del Priore and B. B. Day, "The Engineer and Statistician *Can* Meet," NAVORD Report 4028 (1953).

4. D. E. Hartvigsen, and D. K. Lloyd, "The Application of Statistical Test Designs to Qualification Testing of Rockets in Guided Missiles," *Proc. Western Regional Conference ASQC, San. Francisco, 1957*.
5. D. K. Lloyd, "Multienvironmental Life Testing of Parts and Components in Rockets and Guided Missiles by Statistical Design," *I.R.E. Trans. Reliability and Quality Control, December 1958*.

ADDITIONAL READING

Acheson, M. A., "Quality Acceptance Practices in Specifications," *Proc. 3rd National Symposium on Reliability and Quality Control in Electronics, I.R.E., January 14–16, 1957*, 136–40.

Bailey, N. T. J., "Science, Statistics and Operational Research," *Research* **9,** 202–207 (June 1956).

Ball, L. W., "Management Use of Laboratory Testing To Achieve Reliability," *National Convention Trans., ASQC, 1956*, 663–71.

Batson, H. C., "Applications of Factorial χ^2 Analysis to Experiments in Chemistry," *National Convention Trans. Am. Soc. Quality Control, 1956*, 9–23.

Box, G. E. P., "Integration of Techniques in Process Development," *National Convention Trans., ASQC, 1957*, 687–702.

Brickley, R. L., and Horton, W. H., "Statistical Approach to R and D Problems," *Elec. Mfg.*, **57,** 88–94, 362 (May 1956).

Cleminshaw, C., "Environment-Functional Tests: Key to Systems Reliability," *Ind. Labs.*, **9:**8, 34–34 (August 1958).

Derr, E. H., "Tests to Failure Boost Sparrow I Reliability," *Space/Aeronautics*, **32:**1, 133–41 (July 1959).

Hill, D., Voegtlen, D., and Yueh, J., "Parts vs. Systems: The Reliability Dilemma," *I.R.E. Trans. Reliability and Quality Control*, **PGRQC–6,** 27–38 (February 1956).

Kuzmin, W. R., "Experiments to Expose Marginal Reliability Designs," *Proc. 5th National Symposium on Reliability and Quality Control in Electronics, I.R.E., January 12–14, 1959*, 55–64.

Levenbach, G. J., "Accelerated Life Testing of Capacitors," *I.R.E. Trans. Reliability and Quality Control*, **RQC–10,** 9–20 (June 1957).

Luebbert, W. F., "Achieving Operational Effectiveness and Reliability With Unreliable Components and Equipment," *I.R.E. Convention Record*, Part VI, 41–49 (March 1956).

Zelen, M., "Problems in Life Testing: Factorial Experiments," *Trans. 13th Midwest Quality Control Conference, ASQC, November 1958*, 21–33.

CHAPTER THIRTEEN

STATISTICAL DESIGNS FOR RELIABILITY

13.1 INTRODUCTION

The last chapter discussed the generalities of experimentation and testing, and introduced some of the fundamentals of statistically designed experiments. The availability of these statistical methods provides us with some very efficient procedures for extracting valid data. They are, therefore, an essential part of any reliability program. The designs which have so far appeared in the statistical literature have been developed to a high degree and are extensively covered in theory and practical applications in Refs. 1, 2, and 3. In the last chapter we used two designs of this type for examples (Secs. 12.6.6, 12.6.7). However, although the various standard or classical statistical designs are valuable, they are not appropriate for all the applications we are faced with in the field of reliability. The reason is that the classical designs such as factorials, Latin squares, incomplete blocks, etc. evolved in the fields of agricultural and biological experimentation to fill experimental needs for specific problems. The peculiarities of industrial engineering are such that we need designs in addition to those mentioned, and as a consequence, there are beginning to appear in the literature* designs which will add to our techniques. The sections of this chapter describe some of these more important designs which have application in the field of reliability.

The introduction (Sec. 0.1) considered some of the reasons for unreliability and the situations which produce it. It is not inappropriate to review these reasons in the light of experimental designs. The dynamic complexity of engineering development creates circumstances which are much more a part of industrial experimentation than of other less dyanmic and more controlled areas of scientific investigation. For instance, in

---

* The reader is also referred to the first five articles in *Technometrics* **3**:3, (August 1961) for a series of discussions related to the types of situations and topics covered in this section and extending the subject matter of References 1, 2, 3.

engineering development programs changes frequently occur over which we have no control; schedules and deliveries slip, contracts are canceled, facilities are needed for programs with higher priorities; thus, sometimes there is need for greater flexibility than is generally found in classical statistical designs. In addition, in industrial production, items or observations frequently become available on a sequential basis; again, there is a limited amount of statistical methodology in this area. The sequential occurrence of observations can have both advantages and disadvantages. There are the advantages of being able to utilize the earlier information, make decisions before the whole sample is available, and thus use the information to change the product. On the other hand, especially when we are dealing with an evolutionary development program, there exists the problem of homogeneity of data which are being produced on a sequential basis. A further difference is that sometimes a large number of factors may affect the item and we have the problem of discerning these factors from a small number of expensive items. To complicate this latter situation further, it may only be possible to test an item once, because the test is destructive in nature, making the item unfit for further use. However, new statistical designs and techniques have been evolved and have appeared in the literature to help resolve some of the aforementioned problems, and when they are added to the well-established designs, they contribute greatly to reliability technology.

The five types of testing are described below. They are:

1. continuous experimental designs
2. sensitivity testing
3. random balance designs
4. evolutionary operation
5. response surface experimentation.

13.2 CONTINUOUS EXPERIMENTAL DESIGNS

Section 12.6.7.3 introduced a consideration which is considerably expanded by Garner and Hartvigsen in their paper, "Continuous Development Programs and Experimental Designs" and (Ref. 4). The basic concept is simple. It is to choose segments of complete factorial designs in such a way that the segments can be analyzed independently of each other or can be combined with one another to increase the sample sizes and allow more sensitive comparisons to be made. Such a situation exists in a development program where a certain number of items are available at different times for various phases of testing such as feasibility, verification, and qualification. These phases follow each other, and the decision whether to go on to the next is normally dependent on the outcome of the previous phase(s). In addition, at the beginning of a program we are concerned mostly with large effects of the influencing factors; as these are discovered, our later concern is to discern the magnitude of any lesser effects, which we can do if we can efficiently use all our data including that obtained in earlier testing. In the example we discuss, the segments are fractional factorials, which integrate to form a complete factorial design.

13.2.1 EXAMPLE AND ANALYSIS*

A small rocket engine is being developed which is required to be reliable and safe under a variety of environmental conditions. These environmental conditions consist of the following factors:

Firing temperature	F
Aging	A
Temperature cycling	T
Vibration	V
Humidity	H
Shock	S

It was decided that two levels for each factor be studied. The two levels would be the peripheral limits of the environments which would be experienced in operational use. Since it was reasonably well known that catastrophic failure was not expected at the boundaries, it was not considered necessary to test at any level in between the limits. It is the effect on the variable (chamber pressure) i.e., the amount of its deviation from the nominal under the influence of the environmental factors, rather than whether it will function at all which is in question. If this were not the case and catastrophic failures were to occur at some of the peripheral limits, then intermediate levels should be introduced so that indications of the critical level of the factors could be obtained. If this situation occurred, then major redesign would become necessary and the example given would not apply until later in the program.

As in the usual terminology the higher level of the factor is denoted by the subscript 1 and the lower level (or absence of that factor) by the subscript 0. A minimum of 64 rocket engines were available as follows: 16 for development testing, 16 for proof or verification testing, and 32 for qualification testing. The design for the development test plan was a $\frac{1}{4} \times 2^6$ fractional factorial design. It is shown in Table 13.1 together with *coded* chamber-pressure results.

If we assume that any effects that exist are due to main effects only and that the residual of the sum of squares can be used as an estimate of

* Reprinted by permission of N. R. Garner and D. E. Hartvigsen from an earlier unpublished version of Ref. 4, "Continuous Development Programs and Experimental Designs."

Table 13.1. Development Test Design and Results

			F_0				F_1			
			A_0		A_1		A_0		A_1	
			T_0	T_1	T_0	T_1	T_0	T_1	T_0	T_1
V_0	H_0	S_0	7.7						19.1	
		S_1			9.5		16.6			
	H_1	S_0		9.9						18.2
		S_1				12.9		15.1		
V_1	H_0	S_0		9.7						18.4
		S_1				11.5		14.6		
	H_1	S_0	8.0						19.6	
		S_1			12.2		16.6			

the experimental variation, we can set up a preliminary analysis of variance table based on the results of development tests only (Table 13.2).

Table 13.2 indicates that the firing temperature, F, significantly affects the chamber pressure unless this is caused by a higher-order interaction which is confounded with F. The same comment is true for aging, A. No other factors appear to have any significant effect.

Using one-sided tolerance limits, we see from Table A.9 for a sample size of 8 that we can be 95% confident that 95% of the population of

Table 13.2. Preliminary Analysis of Variance of Development Tests

	Source of variation	d.f.	SS	MS	F-ratio
1	F	1	201.64	201.64	142.78*
2	A	1	33.64	33.64	23.83*
3	T	1	.06	.06	..
4	V	1	.16	.16	..
5	H	1	1.82	1.82	..
6	S	1	.16	.16	..
Preliminary estimate of experimental variation†		9	12.71	1.4122	

* Significant at the 5% level. Where $F_{0.05} = 5.12$ for 1; 9 degrees of freedom.
† Obtained by pooling all effects higher than main effects.

Table 13.3. Sums and Averages of Development Firings

Levels	Sums 0	Sums 1	Averages 0	Averages 1
F	81.4	138.2	10.17	17.27
A	98.2	121.4	12.27	15.75
T	109.3	110.3	13.66	13.79
V	109.0	110.6	13.62	13.82
H	107.1	112.5	13.39	14.06
S	110.6	109.0	13.82	13.62
Total Sum	219.6		13.73	

engines will have a chamber pressure lower than $17.27 + 3.19 \times \sqrt{1.4122} = 21.06$ psia when fired at the hot temperature F_1 (see Sec. 8.7).

If we consider that this is sufficient information so that we can go on to the proof phase without further design change, we do this by testing 16 further engines under the conditions indicated in Table 13.4 and use the combined development data and the proof test data. In the example discussed, the foregoing is the situation. However, if a design change were considered necessary, the results obtained under this second combination would be compared separately with the results performed under the de-

Table 13.4. Development and Proof Test Results

The development test results are those enclosed in parentheses.

			F_0, A_0, T_0	F_0, A_0, T_1	F_0, A_1, T_0	F_0, A_1, T_1	F_1, A_0, T_0	F_1, A_0, T_1	F_1, A_1, T_0	F_1, A_1, T_1
V_0	H_0	S_0	(7.7)			13.4		17.0	(19.1)	
		S_1		9.0	(9.5)		(16.6)			18.8
	H_1	S_0		(9.9)	11.1		16.6			(18.2)
		S_1	8.9			(12.9)		(15.1)	19.4	
V_1	H_0	S_0		(9.7)	12.9		16.2			(18.4)
		S_1	8.8			(11.5)		(14.6)	19.9	
	H_1	S_0	(8.0)			13.4		16.5	(19.6)	
		S_1		8.5	(12.2)		(16.6)			21.1

Table 13.5. Analysis of Variance of Development and Proof Tests

Source of variation	d.f.	S.S.	M.S.	F-ratio
F	1		422.68	565.84†
A	1		83.53	111.82†
T	1		.75	
V	1		.69	
H	1		.75	
S	1		.58	
FA	1		.03	
FT	1		5.69	7.62*
FV	1		.01	
FH	1		.00	
FS	1		.87	
AT	1		.30	
AV	1		2.26	
AH	1		.47	
AS	1		.22	
TV	1		1.09	
TH	1		.05	
TS	1		1.01	
VH	1		.26	
VS	1		.05	
HS	1		1.57	
Proof vs. development tests	1		4.43	5.93*
Experimental variation	9	6.72	.747	
Total	31	534.01		

* Significant at the 5% level for 1,9 degrees of freedom.
† Significant at the 1% level for 1,9 degrees of freedom.

velopment tests to determine the effectivity of the design change. An analysis of variance is performed on these data and is shown in Table 13.5. From the design all main effects and two-factor interactions can be estimated. Investigating the only two-factor interaction, FT, that appears to be significant we find from the $F \times T$ two-way table (Table 13.6) that T_0 vs. T_1 at F_0 contributes a sum of squares of $[(79.1)^2/8] + [(88.3)^2/8] - [(167.4)^2/16] = 5.29$ and T_0 vs. T_1 at F_1 contributes a sum of squares of $[(144.0)^2/8] + [(139.7)^2/8] - [(283.7)^2/16] = 1.15$, indicating that the

Table 13.6

	Sums		Averages	
	F_0	F_1	F_0	F_1
T_0	79.1	144.0	9.89	18.00
T_1	88.3	139.7	11.04	17.46

Table 13.7. Sums and Averages of Development and Proof Firings

Levels	Sums 0	Sums 1	Averages 0	Averages 1
F	167.4	283.7	10.46	17.73
A	199.7	251.4	12.48	15.71
T	223.1	228.0	13.94	14.25
V	223.2	227.9	13.95	14.24
H	223.1	228.0	13.94	14.25
S	227.7	223.4	14.23	13.96
Development............	219.6		13.73	
Proof...................	231.5		14.47	
Total Sum..............	451.1		14.10	

interaction effect is caused by temperature cycling at the cold firing temperature. It is seen that while the average chamber pressure increased in the proof test firings, the amount of variation decreased; i.e., reproducibility improved. Thus if all the motors were fired at F_1 we could be 95% confident that 95% of the rocket engines would have a chamber pressure less than $17.73 + 2.91 \times \sqrt{.747} = 20.2$ psia.*

The conclusions from the data so far generated are:

1. The engine is sensitive to the firing temperature. The chamber pressure increases as the firing temperature increases.
2. Aging increases the chamber pressure.
3. The engine is sensitive to temperature cycling at cold firing temperatures.
4. The chamber pressure of the proof engines has increased slightly over that of the development engines.
5. An estimate of experimental variation is .747 (psia)^2.

Since the critical chamber pressure was 25 psia, which is well above the 99% confidence level of 99.9% of the engines fired at the F_1 upper tolerance limit of 23.4, it was concluded that the next phase of testing, i.e., qualification, could be entered upon without redesign.

The remaining 32 environmental combinations of the complete factorial were applied to the last 32 engines. The complete results including development, proof, and qualification tests are given in Table 13.8.

The summary of the analysis of variance is given in Tables 13.9 and 13.10. Based on the results of analysis of proof and development tests it was decided to analyze only the main effects and FT interaction.

---

* The tolerance factor used in the computation is based on a sample of size 10 in Table A.9, since the standard deviation is based on nine degrees of freedom.

Table 13.8. Complete Set of Coded Data from Development, Proof, and Qualification Tests

			F_0				F_1			
			A_0		A_1		A_0		A_1	
			T_0	T_1	T_0	T_1	T_0	T_1	T_0	T_1
V_0	H_0	S_0	7.7	9.8	13.9	13.4	18.4	17.0	19.1	22.8
		S_1	9.5	9.0	9.5	15.4	16.6	18.2	22.5	18.8
	H_1	S_0	9.3	9.9	11.1	13.1	16.6	16.9	22.4	18.2
		S_1	8.9	11.3	13.3	12.9	18.0	15.1	19.4	22.4
V_1	H_0	S_0	9.6	9.7	12.9	14.6	16.2	18.3	23.2	18.4
		S_1	8.8	9.2	13.0	11.5	17.5	14.6	19.9	22.8
	H_1	S_0	8.0	8.9	13.4	13.4	19.3	16.5	19.6	22.7
		S_1	9.2	8.5	12.2	14.1	16.6	16.5	21.3	21.1

The final conclusions from all the data are as follows:

1. The chamber pressure is highly sensitive to firing temperature. The average at the cold firing temperature is 11.09 psia, and at the hot firing temperature 18.97 psia. However, this is less than originally expected.

Table 13.9. Analysis of Variance Summary

Source of variation	d.f.	S.S.	M.S. = $\frac{\text{S.S.}}{\text{d.f.}}$	F-ratio
F	1		991.46	1135.69
A	1		234.89	269.06
T	1		1.02	..
V	1		.01	..
H	1		.04	..
S	1		.69	..
FT	1		6.70	7.67
Qualification vs. development and proof tests.................	1		55.68	63.78
Experimental variation.........	55	48.04	.873	
Total....................	63	1,138.53		

Table 13.10. Sums and Averages of All Firings

Environment	Sums		Averages	
	0	1	0	1
F	355.0	606.9	11.09	18.97
A	419.6	542.3	13.11	16.95
T	476.9	485.0	14.90	15.16
V	480.4	481.5	15.01	15.04
H	481.8	480.1	15.06	15.00
S	484.3	477.6	15.13	14.93
Qualification tests............	510.8		15.96	
Development and proof tests.	451.1		14.10	

2. The chamber pressure increases with aging. However, the amount is less than originally expected.
3. The qualification engines have a higher chamber pressure than the proof test engines, which in turn have a higher chamber pressure than the development test engines.
4. The chamber pressure becomes slightly sensitive to temperature cycling when the engines are fired at cold temperatures. However, this is not sufficient to affect the reliability of the engine.
5. The chamber pressure is insensitive to all other tested environments.
6. The experimental variation is .873 (psia)^2, i.e., the standard deviation is .934 psia.
7. If these engines were all fired at the worst environmental condition, F_1, 99% of the engines would give less than 22.82 psia 99% of the time.

If this reliability is adequate, the program proceeds to the next phase—mass production. However, it is indicated that causes for difference between engines manufactured in each phase should be investigated and, if necessary, corrections made. For instance, it is quite possible that the differences are due to manufacturing and quality control procedures not being fully developed in the earlier phases of the program. If this is the case, then it is probably unnecessary to take corrective action.

13.3 SENSITIVITY TESTING

Frequently we are faced with the problem of determining the level of a factor which critically affects the performance of a device. This is the level prior to which success can be expected and after which failure occurs, where success and failure are defined on a go, no-go basis. We normally expect (say) environmental factors to affect performance by producing deviations from the normal level of operation under a certain range of conditions such as between two levels of temperature; however, frequently if this range is sufficiently extended a radical change in the response will occur. The reason, generally, is that the physical or chemical properties of the item undergo a significant change, no longer responding in the normal manner. Thus, a diode might cease operating at too low a temperature or a vacuum tube at a certain level of vibration although there was no significant effect on their performance before these critical levels were reached. The classical example usually given in discussions on sensitivity testing is that of explosives and the height at which they will detonate if dropped. The problem is to determine a method of searching for the mean critical level and also to estimate the degree of variation about this average so that confidence and reliability statements can be made. Alternatively, this information might indicate that artificial environments must be created as protection from the critical extremes or that the item must not be used if exposed to a factor above a certain level.

One of the problems with which we are frequently faced in the search for the critical level is that once we have applied the factor at a known level, and if it does not affect the item critically, then if the item is a *one-test* item there is no way of knowing how large a degradation factor towards criticality that level produced. This would not be the case, for example, if a valve froze closed at a certain temperature in a liquid rocket engine test resulting in no ignition, for, when the valve was thawed out it would function quite normally and could be tested again at a known higher temperature, and so on until the critical temperature was established. However, consider the case of a solid propellant rocket engine, which when exposed to low temperature exploded upon ignition: there would be no opportunity to retest the same engine and no way of knowing how far below the critical temperature level we were. If the items are sufficiently numerous, available, or cheap (which is not generally the situation) then one way of obtaining the information is to test groups of the item over a range of levels and measure the proportions in each group which succeed or fail.

The results of such an experiment might be as shown in Fig. 13.1, where the X's represent failures, the O's represent successes, and y_i represents the ith level of factor y (tempera-

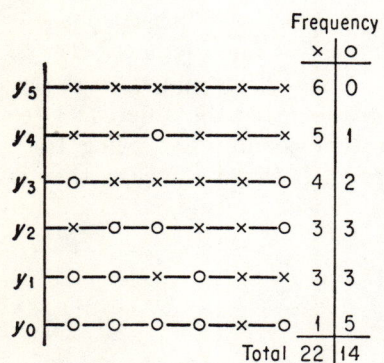

Figure 13.1 An inefficient method of sampling for the critical level.

ture, proportions of mixture, etc.). From the data we can obtain the mean (or median) and standard deviation of the distribution of success in the usual way, transforming or normalizing the data if necessary to enable the usual confidence statements to be made. However, it can be seen that this is not a very efficient experiment as, for example, no information (other than that there are no successes at y_5) is obtained from the six tests at y_5. We would normally expect that successes will occur if we are sufficiently below the critical level and failures when we are sufficiently above the critical level. This suggests that we seek a sampling technique which tends to concentrate the testing close to the critical level. The method which is described has this property, and it is varyingly referred to as *sensitivity testing*, Bruceton analysis, or the "up-and-down" method (Ref. 5).

13.3.1 THE METHOD OF SENSITIVITY TESTING

Two numbers are chosen before commencing the procedure. These are (1) the level h_0 of factor H at which testing will commence and (2) a difference d_0, in level of factor H between two consecutive tests. The testing procedure is as follows:

Test the first item at h_0; if the test is successful, increase the level of H to $h_0 + d_0$ and test the second item. If the results are successful continue this procedure of increasing factor H by a difference d_0 after each test until a failure occurs. When this happens, the next item is tested at a level d_0 below that of the last test. Thus the general rule for each test is: if the last test was successful, increase the level of the factor H by d_0; if the last test was a failure, decrease the level of the factor by d_0. A sequence of tests might appear as shown in Fig. 13.2. The first test was successful, so the next item was tested at level $h_0 + d_0$; this resulted in a failure. The next test was therefore performed at h_0 and was a success, and so on until

Normalized data		Sequence	n_i X's	n_i O's	i
3.2	$h_0 + 4d_0$	× ×	2		
3.0	$h_0 + 3d_0$	× ○ ○ ×	2	2	5
2.8	$h_0 + 2d_0$	× ○ × × × ○ × × ×	7	2	4
2.6	$h_0 + d_0$	× × × × × ○ ○ × ○ × ○ ○ ○ ○ ×	8	7	3
2.4	h_0	○ ○ ○ × × × ○ × ○ ○ ○ ○ ×	5	8	2
2.2	$h_0 - d_0$	× ○ ○ ○ ○ ○	1	5	1
2.0	$h_0 - 2d_0$	○		1	0
		Total	25	25	

Figure 13.2 "Up-and-down" sensitivity sampling.

the sample was exhausted. It is obvious that this procedure automatically concentrates the testing close to the critical level, and thus the mean critical level is approximated by the median unless the distribution is significantly skewed. However, it is also important to have a measure of variation so that confidence limits can be established; i.e., we are far more likely to be interested in a 99 per cent confidence limit as a protection against explosion or some other catastrophic failure than in the 50 per cent confidence limit, i.e., the median. In addition, if the data are transformed so that they follow some known distribution, such as the normal distribution, then much tighter confidence limits can be established. Thus, from previous experimentation it is desirable to have knowledge of the type of distribution followed and to be able to transform the data so that they are normally distributed. Our problem then is to obtain estimates of μ and σ, the parameters of the normal distribution.

In the example given, the data will be considered as a sample from a normal distribution. We proceed by examining the data and finding which of the X's or O's has the lesser total frequency. (We choose the lesser number in order to eliminate the bias caused by choosing h_0 too far away from the critical value, thereby resulting in too many failures or successes before the critical region is reached.) Let N denote this number and let the frequency of this lesser over-all event occurring at the lowest level be n_0, at the next highest level be n_1, and so on up to n_k which is the highest level at which the lesser event occurs. Then

$$\sum_{i=0}^{k} n_i = N$$

Also if h^* is the lowest level of the less frequent event, then the estimate $\bar{x}$ for μ is given by

$$\bar{x} = h^* + d\left[\frac{\sum_{i=0}^{k} i n_i}{N} + \frac{1}{2}\right] \quad \text{if the O's are used}$$

$$\bar{x} = h^* + d\left[\frac{\sum_{i=0}^{k} i n_i}{N} - \frac{1}{2}\right] \quad \text{if the X's are used}$$

The estimate† for σ is given by:

$$s = 1.620 d\left[\frac{N \sum_{i=0}^{k} i^2 n_i - \left(\sum_{i=0}^{k} i n_i\right)^2}{N^2} + 0.030\right]$$

† This formula is accurate when $s/d > 0.533$.

STATISTICAL DESIGNS FOR RELIABILITY

Table 13.11

n_i	i	in_i	$i^2 n_i$
2	5	10	50
2	4	8	32
7	3	21	63
8	2	16	32
5	1	5	5
1	0	0	0
$N = 25$		60	182

$$\bar{x} = 2.0 + 0.2[2.4 + 0.5]$$
$$= 2.58$$
$$s = 1.62 \times 0.2[1.52 + 0.030]$$
$$= 0.502$$

Using Fig. 13.2 as an example we compute Table 13.11 using the O's (in this case the choice is immaterial whether O's or X's are used since both success and failure were experienced an equal number of times).

If confidence intervals are computed, all arithmetic must be performed using the normalized data before converting back into actual data.

13.3.1.1. Standard deviations of the mean and variance. To determine the level of assurance of the estimates of μ and σ we compute their standard deviations, respectively,

$$\sigma_{\bar{x}} = \frac{G\sigma}{\sqrt{N}}, \qquad \sigma_s = \frac{H\sigma}{\sqrt{N}}$$

where G and H are obtained from Fig. 13.3 and are dependent on the magnitude of σ/d. Since we do not generally know the true value of σ it is necessary to use its estimator s, giving

$$s_{\bar{x}} = \frac{Gs}{\sqrt{N}}, \qquad s_s = \frac{Hs}{\sqrt{N}}$$

In the example $\sigma/d \simeq 0.50/0.20 = 2.5$, we find from Fig. 13.3 that $G = 0.92$, $H = 1.92$; thus

$$s_{\bar{x}} = 0.92 \times \frac{0.502}{5} = 0.09$$

$$s_s = 1.92 \times \frac{0.502}{5} = 0.19$$

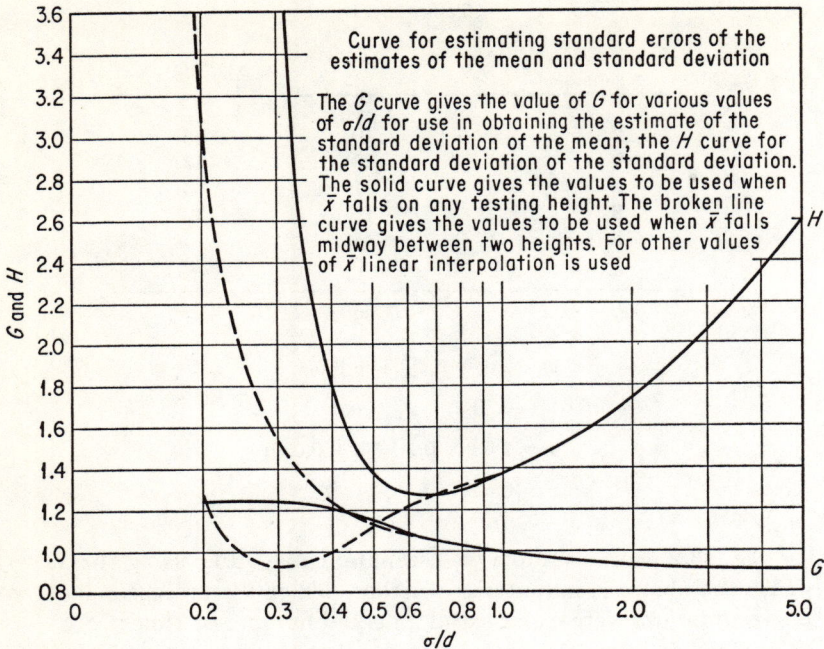

Figure 13.3

13.3.1.2. Confidence limits for μ. $s_{\bar{x}}$ will allow us to put confidence limits about $\bar{x}$ to give the upper and lower bounds for μ. For example, the 95 per cent confidence limits for μ are $2.58 \pm (2.06)(0.09)$, i.e., $\bar{x} \pm (t_{95\%k})s_{\bar{x}}$, where $k = 24$ and is the number of degrees of freedom upon which the estimates are based and $t_{95\%k}$ is obtained from the two-sided t-distribution table.*

13.3.1.3. Percentage limits. The answer to such questions as, "At what level will no more than 5 per cent of the items fail?" is given by $\bar{x} - t_{95\%k}s$, where $t_{95\%k}$ would be obtained from the single-sided t-distribution table with k degrees of freedom. In the example this would be estimated as

$$2.58 - (1.71)(0.50) = 1.72$$

13.3.1.4. Level of assurance of estimate. If we wish to introduce a level of assurance into the answer of the above question we can do so by computing the standard deviation of $\bar{x} - t_{95\%k}s$ which is

$$\sqrt{(s_{\bar{x}}^2 + t_{95\%k}^2 s_s^2)}$$

* The t values used in Secs. 13.2.1.2, 13.2.1.3, and 13.2.1.4 are obtained from Table C, Ref. 1.

STATISTICAL DESIGNS FOR RELIABILITY   389

and associate a one-sided confidence level with the percentage limit estimate. Thus, for example, if we wish to be 90 per cent sure that 95 per cent of the items tested will fail above a certain value, $P_{95\%/90\%}$, then

$$P_{95\%/90\%} = 1.72 - 1.28\sqrt{(0.09)^2 + (1.71)^2(0.19)^2}$$
$$= 1.30$$

13.3.2 CONSIDERATIONS WHEN USING THE TECHNIQUE

Although sensitivity testing is an effective method of sampling, care must be taken in the interpretation of the results. The reason is that the estimators are based on the maximum likelihood method and in the course of derivation several approximations are used which lessen the accuracy of the estimators under certain conditions. For example, when the interval d is being chosen, if σ is approximately known or an estimate of σ is available then d should be put about equal to σ. If d is greater than 3σ then, as is apparent from the curves, accuracy is lost in H and therefore accuracy in s_s. G is an increasing function of d/σ, so that the smaller the value of d the better the estimate of μ. However, if d is small and if the initial choice of test level is poor, then too many observations will be wasted in approaching the region of the median. Also, when d is less than $\sigma/3$ the factor H becomes quite large; therefore, if possible, d is best chosen lying between $\sigma/3$ and $5\sigma/2$. The assumption of normality or the ability to normalize the data is intrinsic in the method; therefore, unless there is strong evidence that this condition exists, it is not prudent to make confidence statements at very high levels of assurance. If it is essential, perhaps for safety reasons, that a (say) 99.9% confidence statement be made, it is better to use the above method to search, to theoretically establish the level, and then to undertake a demonstration of the required confidence and reliability at that point.

13.4 RANDOM BALANCE DESIGNS

A criticism frequently raised against classical statistical experimentation is that it is too "wooden." It is not too well equipped to meet the dynamic demands of industrial operations. This may be true in certain circumstances, but how often are these dynamic demands the result of earlier poor planning and indecision? Nevertheless, the situation is such that we cannot always expect to have our experiments under complete control and protected against outside influences such as changing schedules, loss of hardware, and other such contingencies. Another objection to classical designs is that the factors which influence an item's behavior are so numerous that it is not practical to subject it to all the combinations of those factors which may be expected to occur. The argument against this contention is that we can use fractional factorial designs or some other form of modified classical design and thereby reduce the sample size considerably. This is met with the rejoinder that the design becomes too complicated or is difficult to analyze—particularly if changes occur such as mentioned above. These arguments are valid to various degrees which are dependent on the circumstances surrounding each situation.

Recently designs have been introduced which represent a compromise between the alleged inflexibility of the classical design and the nonstatistical engineering approach to experimentation. The general name for these designs is *random balance* (Ref. 6). These designs eliminate many of the objections raised against the classical designs, but they do so at a price. This price is efficiency of estimation, particularly of confidence statements of interaction effects. Nevertheless, if it is used with due caution and understanding, the random balance technique can be very effective and is a valuable contribution to analytical experimentation.

Its most valuable use is probably as a "screening" tool. Consider the situation at an early stage in a development program where the item is being produced in small numbers with a limited number available, possibly sequentially, for experimentation. In addition, little is known about the effect on the item's performance, response, or reliability, of many factors such as environments, manufacturing, testing conditions, etc. Classical statistical experimentation would find it difficult to furnish a design which could accommodate this situation.

Let us take an example that can be used to illustrate the most simple application of the random balance design.

Consider, the input factors $A, B, C, \cdots, M$ with levels $a_i, i = 1, 2, 3$; $b_j, j = 1, 2, 3, 4, 5$; $c_k, k = 1, 2; \cdots; m_p, p = 1, 2, 3$. Let the response of the rth item be denoted by y_r. Then a random balance design may be represented as shown below:

Output or Response		Input Factors A	B	C	$\cdots$	M
y_1	←	a_3	b_5	c_1	$\cdots$	m_2
y_2	←	a_3	b_4	c_2	$\cdots$	m_1
y_3	←	a_1	b_1	c_2	$\cdots$	m_1
.		.	.	.		.
.		.	.	.		.
.		.	.	.		.
y_r	←	a_2	b_2	c_1	$\cdots$	m_3
.		.	.	.		.
.		.	.	.		.
y_n	←	a_3	b_5	c_2	$\cdots$	m_2

i.e., item 1 received the 3rd level of factor A (temperature)
  item 1 received the 5th level of factor B (conditioning time)
  item 1 received the 1st level of factor C (level of humidity)

. . .
. . .
. . .

  item 1 received the 2nd level of factor M (operator number)

The design is formed by assigning the levels of factor A at random down column A, similarly for factors B through M. It is seen that any matrix within the above $n \times M$ matrix is also a random balance design so that the design can be analyzed even though some of the y_r's may be missing. ("Missing" here does not apply to a "zero" or "infinite" response, either of which might be indicative of catastrophic failure, and must be included in any study of the effects of the factors.) Conversely the design can be extended as more items become available merely by assigning at random the levels of the factors A through M. In addition, since n is independent of M, as many factors as might be of interest can be studied without affecting the number of items in the sample. Thus, it is a very flexible technique.

The initial analysis begins with a graphical presentation of the data in the form of "scatter" diagrams as shown in Fig. 13.4. The response Y

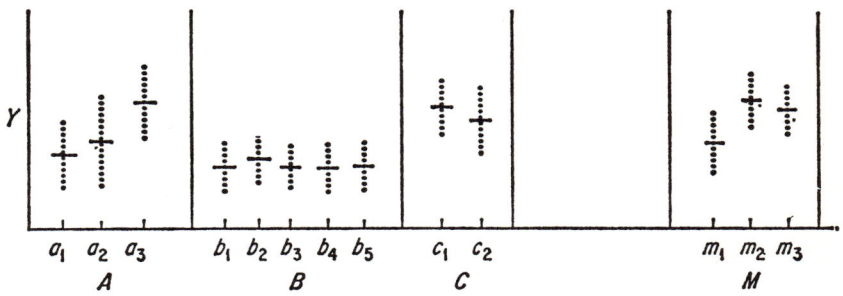

Fig. 13.4 A typical scatter diagram (dashes indicate positions of medians)

is plotted against each level of each factor. This will illustrate whether a factor or particular level has any effect on the response. It will be an approximate comparison, however, since, because of confounding, we do not have a perfectly balanced comparison.

Our first comparison will be by inspection, and rather than computing the arithmetic mean for each level it is quicker and possibly more appropriate to compare medians. Once it is determined which factors appear to have an effect, we can compute the significance level of this effect. (However, it must be borne in mind that because of loss of orthogonality* between comparisons the significance tests are approximate only.)

---

* For a definition of orthogonality see Ref. 1, p. 587.

Each factor being tested is treated in a manner similar to the analysis of a completely randomized block, i.e., by comparing "between treatments" against "within treatments" where generally there are unequal numbers of observations in each classification. For example, to determine whether the factor A had a significant effect we would compute

$$(1) \qquad \frac{(\text{sum of all observations})^2}{\sum_{i=1}^{3} r_i} = \frac{\left(\sum_{i,j} a_{ij}\right)^2}{\sum_{i=1}^{3} r_i}$$

$$(2) \qquad \sum_{i=1}^{3} \left[\frac{\left(\begin{array}{c}\text{sum of all observations,}\\ \text{factor } A \text{ at level } A_i\end{array}\right)^2}{r_i}\right] = \sum_{i=1}^{3} \left[\frac{\left(\sum_j a_{ij}\right)^2}{r_i}\right]$$

(3) $\qquad$ sum of squares of the individual observations $= \sum a_{ij}^2$

where r_i is the number of observations of factor A at level A_i and a_{ij} is the jth observation of factor A at level A_i.

Table 13.12. Analysis of Variance for Factor A

Source of variation	df	Sum of squares	Mean square
Factor A	2	(2) − (1)	[(2) − (1)]/2
Within A	($\Sigma r_i - 3$)	(3) − (2)	[(3) − (2)]/($\Sigma r_i - 3$)
Total	($\Sigma r_i - 1$)	(3) − (1)	

An F-test is used as a test of significance. The F-ratio is obtained from Table 13.12 as the ratio of the mean square of "factor A" to the mean square "within A." The degrees of freedom associated with this F-test are 2 and $(\sum r_i - 3)$. Alternative significance tests based on nonparametric methods might be more appropriate. A nonparametric test is used in the worked example (see Eq. (13.2)).

The standard deviation of a mean is obtained from the expression

$$s_{\bar{a}_i}^2 = \frac{\sum_j a_{ij}^2 - \left(\sum_j a_{ij}\right)^2 / r_i}{r_i(r_i - 1)} \qquad (13.1)$$

When it is determined which of the factors exhibit a significant effect these effects are removed from the data to allow other, lesser effects to be discerned. This is done by subtracting the algebraic effect of those significant factor-levels from the yield. Thus in the above example if A and C are considered to be the only two significant main effects, our next step is to estimate their magnitude. We form a 3×2 array of the means

Table 13.13

	A_1	A_2	A_3
C_1	$\bar{x}_1$	$\bar{x}_2$	$\bar{x}_3$
C_2	$\bar{x}_4$	$\bar{x}_5$	$\bar{x}_6$

as shown in Table 13.13. The observations are first placed in their respective cells. The means of each cell are computed and the average effects of factors A and C are given for each level as follows, when $A_1 C_1$ is taken as the "standard" level:

A_1: $\qquad\qquad\qquad\qquad\qquad 0 = 0$

A_2: $\qquad\qquad [(\bar{x}_2 + \bar{x}_5) - (\bar{x}_1 + \bar{x}_4)]/2 = a_2^*$

A_3: $\qquad\qquad [(\bar{x}_3 + \bar{x}_6) - (\bar{x}_1 + \bar{x}_4)]/2 = a_3^*$

C_1: $\qquad\qquad\qquad\qquad\qquad 0 = 0$

C_2: $\qquad [(\bar{x}_4 + \bar{x}_5 + \bar{x}_6) - (\bar{x}_1 + \bar{x}_2 + \bar{x}_3)]/3 = c_2^*$

The A and C effects can theoretically be eliminated from the response y_r by subtracting out the appropriate effect as determined above by the level of the input factor for any given observation, i.e.,

$$y_1 - a_3^* \qquad\quad \to y_1^1$$
$$y_2 - a_3^* - c_2^* \to y_2^1$$
$$y_3 \qquad\quad - c_2^* \to y_3^1$$
$$\cdots$$
$$y_r - a_2^* \qquad\quad \to y_r^1$$
$$\cdots$$
$$y_n - a_3^* - c_2^* \to y_n^1$$

We now have a second series of y_r's ($\{y_r^1\} \equiv Y_1$) which are plotted as a second series of scatter diagrams and examined once more to see whether any further factors have a significant effect on the yield. This process is repeated until there are no more discernible main effects.

As it stands, the example given would be difficult to analyze much further. It would be difficult to determine significant interactions owing to confounding and absence of observations or inequality of sample sizes appearing in certain factor combinations. The reason is that there are $3 \times 5 \times 2 \times \cdots \times 3$ combinations possible with at the most n of these observable (less if by chance a factor combination is repeated).

We have seen how to abstract the important factors without placing any considerable restriction on the conduct of the tests. Nor are we left with an incomplete design if at any time the configuration of the item is changed or production stopped. Thus, knowing the significant main effects (and interactions in a more sophisticated example), we can now go on to study the effect of these factors under the more controlled conditions of classical statistical experimentation.

13.4.1 MULTIPLE BALANCE DESIGNS

The criticism of the random balance design—that it sometimes results in too much unbalance—is a valid one and has led to modifications in the design to reduce the unbalance. One method of doing this is to construct the design so that an equal number of levels of each factor appear in each column. Another technique is to take subgroups of the factors in traditional statistical designs, assign the items at random for each design, and then compound the several designs. The over-all effect is a random balance design with complete balance within the subgroups. A design of this type is called *multiple balance*. It can be seen that this is a halfway step to the classical designs, and consequently it loses some of the flexibility associated with the simple random balance designs. However, its main advantage is not lost: the number of factors being studied is independent of the number of items available for testing.

Consider the situation where

factor A	has	3 levels	$a_1\ a_2\ a_3$
B		4	$b_1\ b_2\ b_3\ b_4$
C		3	$c_1\ c_2\ c_3$
D		2	$d_1\ d_2$
E		2	$e_1\ e_2$
F		3	$f_1\ f_2\ f_3$
G		3	$g_1\ g_2\ g_3$
H		4	$h_1\ h_2\ h_3\ h_4$

Here we have 8 factors with an assorted number of levels giving a total of $2^2 \times 3^4 \times 4^2 = 5184$ factor combinations possible. For our experiment there are only 36 items available. We might subdivide our factors into three groups—A, B, C; D, E, F; G, H; each group contains no more than 36 combinations. Group ABC has 36 combinations, group DEF has 12 combinations, and group GH has 12 combinations. Thus ABC can be treated as a full factorial design replicated once in the 36 items, DEF as full factorial replicated 3 times, and GH also as a full factorial replicated 3 times.

The three subdesigns have the 36 items assigned at random. Thus item 1 experiences factor levels

$a_1\ b_3\ c_3$ in the ABC subdesign
$d_2\ e_2\ f_1$ DEF
$g_2\ h_3$ GH

and similarly for items 2 through 36. The multiple random balance design can be written as shown in Table 13.14.

Table 13.14. Multiple Balance Design with Three Subdesigns ABC, DEF, GH

Item	A	B	C	D	E	F	G	H
1	a_1	b_3	c_3	d_2	e_2	f_1	g_2	h_3
2	a_2	b_2	c_1	d_2	e_2	f_1	g_2	h_3
3	a_1	b_1	c_1	d_2	e_1	f_3	g_2	h_1
.	.	.	.	.	.	.	.	.
.	.	.	.	.	.	.	.	.
.	.	.	.	.	.	.	.	.
36	a_1	b_1	c_2	d_1	e_2	f_2	g_2	h_3

This design is analyzed in exactly the same manner as the simple random balance design; however, in addition, the factors within the subdesigns can be analyzed in the conventional manner of factorial designs. This means that because of the balance of the subdesign the interaction effects of those factors within a subdesign can be better estimated than can interactions of factors which are members of different subdesigns. In this latter instance generally there will be unbalance in the number of observations for each factor combination. However, the comparisons of factors within a subdesign still will not be "pure" since the levels of the factors external to that subdesign do not balance.

13.4.1.1. Example of multiple balance designs.* In the experiment there were 12 input variables. There were 32 items available for testing and it was decided that each variable would be investigated at two levels only, with the intent that those factors which were found to have an effect would be further studied more thoroughly by the classical statistical designs.

* Reference 7, reprinted by permission of T. A. Budne, *The Application of Random Balance Designs.*

Test run No.	Levels of each variable A B C D E F G H I J K L	Y_1	Y_2	Y_3	Y_4
1	− + + + − + + + − − + +	78	103	103	103
2	+ + − − − − − − − − − −	103	92	100	95
3	+ + + − − + + − + − − +	97	99	99	99
4	+ + − + − + + − − − − +	77	91	99	94
5	+ + + − + − − + + − + +	113	102	102	97
6	− + − − + − + + − + + −	77	90	98	93
7	− + + + + − + + + − − +	66	91	99	94
8	− − + + − − − − + − + −	93	105	105	100
9	− − − − + + + + + + − −	86	99	99	94
10	+ − − − − − − − + − − +	93	82	90	90
11	− − + − − + + + + − + −	87	100	100	100
12	+ − − − − + − − + + + +	100	89	97	92
13	− − − + − + + − − + + +	72	97	105	100
14	+ − + − + + − + − + − −	124	113	113	108
15	− − + − + − + + − − − −	83	96	104	99
16	− + + − + + − − − − + −	88	101	101	101
17	+ − − + − − − + − − + −	96	97	97	97
18	− + + − − − − + + + − +	100	100	100	100
19	− + − + − − − + + − − −	72	84	92	92
20	− − − − − − − + + + + −	86	86	94	94
21	− − + + + + − + + + + +	69	94	102	97
22	+ + + + − − − − + + − −	103	104	104	99
23	+ + + + + + + − + − + +	77	91	91	91
24	+ + − − + + − + − + + +	104	93	101	96
25	− + − + + + − − − + − +	84	96	96	96
26	+ − + − − − + − + − − +	89	91	99	94
27	− + − − − + − + − − − +	92	92	100	100
28	+ − + + − + + − + + + −	78	92	100	95
29	+ + − + + − − − − + + −	87	88	96	96
30	+ − + + + − + − + + − +	81	95	95	95
31	+ − − + + + − − − − + +	99	100	100	100
32	− − − + + − + − − + − −	72	97	97	97

Fig. 13.5 Observed experimental results Y_1 with Y_2, Y_3, Y_4 as successive arithmetic modifications.

Twelve factors each at two levels represent a total of $2^{12} = 4096$ factor combinations. It was decided to subdivide the 12 factors into two groups of 6 variables, each group being arranged as a $\frac{1}{2} \times 2^6 = \frac{1}{32}$ fractional factorial design. The fractional factorial designs were obtained by confounding a 5-factor interaction with a main effect. The items 1 through 32 were randomly assigned to the 32-factor combinations in each group. The first subdesign contains input factors A, B, C, D, E, F, and the second subdesign factors G, H, I, J, K, L. The resultant experiment together with the observations is shown in Fig. 13.5.

The analysis commences with the plotting of 12 scatter diagrams, one for each variable. The diagrams are shown in Fig. 13.6. Inspection of these diagrams indicates the presence of significant main effects. To test for the level of significance we can use the analysis of variance method described

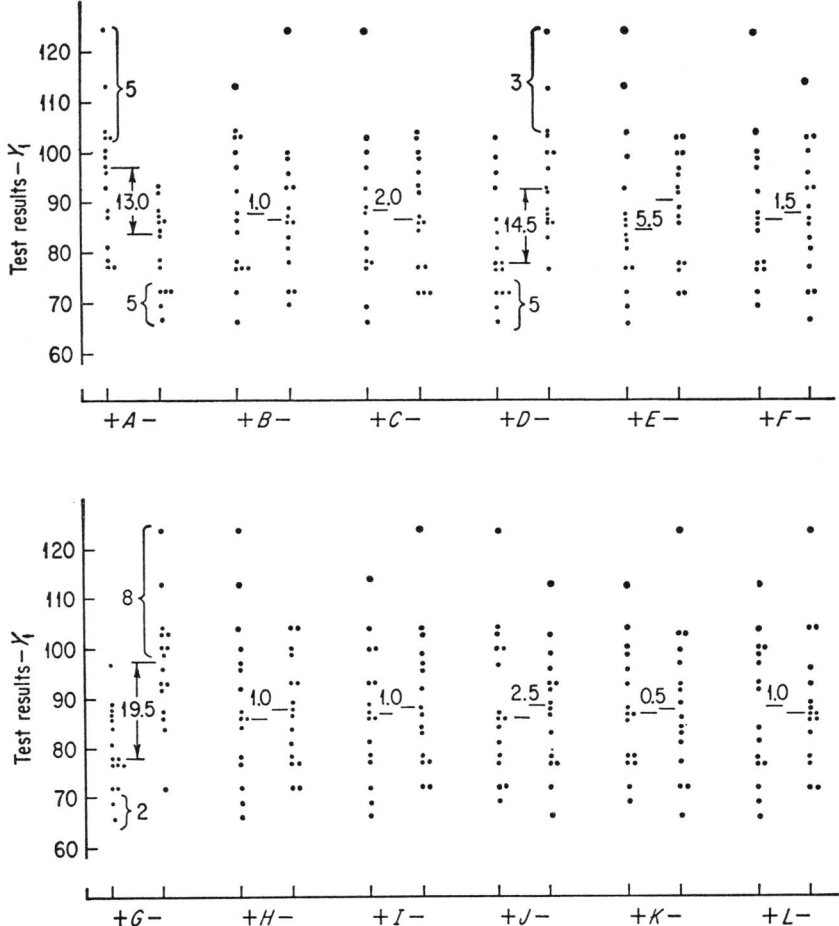

Fig. 13.6 Scatter diagrams for Y_1 (main effects only).

in Sec. 13.4 which in this case reduces to a t-test, or one of the nonparametric tests. A simple nonparametric test exists and is suitable when there are only two levels to the factor and an equal number of observations for each level.* If g is the number of observations of one group higher than any observations of the second group *plus* the number of observations of the second group lower than any observations of the first group (there being at least one observation in both counts), then the probability of the observed or a worse "slippage" is given approximately by

$$P(g) = g/2^g \qquad (13.2)$$

* See J. W. Tukey, "A Quick, Compact, Two-Sample Test to Duckworth's Specifications," *Technometrics*, **1**: 1, 39 (February 1959).

Note that this approximation is independent of n, the number of observations at each level. It can be used when $n \geq 10$. The critical values of g for two-sided 5%, 1%, and 0.1% significance tests are $g \simeq 7$, 10, and 14, respectively.

Factors A, D, and G appear to have significant effects with the differences between the medians equal to 13.0, 14.5, and 19.5, respectively. The values of g for A, D, and G are 10, 8, and 10; therefore, using Eq. (13.2), we find the significance levels of the differences of the factors A, D, and G to be, respectively, 0.01, 0.03, 0.01.

Table 13.15. Grouped Single Observations

	$D(+)$		$D(-)$	
	$A(+)$	$A(-)$	$A(+)$	$A(-)$
$G(+)$	81	78	97	88
	78	72	89	87
	77	72		86
	77	69		83
		69		77
$G(-)$	103	93	124	100
	99	84	113	92
	96	72	104	86
	87		103	
			100	
			93	

Our next step is to estimate the effects of the three significant factors and subtract them out of the observations. We form Table 13.15. It can be seen that unbalance exists by virtue of the unequal sample sizes in the cells so that if the effects for A, D, and G were computed simply as the arithmetic differences between $(+)$ and $(-)$ levels a bias would result. We can reduce the effect of this bias by working with the cell averages instead. Table 13.15 becomes Table 13.16.

Table 13.16. Means of Grouped Observations

	$D(+)$		$D(-)$	
	$A(+)$	$A(-)$	$A(+)$	$A(-)$
$G(+)$	78.25	71.40	93.00	84.20
$G(-)$	96.25	83.00	106.17	92.67

STATISTICAL DESIGNS FOR RELIABILITY

From Table 13.16 we find

$$\sum (A(+) - A(-))/4 = 10.60$$
$$\sum (D(+) - D(-))/4 = -11.87$$
$$\sum (G(+) - G(-))/4 = -12.81$$

Rounding off the effects to whole numbers we modify the observations Y_1 in Fig. 13.5 by adding to them

$$\begin{array}{lll} -11 & \text{when } (+) \text{ appears in column} & A \\ +12 & (+) & D \\ +13 & (+) & G \end{array}$$

This gives us the set of observations Y_2 shown in Fig. 13.5.

The original procedure is repeated. The observations Y_2 are plotted in 12 scatter diagrams, Fig. 13.7, which are again inspected for differences between levels. From inspection of the diagrams C is found to be the only main effect remaining which appears to have a significant effect. Consequently, we now inspect the interactions for effects. Scatter diagrams can be plotted for the interaction effects in an identical manner to the method of plotting main effects. The signs in the two columns corresponding to the interaction being plotted are multiplied together. The sign

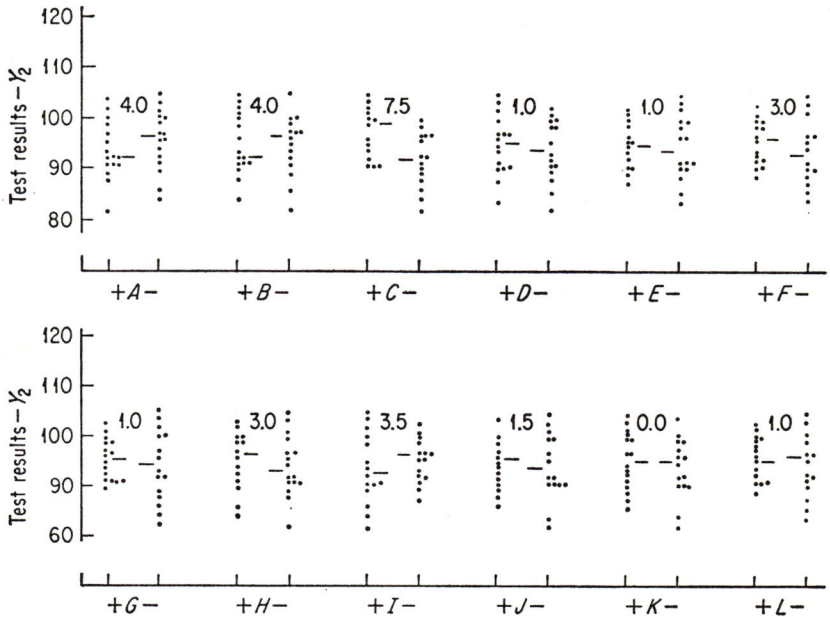

Fig. 13.7 Scatter diagrams of Y_2 (main effects).

of the result $(+)$ or $(-)$ determines the level $(+)$ or $(-)$ of the interaction. Generally, it is neither desirable nor necessary to plot *all* the first-order interactions since there are

$$\binom{n}{2} \equiv \frac{n!}{2!(n-2)!} = \frac{1}{2}n(n-1)$$

ways of choosing first-order interactions from n factors, which in the present example would result in

$$\binom{12}{2} = 66 \text{ different scatter diagrams.}$$

We can avoid having to plot all 66 as follows. Rank the observations Y_2; if the signs $(+)$, $(-)$ of the interactions are well dispersed then indications are that there is no effect from these interactions. However, if there is a tendency for one sign to be grouped towards the top or bottom of the ranking, then there is a strong suspicion of interaction effect. The significance levels can be computed directly from Eq. (13.2). The selected interactions are shown in Fig. 13.8. Two possibly significant interaction effects were observed, JK and AH. The differences between the medians of C, JK, and AH are 7.5, 9.0, and 6.0, respectively.

We expect to proceed by setting up a 2^3 table similar to Tables 13.15 and 13.16 for $C(+), (-)$; $JK(+), (-)$; and $AH(+), (-)$; however,

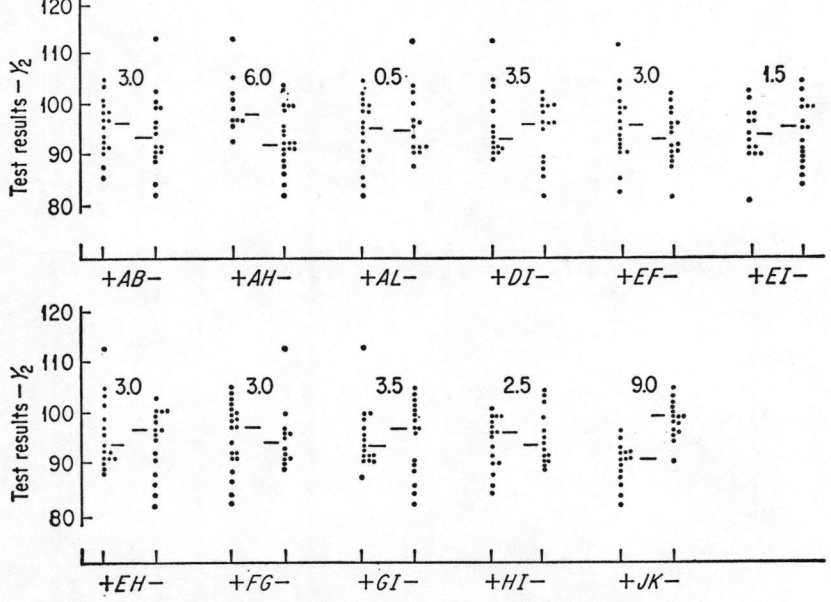

Fig. 13.8 Scatter diagrams of Y_2 (selected interactions).

it is seen that some of the cells are empty. We therefore take the two largest effects C and JK and form a 2^2 array, Table 13.17.

Table 13.17

$C(+)$		$C(-)$	
$JK(+)$	$JK(-)$	$JK(+)$	$JK(-)$
96	113	97	100
94	105	93	99
92	104	92	97
91	103	92	97
91	102	91	96
	101	90	
	100	89	
	100	88	
	99	86	
	95	84	
	91	82	

Again the means of the cells are used in the estimation of the effects of C and JK, giving Table 13.18,

Table 13.18

	$JK(+)$	$JK(-)$
$C(+)$	92.80	101.18
$C(-)$	89.45	97.80

which in turn gives

$$\sum (C(+) - C(-))/2 = 3.36; \quad \sum (JK(+) - JK(-))/2 = -8.36$$

Since the effect of C was much smaller than any of the other previously observed differences it was decided to modify the data only for the JK interaction. This was done by modifying the Y_2 column by adding 8 to the observation when both signs under J, K were identical.

This gives us the set of modified data Y_3 shown in Fig. 13.5. The Y_3 data were similarly plotted as scatter diagrams (which are not shown here, however) and it was found that the difference AH disappeared with the correction for JK. The differences in the medians were much smaller than those previously estimated. The three largest were factor C and interactions EH, CG which had values 3.0, 4.5, and 3.0, respectively, for the differences between the medians. When the $C \times G \times EH$ table was formed it was found that only EH gave a sizable effect of 5.06. Thus, Y_3 was modified by adding (-5) to those observations when both signs under $E, H,$ are identical.

This gives the set of modified data Y_4 shown in Fig. 13.5, which were again plotted as scatter diagrams with the I and C factors seen to have effect. Upon tabulation the effect of C was estimated at 4 units and the effect of I at -5 units. Y_4 was modified accordingly.

Further scatter diagrams of the modified data indicated no more sizable effects, and it was concluded that all the significant main effects and interactions had been found by this analysis. The significant factors and interactions were as shown in Table 13.19.

Table 13.19

Variable	Estimated effect	Regression estimates
G	13	15.9
D	12	12.1
A	-11	-8.7
JK	-8	-10.1
EH	-5	-6.2
I	-5	-5.1
C	4	4.9

The analysis of the data which has just been performed is important in answering the question: Which factors and interactions affect the observations? The method of determining the *magnitude* of the effects leaves something to be desired since, as our analysis shows, unbalance results in waste of information and the introduction of a certain amount of bias. However, knowing *which* factors affect the observations is important, and we can turn to the theory of regression to obtain better estimates of their magnitude. Let the linear relationship be

$$y_i = Ux_{i0} + Gx_{i1} + Dx_{i2} + Ax_{i3} + JKx_{i4} + EHx_{i5} + Ix_{i6} + Cx_{i7}$$

then we can use the method of least squares to obtain the normal equations which when solved give the regression estimates shown in Table 13.19. (It should be noted that in general the contrasts in the random balance designs are nonorthogonal* and therefore the inverse matrices need to be computed in the regression estimation of the effects of G, D, A, JK, EH, I, and C.)

It has been noted (Ref. 8) that a more efficient design could have been used for this particular example in which all of the main effects are unconfounded and are orthogonal. However, the main purpose of the worked example is to show the technique of estimation. These methods are available whether or not balanced designs exist for the problem at hand. Generally, if a balanced design exists it might be preferred—but not if it is available at the expense of not studying all possible significant factors. Random balance is primarily a screening tool and as such has a place in reliability experimentation.

* See S. S. Wilks, *Mathematical Statistics*, Princeton University Press, 1950, §9.6.

During an analysis there are many repeated operations of sorting, modifying, and ordering observations. Digital computing machines are well suited to these types of operations, consequently random balance techniques can be made even more effective under these circumstances.

13.5 EVOLUTIONARY OPERATION

In our normal manufacturing processes, once we have an established operation we are generally concerned with keeping it under control by quality control techniques. We do this by carefully checking the levels of the input variables to make certain they are within specified limits. In addition we measure the output variables to estimate whether the expected variation is due simply to random variation and is within limits. Although it obviously is imperative that this control is imposed, it is also apparent that this leaves our output variables in a form of "status quo." This is satisfactory only if our process is optimum according to some definition. For instance, if our output results in a marginal reliability problem, then it is obviously desirable to somehow improve its reliability. On the other hand, if the output variable is satisfactory, it is possible that the economy or efficiency of the operation might be improved in some way without impairing the quality of the output.

It is possible that the input levels of the process were arrived at by laboratory experimentation; however, even if these levels were obtained by the optimum searching techniques described in Sec. 13.6, it is possible that the laboratory optimum and the manufacturing optimum are not at the same point; i.e., for various reasons optimum experimental input levels are not quite optimum for the manufacturing process. Under these circumstances, then, the only way of arriving at the optimum is from the manufacturing process itself, since laboratory experimentation would only lead us back to the same input levels. The conclusion, therefore, is that in order to search for the process optimum we must introduce a certain amount of variation into the manufacturing process itself, measure its effect, and take the appropriate steps to move in the direction of the optimum until it is finally attained. This procedure, which is appropriately described as an *evolutionary operation* (Ref. 9), is analogous in concept to response surface experimentation but differs in detail for the following reasons.

First, since we are dealing with a manufacturing process which is probably already producing a "fairly satisfactory" item, great care must be taken not to interrupt the productivity. Also it is not psychologically prudent to imply that the technique being introduced is experimental or that the theory behind it is at all complicated. Therefore, the statistical steps that are taken towards the optimum cannot be so large nor so sophisticated as in the laboratory experiments. However, a compensatory effect results from the fact that since the manufacturing process is continuously repeated we can expect to have sufficient observations to obtain good estimates of effects and their standard errors, and we can refrain from changing any variables until we do have sufficient data.

The technique is to introduce a planned amount of variation into each of the input factors of our normal process. This means that now, instead of a single set of input conditions, there are several sets of input conditions by virtue of the various possible factor-level combinations. The output from each of these input combinations is measured several times until the advantage or disadvantage of each one is statistically discernible, i.e., by comparison with the original process output. The favorable levels are then selected and the procedure is repeated until the optimum set of input levels is established.

The operation of running the complete set of input conditions each once is called a *cycle*. The repeated running of cycles until a decision is made to change the input levels is called a *phase*.

The method of evolutionary operation is best described by an example. For ease of illustration we shall discuss the case in which the output variable (response) is a function of two input variables (per cent compound and time). However, the approach can be extended to any number of input variables, and also several outputs might be involved in any decision-making. Thus e.g., performance, per cent successful, cost per unit, etc., could all be measured for each phase and set of input conditions. However —again for ease of illustration—only one type of response is discussed.

13.5.1 EXAMPLE*

In this example the purpose of the evolutionary operation was to determine whether a change in the level of processing time and per cent compound used in the production of a certain type of heat resistant material would improve its ability to withstand a specified amount of heat for a given period of time.

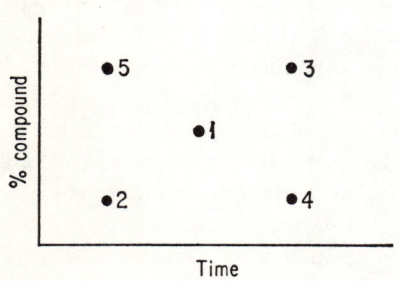

Fig. 13.9 Coordinates of input variables.

The original process was run with the input variables at the levels indicated by 1 in Fig. 13.9. It was decided to introduce both a positive and a negative deviation from the original levels for both time and compound. This resulted in a total of four combinations of these two input factors in addition to the original combination. These were conditions

* Utilizing the methodology described in references 9 and 10.

2, 3, 4, and 5 as shown in the figure. One cycle consisted of the single set of observations from 1, 2, 3, 4, and 5. The parameter being measured is the amount of erosion of the material when subjected to a given amount of heat for a given time. In Table 13.20 the observations have been coded and represent the original thickness of the test sample minus the amount of erosion. Therefore we wish to find those values of the processing time and per cent compound which will minimize the erosion (i.e., maximize the observations). The results of the first four cycles are shown in Table 13.20. Since this is a repetitive operation, it is convenient to set up a standard worksheet (Ref. 10) so that the results are easily interpreted. Figures 13.10, 13.11, and 13.12 are sample worksheets for the first, second, and third cycles.

Table 13.20

Conditions	1	2	3	4	5
Cycle 1	75.9	77.6	76.2	81.4	76.7
Cycle 2	76.8	77.3	80.0	81.5	80.7
Cycle 3	78.1	77.3	79.3	78.7	78.0
Cycle 4	77.5	76.8	78.0	79.8	72.1

Figure 13.10 contains the information from the first cycle; therefore not all the rows are complete. The calculations of the averages and effects are self-explanatory. In the calculations of the standard deviation and the error limits it should be noted that as soon as sufficient observations are obtained (by the end of cycle 3) the range will be used to estimate the standard deviation, s. The error limits are then taken to be ± 2 times the standard error, i.e., the standard deviation multiplied by the root mean square of the coefficients of the effects. This is divided by the square root of the numbers of cycles. Thus the error limits for the effects are $\pm 2s/(n)^{1/2}$. The error limits for a change in the mean are $\pm 4s/(5n)^{1/2}$. The calculation of the standard deviation cannot commence until two cycles have been run. The estimate of σ, i.e., s, is obtained by multiplying the observed range in the cycle by the factor $f_{k,n}$ obtained from Table 13.21 where n is the number of cycles and k is the number of conditions in the cycle. In the given example the estimate of σ, which was used for the first two cycles, was obtained from previous records of the standard process.

As the results are generated by the process outputs and the calculations on the work sheet at the end of each cycle, they are displayed on an "information board." The appearance of the board at the end of any given cycle might be as shown in Fig. 13.13. The purpose of this board is to relay in a simple form the data which will be used in any decision to change the process and to create interest in the effort for product improvement.

From the data on the first three cycles' work sheets we see indications that the process might be improved, although the data are insufficient to

Fig. 13.10. Two-variable evolutionary operation program calculation work sheet, cycle $n = 1$. Response: remaining thickness of sample after test.

Calculation of Averages

Operating conditions	(1)	(2)	(3)	(4)	(5)
(i) Previous cycle sum					
(ii) Previous cycle average					
(iii) New observations	75.9	77.6	76.2	81.4	76.7
(iv) Differences (ii) less (iii)					
(v) New sums	75.9	77.6	76.2	81.4	76.7
(vi) New averages: $\bar{y}_i$	75.9	77.6	76.2	81.4	76.7

Calculation of Standard Deviation

Previous sum s =
Previous average s =
New s = range $\times f_{k,n}$ =
Range =
New sum s =

New average $s = \dfrac{(\text{new sum } s)}{(n-1)}$ =

Calculation of Effects

Time effect $= \tfrac{1}{2}(\bar{y}_3 + \bar{y}_4 - \bar{y}_2 - \bar{y}_5) = 1.65$

% Compound effect $= \tfrac{1}{2}(\bar{y}_3 + \bar{y}_5 - \bar{y}_2 - \bar{y}_4) = -3.05$

Time $\times$ % compound effect $= \tfrac{1}{2}(\bar{y}_2 + \bar{y}_3 - \bar{y}_4 - \bar{y}_5) = -2.15$

Change in mean effect $= \tfrac{1}{5}(\bar{y}_2 + \bar{y}_3 + \bar{y}_4 + \bar{y}_5 - 4\bar{y}_1) = 1.66$

Calculation of Error Limits

For new average $= \dfrac{2}{\sqrt{n}}\sigma^* = \pm 3.2$

For new effects $= \dfrac{2}{\sqrt{n}}\sigma^* = \pm 3.2$

For change in mean $= \dfrac{1.78}{\sqrt{n}}\sigma^* = \pm 2.8$

* Prior estimate $\sigma = 1.6$ used.

STATISTICAL DESIGNS FOR RELIABILITY

Fig. 13.11. Two-variable evolutionary operation program calculation work sheet, cycle $n = 2$. Response: remaining thickness of sample after test.

Calculation of Averages

Operating conditions	(1)	(2)	(3)	(4)	(5)
(i) Previous cycle sum	75.9	77.6	76.2	81.4	76.7
(ii) Previous cycle average	75.9	77.6	76.2	81.4	76.7
(iii) New observations	76.8	77.3	80.0	81.5	80.7
(iv) Differences (ii) less (iii)	−0.9	0.3	−3.8	−0.1	−4.0
(v) New sums	152.7	154.9	156.2	162.9	157.4
(vi) New averages: $\bar{y}_i$	76.4	77.4	78.1	81.4	78.7

Calculation of Effects

Time effect $= \frac{1}{2}(\bar{y}_3 + \bar{y}_4 - \bar{y}_2 - \bar{y}_5) = 1.7$

% Compound effect $= \frac{1}{2}(\bar{y}_3 + \bar{y}_5 - \bar{y}_2 - \bar{y}_4) = -1.0$

Time × % compound effect $= \frac{1}{2}(\bar{y}_2 + \bar{y}_3 - \bar{y}_4 - \bar{y}_5) = -2.3$

Change in mean effect $= \frac{1}{5}(\bar{y}_2 + \bar{y}_3 + \bar{y}_4 + \bar{y}_5 - 4\bar{y}_1) = 2.0$

Calculation of Standard Deviation

Previous sum s =
Previous average s =
New s = range × $f_{k,n}$ = 1.29
Range = 4.3
New sum s = 1.29

New average $s = \dfrac{(\text{new sum } s)}{(n-1)} = 1.29$

Calculation of Error Limits

For new average $= \dfrac{2}{\sqrt{n}} \sigma^* = \pm 2.3$

For new effects $= \dfrac{2}{\sqrt{n}} \sigma^* = \pm 2.3$

For change in mean $= \dfrac{1.78}{\sqrt{n}} \sigma^* = \pm 2.0$

* Prior estimate $\sigma = 1.6$ used.

Fig. 13.12. Two-variable evolutionary operation program calculation work sheet, cycle $n = 3$. Response: remaining thickness of sample after test.

Calculation of Averages

Operating conditions	(1)	(2)	(3)	(4)	(5)
(i) Previous cycle sum	152.7	154.9	156.2	162.9	157.4
(ii) Previous cycle average	76.4	77.4	78.1	81.4	78.7
(iii) New observations	78.1	77.3	79.3	78.7	78.0
(iv) Differences (ii) less (iii)	−1.7	0.1	−1.2	2.7	0.7
(v) New sums	230.8	232.2	235.5	241.6	235.4
(vi) New averages: $\bar{y}_i$	76.9	77.4	78.5	80.5	78.5

Calculation of Effects

Time effect $= \frac{1}{2}(\bar{y}_3 + \bar{y}_4 - \bar{y}_2 - \bar{y}_5) = 1.5$

% Compound effect $= \frac{1}{2}(\bar{y}_3 + \bar{y}_5 - \bar{y}_2 - \bar{y}_4) = -0.5$

Time × % compound effect $= \frac{1}{2}(\bar{y}_2 + \bar{y}_3 - \bar{y}_4 - \bar{y}_5) = -1.5$

Change in mean effect $= \frac{1}{5}(\bar{y}_2 + \bar{y}_3 + \bar{y}_4 + \bar{y}_5 - 4\bar{y}_1) = 1.4$

Calculation of Standard Deviation

Previous sum s = 1.29
Previous average s = 1.29
New s = range × $f_{k,n}$ = 1.54
Range = 4.40
New sum s = 2.83

New average $s = \dfrac{(\text{new sum } s)}{(n-1)} = 1.41$

Calculation of Error Limits

For new average $= \dfrac{2}{\sqrt{n}}\sigma = \pm 1.62$

For new effects $= \dfrac{2}{\sqrt{n}}\sigma = \pm 1.62$

For change in mean $= \dfrac{1.78}{\sqrt{n}}\sigma = \pm 1.45$

Table 13.21. Values of Constant $f_{k,n}$

| Number of cycles = n | \multicolumn{9}{c}{k = number of sets of conditions in block} |
	2	3	4	5	6	7	8	9	10
2	.63	.42	.34	.30	.28	.26	.25	.24	.23
3	.72	.48	.40	.35	.32	.30	.29	.27	.26
4	.77	.51	.42	.37	.34	.32	.30	.29	.28
5	.79	.53	.43	.38	.35	.33	.31	.30	.29
6	.81	.54	.44	.39	.36	.34	.32	.31	.30
7	.82	.55	.45	.40	.37	.34	.33	.31	.30
8	.83	.55	.45	.40	.37	.35	.33	.31	.30
9	.84	.56	.46	.40	.37	.35	.33	.32	.31
10	.84	.56	.46	.41	.37	.35	.33	.32	.31
11	.84	.56	.46	.41	.38	.35	.33	.32	.31
12	.85	.57	.47	.41	.38	.35	.34	.32	.31
13	.85	.57	.47	.41	.38	.36	.34	.32	.31
14	.85	.57	.47	.41	.38	.36	.34	.32	.31
15	.86	.57	.47	.42	.38	.36	.34	.33	.31
16	.86	.57	.47	.42	.38	.36	.34	.33	.32
17	.86	.57	.47	.42	.38	.36	.34	.33	.32
18	.86	.57	.47	.42	.38	.36	.34	.33	.32
19	.86	.58	.47	.42	.38	.36	.34	.33	.32
20	.86	.58	.47	.42	.38	.36	.34	.33	.32

demonstrate this on a statistical basis at the end of the third cycle. The cycles are continued, therefore, until the results become sufficiently substantial to indicate statistically that there is a real effect. At this point a decision can be made to change to a second set of input conditions. If this is done, then a phase is said to have been completed. The next phase might study the same factors at different levels or it might study new factors.

Let us consider how to interpret the results. If the change in any effect is greater than the corresponding error limit, then it may be concluded that the effect of this factor is real. If the response is thought of as mapping out a surface, then the change in the mean will give an indication of the form of the surface in the immediate area of the study. For instance, if the true change in mean is small or zero, then the surface is approximately planar and the maximum response will probably be in the direction of the maximum slope of the plane. The next phase of study would therefore be in this direction. If the change in mean is negative, then the original process would be indicated as being close to the maximum in this region for those factors and responses being studied. If the change in mean effect is positive, then a concave surface is inferred and a move away from this region would be indicated. For a more detailed analytical study of response surfaces the reader is referred to Sec. 13.6.

INFORMATION BOARD

Phase 1   Last cycle completed: 8

Observed measurement of operation		Remaining thickness		Other responses
Requirement		Maximize		
Observation averages	% C O M P O U N D	76.9 77.2	78.4 77.7 80.3 TIME	
Effects with 95% error limits	Time % Compound $C \times T$ Change in mean	+2.3 ± −1.1 ± −0.8 ± +0.4 ±	1.4 1.4 1.4 1.2	
Standard deviation		1.4		
Prior estimate		1.6		

Fig. 13.13. Information board.

Suppose now that the information board indicates that sufficient data have been accumulated so that a decision should or can be made. In this type of study the decision will probably be made by the manager of manufacturing together with specialists from quality control, engineering, and reliability to aid in the interpretation of the results. In the example being discussed, the consequences of being able to increase the response are so important that any decision would be made solely on the single response being measured, with relatively minor concern over secondary responses; i.e., those which do not affect the performance and reliability of the item. Such an example might be taken from the missile field. A solid propellant rocket engine would probably explode if the amount of erosion on a certain part exceeded a specific value. The resultant damage and loss would be very expensive. In these circumstances secondary responses such as cost of production or weight would probably not be considered. However, it is quite apparent that in many other situations the decision to improve the process by changing to one of the superior processing combinations would depend not only on the increase in primary response but on the increase, if any, in cost or weight, etc. Consequently, the information board contains a provision for plotting other responses as well as the prime response.

Returning to the example, it is seen from the information board that processing time has a positive effect and per cent compound has a negative effect (though not significantly so) on the response. Also, both the change in mean and the interaction effects are small, so that it appears that the surface is linear in this area. Therefore, if we wish to move in the direction of the maximum response (i.e., minimum amount of erosion), we might change the standard process condition to combination 4 of Fig. 13.9 and study the region about this point. There is no set rule on determining the spread of the factor levels, but familiarity with the process will suggest the approximate extent of coverage. In this example, the use of the same spread of the time levels about condition 4 as in the previous phase was suggested. However, it was decided to extend the range of coverage of per cent compound in the second phase. The conditions for the second phase are shown by conditions 6, 7, 8, 9, 10 in Fig. 13.14.

The study is undertaken for those levels of combinations of per cent compound and processing time denoted by conditions 6, 7, 8, 9, 10 in the Fig. 13.14. Again the cycles are repeated until the effects, if any, of the factors are discernible and the phase is completed. It is not necessarily expected that each phase will contain the same number of cycles since, as the optimum is approached, the change in effects

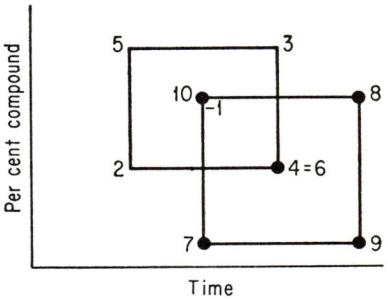

Figure 13.14

will become smaller; and it will consequently require a larger sample size to detect the change.

We need not expect to change the process the moment an effect exceeds an error limit. As pointed out earlier, other responses might be important in addition to the primary response. Also, since any modification or addition to an existing process generally results in an initial increase in cost, even though this expense may be repaid by the consequential improvement in the product or by an even less expensive operation ultimately, it might be wise to let the process "settle down" under its new conditions. Reference 9 considers an example of evaluating the costs of running an evolutionary operation and discusses when the phases might be changed.

It is important to note that the procedure which has just been described is intended to be a continuing operation and part of normal manufacturing operations. Its introduction should be handled with care, however. Once it has shown its value, its "interference" will be less questioned. Once the approximate set of optimum conditions have been established for a given set of factors and before they are investigated more precisely, perhaps by the methods of Sec. 13.6, it is worthwhile to test additional factors. These additional factors might have been tried in earlier phases and rejected as having minor importance; however, once the potential of the "major" factors has been exhausted, it is quite possible that any subsequent improvement in response will be due to changes of the lesser factors and will be important. This would be especially true in marginal reliability areas where every source of improvement should be explored.

It is of considerable importance at the beginning of the evolutionary operation to carefully choose meaningful response(s), that is, to make sure that the measurement we may be taking is realistic and—equally important—that the test being performed is representative of what will be experienced in actual use.

If it is found that there is no change in the response for the factors being considered, there are several possibilities open. One such alternative is to increase the spread of the levels of the factors in case the effect had been small in comparison with the error variation. Another possibility is to study different or additional factors in those cases where the response might be dependent upon other factors or upon their interaction with the original set of input factors.

13.6 RESPONSE SURFACE EXPERIMENTATION

Section 13.5 on evolutionary operation introduced the concept of searching for a set of input conditions which would make the output or response an optimum. Response surface experimentation is also concerned with the determination of the optimum response, but is more sophisticated statistically. The differences between the two methods stem from their different applications. Evolutionary operation is used in normal manufacturing processes, whereas response surface experimentation is generally a laboratory technique. As a consequence the latter methodology does not have the restrictions of the former (i.e., essentially the requirements of simplicity and the noninterruption of productivity) and is therefore able to use more refined and efficient statistical steps to arrive at the optimum set of conditions. The response may be considered as a dependent variable and the input levels as a set of independent variables. When the input variables are allowed to vary, they can be thought of as mapping out a surface for the response. Thus by understanding the topography in the immediate experimental area (i.e., the latter, defined by the values of the input variables) we can determine in which direction the optimum appears to lie. The area of experimentation can thus progress in this direction until the optimum is reached and determined. The basic concept is no different from that of evolutionary operation; however, by estimating the functional relationship of the variables, i.e., the shape of the response surface, we are able to achieve a more detailed knowledge of the area of experimentation and thereby move in a more direct way to the optimum.

13.6.1 GEOMETRICAL REPRESENTATION WITH TWO INPUT VARIABLES

For ease of discussion a response surface generated by two input variables is taken as an example. However, the technique can be extended to any number of input variables.

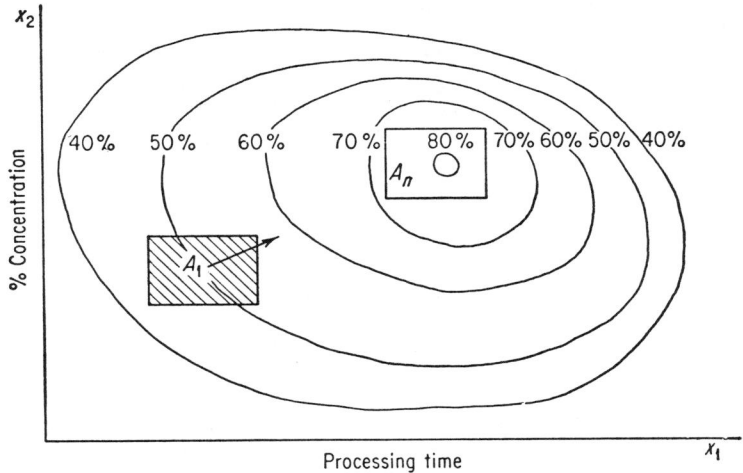

Fig. 13.15 Contour surface per cent yields as a function of x_2, per cent concentration and x_1, time.

Figure 13.15 illustrates how a particular property of a piece of material is dependent on two processing (input) variables. Suppose our initial area of experimentation is at the corners of the shaded area A_1 in the figure. We will then have four responses, one for each of the combinations. We might reasonably assume that we are still some distance from the optimum and therefore that the response surface in the immediate experimental area can be approximated by a plane; i.e., $y = b_0 + b_1x_1 + b_2x_2$. The true response surface is, of course, not known and therefore we test our assumptions (in this case a plane-surface assumption) by means of a statistical goodness-of-fit test as we progress. The plane, which is estimated by a least-squares technique, has a certain inclination to the x_1x_2 plane. Therefore, by finding its line of greatest slope and moving our area of experimentation in that direction we can maximize the rate of our approach to the optimum. This operation is continued until it is found that a plane or linear equation no longer represents the surface. When this occurs, a second-degree equation

$$y = b_0 + b_1x_1 + b_2x_2 + b_{11}x_1^2 + b_{12}x_1x_2 + b_{22}x_2^2 \qquad (13.3)$$

is used to estimate the surface of the response in that area. This would occur, say, when area A_n was reached in Fig. 13.15. If the second-degree equation proves to be a good fit, then by utilizing our knowledge of *the geometry of conical sections* we can study the shape of the surface. We do so by reducing the above second-degree equation (13.3) to its *canonical form*,* i.e., by transferring the origin to the center of the conic section and translating the x_1, x_2 axes along the principal axes X_1, X_2 of the conic giving the equation

$$y - B_0 = B_{11}X_1^2 + B_{22}X_2^2 \qquad (13.4)$$

By inspection of this equation we know much about its form and therefore much about the response surface. At this point it is worth considering Eq. (13.4) in some detail.

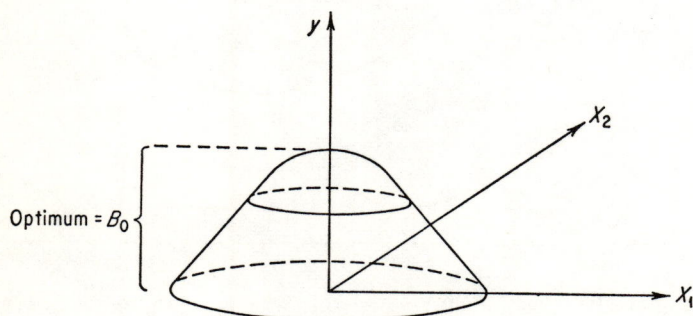

Fig. 13.16 Response surface when $B_{11}, B_{12} < 0$.

When $B_{11}, B_{22} < 0$ then Eq. (13.4) represents a family of concentric ellipses whose center is at $X_1 = 0$, $X_2 = 0$. When this is considered as a response surface, i.e., by plotting y as the third dimension as shown in Fig. 13.16, it can be seen from Eq. (13.4) the maximum value of y is given when $X_1 = X_2 = 0$ and is equal to B_0.

When B_{11} and B_{22} are of different sign, then Eq. (13.4) represents a hyperboloid; that is, the shape of the response surface forms a saddle

* See Sec. 13.6.3.4.

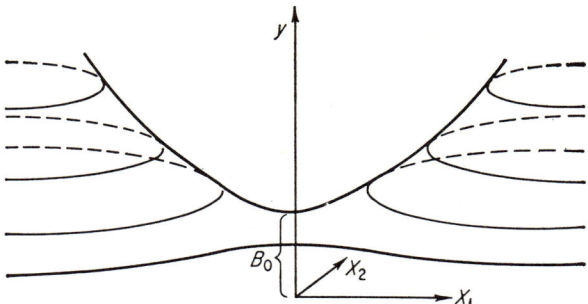

Fig. 13.17 Response surface when $B_{11} > 0$ and $B_{22} < 0$.

with point $y = B_0$ as a minimax. In this case the center of the conic does not give the optimum response for both variables but only the arrival at a maximum for X_2 and a minimum for X_1. The optimum response for y is reached, if one exists, by moving in the direction $\pm X_1$ for $X_2 = 0$.

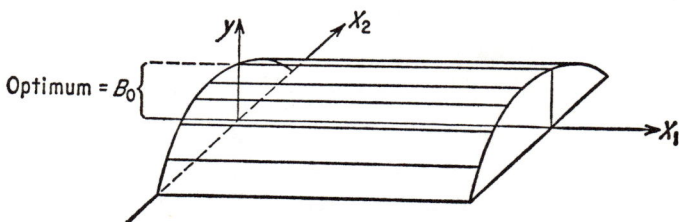

Fig. 13.18 Stationary ridge $B_{11} = 0$, $B_{22} < 0$.

When B_{11} is zero and B_{22} is negative, then Eq. (13.4) gives a set of straight lines parallel to the X_1 axis. There is no unique optimum since any point on the line $X_2 = 0$, $y = B_0$ is an optimum. It can be seen, as shown in Fig. 13.18, that this can be regarded as a particular case of either Fig. 13.16 or 13.17.

There is another surface called a *rising ridge* which occurs when Eq. (13.3) reduces to the form

$$y - B_0 = B_1 X_1 + B_{22} X_2^2$$

This is the equation of a parabola, a curve whose center is at infinity on the X_1 axis. B_1 represents the rate of increase of the response in the direction $+X_1$ for $X_2 = 0$. Fig. 13.19 illustrates this surface.

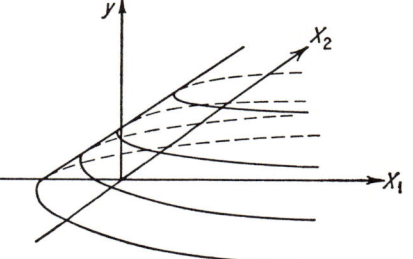

Fig. 13.19 Rising ridge $B_{11} = 0$.

13.6.2 EXAMPLES OF APPLICATION

Before describing the mechanics of the exploration of the surface, let us consider some of the applications. Suppose we have a device which is required to have a high probability of successful operation when called upon. It is known that the first few times it is tested there tends to be a high initial failure rate; it then stabilizes with a lower failure rate for several tests, but then again as test-time accumulates it begins to exhibit a higher failure rate. This type of failure pattern is shown in a variety of equipments: buses, cars, rocket engines, and complex electronic devices. There are various reasons for this behavior. Manufacturing mistakes and human error may cause the earlier "break-in" failures, with the later higher incidence of failure being the result of wearout failures. We attempt, of course, to reduce the probabilities of these failures; nevertheless, it still is important under certain circumstances to discover the point in time at which there will be minimum probability of failure.

Taking the case of a liquid rocket engine, we might be faced with this question. How many starts and how many seconds should we put on an engine in the various calibration, checkout, and acceptance tests so that the next firing has the highest probability of success? This is obviously of prime importance if that next firing is to be an operational firing (as part of either a weapon system or a space vehicle). The operation of a liquid rocket engine is such that a large proportion of failures occur in the initial firings; however, after a number of seconds' operation have been accumulated on the engine, wearout failures will begin to occur on some of its components. That is, the failure distribution of the system is a function of both the number of starts and the number of seconds. We must find out not only the number of firings for the optimum but also the average amount of time for each firing. In this example we might therefore set up a test program in which one input variable equaled the number of starts on the engine and the second input variable represented the intended length of a firing. Each engine would experience firings of a given duration; that is, the first engine would experience 20-second firings only, the second engine would experience 40-second firings only, etc. We might take as our response (to be maximized) y = total number of starts $\times$ cumulated number of operating seconds $\div$ the number of failures. We could therefore use the methodology of this section to estimate the surface generated by this function.

Another example might be the tensile strength of a piece of solid propellant, which is dependent on the concentration of its chemicals and its curing time. If, in its processing, it has too small a concentration of one of its chemical ingredients and/or too short a processing time, then it is too "soft." Under these circumstances it would be unsatisfactory because the propellant would slump in storage. On the other hand, too great a concentration or processing time would result in its being too brittle, subject to cracks in handling and transportation and, therefore, possible failure during firing. In this example, the response might represent the percentage of successes on samples of the propellant which are subjected to conditions representing tests for slump and brittleness, i.e., for both properties, either being unsatisfactory.

There is no reason why more than one type of response should not be studied. For example, we may wish to maximize a certain type of performance subject to the requirements that the reliability should not be decreased or that the weight be kept below a given value. Or, more ambitiously, we might wish to optimize all of these responses; however, it is difficult to maximize more than one response concurrently unless there is a known functional relationship or the relative importances of the several responses are established. This essentially has the property of reducing the multiple responses to one response and therefore one optimum. If the relationship or relative importance of the responses cannot be determined, then one response should be chosen as the principal response which is then optimized with certain decisions or restrictions being based on the secondary responses.

13.6.3 THE METHODOLOGY OF EXPERIMENTATION

If y is the response and x_1, x_2, $\cdots$ are the input variables or factors, then the true response surface can be described by the functional relationship

$$y = \phi(x_1, x_2, \cdots)$$

where ϕ is some function of the x_i's. This equation may be written in its Taylor series about the origin of x_i's

$$y = \beta_0 + \sum \beta_i x_i + \sum_{i \leq j} \beta_{ij} x_i x_j + \sum_{i \leq j \leq k} \beta_{ijk} x_i x_j x_k + \cdots \quad (13.5)$$

It is considered that in the local area of experimentation, the surface can be represented by either a plane or a second-degree equation. The equations of the plane or conical surface are both estimated by the method of least squares from observations from an experimental design. Many designs have been proposed which are particularly suited to this form of experimentation. However, it is not within the scope of this book to discuss any but the simplest type of design; the reader should consult the references (1, 2) for explicit details.

The procedures used in the search for the optimum are fairly distinct according to whether the experimental area can be represented by a linear equation or a quadratic. It is convenient, therefore, to describe the methods separately, though in practice they are supplementary.

13.6.3.1. The linear equation and method of steepest ascent.
Unless it is known from experience that the area of experimentation is expected to contain the optimum conditions, the first area of experimentation is assumed to be approximated by a $(p - 1)$-dimensional plane and estimated by Eq. (13.6) from n experimental points (responses)

$$\hat{y} = b_0 + \sum_{i=1}^{p-1} b_i x_i \qquad (13.6)$$

where the b_r are estimates of β_r obtained by minimizing $(y - \beta_0 - \sum \beta_i x_i)^2$ and solving the normal* equations.

It is next determined whether Eq. (13.6) represents a good fit, i.e., that the surface is linear. We first compute the "mean square of the lack of fit."

$$\sum (y_E - y_O)^2 / (n - p) \qquad (13.7)$$

where y_E is the expected value of $\hat{y}$ from (13.6) for a given set of input conditions and y_O is the observed response from the experiment for that same set of input conditions. Expression (13.7) contains two sources of variation. One is due to the lack of fit and the other due to sampling error. Therefore, if an estimate of the sampling error is available, we can determine by means of an F-test whether expression (13.7) contains any significant variation due to lack of fit, or whether its magnitude may be considered due solely to sampling error. An estimate of the error can be provided for in the experiment by adding to the design, say, c observations at its center and computing the sum of squares, S_c^2, for this set of c observations. An unbiased estimate of the error variance is therefore $S_c^2/(c - 1)$. It should be noted that these additional points do not change the estimated values of the coefficients b_i which multiply x_i, although b_0 is changed. (The reader should verify this fact by solving the normal equations first when center points are absent from the design and second when they are present.) We can form the F-ratio as follows:

$$F(c - 1, n - p) = \frac{(c - 1) \sum (y_E - y_O)^2}{(n - p) S_c^2}$$

If this value is greater than the tabulated value of the F-distribution for $(c - 1)$, $(n - p)$ degrees of freedom at a prechosen significance level, this is an indication that the surface is not a good fit, and we should investigate equations of higher degree using the above F-test until an adequate fit is indicated.

If the error variance is known, say, σ_0^2 (and therefore does not need to be estimated by making further observations), then we may use a χ^2

---

* See Ref. 1, Appendix 11B, for discussion of the method of least squares.

(chi-square) test for goodness of fit, where

$$\chi^2_{n-p} = \frac{\sum (y_E - y_O)^2}{\sigma_0^2}$$

If $\chi^2_{n-p}/(n - p)$ is larger than the entry in Table A.8 corresponding to $n - p$ degrees of freedom and the pre-chosen significance level, we conclude that the fitted equation is not adequate. If the fit is good, then $(b_1, b_2, b_3, \cdots, b_n)$ are the estimated slopes of the plane and we can proceed to move up the path of steepest ascent in the following manner:

b_i represents the rate of change of response with respect to x_i in coded units and k_i represents the number of original units in the coded units, then b_i/k_i is the rate of change of response with respect to factor x_i in its original units. Table 13.22 is formed.

Table 13.22

	x_1	x_2	$\cdots$	x_n
Change in response per coded unit change in x_i.	b_1	b_2	$\cdots$	b_n
Change in original units of x_i for change in response	$b_1 k_1$	$b_2 k_2$	$\cdots$	$b_n k_n$
Change per unit change of x_1.	1	$b_2 k_2/b_1 k_1$	$\cdots$	$b_n k_n/b_1 k_1$

If the coordinates of the center of the experimental area are $x_1^*, x_2^*, \cdots, x_n^*$, in original units then the coordinates of the steps up the plane are

first step: $\quad x_1^* + 1, \quad x_2^* + b_2 k_2/b_1 k_1, \quad \cdots, \quad x_n^* + b_n k_n/b_1 k_1$

second step: $\quad x_1^* + 2, \quad x_2^* + 2b_2 k_2/b_1 k_1, \quad \cdots, \quad x_n^* + 2b_n k_n/b_1 k_1$
(etc.)

Therefore by substituting these values, when coded, into the equation

$$\hat{y} = b_0 + b_1 x_1 + b_2 x_2 + \cdots + b_n x_n$$

we obtain the predicted value of $\hat{y}$ for each of the steps, all of which obviously lie on the plane. We now wish to determine how far up the plane our extrapolations will agree with an observed response at that point. Since we have several predicted responses for several sets of the assumed values of the inputs, the next step is to test whether one set of the assumed x's (inputs) does actually give a response close to the predicted response. If there is poor agreement, then we move back along the line of steepest ascent, taking observations for each assumed set of inputs, until it is determined where the linear relationship begins to falter as we move away from the initial levels. Care must be taken not to test the predicted against actual response at too great a distance from the initial levels; otherwise, the investigation might "step over" the optimum response. There is no way of determining how large the steps should be, but familiarity with the subject matter will provide intelligent estimates. It might be suggested that a tendency to conservatism is in order. If sufficient back-tracking has been necessary to determine where loss of linearity from the original equation occurred, then there will be a number of observations available whereby another linear equation can be estimated giving a new set of b_i's and thus a new path of steepest ascent. This procedure is repeated until a linear equation no longer fits the data. When this occurs, then a higher-order surface is fitted to the data.

13.6.3.2. An example of the method of steepest ascent.* An experiment was being carried out in which the efficiency of the output was under 50%. It was thought that this efficiency could be increased to at least 70% or even 80%. The efficiency was thought to be dependent on five input factors, S, R_1, P, D, R_2. In the experiments the factors were each varied at two levels. S represented the amount of solvent, R_1 the ratio of the first chemical to the base, P the purity of the solvent, D drying time, R_2 ratio of the second chemical to the base. Table 13.23 shows the limits of the levels, where (-1), $(+1)$ denote lower and upper levels, respectively.

Table 13.23

	-1	$+1$	
S	500	550	Solvent
R_1	9.0	9.5	Ratio of first chemical to base
P	95	98	Purity of solvent
D	4	5	Drying time
R_2	3.0	3.5	Ratio of second chemical to base

Since it was considered that the process was still far from the optimum, it was decided that for the first area of experimentation a plane would approximate the response surface. The design chosen to sample the 2^5 combinations was a $\frac{1}{4}$ fractional factorial providing $\frac{1}{4} \times 2^5 = 8$ observations. It was not expected that there would be any significant effects other than main effects. However, the design was chosen so that the one higher-order interaction which might possibly have a real effect was not confounded with any main effects.

The design is shown in Table 13.24 where the relationship between the natural variable and x_i (the coded variable) is

$$x_1 = (S - 525)/25 \qquad x_4 = (D - 4.5)/0.5$$
$$x_2 = (R_1 - 9.25)/0.25 \qquad x_5 = (R_2 - 3.25)/0.25$$
$$x_3 = (P - 96.5)/1.5$$

* Reference 11. Reprinted by permission of G. E. P. Box and K. B. Wilson.

STATISTICAL DESIGNS FOR RELIABILITY

Table 13.24*

	b_0	b_1	b_2	b_3	b_4	$-b_5$	b_{12}	b_{13}		
	x_0	x_1	x_2	x_3	x_4	x_5	x_1x_2	x_1x_3	y_O	y_E
Experiment (1)	1	-1	-1	-1	-1	-1	1	1	34.4	36.0
Experiment (2)	1	-1	-1	1	1	1	1	-1	51.6	49.6
Experiment (3)	1	-1	1	-1	1	1	-1	1	31.2	33.2
Experiment (4)	1	-1	1	1	-1	-1	-1	-1	45.1	43.6
Experiment (5)	1	1	-1	-1	1	-1	-1	-1	54.1	52.6
Experiment (6)	1	1	-1	1	-1	1	-1	1	62.4	64.6
Experiment (7)	1	1	1	-1	-1	1	1	-1	50.2	48.2
Experiment (8)	1	1	1	1	1	-1	1	1	58.6	60.2

* This table is reprinted from Table 11.21 of Owen L. Davies, *Design and Analysis of Industrial Experiments*, 1st Edition, 1954, published by Oliver and Boyd, Ltd., Edinburgh, by permission of the author and publishers.

The numbers in the y_O column are the observed responses. The numbers in the y_E column are the expected number and are obtained from Eq. (13.9). This design is a result of confounding† the following identities:

$$\text{main effect } x_4 \equiv \text{ interaction } x_1x_2x_3$$
$$\text{main effect } x_5 \equiv \text{ interaction } -x_2x_3 \tag{13.8}$$

If the design were considered to be a full factorial in x_1, x_2, x_3, then the following regression equation in x_1, x_2, x_3,

$$y = \beta_0 + \beta_1 x_1 + \beta_2 x_2 + \beta_3 x_3$$
$$+ \beta_{12}(x_1x_2) + \beta_{13}(x_1x_3) + \beta_{23}(x_2x_3) + \beta_{123}(x_1x_2x_3)$$

contains both the main effects and interaction effects of x_1, x_2, and x_3. The method of least squares is used to estimate the β's which in turn estimate the coefficients of the following regression equation of x_1, x_2, x_3, x_4, and x_5, by virtue of the confounding identities defined in (13.8).

$$\hat{y} = b_0 + b_1 x_1 + b_2 x_2 + b_3 x_3 + b_{12}(x_1x_2) + b_{13}(x_1x_3) - b_5 x_5 + b_4 x_4$$

However, unless b_{12} and b_{13} are comparable in size with the linear term coefficients, they will be ignored since they are associated with interaction effects which are not expected to be significant at the present distance of experimentation from the optimum. Thus, solving the normal equations and ignoring the coefficients of the higher order interactions implicit in

† For a discussion of the subject of confounding see References 1, 2 and 3.

the estimates of the b_i's, we have the following relationships:

$$\beta_0 \to b_0 = 48.5 \qquad \beta_{12} \to b_{12} = 0.2$$
$$\beta_1 \to b_1 = 7.9 \qquad \beta_{13} \to b_{13} = -1.8$$
$$\beta_2 \to b_2 = -2.2 \qquad \beta_{23} \to -b_5 = 0.4$$
$$\beta_3 \to b_3 = 6.0 \qquad \beta_{123} \to b_4 = 0.4$$

assumed to be zero

giving the equation of the plane response surface as

$$\hat{y} = 48.5 + 7.9x_1 - 2.2x_2 + 6.0x_3 + 0.4x_4 + 0.4x_5 \qquad (13.9)$$

When we compute $\sum (y_O - y_E)^2$, a value = 26.5 is obtained. This is large, since the error variance was known from prior data to equal 0.5, and is due to the fact that the interaction of x_1 and x_3 *does appear to have a real effect*. If the effect due to $x_1 x_3$ and $x_1 x_2$ had been included in the computation of $\sum (y_O - y_E)^2$, the value obtained would become 0.06 indicating a very close fit. Had the closeness not been determined by inspection, the statistical test for lack of fit would have been the chi-square test rather than the F-test, since we are testing against a known variance (see discussion in 13.6.3.1).

It is probable that the interaction effect $x_1 x_3$ is real. However, it is small in comparison with main effects of x_1, x_2, x_3; consequently, it was ignored, and a linear equation was used to compute the line of steepest ascent.

We are now ready to move along the line of steepest ascent (Table 13.25).

After the 8th step of predicted ascent, it was decided to perform an experiment, i.e., No. (9), at the same values of x_i as step 5 to see whether the linear relationship still held. The result of the experiment was an observed response of 77% compared with the predicted value of 74.8%, indicating a close relationship at this level. Therefore, a further experiment, No. (10), was performed at step 7 giving an observed result of 80% compared with a predicted 85.2%. Since the response has to be less than 100%, it was expected that the response would start falling away from the plane surface, as it appears to be doing in the area of step 7. Consequently, to see whether we had reached a stationary area or were on a second, less steep plane, further experiments were conducted at steps 6 and 8. The observations indicated that a curved surface was evident and quadratic terms should be included in the next estimating equation. In a second experiment designed to provide the observations for this estimation it was decided that the center of the design should be at the coordinates of step 7 and that the spread of the levels for factors x_1, x_2, and x_3 would remain the same but that they would be increased slightly for factors x_4 and x_5.

Table 13.25

	x_1	x_2	x_3	x_4	x_5	Predicted y	Observed
Base level...........................	525	9.25	95.5	4.5	3.25	..	..
No. of original units in coded unit, k_i....	25	0.25	1.5	0.5	0.25	..	..
Change in response per coded unit, b_i....	7.9	−2.2	6.0	0.4	0.4	..	..
Change in original units of x_i, $b_i k_i$.......	197.5	−0.55	9.0	0.2	0.1	..	..
Change in level per 10 units of x_1........	10	−0.028	0.456	0.011	0.005	..	..
Value of x_i's for 1st step of ascent.......	535	9.22	96.0	4.5	3.25	53.9	..
Value of x_i's for 2nd step of ascent.......	545	9.19	96.4	4.5	3.26	59.0	..
Value of x_i's for 3rd step of ascent.......	555	9.17	96.9	4.5	3.26	64.3	..
Value of x_i's for 4th step of ascent.......	565	9.14	97.3	4.5	3.27	69.3	..
Value of x_i's for 5th step of ascent.......	575	9.11	97.8	4.6	3.27	74.8	77
Value of x_i's for 6th step of ascent.......	585	9.08	98.2	4.6	3.28	79.9	81
Value of x_i's for 7th step of ascent.......	595	9.06	98.7	4.6	3.28	85.2	80
Value of x_i's for 8th step of ascent.......	605	9.03	99.1	4.6	3.29	90.3	76

The optimum was then determined by the methods which are now described.

13.6.3.3. The quadratic equations. The method of steepest ascent will generally bring the region of experimentation close to an optimum or stationary area but will not in itself permit the optimum to be found. At this time it is usually necessary to "cover" the area of experimentation with more observations, since the number of coefficients to be estimated in the higher-order equation increases rapidly with the number of factors in the study, as shown in Table 13.26.

Table 13.26

No. of factors in experiment	No. constants to be estimated		No. of observations in a factorial design (each factor 2 levels)			
	Degree of fitted equation		Degree of Factorial			
	1 (Plane)	2 (Quadratic)	Full	1/2	1/4	1/8
2	3	6	4	2	1	..
3	4	10	8	4	2	1
4	5	15	16	8	4	2
5	6	21	32	16	8	4
6	7	28	64	32	16	8

It can be seen that even a full factorial design will not provide sufficient observations for estimation of the regression equation in certain cases. Box et al. (Ref. 1) have suggested that the additional observations be obtained at points which are equidistant from the center of the factorial design with the coordinates of the original observations. Figure 13.20 illustrates an example for two factors. Such designs are called "orthogonal composite designs." The literature (Ref. 1, 2) contains examples of designs suitable for response-surface experimentation including experiments with factors at more than two levels.

When we have obtained a sufficient number of observations, we again estimate by the method* of least squares the second-degree approximation to $y = \phi(x_1, x_2, \cdots, x_n)$.

$$\hat{y} = b_0 + \sum_i b_i x_i + \sum_{i \leq j} b_{ij} x_i x_j \tag{13.10}$$

This equation is tested for its goodness of fit by the F-test method of Sec. 13.6.3.1. We assume for the purposes of this discussion that a second-order equation can be found to satisfy the surface in the local area of the

---

* For a discussion of the method of least squares for fitting nonlinear regression functions, see H. O. Hartley, *Technometrics* **3**:2, 269–280 (May 1961).

optimum of the experiment. If this is not the case, then cubic or even higher-order equations may be estimated; however, their interpretation becomes increasingly difficult. An understanding of the subject matter will indicate whether higher-order equations are appropriate; however, the first review of the situation would be as to the "area" of the experiment. For instance, a second-order equation may not give a good fit because the difference between the levels of the factors might be too large, or perhaps all the factors affecting the response were not being taken into account. These possibilities should be considered, otherwise the technique loses much of its appeal.

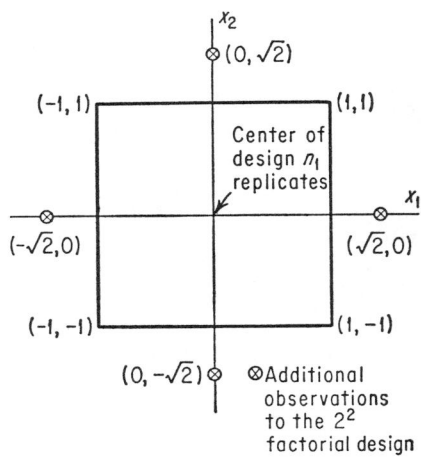

Figure 13.20

Equation (13.10) is then reduced to its canonical form (Sec. 13.6.3.4).

$$\hat{y} = B_0 + \sum B_{ii}X_i^2 \tag{13.10a}$$

The properties of the surface defined by this equation are extensions of the properties discussed for the two-dimensional case of Sec. 13.6.1.

Thus if all the B_{ii} are negative, the estimated surface will have a maximum equal to B_0 at the center of the ellipsoids.

If some of the B_{ii} are positive, the surface is defined by elliptic hyperboloids indicating the presence of minimaxes.

If one or more of the B_{ii} approach zero, this indicates the presence of ridges in the directions of the corresponding X_i axes.

Once the coefficients b_i have been estimated in Eq. (13.10), the major portion of the remaining work, which we shall now describe, consists of the reduction of Eq. (13.10) to its canonical form Eq. (13.10a).

13.6.3.4. Reduction to canonical form. The linear terms in Eq. (13.10) are removed by transferring the origin to the center of the surface. The center is given by differentiating the right-hand side of Eq. (13.10) with respect to each x_i in turn and putting the resulting expressions equal to zero.

$$0 = b_1 + 2b_{11}x_1 + b_{12}x_2 + b_{13}x_3 + \cdots + b_{1n}x_n$$
$$0 = b_2 + b_{21}x_1 + 2b_{22}x_2 + b_{23}x_3 + \cdots + b_{2n}x_n$$
$$\vdots$$
$$0 = b_n + b_{n1}x_1 + b_{n2}x_2 + b_{n3}x_3 + \cdots + 2b_{nn}x_n$$

Denote the solutions to these equations by x_{1s}, x_{2s}, etc. Then these values represent the center of the surface, and in addition they give the stationary value for y, the response. The origin is now transferred to the center of the surface by the transformations.

$$x_1^* = x_1 - x_{1s}, \qquad x_2^* = x_2 - x_{2s}, \qquad \cdots$$

which when substituted in (13.10) gives

$$\hat{y} - B_0 = \sum_{i<j} b_{ij} x_i^* x_j^* + \sum b_{ii} x_i^{*2} \qquad (13.11)$$

where $B_0 = b_0 + \frac{1}{2} \sum b_i x_{is}$ (i.e., the stationary value).

The cross-product terms in (13.11) are now removed by rotating the transferred coordinate axes about the new origin so that they become the principal axes of the quadratic surface. The new axes are also orthogonal. The result is Eq. (13.10a), repeated here for the reader's convenience.

$$\hat{y} - B_0 = B_{11} X_1^2 + B_{22} X_2^2 + \cdots + B_{nn} X_n^2$$

where B_{ii}, $i = 1, \cdots, n$ are the roots of the characteristic equation

$$0 = \begin{vmatrix} (b_{11} - B) & \frac{1}{2} b_{12} & \cdots & \frac{1}{2} b_{1n} \\ \frac{1}{2} b_{12} & (b_{22} - B) & & \frac{1}{2} b_{2n} \\ \vdots & \vdots & & \vdots \\ \frac{1}{2} b_{1n} & \frac{1}{2} b_{2n} & \cdots & (b_{nn} - B) \end{vmatrix} \qquad (13.12)$$

and X_i are linear expressions in $x_i^* = (x_i - x_{is})$ such as

$$\begin{aligned} X_1 &= m_{11} x_1^* + m_{12} x_2^* + \cdots + m_{1n} x_n^* \\ &\quad \vdots \\ X_n &= m_{n1} x_1^* + m_{n2} x_2^* + \cdots + m_{nn} x_n^* \end{aligned} \qquad (13.13)$$

where the m_{ij} have to be determined and are solutions of the n sets of equations, $i = 1, 2, \cdots, n$

$$\begin{aligned} (b_{11} - B_{ii}) m_{i1} + \tfrac{1}{2} b_{12} m_{i2} + \cdots + \tfrac{1}{2} b_{1n} m_{in} &= 0 \\ \tfrac{1}{2} b_{12} m_{i1} + (b_{22} - B_{ii}) m_{i2} + \cdots + \tfrac{1}{2} b_{2n} m_{in} &= 0 \\ &\vdots \\ \tfrac{1}{2} b_{1n} m_{i1} + \tfrac{1}{2} b_{2n} m_{i2} + \cdots + (b_{nn} - B_{ii}) m_{in} &= 0 \end{aligned} \qquad (13.14)$$

where B_{ii} are solutions of B from Eq. (13.12).

STATISTICAL DESIGNS FOR RELIABILITY

It will be seen that the above set of equations is constructed by placing the elements of the determinant of the characteristic equation (13.12) as coefficients of the m's in the ith row of Eqs. (13.13), where the elements in Eq. (13.12) have B_{ii} replacing B. The solutions of Eqs. (13.14) do not allow the m's to be found explicitly but only proportionally to their true values, which are found by the additional sets of conditions for orthogonality

$$m_{i1}^2 + m_{i2}^2 + \cdots + m_{in}^2 = 1$$

for $i = 1, 2, \cdots, n$.

Since the transformation is orthogonal, Eq. (13.15) permits us to express x_i^* in terms of X_i

$$\begin{aligned} x_1^* &= m_{11}X_1 + m_{21}X_2 + \cdots + m_{n1}X_n \\ & \vdots \\ x_n^* &= m_{1n}X_1 + m_{2n}X_2 + \cdots + m_{nn}X_n \end{aligned} \quad (13.15)$$

13.6.3.5. An example of determining the optimum from a quadratic surface.* The processing of a certain type of material consisted of mixing a given weight of powder with a solvent, casting into a given configuration, and drying out the solvent in an oven. The variables concerned in this operation were: T, the temperature of the oven; c, the concentration of solvent; and t, drying time. The response being considered was to be optimized against these three variables. Earlier experiments had established the approximate process values to be $T = 167°C$, $c = 27.5\%$, and $t = 6.5$ hours.

It was decided to run a 3-factor factorial experiment, each variable at two levels, providing eight of the minimum of ten observations necessary for the estimation of coefficients of a quadratic surface (see Table 13.26). In addition, seven other combinations were observed as shown in Table 13.28.

If the variables are coded by the following relationships

$$x_1 = \frac{T - 167}{5}, \quad x_2 = \frac{c - 27.5}{2.5}, \quad x_3 = \frac{t - 6.5}{1.5}$$

then the relationships between the design levels and the original units are as shown in Table 13.27.

---

* Reference 12. Reprinted with permission of G. E. P. Box and Owen L. Davies, Editor of *Design and Analysis of Industrial Experiments*, published by Oliver and Boyd, Ltd., Edinburgh, by permission of the author and publishers. 1st Edition, 1954.

Table 13.27

Factor levels in units of the experiment		−2	−1	0	1	2
Original units	T = temperature	157	162	167	172	177
	c = concentration	22.5	25	27.5	30	32.5
	t = time	3.5	5	6.5	8	9.5

The responses are shown against the coded levels of the input factors in Table 13.28.

Table 13.28. Coded Levels of Experimental Variables and Results Obtained

	Experiment	x_1	x_2	x_3	Response
2^3 Factorial	1	−1	−1	−1	45.9
	2	−1	−1	1	53.3
	3	−1	1	−1	57.5
	4	−1	1	1	58.8
	5	1	−1	−1	60.6
	6	1	−1	1	58.0
	7	1	1	−1	58.6
	8	1	1	1	52.6
Additional points to form a central composite design.	9	0	0	0	56.9
	10	2	0	0	55.4
	11	−2	0	0	46.9
	12	0	2	0	57.5
	13	0	−2	0	55.0
	14	0	0	2	58.9
	15	0	0	−2	50.3

The second-degree equation

$$\hat{y} = b_0 + \sum_{i=1}^{3} b_i x_i + \sum_{i=1 \leq j=1}^{3,3} b_{ij} x_i x_j$$

was fitted* to the data, giving the estimated equation of the response surface in the area of experimentation as

$$\hat{y} = 57.71 + 1.94x_1 + 0.91x_2 + 1.07x_3 - 1.54x_1^2 - 0.26x_2^2$$
$$- 0.68x_3^2 - 3.09x_1x_2 - 2.19x_1x_3 - 1.21x_2x_3 \quad (13.16)$$

* For details of estimating the coefficients see Ref. 1, ¶11.72

The sum of squares obtained from the difference between the observed values of y and the expected values was 24.6. This was associated with 5 degrees of freedom. A prior estimate of the experimental error was available and equal to 1.5; therefore, it was concluded that the second-degree equation represented a sufficiently good fit, since $\chi_5^2 = 24.6/(1.5)^2 = 11$ is not quite significant at the 5% level (see Table A.8).

The center of the system was obtained by partial differentiation of Eq. (13.16) for x_1, x_2, and x_3 and by solving these expressions when equated to zero.

$$0 = 1.94 - 3.08x_1 - 3.09x_2 - 2.19x_3$$

$$0 = 0.91 - 3.09x_1 - 0.52x_2 - 1.21x_3$$

$$0 = 1.07 - 2.19x_1 - 1.21x_2 - 1.36x_3$$

giving

$$x_{1s} = 0.061, \quad x_{2s} = 0.215, \quad x_{3s} = 0.499, \quad \hat{y}_s = 58.14$$

The canonical coefficients B_{ii} are obtained from

$$0 = \begin{vmatrix} (-1.54 - B) & -1.54 & -1.10 \\ -1.54 & (-0.26 - B) & -0.60 \\ -1.10 & -0.60 & (-0.68 - B) \end{vmatrix}$$

giving

$$B_{11} = -3.19, \quad B_{22} = -0.07, \quad B_{33} = 0.78$$

The m_{ij} for $i = 1, 2, 3, j = 1, 2, 3$ are obtained by three sets of equations obtained from (13.14). Their solutions produce the following orthogonal relationship between X_i and x_i:

	$(x_1 - 0.061)$	$(x_2 - 0.215)$	$(x_3 - 0.499)$
X_1	0.751	0.488	0.444
X_2	0.307	0.338	-0.890
X_3	0.585	-0.804	0.104

Thus, e.g.,

$$X_1 = 0.751(x_1 - 0.061) + 0.488(x_2 - 0.215) + 0.444(x_3 - 0.499)$$

or

$$(x_1 - 0.061) = 0.751X_1 + 0.307X_2 + 0.585X_3$$

Equation (13.16) when written in its canonical form is

$$\hat{y} - 58.14 = -3.19X_1^2 - 0.07X_2^2 + 0.78X_3^2 \qquad (13.17)$$

Inspection of this equation reveals that the coefficient of X_2^2 is close to zero and that since the coefficients of B_{11} and B_{33} are opposite in sign, we have a saddle point. However, the coefficient B_{33} is also small so that the surface is a flattened saddle. The minimax is at $X_1 = 0$, $X_2 = 0$, $X_3 = 0$. To move along the X_1 axis would result in a sharp decrease in response. However, we would expect to increase the response by moving in either direction along X_3. Consequently, four additional experiments, 16 through 19 (Table 13.29), were conducted in these directions and the canonical form recalculated (Eq. (13.18)).

Table 13.29. Levels of Experimental Variables and Results Obtained

	Experiment	x_1	x_2	x_3	Response
Additional points	16	2	−3	0	59.4
	17	2	−3	0	61.5
	18	−1.4	2.6	0.7	59.5
	19	−1.4	2.6	0.7	58.5

$$\hat{y} - 59.23 = -3.51X_1^2 - 0.25X_2^2 + 0.24X_3^2 \qquad (13.18)$$

The additional observations resulted in a minor change to the maximum. Therefore, since the effects of X_2 and X_3 were small, it was concluded that the maximum was given by $X_1 = 0$ and was equal to approximately 60. This response could be obtained for any point on the $X_1 = 0$ plane through the X_2X_3 axes. Thus any set of x_1, x_2, x_3 which satisfied Eq. (13.19) would result in a maximum.

$$0 = 0.751(x_1 - 0.061) + 0.488(x_2 - 0.215) + 0.444(x_3 - 0.499)$$

$$(13.19)$$

or in terms of the original units:

$$0 = 0.150T + 0.195c + 0.296t - 32.76 \qquad (13.20)$$

It must be remembered that Eq. (13.20) can be expected to hold only in the region of the experiment, so that values of T, c, t outside the limits 157–177, 20–34, 3.5–9.5, respectively, and which also satisfied Eq. (13.20) would not necessarily be expected to give a response equal to the maximum 60.

In most cases of this type secondary considerations would be used to establish which set of T, c, t would be chosen. Thus time t, might be a limiting factor in a process to be introduced into a production line. Under these circumstances, the minimum of 3.5 might be chosen, with a second of the remaining variables, c or T, picked to satisfy some other particular requirement.

The methodology described in this section is important not only in finding the optimum combination of the various factors which together produce a response but also in understanding the basic theory of the process itself (Ref. 13). This is why the mathematical form of the response surface is introduced into the methodology. For instance, evolutionary operation will lead to an optimum set of input variables, and the fact that it does so relatively slowly is immaterial; however, its methodology does not permit the functional relationship to be assessed. The method of response-surfaces exploration is linked with the fundamental assumption that a functional relationship exists which can be approximated and estimated.

REFERENCES

1. O. L. Davies, *Design and Analysis of Industrial Experiments*, Oliver and Boyd London and Edinburgh, 1954.
2. W. G. Cochran and G. M. Cox, *Experimental Designs*, 2d ed., John Wiley & Sons, New York, 1957.
3. O. Kempthorne, *The Design and Analysis of Experiments*, John Wiley & Sons, New York, 1957.
4. N. R. Garner and D. E. Hartvigsen, "Experimental Designs in Development, Proof and Qualification Testing of Rocket Motors," *Proc. Western Regional Conference ASQC, August 1958, San Diego*.
5. "Statistical Analysis for a New Procedure in Sensitivity Experiments," Applied Mathematics Panel, National Defense Research Committee, AMP Report No. 101, 1R, SRG-P No. 40, July 1944.
6. F. E. Satterthwaite, "Random Balance Experimentation," *Technometrics*, **1:**2, 111-37 (May 1959).
7. T. A. Budne, "The Application of Random Balance Designs," *Technometrics*, **1:**2, 139-55 (May 1959).
8. J. S. Hunter, *Technometrics*, **1:**2, 180-184 (May 1959).
9. G. E. P. Box, "Evolutionary Operation: A Method for Increasing Industrial Productivity," *Applied Statistics*, **6:**2, 3-23 (1957).
10. G. E. P. Box and J. S. Hunter, "Condensed Calculations for Evolutionary Operations Programs," *Technometrics*, **1:**1, 77-95 (February 1959).
11. G. E. P. Box and K. B. Wilson, "On the Experimental Attainment of Optimum Conditions," *J. Roy. Stat. Soc.*, Series B, **13:**1, 18 (1951).
12. G. E. P. Box, "Exploration and Exploitation of Response Surfaces: Some General Considerations and Examples," *Biometrics*, **10:**1, Sec. 8.
13. G. E. P. Box and P. V. Voule, "The Exploration and Exploitation of Response Surfaces: An Example of the Link Between the Fitted Surface and the Basic Mechanism of the System," *Biometrics*, **11,** 287-322 (1955).

ADDITIONAL READING

Anscombe, F. J., "Quick Analysis Methods For Random Balance Screening Experiments," *Technometrics*, 1:2, 195–209 (May 1959).

Box, G. E. P., "A Basis for Selection of a Response Surface Design," *J. Am. Stat. Assoc.*, 54:287, 622–54 (September 1959).

Dickinson, A. W., "Computer Program Abstracts, Response Surface Evaluation (025)," *Chem. Eng. Progress*, **55,** 86, 88 (September 1959).

Hartley, H. O., "Smallest Composite Designs for Quadratic Response Surfaces," *Biometrics*, 15:4, 611–24 (December 1959).

Hunter, J. S., "Some Statistical Principles Underlying Evolutionary Operations," *Proc. 2d Stevens Symposium*, 63–75, and *Chemical Division and Metropolitan Section, ASQC, January 1958*.

Hunter, J. S., "Statistical Methods for Determining Optimum Conditions," *National Convention Trans. ASQC, 1956*, 415–28.

Koehler, T., "Evolutionary Operations, Some Actual Examples," *Proc. 2d Stevens Symposium*, 5–8.

Lloyd, D.K., "Long Range Service Life Analysis (LRSLA) Estimating Procedure," *J. Spacecraft and Rockets*, Vol. 14 (June 1977).

Satterthwaite, F. E., "New Developments in Experimental Design," *Proc. Rutgers Quality Control Conf., ASQC, 1956*, 55–57.

Shainin, D., "Use Logical Research Methods For Component Specification," *Automatic Control*, **6,** 28–32 (June 1957).

Vaswani, R., "Sequential Decisioning Technique for Optimization of Complex Systems," *J. Ind. Eng.*, 7:4, 174–78 (July–August 1956).

Whidden, P., "Design of Experiment in Metals Processing," *National Convention Trans., ASQC, 1956*, 677–83.

Whitewell, J. C., "Practical Applications of Evolutionary Operations," *National Convention Trans., ASQC, 1959*, 603–616.

Youden, W. J., "Engineering vs. Classical Test Patterns," *Ind. Eng. Chem.*, **48,** 59A–60A (August 1956).

SECTION III
EXAMPLES OF RELIABILITY EVALUATION AND DEMONSTRATION PROGRAMS

Three examples are presented in this Section. Each relates to a different type of equipment. Their situations and restrictions differ considerably, so that in each case variation is found in the solution of their problems. Also, except in the second case, no equipment was available specifically for reliability tests; consequently, only development tests could provide the reliability information.

The first example is that of a large solid propellant rocket engine undergoing engineering development. There are two features which determine the extent and character of the development and reliability program. First, a solid propellant rocket engine can be test-fired only once, and this, together with the high cost of its materials and manufacture, means that the number of tests of full-scale engines is therefore strictly limited. Second, the nature of a solid propellant rocket engine is such that it is a simple but highly integrated system; that is, a design change of a single component will frequently require associated changes in design of many of the components in the remainder of the system. Both features result in component design changes being made in groups; consequently, the program consists of distinct but overlapping phases as the design changes are made and incorporated in the full scale engine.

The second example describes how, at the earliest time possible, it is desired to improve and accept an already well-developed device when it has demonstrated a specified level of reliability. The device is a turbogenerator subsystem for which single units can be retested many times.

The third example is that of a large liquid propellant rocket engine which is undergoing continuous system development by both component and complete system testing and modification. The system is complex, but there is much weaker mutual interdependence between the components of this system than in the case of the solid rocket engine; that is, components may be developed to a high degree in separate tests prior to tests within the engine system. Also, relatively major adjustments may be made without affecting performance of other components. Another important feature of the test program is that an engine assembly may be test-fired several times without exhibiting wearout-type failures. The duration of each test-firing is of major significance in terms of reliability measurement.

The general approach to solving each of the examples is similar. Thus provisions are made for defining the configurations, establishing the applicability of any one test, defining the criteria for judging or classifying the test results, and protecting the integrity of the data. The methods by which the data are evaluated are considerably different, owing to the restrictions and circumstances of testing of each equipment as well as to the way the data may be interpreted for their measure as an estimate of reliability.

Each example is complete in itself and discusses specific applications of general considerations mentioned in earlier chapters.

CHAPTER FOURTEEN

A RELIABILITY EVALUATION PROGRAM FOR A LARGE SOLID PROPELLANT ROCKET ENGINE DURING ENGINEERING DEVELOPMENT

14.1 OBJECTIVES AND PROBLEMS

The engine in question represents a major advancement in size and design over any previous solid propellant rocket engine. Consequently, there is little previous experience upon which reliability information is available. In addition, owing to the expense and nature of the engine, no or few test firings are made specifically for reliability demonstration purposes* and only one test firing can be made per engine.

The objectives of this reliability evaluation program are therefore to define and measure, during a research and development program, charac-

---

* In a development program of a rocket engine there usually exist certain points at which the progress is monitored. For instance, the PFRT (Pre-Flight Rating Test) program would be one such point at the culmination of the initial R and D (Research and Development) program, demonstrating with a sample of engines of an essentially fixed design the ability of the engine to enter into the flight phase where further R and D work will be undertaken. After a period in this phase, there results from the flight and static tests a more refined configuration and stabilized production process. The design is frozen and a Qualification Test Program is undertaken in order to demonstrate the suitability of the engine and the production facilities for the mass production phase. The periods in the program in which several engines of a fixed design are test-fired probably produce the most valid and representative data at the time. However, these periods represent only a moderate fraction of the total number of engines fired. Consequently, the PFRT and Qualification Test programs in themselves cannot demonstrate the high reliability requirements expected of solid propellant rocket engines. Therefore, data from the other sources, R and D static firings and R and D flight firings must also be included to build up the sample size. This chapter presents a method of reliability evaluation which utilizes the data from the beginning through the end of the PFRT program only. The method, however, can be applied with minor modifications (mostly administrative in nature) to the R and D Flight and through the Qualification Test programs.

teristics which are indicative of the engine's ability to reliably perform its function of propelling a missile of one or more stages into a certain target area.

To accomplish these objectives three major practical and statistical problems must first be considered:

1. Relating engine reliability requirements in a Pre-flight R&D program to the weapon system's operational reliability requirements.
2. Obtaining valid reliability data in an R&D program when the objectives of the tests vary and the configuration of the engine undergoes continuous modification.
3. Making efficient reliability estimates based upon the results of a limited number of full-scale engine test firings.

14.2 DISCUSSION OF PROBLEMS

1. The ultimate intention for a given missile is that it will fall within a certain target area. This intention can be projected into single-stage engine reliability requirements as implying that for successful operation of the missile there should be no catastrophic failure which would result in mission abort nor any unsatisfactory performance which would result in a significant target miss. On the other hand, operation of a given engine which exhibits no catastrophic failure, and for which the performance parameters all lie within specified limits, would, assuming successful operation of the remainder of the missile, produce a scatter of hits in and around the target area. Initially, only developmental test-stand firings can be utilized for reliability evaluation during the program, since flight tests will not yet have been conducted. Thus, the effect of performance interactions of an engine with the remainder of the systems in the missile cannot be comprehensively known during this period. The initial ground rules which interpret the success or failure criteria of the weapon system into the success or failure criteria of the engine may, therefore, be based on engineering and arbitrary judgments. However, provided that the ground rules are clearly stated, understood, and rigorously applied, reliability estimates can be made which are valid within the framework of those ground rules. Therefore, if reliability is defined as the probability of a successful operation of the weapon system, then the reliability of the engine, as a functioning subsystem of that weapon system, can be interpreted as the probability of successful operation of the engine. This in turn can be regarded as the probability of performing within engine model specifications. Thus, the relationship between the weapon system and engine is defined by the specifications which, of course, are revised when flight data become available.

2. The numerical value of reliability which the engine may be expected to display or contractually demonstrate would depend on the "state of the art" of the engine, the number of units available for testing, and the reliability requirement of the weapon system.

The applicability of engineering development tests for reliability evaluation is dependent on the validity of the data and the intention and circumstances of the tests generating the data. Generally, the nature of developmental testing differs from reliability testing. Reliability testing would usually involve a large homogeneous sample of engines representative of the final configuration. It is generally not practical to produce a large number of expensive engines specifically to demonstrate the system's reliability. Even if this were done, by the time production and testing were completed the reliability results would refer to an obsolescent configuration, since simultaneous engineering testing would probably have resulted in further design improvements. Consequently, in this program reliability must be estimated from the results of R&D engineering tests. However, engineering testing is usually performed with small sample sizes on changing configurations. Therefore, it is necessary to take into account the objectives of an engineering test, to ascertain how these objectives differ from those of a reliability test and to see whether it is possible to reconcile the two. This must be done without restricting the exploratory nature of development testing.

This is accomplished by "screening" the engineering tests for reliability use by means of a "declaration policy" and determining the degree of representation of the engine towards its flight configuration. Thus, even though the item is evolving during the R&D test period, representative data can be obtained and used to determine a valid estimate of reliability.

The foregoing discussion implies that only full-scale engine tests will be used for reliability evaluation. A development program for state-of-the-art solid-propellant engines involves a great deal of experimental testing with subscale engines. Initial feasibility studies for new or modified propellant formulations are best and most economically undertaken in small and subscale engines in order to provide evidence that the propellant will satisfy internal ballistic requirements in the full-scale engines. In many instances, accurate scaling predictions of internal ballistic performance can be established. Further testing with subscale engines is undertaken in order to determine properties of charge and case designs, insulation materials, movable nozzle designs, etc. for the full-scale engine. However, there exist differences between subscale and full-scale engines which cannot be completely resolved by the use of scaling or correlation factors. For example, these differences might be due to the lack of sufficient knowledge of the mechanical properties of scaled-up propellant charges, such as of stress magnitudes and patterns, which are intimately related to problems of propellant cracking and propellant-liner separation. Also, any problems specifically associated with manufacturing and quality control processes of the full-scale engine can only be exhibited by the full-scale unit itself.

Thus, although the engineering information obtained from subscale engine tests is essential for the development of functionability and performance of the full-scale engine, the best evidence to ascertain that the end-product requirements will be met is obtainable only from tests of full-scale engines sufficiently representative of final design configuration.

3. Since there are only a limited number of full-scale engines representative of final configuration available for testing, the statistical method of making reliability evaluations becomes of great importance. The statistical technique employed in this program utilizes the methods of Sec. 9.2.2.

14.3 DESCRIPTION AND CONDUCT OF THE PROGRAM

The program is divided into four parts:
1. a method of system apportionment into *Principal Subsystems*
2. a declaration policy
3. a method of classifying R&D test results for reliability evaluation
4. a technique of estimating reliability.

14.3.1 SYSTEM APPORTIONMENT

To compensate for the relatively small number of tests of full-scale engines, it is essential that all representative data be utilized. This is done by apportioning the engine into three Principal Subsystems, each engine test being evaluated in terms of the behavior of the Principal Subsystems. Thus, the fact that an engine is not fully representative of the final configuration in any one test will not prevent the evaluation of those Principal Subsystems which are operating in a configuration or manner representative of flight status. The three Principal Subsystems are:

A. The Propellant Charge-Ignition Subsystem
B. The Case-Liner-Internal Insulation Subsystem
C. The Thrust Vector Control Subsystem.

Engine reliability estimates are made from the Principal Subsystem test results (tested within the environment of the full-scale engine) and can begin with the first test firing.

14.3.2 APPLICABILITY OF PRINCIPAL SUBSYSTEMS

The Principal Subsystems tested in a full-scale engine firing will be classified as *applicable* or *inapplicable* for purposes of reliability evaluation. In order to determine which of the subsystems being tested in any full-scale test firing are sufficiently representative of flight configuration to be useful for reliability evaluation, it is necessary to set up criteria which determine the subsystem's applicability (Table 14.1).

Table 14.1

Principal Subsystem	Required characteristics for applicability*
Subsystem A Propellant Charge-Igniter	W_p = propellant charge weight lies within $\pm$ __ % of current model specification limits
Subsystem B Case-liner-internal insulation	(1) Engine configuration includes flight-weight case and end-closure, liner and insulation, as specified by current weight and balance status report. (2) Propellant weight as specified for Subsystem A.
Subsystem C Thrust vector control	(1) Engine configuration includes flight-weight movable nozzles. (2) Flight-weight nozzle actuator subassembly. (3) Propellant weight as specified for Subsystem A. (4) Nozzles must be intended to actuate during test firing. (5) Predicted action time not less than current model specification limit unless thrust-termination operation is being tested.

* The reader will note that for the purposes of reliability evaluation in the development phase of a solid propellant rocket engine program, the required characteristics for applicability can be stated in very general terms. The fundamental engine design criteria and the schedules are the primary sources for directing the development engineering effort toward a well-defined design configuration. In fact, the engine so developed has its first reasonably complete definition in the Model Specification document generated for the PFRT engine configuration. Even then, the internal ballistic performance is usually not known to a satisfactory precision until the completion of the PFRT program.

Those Principal Subsystems which are applicable will be evaluated and categorized as *success*, *failure*, or *exclusion*. However, a result may be excluded *prior* to the test; the circumstances which permit this are listed in Table 14.2.

Table 14.2. Ground Rules for Pretest Exclusion

	All Principal Subsystems (A, B, C)
Pretest exclusion (subject to approval of prime contractor/program manager)	(1) The Principal Subsystem does not have required characteristics for applicability. (2) Owing to the intention or circumstances of the test, a particular Subsystem may be excluded because of stated uncertain characteristics of operation relating to performance and/or possibility of malfunction; e.g., if inspection shows propellant voids or cracks of a sufficient degree so that malfunction would be expected, then the Propellant Charge-Ignition Subsystem may be considered for exclusion. (3) If internal ballistic performance is predicted to be outside current model specification limits, but the test firing is approved, then performance will be excluded. A maximum of ―― exclusions is allowable under this ground rule prior to the formal PFRT (Pre-Flight Rating Test) program. No exclusions under this ground rule are allowable during the formal PFRT program. (4) Provisional exclusion: A Subsystem will be excluded if there is a failure of an experimental part which has been so listed on the declaration form prior to the test, and which is being tested for the first time in a full-scale engine. However, if the Subsystem fails because of the failure of a nonlisted part, it will be classified as a failure. (5) Provisional exclusion: A Subsystem will be excluded if there is a failure of an obsolete part which has been so listed on the declaration form prior to the test. An obsolete part is defined as a part used in a test configuration for reasons of expediency, but for which there already exists a Significant or Reliability Design Change. (See Sec. 14.4.1.)

Note: Reasons for all exclusions must appear on the declaration form prior to the test.

14.3.3 DECLARATION POLICY

In order to establish and protect the integrity of the data, it is necessary to determine the intentions of a test and, by applying the ground rules for applicability, to establish the utility of the test results for reliability purposes. This is done by means of the *declaration form* which must be completed *prior to each full-scale test*. It is then submitted to the program manager for approval. The form as shown in Fig. 14.1 is self-explanatory, but it is appropriate to note that when exclusions are declared, the reasons should be given, together with any substantiating information and/or

RELIABILITY EVALUATION OF LARGE SOLID ROCKET ENGINES

DECLARATION FORM

A. Engine Serial No. _FWE-23_ Engine Type _XM-83_ STAGE II Test Stand _G-8_
Test No. _3.5FS-12_ Date of Test _4 JAN 196-_ Model Spec. Ref _SPR-4003_

B. Which of the following Principal Subsystems are applicable for reliability evaluation?

	Applicable	Not Applicable
(1) Propellant Charge-Ignition (A)	X	
(2) Case-Liner-Internal Insulation (B)		X
(3) Thrust Vector Control (C)	X	

C. If any of the Principal Subsystems are not applicable state reasons. _NON FLIGHT TYPE INTERNAL INSULATION_

D. Which parts, components, etc., are omitted from the test configuration and state reasons: _EXTERNAL INSULATION FROM NOZZLES_

E. Which performance parameters should be excluded from reliability evaluation? (Development firings only) State reasons: _____

F. Which components or parts are declared experimental or obsolete (for provisional exclusion)? _1) NOZZLE THROAT INSERTS DECLARED EXPERIMENTAL (NEW PART NO. 3-157192; PREVIOUS PART NO. 3-157087) THIS IS FIRST TEST TO DETERMINE WHETHER USE OF — ALLOY WILL RESULT IN LESS EROSION. 2) SCRAPER RING NOZZLES #1+2 DECLARED OBSOLETE (PREVIOUS PART NO. 3-156863 IS BEING USED INSTEAD OF CURRENT PART NO. 3-156921) PART NO. 3-156921 WAS INTRODUCED AS A RELIABILITY DESIGN CHANGE IN TEST NO. 3.5FS-10. USE OF OLDER PART MAY RESULT IN EXCESSIVE TORQUE OR STALLING OF NOZZLES. OLDER PART USED DUE TO LACK OF AVAILABILITY OF NEWER PARTS AT TIME OF ASSEMBLY._

G. Comments: _THIS IS SIXTH DEVELOPMENT TEST IN WHICH HLK PROPELLANT HAS BEEN USED._

H. Nozzle Program: Number of Cycles, angles of deflection, and period of operation _SIX CYCLES AT ±3° INITIATED AT 5, 15,...65 SECONDS ALTERNATING WITH SIX CYCLES AT ±6° INITIATED AT 10, 20,...60 SECONDS FROM 0 TIME_

Signatures and Approvals:
Program Management _____ Test Engineer _J. Doe 1/4/6-_
Date _____ Reliability _____

Fig. 14.1 Sample declaration form as submitted to Reliability just before test-firing.

references to failure reports, inspection reports, engineering change orders, etc. For example, if a part is provisionally excluded by reason of obsolescence, the engineering change order number or the new part (drawing) number, etc., should be given on the form. The program manager might establish a limit to the number of provisional exclusions.

14.3.4 TEST RESULT CLASSIFICATION

Each principal subsystem will be classified as having succeeded or failed depending on whether its performance in operational use would have resulted in a successful or failing flight. The exception to this would occur when for causes external to the Subsystem, the Subsystem was not given the opportunity to succeed or fail. The detailed ground rules are given in Table 14.3.

Table 14.3. Ground Rules for Classification of Test Results

	Propellant Charge-Ignition Subsystem (A)
Success	(1) Subsystem A does not fail and is not excluded, and performance is within the current model specification limits. (2) It operates without failure with performance outside current model specification limits, provided that this intention is so stated on the declaration form, and approved, prior to test. (See limitation Table 14.2, Item (3).)
Failure	(1) The igniter fails to operate or fails to ignite the propellant. (2) The ignition delay is greater than the maximum value specified by the current model specification limits. (3) The ignition peak pressure/thrust is greater than the equilibrium chamber pressure/thrust. (4) Rough combustion; i.e., chamber pressure/thrust peaks $\geq 10\%$ of equilibrium chamber pressure/thrust. (5) Engine blow-up attributable to propellant or igniter. (6) The performance values lie outside the current model specification limits when the performances were declared to be within the current model specification limits.
Post-test exclusion	(1) Tests in which failure occurs owing to causes external to Subsystem A, e.g., other Principal Subsystem failures, test operator error, instrumentation or facility malfunction, provided that failure of Subsystem A has not already occurred. (2) Failure of the Subsystem owing to the failure of an obsolete or experimental part so listed prior to the test on the declaration form. (3) Tests in which Subsystem A did not fail but did not have the opportunity to satisfy the declared intention of the test.

RELIABILITY EVALUATION OF LARGE SOLID ROCKET ENGINES

	Case-Liner-Internal Insulation Subsystem (B)
Success	(1) Subsystem B does not fail and is not excluded.
Failure	(1) Case or end-closure burn-through during normal operation.
Post-test exclusion	(1) Tests which fail owing to causes external to Subsystem B, e.g., other Principal Subsystem failures, test operator error, instrumentation or facility malfunction, provided that failure of Subsystem B has not already occurred. (2) Failure of Subsystem B owing to the failure of an obsolete or experimental part so listed prior to the test on the declaration form. (3) Tests in which Subsystem B did not fail but did not have the opportunity to satisfy the declared intention of the test.

	Thrust Vector Control Sul system (C)
Success	(1) Subsystem C does not fail, is not excluded, and performs as programmed for the duration of the test.
Failure	(1) Movable nozzles stick, jam, or do not deflect as programmed. (2) Gas leakage or burn-through in any part of the movable nozzle assembly occurs. (3) Actuators malfunction.
Post-test exclusion	(1) Tests which fail owing to causes external to Subsystem C, e.g., other Principal Subsystem failures, test operator error, instrumentation or facility malfunction, provided that failure of Subsystem C has not already occurred. (2) Failure of Subsystem C owing to the failure of an obsolete or experimental part so listed prior to the test on the declaration form. (3) Tests in which Subsystem C did not fail but did not have the opportunity to satisfy the declared intention of the test.

Notes: It may be required that the engine contractor be held individually responsible for attaining a numerical reliability requirement. To prevent controversy of a legal nature the contractor's numerical reliability requirement should not encompass: (1) equipment developed by other *associate contractors** which is tested in conjunction with the engine contractor's system; nor (2) interface attachments, the function of which is a joint responsibility between two or more associate contractors. When, however, equipment is furnished by *vendor* or *subcontractor* to the engine contractor, and does not fall in category (2), the equipment is considered to be the engine contractor's responsibility with respect to meeting any numerical reliability requirements.

* An associate contractor is one of several contractors each of whom develops a separate major subsystem. However, no one associate contractor has complete responsibility for the entire weapon system.

14.4 RELIABILITY REPORTING AND ESTIMATION

The *reliability report form* (Fig. 14.2 is a sample reliability report form) is completed as soon as possible after the test data are reduced. The data are then evaluated by the contractor's reliability group on the basis of the ground rules and the information contained on the declaration form. This is done for each Principal Subsystem after each full-scale test. A short description of the failure, or the reason for exclusion when any Principal Subsystem is so classified, should appear in the remarks column. When a failure occurs which cannot be assigned to any particular subsystem, this fact should be noted as well. References to failure reports and to corrective action reports, etc., should also be given as necessary. The reliability report forms covering a particular calendar month can then be gathered in chronological order and summarized as illustrated in the example to follow. The point estimate and $\gamma\%$ lower confidence limit of reliability can then be obtained by the methods given in Sec. 14.4.2 and reported monthly.

14.4.1 REPRESENTATIVE AND CURRENT DATA

Because of the low number of tests available for evaluation, each Principal Subsystem must necessarily display a high reliability, and few failures must, therefore, occur. However, at the beginning of the program several failures may happen, thus establishing such a high cumulative failure rate that, where there are no provisions for discarding data, an exorbitantly large number of successful tests would be needed before the earlier failure rate would be "absorbed." This is not feasible, and in a development program which is expected to improve the product, it is not realistic to handle the data this way. If these failures were random or failures due to unassignable causes, then it would not be legitimate to discard the earlier data, as this failure pattern would be the manifestation of the inherent reliability and, as such, would indicate that the Subsystem was not sufficiently reliable. However, the earlier failures are not generally random; i.e., they do have assignable causes, and in a development program are subject to analysis and corrective action.

This corrective action is called a *Reliability Design Change*, which is defined as a modification to correct a previously observed failure in the full-scale engine, and must be intended to appear on all subsequent engines. Data generated after the Reliability Design Change will be regarded as homogeneous, and earlier data discarded as being no longer representative of the current design. Data produced after the Reliability Design Change will be called *Current Data*. The decision as to what constitutes a Reliability Design Change is subject to the approval of Program Management and relates only to that particular Subsystem for which corrective action has been taken. In addition, during the development program there may occur

RELIABILITY REPORT FORM

Date of Test __4 JAN 196-__ Test No. __3.5 FS-12__ Test Stand __G-8__ Test Report No. __36__

Engine Type __STAGE II XM83__ Engine Serial No. __FWE-23__ Model Spec. No. __SPR-4003__

Engine Test Data

Total Impulse, lbf-sec _____	Ignition Delay, sec _____
Average Thrust, lbf _____	Ignition Peak Thrust, lbf _____
Specific Impulse, lbf sec/lbm _____	Useful Propellant Weight, lbm _____
Max-Min (Thrust Time Curve), lbf _____	Engine Weight, lbm _____
Rate of Thrust Rise, lbf/sec _____	Propellant Mass Fraction _____
Rate of Thrust Decline, lbf/sec _____	Firing Temperature, F _____
Action Time, sec _____	Other Environments _____ Rel Humidity 60%
Average Chamber Pressure, psia _____	ENGINE Temp cycled twice between --

Test Result Classification
(to be completed by reliability group)

| Principal Subsystem | Exclusion | | Failure | | Success | | Remarks |
	Pre-test	Post-test	Perf Excl	Perf Not Excl	Perf Excl	No Perf Fail	
A			X				Ignition delay beyond model spec. limits.
B	X						See declaration form (Fig. 14.1)
C						X	Erosion <2%. Torque values acceptable.

Additional Remarks: _____

Responsible Engineer _____ Department and Group _____ Date _____

Fig. 14.2 Sample reliability report form.

design changes which are not necessarily for reliability purposes; e.g., for weight and performance improvements. If these changes result in a *Significant Design Change*, according to mutual agreement between Program Management and the contractor, then data produced after the Significant Design Change will also be considered Current Data.

14.4.2 ENGINE RELIABILITY ESTIMATION

14.4.2.1. Technique of estimation. Engine reliability will be computed from the Principal Subsystem test results. After each test it will generally be possible to classify each of the Principal Subsystems as having succeeded, failed, or been excluded. The exception to this classification can occur when there is a failure but it is not known which Principal Subsystem(s) failed. While it is desirable to be able to classify completely each test, an unassignable failure still represents a system failure and must be incorporated into any system reliability estimate. In the event of this occurrence an unassignable failure will be arbitrarily assigned to that Principal Subsystem experiencing the most failures in the sample from which the system estimate is made. If later analysis permits unassignable failures to be assigned correctly, the estimates will then be recomputed.

All engine reliability estimates (as described in the example below) will be based on Current Data only. Thus, when counting applicable tests, the count should not be extended any further back than the last Reliability Design Change or Significant Design Change for each Principal Subsystem.

The general procedure is as follows: The number of applicable tests on each Principal Subsystem are determined. The smallest of these three numbers, N, gives the number of *equivalent* engine tests performed. To compute the reliability for the engine it is necessary to count the number of known failures which occurred in the last N tests of each Subsystem. In addition, those failures which have not been assigned or attributed to any Principal Subsystem in particular are arbitrarily assigned to the individual Principal Subsystem as stated previously. Thus, the failure arrangement for a given number of equivalent engine tests is obtained and can be written $(N; f_1, f_2, f_3)$, where f_1, f_2, f_3 are the number of failures of the separate Subsystems; and $f_1 + f_2 + f_3$ is therefore the total number of assignable and unassignable failures during the period of N equivalent engine tests. The minimum reliability for (say) a 90 per cent confidence can then be obtained from Fig. A.10 (p. 565) for $(N; f_1, f_2, f_3)$.

14.4.2.2. An example of estimation. a. An example will illustrate the method of engine reliability estimation. The method of tabulating the results as shown in Table 14.4 below will be found convenient. The table is simply a summary, listed in chronological order, of the test result classifications obtained from the reliability report forms. During each test all

Principal Subsystems are physically present but may or may not be applicable for reliability evaluation according to the ground rules (*cf.* Table 14.1). The result of the test for each of A, B, and C will be success (S), failure (F), or exclusion (which is indicated by a blank in the table).

Table 14.4

Test No. (Chronological)	Reliability Report No.	Date of Test	Results		
			A	B	C
1			S		
2				F	
3			S		S*
4			F	S*	
5			S*	S*	F*
6			S*		
7			S*	S*	S*
8			S*	S*	F*
9			S*	S*	S*

Subsystem	Total no. of times tested	Total failures	Failures in last 5 tests*
A	8	1	0
B	6	1	0
C	5	2	2

Test No.	Explanation of tabulation
1	Subsystem A was applicable and a success. B, C were excluded.
2	Subsystem B was applicable and a failure. A, C were excluded.
3	Subsystems A and C were applicable and successes. B was excluded.
4	Subsystems A and B were both applicable. A failed, B was successful. C was excluded.
5	A, B were both successful; C a failure.
6	A was a success. B and C were both excluded.
7	All Subsystems successful.
8	All Subsystems applicable but a failure occurred in C.
9	A, B, and C were all successful.

* See 14.4.2.2b.

b. In the above example, the Principal Subsystem experiencing the fewest number of applicable, non-excluded tests is C with 5; thus, a maximum of the equivalent of five complete engines have been tested. Counting back from Test No. 9 for the last 5 tests of each Principal Subsystem, it is found that A was present and applicable in Tests No. 9, 8, 7, 6, and 5; Subsystem B in Tests No. 9, 8, 7, 5, and 4; Subsystem C in Tests No. 9, 8, 7, 5, and 3. Thus, only the results of the aforementioned tests for each Subsystem should be utilized in obtaining the engine reliability estimate; i.e., only those results marked by an asterisk in the table. During this period of testing the five equivalent complete engines, A has no failures. B has no failures. C has two failures (Tests no. 5 and 8).

c. Thus, in the notation of Sec. 14.4.2.1, the reliability information can be written (5; 0, 0, 2). This indicates that there was the equivalent of a maximum of 5 complete engines tested, and no failures occurred in the last five tests of Principal Subsystems A and B. However, 2 failures occurred in Principal Subsystem C.

The 90 per cent confidence limit for reliability, based on the number above, is 24.7 per cent and is obtained from Fig. A.10.

The point estimate of reliability is obtained directly from the product rule and, hence, is

$$(5 - 0)(5 - 0)(5 - 2)/5^3 = 3/5 = 60\%$$

14.5 CONCLUSION

This chapter has shown how the reliability of a solid propellant rocket engine can be estimated from engineering development tests without restricting the exploratory nature of the tests. These estimates can be obtained even though the configuration and/or hardware being evaluated has not reached its final design.

The concepts used and the approach taken to solve the particular problem discussed in this chapter may be generalized to apply and to be utilized for systems in which the following situations occur.

1. It is desired to evaluate the reliability of the system against its end use.
2. The system is in a state of continuous development towards an end configuration.
3. The intentions of the tests vary as the system evolves and there are no or few tests specifically for reliability evaluation.
4. There is a limited number of systems available for testing and/or there is a limited number of times (perhaps only once) that a system can be tested.
5. The reliability is estimated by attribute (Success-Failure) data only.

CHAPTER FIFTEEN

A RELIABILITY EVALUATION PROGRAM FOR A TURBO-GENERATOR DEVICE

15.1 OBJECTIVES AND PROBLEMS

Previous experience had shown that the turbo-generator equipment under consideration had a fairly high value of reliability. However, it was in competition with a different type of equipment, electrochemical, which had a known, even higher, reliability but which was not completely satisfactory on account of weight. It was decided that if, with a certain limited amount of development work, the turbo-generator system could be made to be more reliable and to demonstrate a certain stated value of reliability, it would be chosen to replace the existing equipment.

To accomplish this objective the following problems had to be considered:

1. Obtaining valid reliability data during a program when the objectives of the tests varied and the configuration and design of the device changed slightly.
2. Demonstrating at the earliest time possible that the reliability requirements had been satisfied and, in addition, establishing a statistical monitoring method compatible with the demonstration method.

15.2 TEST CONDITIONS AND GROUND RULES

The configuration being well defined, it was considered that all development tests could be used for reliability evaluation, with the exception of certain tests conducted to explore the effect of peripheral and extreme environments on performance. In order to control the integrity of the data, a declaration form was initiated prior to each test. This form stated whether the test was applicable for reliability purposes. In addition, it contained the basic information regarding the test and the device such as date, equipment serial number, conditions of test, including use of experimental hardware, procedures, and environmental conditions.

The ground rule for success was "the ability to perform within specification limits for the time and under the conditions of intended customer use." Failure was defined as the inability to perform successfully. The only exceptions were for those tests which were stated to be inapplicable on the declaration form and those which failed (or were not given the opportunity to succeed) for some reason *external* to the device. Test results of this nature were classified as *exclusions* and were not incorporated into the reliability estimate.

The test conditions except when otherwise noted on the declaration form were considered to be sufficiently representative of customer use. This also applied to the operating duration of each test, which was always intended to be equal to the length of actual operation. Previous experience in the earlier tests indicated that no simple statistical time-to-failure distribution described the failure pattern. It was also known that each device could receive a large number of tests before any wearout failures appeared. Therefore, it was considered that each test could be regarded as being independent of any other, provided that the equipment had not accumulated too many tests. Consequently, the ordinary binomial sampling methods could be used for reliability evaluation (*cf.* Sec. 10.5). No equipment which was used in the reliability evaluation program was permitted to have more than a certain number of tests. Those equipments which had accumulated more than this number and were still being tested were automatically excluded from reliability evaluation or demonstration, although they served to provide engineering and time-to-wearout information.

15.3 STATISTICAL CONSIDERATIONS

The character of the testing was such that the test results were generated in sequence so that a sequential sampling plan was indicated. As the reliability requirement was very high, it was expected that it would take a year's testing (about 2500 tests) to both improve the reliability from its existing level and also demonstrate the requirement.

It was also considered necessary to modify the usual sequential plan (*cf.* Sec. 10.4)—for two reasons: (1) it was expected that only improvements in the reliability would occur and should be sampled for, and (2) since the population was not homogeneous, a method of eliminating the earlier and higher failure rate data was required. These two problems were resolved, respectively, (1) by making the usual two-sided sequential sampling plan into a one-sided plan, and (2) by introducing a test of significance into the procedure to determine the probability of a difference in failure rate between two periods of data. It is worthwhile to explore both these points in greater detail.

The sequential sampling plan has the advantage not only in being appropriate to the manner in which the data were generated but also in needing fewer tests, on the average, to demonstrate a specified reliability with given confidence than does a fixed-sample size sampling plan.

A sequential plan usually incorporates both acceptance and rejection lines; however, in this situation a rejection line was not pertinent. There was no possibility of rejection since if the equipment were not able to pass the acceptance boundary, development activity would be intensified, as no third alternative device existed which could possibly be made competitive in the time available. In addition it was desirable to obtain considerable data on the device, since there was a strong possibility that it could be used on a future space system where weight-saving considerations would be of utmost importance. The principal risk on the existing weapon system was that of accepting a device less reliable than was required, i.e., the probability, β, of accepting a device with reliability equal to or less than the minimum specified should be low, say, 5%. Consequently, a sequential binomial plan utilizing only the acceptance boundary was used. The mathematical aspects of this plan are discussed in Appendix 15A.

Another important consideration was that the method of periodic interim reliability evaluation reporting should be consistent with the demonstration of the end requirements. This was resolved by generating a "fan" of lines in exactly the same manner as the acceptance line was established. The details of the method of computing the lines are described in Appendix 15A. The manner in which this fan is used for periodic reporting is described in Sec. 15.4.3. Any reliability number evaluated by this procedure would be regarded as an "index," having more meaning as a comparison against the end sampling requirements than as an absolute estimate.

The second consideration—that of testing for a change in the failure rate—is also very important. Normally, when testing statistically for an improvement or a change in the output of an item, we would take two or more sets of homogeneous data and compare means using some standard statistical tests such as the t-test or the F-test (assuming normal distributions). When we do this, we are actually testing the difference between items as represented by their data, whereas in our present situation we are testing between groups of data which represent accumulation of changes in the item over periods of time. From a strict statistical viewpoint we should test the design for each change before pooling data; however, it is impossible to differentiate between individual changes with the limited data on any one design, and so we state the null hypothesis that *it is the reliability of the item, not the item itself, which remains unchanged* until the data prove otherwise. In a sense we demonstrate a confidence in the design engineer such that, although minor design changes are made, because of his experience, no reduction in reliability is expected. This attitude is almost forced upon us because it can be argued that in a complex system each test or system tested is different from any other, depending on how strict or loose the criteria for a difference are. However, if there is a major change in the item, then the reliability of the essentially new item should be demonstrated separately. The difference between a major and minor change is often a matter of judgment, although, if a change, which is initially regarded as minor, has a significant effect on the reliability, this will be discovered by the statistical techniques described in Sec. 15.4.1. The data generated prior to the point of a statistically significant change in reliability can be regarded as being no longer representative of the current configuration. All data after the significant change point are regarded as "current data."

This step is necessary since the difficulty with using the usual sequential analysis in a development program is that the lower reliability exhibited by the results of earlier tests would indicate that the equipment was unsatisfactory; i.e., it would appear very unlikely to meet the acceptance criteria. A development program is intended to improve inadequate equipment, but the expected and more numerous earlier failures would build up such a high failure rate that unless there were some provision for discarding

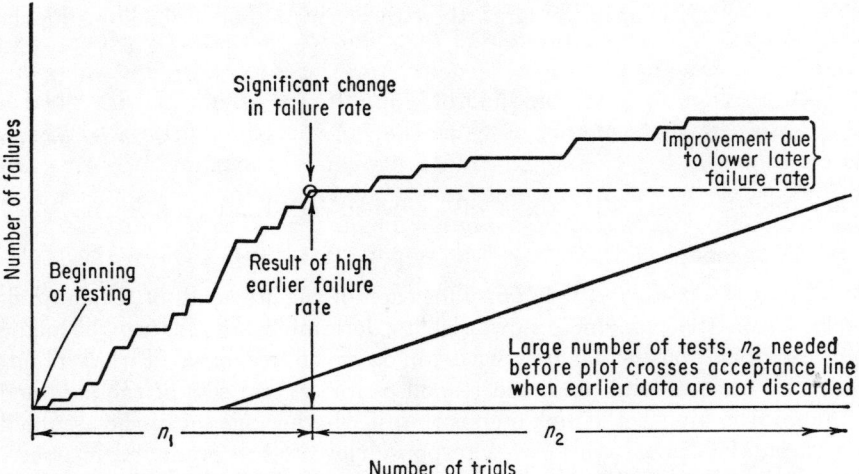

Fig. 15.1 The consequences of not discarding non-representative data. To compensate for the low reliability of items tested before n_1 either (a) a product of much higher than required reliability has to be produced in the next n_2 trials or (b), far more trials of a product with satisfactory reliability are required to intercept the acceptance line than are really necessary. Thus, in both cases, it is costly to use non-representative data.

the earlier data we would not be able to absorb the earlier failure rate (Fig. 15.1), which, as the development program progresses, would become less and less representative.

The technique of defining "current data" by means of the "reliability design change" (*cf.* Sec. 14.4.1) was not considered appropriate, since, when a very high reliability level already exists, the modifications can only be minor design changes. Also, the number of tests available is not restricted so that the data can be allowed to "speak for themselves" and we can search for a significant change in the failure rate as shown by the data.

15.4 STATISTICAL ANALYSIS OF THE DATA*

As the test results were received they were plotted chronologically in the usual manner of sequential analysis, as failures versus trials. Typical plots are shown in Figs. 15.1 through 15.5. The data are reviewed on a

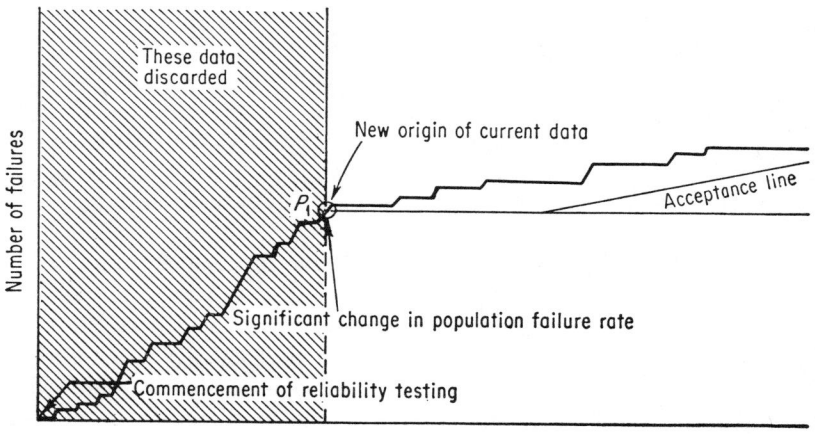

Fig. 15.2 Changing failure rates and discarding data.

monthly basis. They were first examined to see whether all or part of the month's data were homogeneous compared with existing current data. If so, they were then pooled and added to the existing current data. On the other hand, if a significant change in the failure rate was found, the exact trial at which the change statistically occurred was established and this was considered as the new origin of the current data (Fig. 15.2). The methods for finding the new origin are now described.

* The reliability numbers used in this section are not high in the sense implied in the preceding sections of this chapter. They were purposely chosen of the stated magnitude for ease of graphical and computational illustration.

15.4.1 THE STATISTICAL TEST FOR A SIGNIFICANT CHANGE*

Suppose we have a plot of data as shown in Fig. 15.3. Point P appears by inspection to be where the failure rate changes. We now wish to test for a significant difference between the respective true failure rates in the

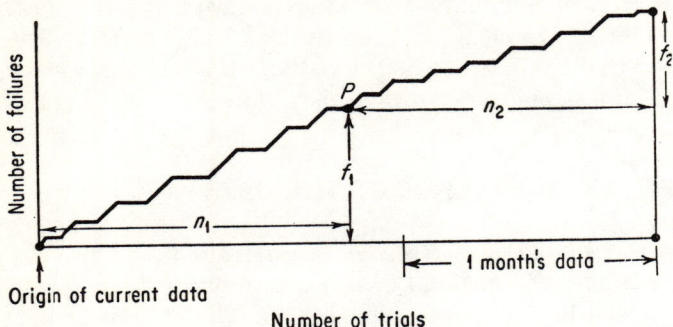

Fig. 15.3 Notation for testing for a change in failure rate.

populations formed by the division point P. Let there be f_1 failures in the n_1 trials between the origin of the current data and P, and f_2 failures in the n_2 trials between the point P and the end of the data. If $h_1 = f_1/n_1$ and $h_2 = f_2/n_2$ (relabeling, if necessary, h_1, h_2 so that $h_1 > h_2$) the test is made by computing the probability $P(h_1, h_2)$ of a difference as large as, or larger than $|h_1 - h_2|$; i.e.,

$$P(h_1, h_2) = 2 \sum_{r=f_1}^{n_1} \binom{n_1}{r}\binom{n_2}{f_1 + f_2 - r} \bigg/ \binom{n_1 + n_2}{f_1 + f_2}$$

If this value is less than $\alpha = 5$ per cent or 1 per cent or any other chosen level of significance, then we "know" there is a significant change between the failure rates of the earlier and later populations. Evidently, there may be several points P, each of which divides the data into two groups representing populations with significantly differing failure rates. We now arbitrarily define the origin O of the new current data as the most recent point P (i.e., the farthest to the right in Fig. 15.3) such that $P(h_1, h_2) < \alpha$.

15.4.1.1. An example. P_1 was first investigated to see whether it represented a change point. At this point the two observed failure rates between O and N are $\frac{43}{92}$ and $\frac{13}{59}$. The probability of getting this or a worse split is 0.00335, which is highly significant (working at a 1% α risk.) This is obviously not the latest point, so P_3 is investigated. The observed failure rates are $\frac{43}{98}$ and $\frac{13}{53}$. The respective probability is 0.02815, which is not significant at the 1 per cent level. The latest change, therefore, lies within

---

* See Appendix 15B for details of computation.

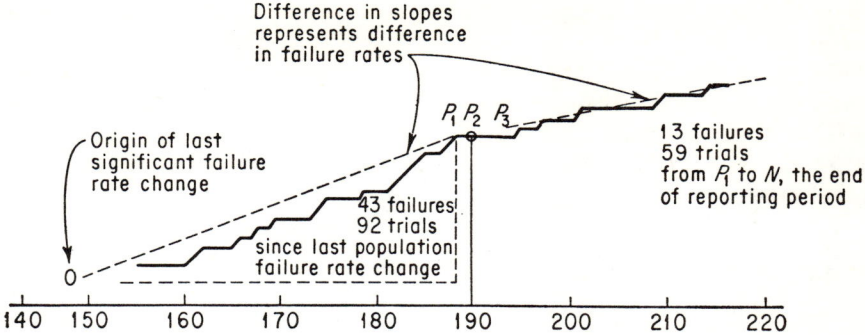

Fig. 15.4 Testing to see if P_1 represents a significant change point.

P_1 and P_3. (It is obvious from Fig. 15.4 that significant change points would not lie to the right of P_3, since the observed failure rates would become more nearly equal.) The probability at P_2 (i.e., $\frac{43}{95}$, $\frac{13}{56}$) is 0.01013 which is not quite significant; and by calculation it was found that the latest change in population failure rates is at trial 190, one trial before P_2. Had no change been discovered, the data would have been regarded as being homogeneous. As the data for each period are added, the aforementioned test is made to determine the latest change. It frequently occurs that a trend is indicated in one period but insufficient data are available at that time, and it is not until the following period that the change can be substantiated. It is possible, but not likely, that two or more real changes can be discovered in one period: in these circumstances, the change in failure rate would be very great.

A note of caution must be inserted here. The test for a change in the failure rate should be performed a limited number of times. If it were to be performed, say, after each test, the significance level of the statistical test would be compromised by the repeated sampling. However, if we apply our significance test at infrequent intervals, i.e., a limited number of times, then the significance level should not be unduly compromised. The exact statistical consequences of this procedure are not known; however, it is considered that the merits of being able to determine a unique point for the origin outweigh the slight statistical compromise.

15.4.2 RELIABILITY DEMONSTRATION

Having discovered which part of the data are current and homogeneous, a criterion is needed to decide when they indicate sufficient reliability. The one-sided sequential acceptance plan fulfills this requirement. The origin of the sequential plan is placed at the origin of the current data, and when the plot of the data crosses the acceptance line the device is said to have demonstrated the numerical reliability requirements. There may be some additional requirements—environmental tests, etc.—which are required in sufficient proportions in the current data. If these additional requirements have not been met, then testing is continued under environmental conditions until the proportions satisfy the required proportions. Equations (15A.2) and (15A.3) determine the acceptance line from the reliability requirements.

15.4.3 RELIABILITY REPORTING

Reliability reporting is accomplished by use of a measure exactly analogous to the acceptance demonstration. A convenient method is to construct a transparent template which has a *fan* of lines drawn on it, determined by Eqs. (15A.1) and (15A.2) corresponding to an intermediate sequence of reliability numbers and the same confidence level as the one used for the acceptance line. The origin of the transparency is placed at the origin of the latest current data; the fan line corresponding to the highest reliability number attained by the data trace (interpolating between the fan lines as necessary) is the reported reliability. Thus, in Fig. 15.5, the reliability would be reported as 76% with 90% confidence. We do not report any lower reliability unless there is a change in population. The reason for this is to reconcile the underlying philosophy of the se-

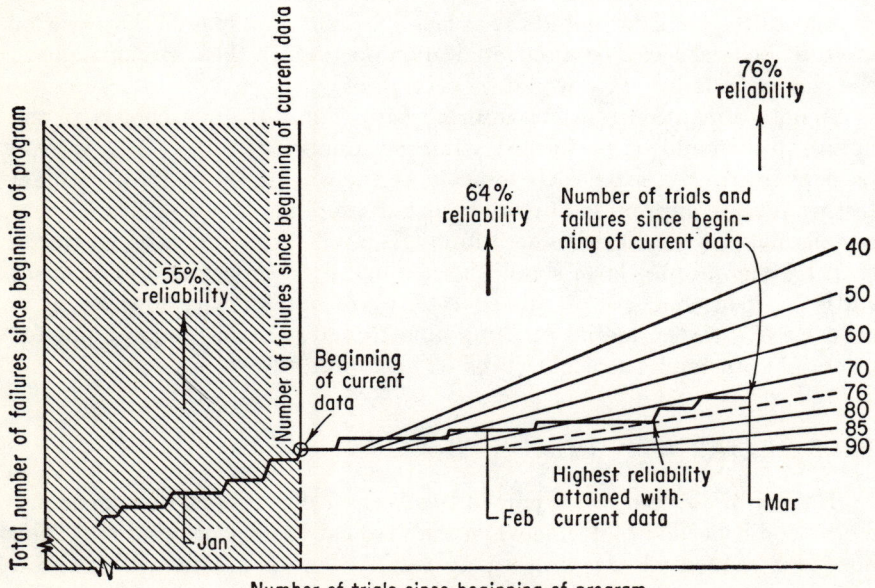

Fig. 15.5 Using the reliability reporting fan.

quential test with the present procedure, since in the standard sequential test we would have stopped sampling when the 76 per cent reliability line had been crossed, had 76 per cent been the final required reliability. Therefore, at this test the product would have been accepted. Generally, the reliability shows an increase so that the trace tends to curve towards the abscissa axis. This fan is used every reporting period.

15.5 CONCLUSIONS AND COMMENTS

This chapter has illustrated how existing statistical techniques can be modified, if necessary, and applied to provide reliability measurements suitable for particular problems. Although a sequential binomial sampling plan was used in the given example, the sequential plan could just as well have been based on an exponential (*cf.* Sec. 10.9) or normal distribution,* etc., provided that there was good evidence to indicate that one of these distributions described the failure pattern of the equipment. For the case in which the exponential distribution of time-to-failure applies (as is often assumed for electronic parts) the acceptance boundaries and the fan lines could be generated by a method analogous to that shown in Appendix 15A. The ordinate and abscissa would be, correspondingly, *total life* and *failures* (*cf.* Sec. 10.9). However, the details are not presented here.

REFERENCES

1. J. C. P. Miller, *Table of Binomial Coefficients*, Cambridge University Press (1954).
2. A. Hald, *Statistical Tables and Formulas*, John Wiley & Sons, Inc., New York, 1952.
3. A. Hald, *Statistical Theory with Engineering Applications*, John Wiley & Sons, Inc., New York, 1952.
4. *Tables of the Cumulative Binomial Probability Distribution*, Harvard University Press, Cambridge, Mass., 1953.

ADDITIONAL READING

Fairfield, J. H., "A Rapid Method of Comparing Two Percentages," *Industrial Quality Control*, **16**:5, 20–21 (November 1959).

Page, E. S., "On Problems in Which a Change in a Parameter Occurs at an Unknown Point," *Biometrika*, **44**, Parts I and II, 248–52 (June 1957).

Ruther, F. J., "Reliability Via the Component Part," *Military Electronics*, **5**:2, 12–15 (August 1958).

---

* Sequential sampling plans based on the normal distribution are described in A. Wald, *Sequential Analysis*, John Wiley & Sons, Inc., New York (1947), Chapters 7–9.

APPENDIX 15A
ONE-SIDED SAMPLING ACCEPTANCE AND REPORTING LINES FOR RELIABILITY ESTIMATION

The usual sequential binomial sampling plan contains both acceptance and rejection lines (cf. Sec. 10.4). Since our purpose is to avoid a decision to reject, we now wish to construct a plan with no rejection boundary; i.e., only the acceptance boundary, as shown in Figure 15A.1. This can be accomplished by making use of the equations in Sec. 10.4, in which the "producers risk" α is set equal to zero. The details are not given here, but may be seen in the noted reference.*

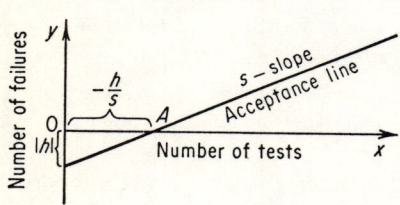

Figure 15A.1

The properties of the resulting plan are as follows. The OC function $L(p)$ becomes equal to unity for values of $p \leq s$; for $p > s$, $L(p)$ is related to p by Eq. (15A.1):

$$p = \frac{(L(p))^{-s/h} - 1}{(L(p))^{-1/h} - 1} \tag{15A.1}$$

where h is the intercept of the acceptance line on the failures axis, and s is the slope of the acceptance line. Thus, the probability of reaching a decision (of acceptance only) is unity when $p \leq s$. However, there is a positive probability of not reaching a decision at all when $p > s$.

One relation to determine the two constants s and h may be obtained from the acceptance specification, namely, to demonstrate reliability

---

* Statistical Research Group, Columbia University, Report No. SRG 255, *Sequential Analysis of Statistical Data: Applications*, Columbia University Press, New York, (1945), Secs. B.0933, B.0935.

$\hat{R}_L \equiv 1 - p^*$ with confidence $\gamma \equiv 1 - \beta$. Thus, from Eq. (15A.1)

$$p^* = \frac{\beta^{-s/h} - 1}{\beta^{-1/h} - 1} \qquad (15A.2)$$

For example, the acceptance criteria might be to demonstrate 98 per cent reliability with 95 per cent confidence, in which case $p^* = 0.02$ and $\beta = 0.05$.

Another relation between the constants s and h is needed in order to determine them uniquely. This can be done in a variety of ways. The way in which it was actually accomplished in the reliability program described in Chapter 15 was to specify that when $p = p^*$, the customer should run a certain risk, $\Delta\beta$, that the device would be considered to meet the reliability requirements by an initial run of consecutively successful trials. Thus, the second relation would be

$$\Delta\beta = (1 - p^*)^{-h/s} \qquad (15A.3)$$

(since $-h/s$ is the number of trials to the point where a decision of acceptance can be first reached (*cf.* Fig. 15A.1)). It was arbitrarily decided that $\Delta\beta$ should be about $\frac{1}{4}$ of the total risk of acceptance when $p = p^*$; i.e., $\beta/4$.

The two equations (15A.2) and (15A.3), may now be solved to determine the acceptance boundary by setting $p^* = 0.02$, $\beta = 0.05$, and $\Delta\beta = 0.0125$. Furthermore, the fan lines may also be determined in the same way by choosing an intermediate sequence of values of p^*. Table 15A.1 gives the intercepts on the trials axis, $-h/s$, and the slopes, s, of the fan lines corresponding to various values of p^*.

Table 15A.1. Values of Constants to Determine Fan Lines

p^*, failure rate	0.40	0.30	0.20	0.15	0.10	0.05	0.02
Reliability (%)	60%	70%	80%	85%	90%	95%	98%
$-h/s$, trials intercept	8.58	12.29	19.64	26.96	41.59	85.43	216.90
s, slope	0.2615	0.1909	0.1241	0.0920	0.0607	0.0300	0.01192

APPENDIX 15B
METHOD OF COMPUTATION FOR CHANGE IN FAILURE RATE

The formula for computing the probability that an observed split of failures among two groups of trials indicates a significant change in population failure rates is given in Sec. 15.4.1. The formula, repeated here for convenience, is

$$P(h_1, h_2) = 2 \sum_{r=f_1}^{n_1} \binom{n_1}{r}\binom{n_2}{f_1 + f_2 - r} \bigg/ \binom{n_1 + n_2}{f_1 + f_2} \quad (15B.1)^*$$

The probability $P(h_1, h_2)$ may be computed directly by making use of tables of binomial coefficients (Ref. 1), or tables of logarithms of binomial coefficients (Ref. 2, Table XIV), or, if necessary, by using tables of logarithms of factorials (Ref. 2, Table XIII) which define the binomial coefficients in Eq. (15B.1). Methods of approximating $P(h_1, h_2)$ are also given in Ref. 3, pp. 705–711.

In this appendix, an example of the exact calculation of $P(h_1, h_2)$ is given that makes use of tables of the binomial distribution function (Ref. 4). To use these tables we first multiply and divide the terms in Eq. (15B.1) by appropriate powers of p and $1 - p$, to obtain

$$P(h_1, h_2) = 2 \frac{\sum_{r=f_1}^{n_1} \binom{n_1}{r} p^r (1-p)^{n_1-r} \binom{n_2}{f_1 + f_2 - r} p^{f_1+f_2-r}(1-p)^{n_2-(f_1+f_2-r)}}{\binom{n_1 + n_2}{f_1 + f_2} p^{f_1+f_2}(1-p)^{n_1+n_2-f_1-f_2}}$$

(15B.2)

* Since

$$\binom{n_2}{f_1 + f_2 - r} \equiv 0$$

when $r > f_1 + f_2$, the upper limit of the summation could be given as: **min.**$(n_1, f_1 + f_2)$.

Table 15B.1. Calculation of $P(h_1, h_2)$

$(h_1 = \tfrac{11}{24}, h_2 = \tfrac{14}{60}; p = \tfrac{25}{84} \simeq 0.30)$

(1)	(2)	(3)*	(4)	(5)*	(6)
		$\Delta_r(2) = \binom{24}{r}(0.30)^r(0.50)^{24-r}$		$-\Delta_r(4) = \binom{60}{25-r}(0.30)^{25-r}(0.70)^{60-(25-r)}$	
r	$B(24, 0.30; r)$		$B(60, 0.30; 26-r)$		$(3) \times (5)$
11	0.07424	0.04285	0.83789	0.06215	0.002663
12	0.03139	0.01989	0.90004	0.04319	0.000859
13	0.01150	0.00787	0.94323	0.02729	0.000215
14	0.00363	0.00265	0.97052	0.01560	0.000041
15	0.00098	0.00075	0.98612	0.00801	0.000006
16	0.00023		0.99413		
					$\Sigma = 0.003784$

$\binom{84}{25}(0.30)^{25}(0.70)^{59} = B(84, 0.30; 25) - B(84, 0.30; 26) = 0.55998 - 0.46524 = 0.09474$

$$P(h_1, h_2) = \frac{2 \times (0.003784)}{0.09474} = 0.0799 \simeq 8\%$$

(not significant at 5% level)

* $\Delta_r(j)$ = entry in row r, col. j, minus entry in row $r+1$, col. j.

The reason for doing this is that the Ref. 4 tables enable one to calculate terms of the form, e.g.,

$$\binom{n_2}{f_1 + f_2 - r} p^{f_1+f_2-r}(1-p)^{n_2-f_1-f_2+r}$$
$$= B(n_2, p; f_1 + f_2 - r) - B(n_2, p; f_1 + f_2 - r + 1)$$

by simply taking differences in the tables.* One must first select some fixed value of p in order to make use of the tables. It is generally best to use

$$p \simeq \frac{(f_1 + f_2)}{(n_1 + n_2)}$$

since this will ensure that the denominator in Eq. (15B.2) is nearly at a maximum and, hence, will have the maximum number of significant digits. The procedure is best illustrated by a numerical example, with the calculations as set-up in Table 15B.1.

* $B(n, p; x) \equiv \sum_{j=x}^{n} \binom{n}{j} p^j (1-p)^{n-j}$

CHAPTER SIXTEEN

A RELIABILITY EVALUATION PROGRAM FOR A LARGE LIQUID ROCKET ENGINE DURING ENGINEERING DEVELOPMENT

16.1 OBJECTIVES AND PROBLEMS

The purpose of this program is to measure the reliability of a large liquid propellant rocket engine as it evolves from a prototype configuration into an engine suitable for production for operational use.

The environment of the engine is severe, so that although much of the development testing can be performed with separate component tests, the major portion and final development work has to be performed utilizing complete engine system tests. This means that the system configuration is continuously changing and that many tests are upon systems which contain experimental hardware or employ non-standard experimental procedures. A further problem is that the particular fuel used is expensive, and test facilities have limited fuel tank capacities, so that a large portion of the firings are for less than the rocket engine's full-rated duration.

The main objectives are:

1. relating the engine's performance in an R&D program to its operational requirements
2. obtaining valid reliability data without restricting the exploratory nature of engineering testing
3. making valid reliability estimates based upon the results of tests some of which differ in duration from operational duration.

16.2 DISCUSSION OF PROBLEMS

The objectives listed under (1) and (2) are resolved in the same general manner as the similar problems for a solid propellant rocket engine described in Chap. 14. That is, the engine model specifications are used as the criteria in interpreting the weapon system requirements in terms of engine requirements. These limits are derived by "flying" the missile's systems on an electronic computer. Thus an operation of an engine is defined as a success only if it performs in a manner which would have resulted in a successful missile flight; if the operation would have resulted in a failed flight, i.e., a hit at a distance greater than a specified distance from the target, then the engine test is regarded as a failure unless it should be excluded. The detailed test-firing ground rules are given in Table 16.1.

16.2.1 RELIABILITY GROUND RULES

It was considered that most development test results could be used for reliability purposes. The exceptions would be for deliberate malfunction tests, tests-to-failure, safety limits tests, and the first checkout or calibration tests, etc., or tests intended for less than 10 seconds. In all cases such an intention would be stated prior to the test on a pretest declaration form. All tests falling into these categories would remain excluded no matter what the outcome of the test might be. An example of a declaration form is illustrated in Fig. 16.1. The 10-second minimum limit was chosen because it was considered that the engine would not otherwise have time to reach its steady-state operating conditions and consequently would not exemplify a condition representative of required operation.

There are further sets of reasons why certain tests should be excluded from any reliability evaluation. During a development program the system is undergoing continuous change and modification. The reason is that although the majority of component development can take place in separate component tests, the final work and demonstration of compatibility with the system must take place within the engine system test. Consequently, the system will often contain components which are *experimental*, and a failure of the system due only to the failure of the experimental component would not necessarily be a true failure of the system. Under these circumstances the test result should be excluded. On the other hand, to exclude the test completely because it contained an experimental component would represent a waste of valuable information. If the test were a success, then it would mean that success information on the remainder of the system was not being utilized. This situation is resolved by allowing post-test exclusion for failures on a component which had been declared experimental and so listed on the declaration form. The same argument also applies to *obsolete*, *worn-out*, or *damaged* components which might have been used in the system configuration for purposes of expediency.

It can be seen that these provisions can easily be abused unless some form of control is maintained. One control consists of the requirement that only components satisfying one of the above four definitions can be listed on the declaration form. A second control limits the number of

```
          DECLARATION FORM FOR RELIABILITY EVALUATION PROGRAM
DATE OF TEST: _____

ENGINE                        TEST                    TEST
  MODEL: _____        STAND: _____    NUMBER: _____
ENGINE SERIAL NUMBER: _____
1. Is test applicable for Reliability Evaluation: [___] Yes; [___] No, Why?
        [___] Calibration or First Checkout Test    [___] Item 2
        [___] Test-To-Failure                       [___] Item 4
        [___] Safety Limits                         [___] Exceed Limits of Item 5
        [___] Malfunction Tests                     [___] Other
2. Intended Duration: _____
3. Environmental Conditions of Test: _____
4. Intended Performance Outside Model Specification Performance: ____

5. Components with deviations from master list; i.e., experimental or obsolete
   as well as worn-out and damaged:
   a. Major: _____   Why? _____
   b. Minor: _____   Why? _____
   c. Minor: _____   Why? _____
   d. Minor: _____   Why? _____

      Signature and Department of Person Initiating Form _____

                 To Be Completed By the Reliability Group

   Result of Test:    Success   [___]       Actual Time of Test _____
                      Failure   [___]
                      Exclusion [___]
   Remarks: _____
   _____

         Signature of Responsible Person in Reliability: _____
                                              Date: _____
```

Figure 16.1

components provisionally excluded to one major and three minor components, as defined in the ground rules. Not all components falling into one of these provisional excluded categories need be declared; however, a component failure will exclude the test *only* if it has been listed. If it has not been listed and it fails, then the test is regarded as a system failure. This might infer that the test or development engineer must take his chances by being restricted to four particular components from many which might fail; however, in this situation he could exclude the whole test since the configuration is too experimental. However, the fact that the engine system contains more than one major and three minor excludable components does not necessarily imply that the test has to be excluded.

Table 16.1. Ground Rules for Reliability Program

Success

Generally a firing which would have resulted in a satisfactory missile flight with respect to the engine subsystem.

    a. The engine exhibits successful ignition, start, stable steady-state operation, and controlled shutdown within model specifications.

    b. The engine operation time equals the declared intended duration within a shutdown tolerance of $\pm 5\%$ where the intended duration must be at least 10 seconds. If the run duration exceeds the above tolerance owing to conditions external to the engine, the run may be classified as a success by special agreement with the program management, if all other requirements for the classification success are met.

Failure

An engine test firing shall generally be classified as a failure if the firing would have resulted in a failed flight unless the test firing can be classified as an exclusion.

    a. Performance parameters fail to meet the model specification requirements.

    b. An unscheduled shutdown of the engine occurs owing to a condition originating within the engine, i.e., in order to prevent damage.

Exclusion

An engine test firing can be classified as exclusion either before or after the test is conducted. Those test firings which are classified as exclusions before the test is conducted must remain so classified irrespective of the outcome of the test. All reasons used as a basis for classifying a test as an exclusion must appear on the pretest declaration form.

Pretest Exclusion

    a. If the test firing is to perform a malfunction test, test-to-failure, safety-limits test, first checkout or calibration test, or test intended for less than 10 seconds, then the test will be excluded.

    b. If the engine configuration contains more experimental, obsolete, worn-out or damaged components than can be provisionally excluded within the limits of one major component and three minor components, it may, at the discretion of the project engineer, be excluded as being nonrepresentative of the most reliable configuration, i.e., as defined by the component master list (Sec. 16.2.2).

Post-test Exclusion

    a. A test firing will be excluded if it fails solely because of the use of new test procedures not covered by the current model specifications. However, if the outcome of the test is such as to result in meeting the requirements for success, then it will be so classified.

    b. A test firing will be excluded if it fails solely because of the failure of an experimental, obsolete, worn-out, or damaged component which has been provisionally excluded. If the outcome of the test is such as to result in meeting the classification for success, it will be so classified. (The definitions for experimental, obsolete, worn-out, and damaged components are contained under the discussion of the master list (Sec. 16.2.2)). The number of components that can be listed on the pretest declaration form and used as a basis for a post-test exclusion is limited to one major component and three minor components, where a major component is defined as a thrust chamber assembly, turbopump assembly, or gas generator assembly. All other components are considered minor.

    c. A test firing will be excluded if the failure is due to conditions originating externally to the engine system—e.g., facility malfunction, instrumentation or test operator error—thereby preventing the objectives of the test from being fulfilled.

16.2.2 THE MASTER LIST AND EXCLUDABLE COMPONENTS

It is therefore necessary to define *experimental*. Perhaps this is not the best word, for in its popular meaning it might be understood to imply a design which is not the final design. If we use the word experimental in this sense, then probably the majority of the components in the engine will receive modifications before the final design is established, and this will not allow us to evaluate the reliability of the current engine since too much of it will be excluded or too many tests will be excluded. Therefore we define experimental by means of our third control: the *component master list*. This master list is made up at that point in time during the development program when the first engines are tested. It will include those engine components which in the judgment of the responsible component development engineers are the most reliable at that particular point in time. Only one design for each component can be selected for inclusion, and each component must be represented on the list. As the development program progresses, the engineer will have the option of replacing any design on the master list if he considers that the replacement will be more reliable. The master list must be kept current at all times. An experimental component is therefore defined as one which has not yet appeared on the master list; once it has appeared, it can no longer be excluded as an experimental part. A component which has appeared on the master list but has been replaced by a more reliable part is defined as an *obsolete* component. A *worn-out* component is one which has been used on an engine more than a specified number of firings or a specified operating time. A *damaged* component is one which has experienced excessive conditions, such as "rough" handling, dropping or severe, beyond-specification limits performance, and which might fail because of its suspect condition. Such a component would be used only in emergencies.

It is necessary to go to these lengths of definition, for when we know the current most reliable configuration we can evaluate its ability to perform against the end performance requirements.

16.2.3 SCOPE OF RELIABILITY EVALUATION

The tests which are used for reliability evaluation include those conducted at the engine contractor's test site for R&D purposes, acceptance test firings, and qualification test firings, as well as those conducted in the field as part of missile system tests, and flight tests. While it is apparent that the weapon system flight tests are the most realistic, they are relatively limited and furthermore if a flight failure occurs, it might not even be possible to assign the cause of failure to a specific system, such as, the propulsion system, due to lack of conclusive evidence. For instance, the missile would be frequently non-recoverable and camera or telemetry coverage of the event might not be complete. Also a flight test is a one-shot result; consequently, most of the engine test results come from R&D static tests and the more formal production acceptance and qualification tests. The R&D tests will be on systems containing some parts that are experimental, obsolete, and so on, whereas the more formal engine tests will be on engines composed almost exclusively of components on the master list. Except for the flight tests, the length of the test firings will not necessarily be full duration; however, there will probably be several firings from each engine. The variation in test durations is due to test facility limitation, expense, or the imposition of restrictions to prevent accumulation of too much time on the engine before its operational use.

16.3 STATISTICAL ANALYSIS OF DATA

It can be seen, therefore, that the data are quite mixed. This problem is partially solved by segregating the data and making comparisons between, say, flight results and acceptance test results. The declaration policy represents a further control. But a major problem remains: the different lengths of firing durations and the effect on failure rate. It is well known that the starting and shutdown of this type of equipment has a much more detrimental effect on the system than the simple accumulation of time; therefore we cannot say, for example, that starting an engine and running it for one minute is equivalent to one-third of the operation of starting the engine and running it for three minutes. But neither is it proper to equate a short-duration run to a full-duration run. Therefore, some *weighting factor* lying between zero and one must be sought which can be associated with the firings of less than full duration. This will allow us to talk in terms of failures per *equivalent full-duration* test. It can be seen that this represents an important consideration, otherwise severe bias would result in any reliability estimate. If we were to ignore the length of duration, that is, to give a short-duration firing equal weight with a full-duration run, then any reliability estimate would be biased on the high side of the actual value; alternately, if we established a weight equal to the ratio of the length of time of the test to the full-rated duration, then we would be underestimating the reliability.

It was decided that the test data themselves should furnish the weighting factor, w_i, where w_i is some value lying between 0 and 1 which would be associated with any test firing whose intended duration was T_i. The discussion of the statistical method of reliability estimation for this equipment can now be subdivided into two parts. The first part is the method of estimating the weighting factors. The second part is the question of how the weighting factors should be applied to the data to obtain valid, unbiased reliability estimates.

16.3.1 WEIGHTING FACTORS

The weighting factor w_i associated with any test firing of intended duration T_i is defined by

$$w_i = \frac{p_i + \epsilon(1 - p_i)}{p_k + \epsilon(1 - p_k)} \qquad (16.1)$$

where p_i is the probability of failure of the engine due to failures produced by starting and operating the engine for time T_i; p_k is the probability of failure for a full duration firing. The quantity ϵ is the conditional probability of shutdown failure, given that failure had not occurred prior to shutdown. Thus, in words, Eq. (16.1) states that the weighting factor w_i is the ratio of the probability of failure caused by start, operating, and shutdown failures in time T_i to the probability of failure caused by start, operating, and shutdown failures in time T_k, which is defined as full duration.* It is noted that shutdown failures are treated a little differently from start and operating failures—for two reasons. First, we wish to differentiate shutdown failures from the other "ordinary" failures which happen to occur at the time of shutdown, and second, shutdown failures are not necessarily given the opportunity to occur owing to the fact that an ordinary failure had already terminated the test.

In order to estimate w_i we need the following set of definitions:

T_i = the intended duration of firing of group i
N_i = the number of firings in group i
f_i = the number of shutdown failures in group i
f_{ij} = the number of ordinary failures in group i occurring within the time interval (T_{j-1}, T_j)

A hypothetical set of test results could then be arranged in accordance with the foregoing definitions into Table 16.2. For ease of computation a further set of definitions are established:

$H_i = N_i - f_i - \sum_{j=1}^{i} f_{ij}$ (i.e., the number of successes in group i)

$F_j = \sum_{r=j}^{k} f_{rj}$ (i.e., the number of ordinary failures in column j)

$G_i = \sum_{j=i}^{k} (F_j + H_j + f_j)$ \hfill (16.2)

---

* w_i can equivalently be defined as the conditional probability of failure at or before time T_i, given that failure occurs at or before T_k.

Table 16.2

Let $k = 6$, $T_k = 140$

Group i	Intended duration T_i	No. of firings N_i	No. failures occurring in time intervals (T_{j-1}, T_j)							H_i
			(0, 10)	(10, 20)	(20, 40)	(40, 60)	(60, 100)	(100, 140)	
1	10	162	f_{11} | f_1 — 20 | 2	...	...	...	...	...	140
2	20	27	f_{21} 1	f_{22} | f_2 — 0 | 0	...	...	...	...	26
3	40	26	f_{31} 0	f_{32} 0	f_{33} | f_3 — 0 | 1	...	...	...	25
4	60	1	f_{41} 0	f_{42} 0	f_{43} 0	f_{44} | f_4 — 0 | 0	...	...	1
5	100	10	f_{51} 0	f_{52} 0	f_{53} 0	f_{54} 0	f_{55} | f_5 — 0 | 0	...	10
6	140	126	f_{61} 12	f_{62} 2	f_{63} 5	f_{64} 2	f_{65} 4	f_{66} | f_6 — 1 | 3	97
		F_j	33	2	5	2	4	1	

470 RELIABILITY EVALUATION OF LARGE LIQUID ROCKET ENGINES

The maximum likelihood estimators $\hat{\epsilon}$ and $\hat{p}_i$ can now be computed by the formulas

$$\hat{\epsilon} = \frac{\sum_{r=1}^{k} f_r}{\sum_{r=1}^{k} H_r + \sum_{r=1}^{k} f_r}$$

$$\hat{p}_i = \hat{p}_{i-1} + \frac{F_i(1 - \hat{p}_{i-1})}{G_i} \qquad i = 1, 2, \cdots, k$$

$$\hat{p}_0 \equiv 0$$

(16.3)

These values are then substituted in Eq. (16.1) to provide $\hat{w}_i$, the estimator of w_i. The quantity $\hat{\epsilon}$, as defined in (16.3), is simply the ratio of the total of shutdown failures to the total number of firings which did not experience ordinary failure. The expressions in (16.3) are derived in Appendix 16A.

16.3.2 AN EXAMPLE OF COMPUTING WEIGHTING FACTORS

We can utilize the data in Table 16.2 as an example. Using the relationship expressed in (16.2) and (16.3), we can organize the computations as shown in Table 16.3. The weighting factors should be estimated from a

Table 16.3

	F_i	H_i	f_i	G_i	$\hat{p}_i$	$\hat{p}_i + \hat{\epsilon}(1 - \hat{p}_i)$	$\hat{w}_i$
$i = 1$	33	140	2	352	0.0938	0.1116	0.551
$i = 2$	2	26	0	177	0.1040	0.1216	0.600
$i = 3$	5	25	1	149	0.1341	0.1511	0.746
$i = 4$	2	1	0	118	0.1487	0.1654	0.817
$i = 5$	4	10	0	115	0.1784	0.1946	0.961
$i = 6$	1	97	3	101	0.1865	0.2025	1.000

$$\hat{\epsilon} = 0.01967$$

sufficiently large body of data so that enough failures are in the various cells to obtain "stable" values. Nine months' data might be sufficient, or a minimum of (say) 30 failures might be required. The appropriate requirements would depend on the failure rate of the equipment as well as the rate of testing. Normally the weighting factors will vary somewhat during any development program; therefore, they should be computed periodically. If monthly reliability reports are issued, then the weighting factors are computed immediately before applying them to the data. Once a weighting factor has been computed and associated with a test, it should remain associated with that test even though computation of weighting factors from later data indicates that the weighting factor for a given duration has changed by the later date. This has the advantage of convenience in both application and computation.

16.3.3 AN EXAMPLE OF COMPUTING RELIABILITY USING WEIGHTING FACTORS

The weighting factors have been defined and are estimated from a broad base of data for the purpose of applying them to a smaller, more current group of data (see Secs. 16.3.5, 16.4). The reason for the particular definition (16.1) may be seen by applying the estimated weighting factors *to the same group of data* from which they were obtained.

The method of application is to estimate the reliability R as

$$\hat{R} = 1 - \frac{\text{total number of failures}}{\sum_i N_i \hat{w}_i} \qquad (16.4)$$

For the data of Table 16.2 and making use of the computations given in Table 16.3, we would obtain

$$\hat{R} = 1 - (53/261.3) = 0.7972 \simeq 79.7\%$$

Note that, numerically, this value of $\hat{R}$ is extremely close to the quantity

$$1 - (\hat{p}_k + \hat{\epsilon}(1 - \hat{p}_k)) = 0.7975,$$

(where $k = 6$), which is by definition an estimate of the probability R of no failure (ordinary or shutdown) on or before the full duration T_k. In fact, if we assumed that the weighting factors were known *exactly* (i.e., not merely estimated) we would have

$$\hat{w}_i = \frac{p_i + \epsilon(1 - p_i)}{p_k + \epsilon(1 - p_k)}.$$

Then, from Eq. (16.4), the expected value of $\hat{R}$ would be

$$E(\hat{R}) = 1 - \frac{\sum_i N_i(p_i + \epsilon(1 - p_i))}{\sum_i N_i \frac{(p_i + \epsilon(1 - p_i))}{(p_k + \epsilon(1 - p_k))}} \qquad (16.5)$$

$$= 1 - (p_k + \epsilon(1 - p_k)) \equiv R \qquad (16.6)$$

Thus, $\hat{R}$ would be an *unbiased* estimator for R.

EXERCISE: Show that the quantity

$$\sum_i N_i(p_i + \epsilon(1 - p_i))$$

is the expected value of the total number of failures.

EXERCISE: Show that, under the assumption that the weighting factors are known exactly,

$$\hat{R}' = 1 - \frac{\sum_i \frac{\text{total number of failures in } i\text{th group}}{\hat{w}_i}}{\sum_i N_i}$$

is also an unbiased estimator of R. Compare numerically the estimators $\hat{R}$ and $\hat{R}'$ using the data of Tables 16.2 and 16.3.

EXERCISE: Consider the following results (*cf.* Table 16.2):

			(0, T_1)		(T_1, T_2)	
(Short-Duration)	T_1	N	$f_{11} = f$	$f_1 = 0$		
(Full Duration)	T_2	N	$f_{21} = f$		$f_{22} = g$	$f_2 = 0$

Show algebraically that $\hat{R} = \hat{R}' = 1 - \hat{p}_2$ (note that $\hat{\epsilon} = 0$).

16.3.4 USE OF THE WEIGHTING FACTORS AND RELIABILITY EVALUATION

The test results are compiled each month by the reliability group from each test site in chronological order. They would appear as in Fig. 16.2, which is a sample sheet. This list consists of both R&D and production engines identifiable by their serial numbers. Since these engines are of the same configuration, they can be grouped together. Flight data and other field tests are upon engines of a slightly earlier development configuration, and in addition, they will have received different handling, been operated by different personnel, and of course, the flight tests will have experienced the actual operation. Consequently, these data should be analyzed similarly but separately from the production and R&D tests and the effect of the different factors estimated by comparing their reliability estimates.

Page No. _____   Rocket Engine _____

CHRONOLOGICAL LIST OF TEST DATA FOR RELIABILITY EVALUATION PROGRAM, MONTH ENDING X-X-6X

List No.	Test No. Stand	Date Eng.S/N	Type Test	Int.Dur. Act.Dur.	Weight Factor	Result	Remarks
341	2.1-002 α 3	x-x-6x 916	R and D	160 161	1.00	S	All performance specifications met.
342	3.2-051 α 1	x-x-6x 1408	P	60 60	.82	S	All performance specifications met.
343	2.3-110 β 1	x-x-6x 1100	R and D			E	Declared malfunction test. See Declaration Form 343.
344	2.3-111 β 1	x-x-6x 1100	R and D	140 65	1.00	F	Rough combustion cutoff. See Failure Report No. FR 95.
345	2.4-021 δ 3	x-x-6x 910	R And D	100 101	.96	S	All performance specifications met.
346	3.2-052 α 1	x-x-6x 1408	P	40 40	..75	S	All performance specifications met.
347	2.1-003 α 3	x-x-6x 916	R and D			E	Test-to-failure. See Declaration Form 347
348	2.1-004 α 3	x-x-6x 916	R and D	60 61	.82	S	All performance specifications met.

Legend: R and D: development tests.
P: production acceptance tests.
S: success, F: failure, E: exclusion

Summary: 6 applicable tests out of 8 starts.
1 failure out of 5.35 equivalent full-duration tests.

Figure 16.2

16.3.5. TYPE OF STATISTICAL ESTIMATE

It was considered that a moving average reliability estimate would be the most informative and useful type of reliability number. A sample size of the latest 100 equivalent full duration test results was considered sufficiently large to provide a good level of assurance, but not too large so that too many of the earlier, unrepresentative development data were incorporated. A sequential plan was not used since a constant sample size was required, because the total propulsion system consisted of several other engines of which the Fig. 16.2 represents data from only one. It was necessary to combine the estimated reliability from several engines to obtain the propulsion system reliability estimate; this was much easier as well as statistically acceptable if all the data were based on equal size samples, i.e., the last 100 equivalent full duration tests from each engine.

16.4 RELIABILITY REPORTING

The computation of reliability is quite straightforward. With each test firing, which is not excluded, there is associated a weighting factor (see Fig. 16.2), which is dependent only upon the intended duration of the test. As of the last test of the reporting period the weighting factors are added in reverse chronological order until a total of at least 100 but less than 101 is reached. The total number of tests classified as failure and encompassed by the latter sequence of runs is then counted. If f is this number, then the point estimate of reliability is $(100 - f)/100$. If a confidence limit is required, this can be obtained from Figs. A.1–A.5 for the appropriate confidence level. This number is reported monthly and will illustrate the growth of the engine's reliability within a known framework of controls and criteria.

16.5 RELIABILITY DEMONSTRATION

In addition to the reporting requirements, there is also the requirement for demonstrating a contractual reliability goal. This might be stated in the form that the entire propulsion system, consisting of, say, three stages, must demonstrate as of July 196x a minimum of 90% reliability at 90% confidence. This implies (Fig. A.3) that there should be no more than a total of six failures in 100 tests of each of the three stages. In the planning stages of the program, if all three engines required equal development, these six failures might have been apportioned equally between each engine, and therefore each would have a point estimate reliability goal of 98%. If each stage engine were being made by a different contractor, then the 98% point estimate or a minimum reliability of 95% at a 90% confidence might be the reliability goal for each contractor. However, if all stages are made by the same contractor, then it is unnecessary to specify a contractual requirement for each stage; in fact, the contractor obviously has a smaller chance of meeting a requirement for each and every stage compared with satisfying the total requirement for the complete propulsion system.

There may be additional side requirements imposed upon the last 100 equivalent full duration tests used in the demonstration sample. For instance, a specified proportion of the tests may be under environmental conditions, a certain number of tests must be full-duration firings, and a restriction on the number of tests per engine should also be imposed. If possible, these factors should be introduced in the form of a statistical design. This is not always possible owing to hardware, schedule, or test facility limitations; however, a balanced statistical demonstration of reliability in this manner will provide additional confidence in the results. Alternately, if the goal is not passed, then the causation factor can be determined. The development program should allow for the introduction of environmental testing and normal operating and production procedures as early as possible so that their effects can be measured and appropriate steps taken.

ADDITIONAL READING

Barton, H. A., and Thatcher, A. G., "Rocket Engine Reliability," *Ordnance*, **41**:220, 722–26 (January–February, 1957).

APPENDIX 16A

MAXIMUM LIKELIHOOD EQUATIONS FOR p_i, ϵ

The probability of a failure equals the probability of the *union* of two mutually exclusive events: (1) an ordinary, nonshutdown failure (2) a shutdown failure. Figure 16A.1 illustrates the failure model and its appropriate parameters.

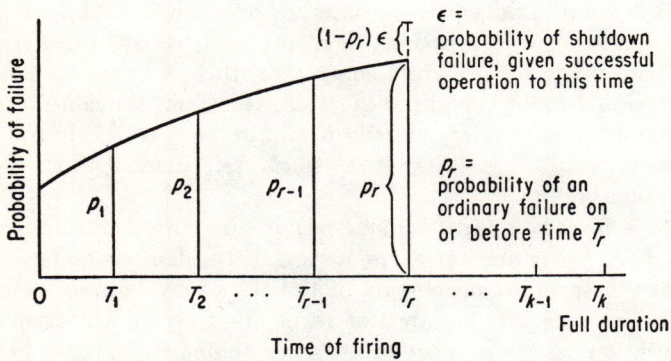

Figure 16A.1

The times shown, $T_1, T_2, \cdots, T_{r-1}, T_r, \cdots, T_k$, are the intended durations of operation for the corresponding p_i as well as being the discrete time intervals for classifying the failure times. p_i is defined as the probability of ordinary failure at or before time T_i for $i = 1, \cdots, k$. The quantity ϵ is defined as the conditional probability of shutdown failure (given that the engine has not failed prior to shutdown). Hence, the probability of a failure at or before time T_i is $p_i + \epsilon(1 - p_i)$. We now wish to estimate p_i and ϵ by the method of maximum likelihood.

First, consider only the group of N_r firings intended for time T_r. If f_r failures are observed at shutdown and f_{rj} are the numbers of failures

RELIABILITY EVALUATION OF LARGE LIQUID ROCKET ENGINES   477

occurring in the time interval $T_j - T_{j-1}$ for $j = 1, \cdots, r$, then

$$H_r = N_r - f_r - \sum_{j=1}^{r} f_{rj}$$

is the number of successes in this group.

The likelihood function is given by the probability of the observed event's occurring (see Secs. 7.4 and 6.3); i.e.,

$$L_r = \frac{N_r!}{H_r! f_r! \prod_{j=1}^{r} f_{rj}!} [(1 - p_r)(1 - \epsilon)]^{H_r} [(1 - p_r)\epsilon]^{f_r} \prod_{j=1}^{r} (p_j - p_{j-1})^{f_{rj}} \quad (16A.1)$$

The likelihood function L for all groups of firings is

$$L = \prod_{r=1}^{k} L_r \quad (16A.2)$$

Taking logarithms of both sides of (16A.2), where $\mathcal{L} \equiv \log L$, we have

$$\mathcal{L} = \text{constant} + \sum_{r=1}^{k} (H_r + f_r) \log (1 - p_r) + \sum_{r=1}^{k} f_r \log \epsilon$$

$$+ \sum_{r=1}^{k} H_r \log (1 - \epsilon) + \sum_{r=1}^{k} \sum_{j=1}^{r} f_{rj} \log (p_j - p_{j-1}) \quad (16A.3)$$

or

$$\mathcal{L} = \text{constant} + \sum_{r=1}^{k} (H_r + f_r) \log (1 - p_r) + \sum_{r=1}^{k} f_r \log \epsilon$$

$$+ \sum_{r=1}^{k} H_r \log (1 - \epsilon) + \sum_{j=1}^{k} F_j \log (p_j - p_{j-1}) \quad (16A.4)$$

where $F_j = \sum_{r=j}^{k} f_{rj}$.

The parameters p_i and ϵ which maximize $\mathcal{L} \equiv \log L$ will also maximize L. Thus we differentiate $\mathcal{L}$ with respect to p_i and ϵ, obtaining:

$$\left. \frac{\partial \mathcal{L}}{\partial p_i} = -\frac{H_i + f_i}{1 - p_i} + \frac{F_i}{p_i - p_{i-1}} - \frac{F_{i+1}}{p_{i+1} - p_i} \right\}$$

when $i = 1, 2, \cdots, k - 1$, $\quad (16A.5)$

and

$$\frac{\partial \mathcal{L}}{\partial p_k} = -\frac{H_k + f_k}{1 - p_k} + \frac{F_k}{p_k - p_{k-1}}$$

and

$$\frac{\partial \mathcal{L}}{\partial \epsilon} = -\frac{1}{1 - \epsilon} \sum_{r=1}^{k} H_r + \frac{1}{\epsilon} \sum_{r=1}^{k} f_r \quad (16A.6)$$

If we now set

$$\frac{\partial \mathcal{L}}{\partial \epsilon} = 0, \qquad \frac{\partial \mathcal{L}}{\partial p_i} = 0, \qquad i = 1, 2, \cdots, k$$

we obtain the set of equations which are to be solved for $\hat{p}_i, \cdots, \hat{p}_k$ and $\hat{\epsilon}$ (the quantities p_i, ϵ are written with "hats" to denote that they are estimates of the true values):

$$\left.\begin{array}{r}\dfrac{F_i}{\hat{p}_i - \hat{p}_{i-1}} - \dfrac{F_{i+1}}{\hat{p}_{i+1} - \hat{p}_i} = \dfrac{H_i + f_i}{1 - \hat{p}_i}, \quad \begin{array}{l} i = 1, 2, \cdots, k-1 \\ \hat{p}_0 = 0 \end{array} \\[2ex] \dfrac{F_k}{\hat{p}_k - \hat{p}_{k-1}} = \dfrac{H_k + f_k}{1 - \hat{p}_k} \\[2ex] \dfrac{1}{\hat{\epsilon}} \sum_{r=1}^{k} f_r = \dfrac{1}{1 - \hat{\epsilon}} \sum_{r=1}^{k} H_r \end{array}\right\} \quad (16\text{A}.7)$$

It is easily verified that the solutions given by

$$\left.\begin{array}{c} \hat{\epsilon} = \dfrac{\sum\limits_{r=1}^{k} f_r}{\sum\limits_{r=1}^{k} H_r + \sum\limits_{r=1}^{k} f_r} \\[4ex] \hat{p}_i = \hat{p}_{i-1} + \dfrac{F_i}{G_i}(1 - \hat{p}_{i-1}) \quad \begin{array}{l} i = 1, 2, \cdots, k \\ \hat{p}_0 = 0 \end{array} \end{array}\right\} \quad (16\text{A}.8)$$

where $G_i = \sum\limits_{j=i}^{k} F_j + H_j + f_j$

satisfy the system of Eqs. (16A.7).

APPENDIX 16B

VARIANCES AND COVARIANCES OF THE $\hat{p}_i$, $\hat{\epsilon}$

To obtain estimates of the variances and covariances of the $\hat{p}_i$ and the $\hat{\epsilon}$, we proceed as in Sec. 7.5.4. From Eqs. (16A.5) and (16A.6) we have

$$\mathcal{L}_{ii} \equiv \frac{\partial^2 \mathcal{L}}{\partial p_i^2} = -\frac{H_i + f_i}{(1-p_i)^2} - \frac{F_i}{(p_i - p_{i-1})^2} - \frac{F_{i+1}}{(p_{i+1} - p_i)^2} \qquad (16\text{B}.1)$$

for $i = 1, 2, \cdots, k-1$; and

$$\mathcal{L}_{kk} \equiv \frac{\partial^2 \mathcal{L}}{\partial p_k^2} = -\frac{H_k + f_k}{(1-p_k)^2} - \frac{F_k}{(p_k - p_{k-1})^2} \qquad (16\text{B}.2)$$

also

$$\mathcal{L}_{\epsilon\epsilon} \equiv \frac{\partial^2 \mathcal{L}}{\partial \epsilon^2} = -\frac{1}{(1-\epsilon)^2} \sum_{r=1}^{k} H_r - \frac{1}{\epsilon^2} \sum_{r=1}^{k} f_r \qquad (16\text{B}.3)$$

The mixed partial derivatives are

$$\left. \begin{aligned} \mathcal{L}_{ij} &\equiv \frac{\partial^2 \mathcal{L}}{\partial p_i \, \partial p_j} = \frac{\partial^2 \mathcal{L}}{\partial p_j \, \partial p_i} \equiv \mathcal{L}_{ji} = 0, \quad &&\begin{array}{l} j < i-1 \\ j > i+1 \end{array} \\ &= \frac{F_i}{(p_i - p_{i-1})^2}, &&j = i-1 \\ &= \frac{F_{i+1}}{(p_{i+1} - p_i)^2}, &&j = i+1 \end{aligned} \right\} \qquad (16\text{B}.4)$$

and $\quad \mathcal{L}_{i\epsilon} = \mathcal{L}_{\epsilon i} \equiv \dfrac{\partial^2 \mathcal{L}}{\partial \epsilon \, \partial p_i} = 0, \quad i = 1, 2, \cdots, k$

For example, when $k = 2$, the matrix of partial derivatives becomes (taking p_1, p_2, ϵ in order):

$$\mathbf{A} = \begin{bmatrix} -\dfrac{H_1 + f_1}{(1 - p_1)^2} - \dfrac{F_1}{p_1^2} - \dfrac{F_2}{(p_2 - p_1)^2} & \dfrac{F_2}{(p_2 - p_1)^2} & 0 \\ \dfrac{F_2}{(p_2 - p_1)^2} & -\dfrac{H_2 + f_2}{(1 - p_2)^2} - \dfrac{F_2}{(p_2 - p_1)^2} & 0 \\ 0 & 0 & -\dfrac{H_1 + H_2}{(1 - \epsilon)^2} - \dfrac{f_1 + f_2}{\epsilon^2} \end{bmatrix}$$

To obtain estimates of the variances and covariances, it is easiest to replace the p_i, ϵ in the matrix $\mathbf{A}$ by the estimators $\hat{p}_i$, $\hat{\epsilon}$; thus $\mathbf{A}$ becomes a matrix of numbers. The next step is to find the negative of the inverse of $\mathbf{A}$; the elements of the resulting matrix are (in the same order) the desired estimates of the variances and covariances.

As an example, we take the data of Sec. 16.3.2. Using Eqs. (16B.1) through (16B.4), we obtain

$$\mathbf{A} = \begin{bmatrix} -23147. & 19223. & 0 & 0 & 0 & 0 & 0 \\ 19223. & -24774. & 5519. & 0 & 0 & 0 & 0 \\ 0 & 5519. & -14936. & 9383. & 0 & 0 & 0 \\ 0 & 0 & 9383. & -13919. & 4535. & 0 & 0 \\ 0 & 0 & 0 & 4535. & -19791. & 15242. & 0 \\ 0 & 0 & 0 & 0 & 15242. & -15393. & 0 \\ 0 & 0 & 0 & 0 & 0 & 0 & -15819. \end{bmatrix}$$

The inverse $\mathbf{A}^{-1}$ is not too difficult to compute[*] but it is somewhat tedious; the details are omitted here. The final result is the variance-covariance matrix $\mathbf{B} = -\mathbf{A}^{-1}$:

$$\mathbf{B} = \begin{bmatrix} .00024159 & .00023888 & .00023085 & .00022698 & .00021904 & .00021689 & 0 \\ .00023888 & .00028764 & .00027797 & .00027331 & .00026374 & .00026115 & 0 \\ .00023085 & .00027797 & .00044373 & .00043628 & .00042102 & .00041689 & 0 \\ .00022698 & .00027331 & .00043628 & .00053376 & .00051508 & .00051003 & 0 \\ .00021904 & .00026374 & .00042102 & .00051508 & .00070987 & .00070290 & 0 \\ .00021689 & .00026115 & .00041689 & .00051003 & .00070290 & .00076096 & 0 \\ 0 & 0 & 0 & 0 & 0 & 0 & .00006322 \end{bmatrix}$$

The diagonal elements of the matrix $\mathbf{B}$ are, in order, estimates of Var $\hat{p}_1$, $\cdots$, Var $\hat{p}_6$, Var $\hat{\epsilon}$. The remaining elements of $\mathbf{B}$ are estimates of the covariances of the corresponding estimates. For example, in row 1, column 2, we have[†] $\langle \text{Cov}\ (\hat{p}_1, \hat{p}_2) \rangle = 0.00023888$. Note that the matrix $\mathbf{B}$

---

[*] See J. G. Kemeny *et al.*, *Finite Mathematical Structures*, Prentice-Hall, Inc., Englewood Cliffs, N. J., 1959, pp. 240–243.

[†] "$\langle\ \rangle$" denotes a numerical estimate of the quantity within the brackets.

is symmetric, as it should be, since Cov $(\hat{p}_i, \hat{p}_j) =$ Cov $(\hat{p}_j, \hat{p}_i)$. Note also that $\langle \text{Cov}(\hat{\epsilon}, \hat{p}_j) \rangle = 0$; i. e., $\hat{\epsilon}$ and $\hat{p}_j$ are uncorrelated.

If it is desired to calculate an estimate of the correlation coefficient ρ_{ij} between $\hat{p}_i, \hat{p}_j$, then the following formula is used:

$$\langle \rho_{ij} \rangle = \frac{\langle \text{Cov}\,(\hat{p}_i, \hat{p}_j) \rangle}{(\langle \text{Var}\,\hat{p}_i \rangle \langle \text{Var}\,\hat{p}_j \rangle)^{1/2}} \quad (16\text{B}.5)$$

For example

$$\langle \rho_{12} \rangle = \frac{0.00023888}{[(0.00024159)(0.00028764)]^{1/2}} = \frac{0.00023888}{0.00026361}$$

or
$$\langle \rho_{12} \rangle = 0.9062$$

As another example

$$\langle \rho_{56} \rangle = \frac{0.00070290}{[(0.00070987)(0.00076096)]^{1/2}} = \frac{0.00070290}{0.00073497}$$

or
$$\langle \rho_{56} \rangle = 0.9564$$

Suppose we consider ρ_{16}, i.e., the correlation coefficient for "widely spaced" $\hat{p}_i$. We have

$$\langle \rho_{16} \rangle = \frac{0.00021689}{[(0.00024159)(0.00076096)]^{1/2}} = 0.5058$$

The above result indicates that the correlation coefficient tends to decrease as the times at which the time-to-failure distribution is estimated become more widely separated.

The variances of the weighting factors w_i can be estimated as follows. Since

$$\hat{w}_i = \frac{\hat{p}_i + (1 - \hat{p}_i)\hat{\epsilon}}{\hat{p}_k + (1 - \hat{p}_k)\hat{\epsilon}} = U_i(\hat{p}_i, \hat{p}_k; \hat{\epsilon}) \quad (16\text{B}.6)$$

Then

$$\langle \text{Var}\,\hat{w}_i \rangle \simeq \left\langle \left(\frac{\partial U_i}{\partial \hat{p}_i}\right)^2 \right\rangle \langle \text{Var}\,\hat{p}_i \rangle + \left\langle \left(\frac{\partial U_i}{\partial \hat{p}_k}\right)^2 \right\rangle \langle \text{Var}\,\hat{p}_k \rangle$$

$$+ \left\langle \left(\frac{\partial U_i}{\partial \hat{\epsilon}}\right)^2 \right\rangle \langle \text{Var}\,\hat{\epsilon} \rangle + 2\left\langle \frac{\partial U_i}{\partial \hat{p}_i} \right\rangle \left\langle \frac{\partial U_i}{\partial \hat{p}_k} \right\rangle \langle \text{Cov}\,(\hat{p}_i, \hat{p}_k) \rangle \quad (16\text{B}.7)\text{*}$$

* The terms containing Cov $(\hat{p}_i, \hat{\epsilon})$, Cov $(\hat{p}_k, \hat{\epsilon})$ are left out since these are zero in this case (see Sec. 9.5.1).

The quantities $\langle \text{Var } \hat{p}_i \rangle$, $\langle \text{Cov } (\hat{p}_i, \hat{p}_k) \rangle$, etc., are available from the matrix **B**. The values of the partial derivatives in Eq. (16B.7) are as follows:

$$\left\langle \frac{\partial U_i}{\partial \hat{p}_i} \right\rangle = \frac{1 - \hat{\epsilon}}{\hat{p}_k + (1 - \hat{p}_k)\hat{\epsilon}} \quad (16\text{B}.8)$$

$$\left\langle \frac{\partial U_i}{\partial \hat{p}_k} \right\rangle = \frac{(1 - \hat{\epsilon})(\hat{p}_i + (1 - \hat{p}_i)\hat{\epsilon})}{(\hat{p}_k + (1 - \hat{p}_k)\hat{\epsilon})^2} \quad (16\text{B}.9)$$

$$\left\langle \frac{\partial U_i}{\partial \hat{\epsilon}} \right\rangle = \frac{\hat{p}_k - \hat{p}_i}{(\hat{p}_k + (1 - \hat{p}_k)\hat{\epsilon})^2} \quad (16\text{B}.10)$$

For example, let us find $\langle \text{Var } \hat{w}_1 \rangle$. From Table 16.3, $\hat{p}_1 = 0.0938$, $\hat{p}_6 = 0.1865$, $\hat{\epsilon} = 0.01967$, thus

$$\left\langle \left(\frac{\partial U_1}{\partial \hat{p}_1}\right)^2 \right\rangle = (4.841)^2 = 23.44 \quad (16\text{B}.11)$$

$$\left\langle \left(\frac{\partial U_1}{\partial \hat{p}_6}\right)^2 \right\rangle = (-2.668)^2 = 7.118 \quad (16\text{B}.12)$$

$$\left\langle \left(\frac{\partial U_1}{\partial \hat{\epsilon}}\right)^2 \right\rangle = (2.261)^2 = 5.112 \quad (16\text{B}.13)$$

and

$$\left\langle \frac{\partial U_1}{\partial \hat{p}_1} \right\rangle \left\langle \frac{\partial U_1}{\partial \hat{p}_6} \right\rangle = (4.841)(-2.668) = -12.92 \quad (16\text{B}.14)$$

Using the above quantities in Eq. (16B.7), together with the required variances and covariance values from the matrix **B**, we finally obtain:

$$\langle \text{Var } \hat{w}_1 \rangle \simeq 0.005798$$

EXERCISE: Find $\langle \text{Var } \hat{w}_2 \rangle, \cdots, \langle \text{Var } \hat{w}_5 \rangle$. Note that $\langle \text{Var } \hat{w}_6 \rangle \equiv 0$.

Answer: 0.006086, 0.005853, 0.004937, 0.001451.

Since $p_k + (1 - p_k)\epsilon$ is the probability that a device will fail at or before time T_k (the full-rated duration), we can find an upper confidence limit on this quantity by the formula (*cf.* Sec. 7.4.4)

$$1 - \hat{R}_L = \hat{p}_k + (1 - \hat{p}_k)\hat{\epsilon} + K_{1-\gamma}\sqrt{\langle \text{Var } (\hat{p}_k + (1 - \hat{p}_k)\hat{\epsilon}) \rangle} \quad (16\text{B}.15)$$

where

$$\langle \text{Var } (\hat{p}_k + (1 - \hat{p}_k)\hat{\epsilon}) \rangle = (1 - \hat{\epsilon})^2 \langle \text{Var } \hat{p}_k \rangle + (1 - \hat{p}_k)^2 \langle \text{Var } \hat{\epsilon} \rangle \quad (16\text{B}.16)$$

In the example we have been using, $k = 6$; thus

$$\langle \text{Var } (\hat{p}_6 + (1 - \hat{p}_6)\hat{\epsilon}) \rangle = 0.0007732$$

RELIABILITY EVALUATION OF LARGE LIQUID ROCKET ENGINES 483

(from Table 16.3 and the matrix **B**); also, $\hat{p}_6 + (1 - \hat{p}_6)\hat{\epsilon} = 0.2025$ and therefore, for $\gamma = 0.90$ ($K_{0.10} = 1.282$ is the standard normal deviate exceeded with probability 0.10),

$$1 - \hat{R}_L = 0.2025 + 1.282(0.02781)$$

or $$1 - \hat{R}_L = 0.2382 \qquad (16B.17)$$

The result given by (16B.17) is equivalent to saying that

$$\hat{R}_L = 0.7618 \qquad (16B.18)$$

is a 90 per cent lower confidence limit for the probability of successfully meeting the full-rated duration requirement.

It is interesting to compare the above estimate and confidence limit with those obtained from Eq. (16.4). We have

$$1 - \hat{R} = \frac{\text{total number of failures}}{\sum N_i w_i} \qquad (16B.19)$$

$$= \frac{53}{261.3} = 0.2028 \qquad (16B.20)$$

The simple binomial variance is given by

$$\frac{\hat{R}(1 - \hat{R})}{\sum N_i w_i} = 0.0006188$$

whose square root is 0.02488. Thus an upper 90 per cent confidence limit on probability of failure is

$$1 - \hat{R}'_L = 0.2028 + 1.282(0.02488) = 0.2347 \qquad (16B.21)$$

or $$\hat{R}'_L = 0.7653 \qquad (16B.22)$$

is a 90 per cent lower confidence limit for the probability of successfully meeting the full duration requirement. The two confidence limits $\hat{R}_L$ and $\hat{R}_L'$, given by Eqs. (16B.18) and (16B.22) respectively, are quite close, as are the point estimates for reliability.

We can also find a binomial confidence limit using Fig. A.3 (p. 558). Based on 53 failures in 261 equivalent full duration trials, the lower 90 per cent confidence limit on reliability is $1 - 0.24 = 0.76$.

CHAPTER SEVENTEEN
SOFTWARE RELIABILITY

17.1 INTRODUCTION

This chapter discusses the newly developing technology needed to control and measure computer software reliability. The importance of this topic is indicated by the increasing ratio of software to hardware costs observed in military systems, which apparently is also the case for civil and commercial systems.

More important is the increasing reliance upon larger and more complex computer programs which are used to control and make decisions on the status of safety-critical systems such as nuclear reactors, transportation systems, medical life support equipment and military weapon systems.

Section 17.2 discusses the background of the software reliability problem, including the growing costs of unreliable software, some of the traditional views of the software specialists toward "system reliability," and some of the important differences between hardware and software reliability.

Section 17.3 considers the nature of errors which can ultimately cause failures of the software subsystem and which, unless redundant or "fault-tolerant" design techniques are used, will almost surely cause system failure. Since errors do occur, methods of reporting and correction are discussed.

Section 17.4 covers techniques of software reliability evaluation in some detail, in keeping with the emphasis given to quantitative methods in this book. Three categories of evaluation are defined: *reliability prediction, reliability estimation,* and *reliability measurement,* each leading to a different set of models and analysis techniques. Usually all three types will be used by the software engineer, since each is largely complementary to the others, and can be applied at different phases of development.

Finally, a selected bibliography is given which incorporates journal articles, and government and government-sponsored reports which represent the current trends in design and production of reliable software.

17.2 THE SOFTWARE RELIABILITY PROBLEM

17.2.1 THE COST OF UNRELIABLE SOFTWARE

Recent studies of United States Air Force command and control data processing systems have shown that computer software costs have been increasing as computer hardware costs have gone down. In 1972 the Air Force spent between $1 and $1.5 billion on software (about 4 to 5 percent of the total Air Force budget). In comparison, the hardware cost was $300–$400 million (Ref. 1). Figure 17.1 shows the likely trends in the relative costs of hardware and software, going to perhaps a relative software cost of 90 percent in the 1980's.

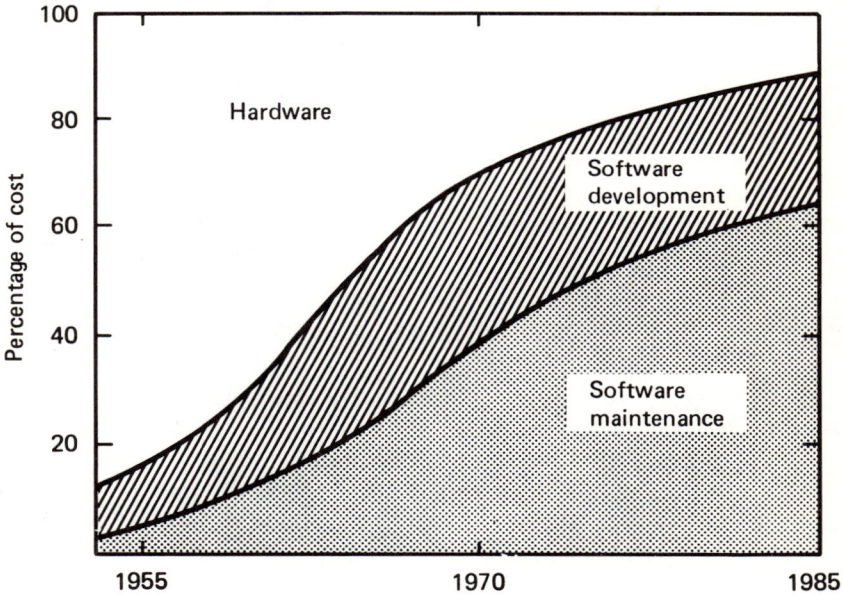

Fig. 17.1 Hardware-software cost trends.

The data are probably representative of other United States Military Services, and may also indicate the trend in other large scale government and civil software systems, such as literature abstract services or airline reservations systems.

A major portion of software costs occur during the operations and maintenance phase (also shown in Fig. 17.1), a period that may extend for many years beyond the original development phases. Figure 17.2 depicts typical phases in the software life-cycle. According to one study (Ref. 2) almost 40 percent of the software effort in Great Britain goes into maintenance (1973). Another survey in 1972 (Ref. 3) showed that some American data processing organizations spend

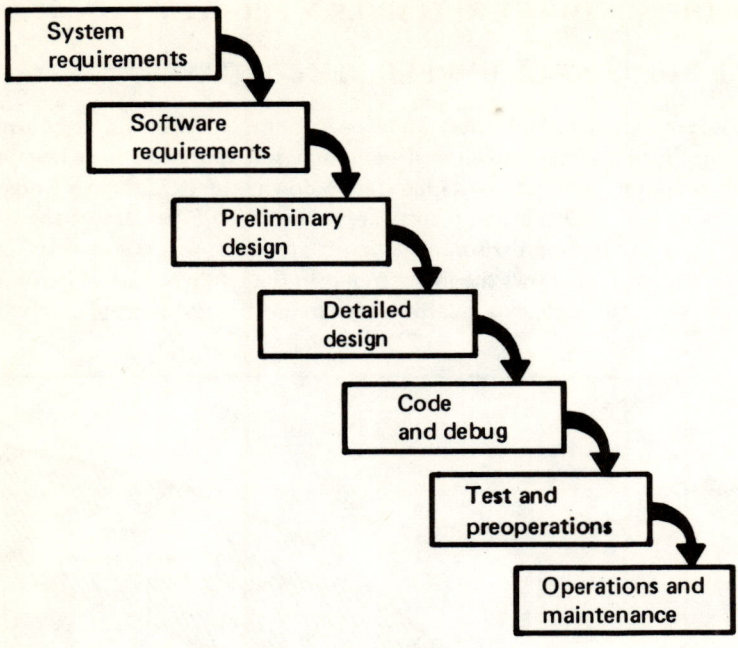

Fig. 17.2 Phases in the software life cycle.

in the range of 30—80 percent of their data processing budgets on maintenance. Primary efforts during the operations and maintenance phase are to correct errors which prevent the software from fulfilling its specified requirements, or to accommodate changed requirements, some planned, but others whose need becomes apparent only when the system is operational in the user environment. Other bases for software modifications are to improve performance or processing efficiency, adapt to changes in data base restructuring or new hardware, or simply to improve maintainability by, for example, rewriting its documentation or adding comments to the source code listings. Even maintenance for the latter reasons may introduce more errors, Reference 4 reports on a study indicating that if 10 or fewer source statements are changed, there is about a 50 percent chance of running correctly on the first attempt, and if 50 statements are changed, the probability drops below 20 percent.

The cost of correcting the error is not only in the manhours devoted to determining the actual fault, and the tests to verify the change, but also in the possible loss of capability of the system to perform functions expected of it while the correction is being made. From one viewpoint, the latter can be considered a monetary loss because payment may have been made to the software producer to provide the given capability, say, 95 percent of the time. If the capability were present only 75 percent of the time, then the user could have paid less for the software on the basis that 75 percent capability was the actual requirement.

Of course a direct monetary loss to the user, say an airline, may ensue since in the case of a software malfunction the airline may then have to hire more clerks to manually perform the unavailable functions, lose business temporarily to other airlines, and so on.

The above described losses are monetary, but other failures result in costs to the populace which cannot easily be measured (for example, several generations of human beings could suffer from the effects of radiation in the event of an uncontained nuclear power plant explosion). Of course such losses could be consequences of either software or hardware failure. However, there is a growing recognition that many computer programs have become more complex, and therefore error prone, than corresponding hardware systems.

Discovering errors early in the software life cycle can yield a large payoff (Ref. 5). As can be seen in Fig. 17.3, if cost of an error discovered during coding is set at a unit value, this is still 5–10 times the cost of correcting the error if its

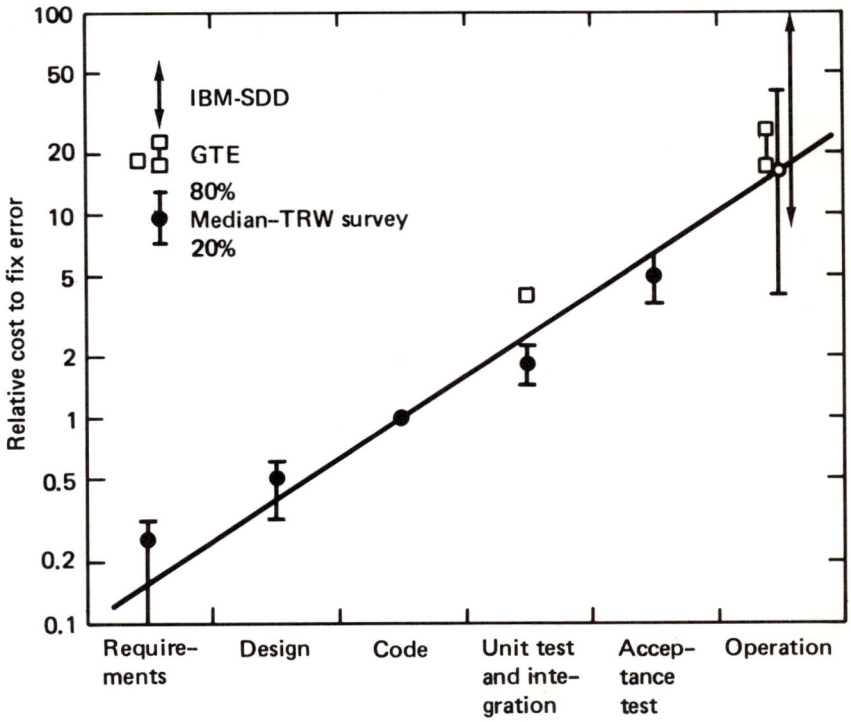

Fig. 17.3 Software validation: the price of procrastination.

cause were found during the requirements specification phase. Toward the end of development, during acceptance test, the relative cost is about 5, and during operations the relative cost is about 20. Thus, comparing extreme ends of the software life cycle, one manhour spent in finding and correcting a requirements specification error during the requirements phase would be multiplied by 100 were the error to be discovered during user operations.

In Sec. 17.3 methods will be discussed for preventing or detecting errors during software development, which could therefore considerably reduce costs of errors during operations and maintenance. However, the above-described leverage on the early part of the software life cycle is offset by the costs of developing and applying new or modified techniques and automated support tools. This tradeoff is one of the key problems in software production and represents the basis for the newly developing technologies of "software engineering."

17.2.2 TRADITIONAL VIEWS OF SOFTWARE RELIABILITY

Until recently, it appears to the authors, most software specialists took the view that a computer program is either 0 or 100 percent reliable, since program instructions once formed correctly cannot degrade or become faulty. According to this view, if a single fault* were present in a program, the intended function of the program could not be carried out in all respects and therefore the program had a reliability of zero.

However, experience with many systems (particularly as they became more complex) showed that software can have faults and still function acceptably nearly all of the time. This performance might be adequate if the user were willing to tolerate the absence of some function for the interval of time between the first appearance of the problem and its eventual correction. On the other hand, there are many applications, such as real-time processing of medical data for a patient under intensive care, or assessment of safety-critical parameters for a nuclear reactor, where the loss of some functions would be intolerable for the time it takes to correct the software error* which caused the problem.

A software failure may not be immediately apparent after it occurs. An example is when the value of a computed quantity is numerically in error by more than a prescribed tolerance, subsequently (perhaps hours later) causing an output to be obviously wrong. A failure may not actually occur for some time (months or even years) after the software is in use because (1) the particular logic path containing the faulty sequence of instructions has not yet been executed, or (2) the same logic path in its execution will manifest the error for some but not for all input data. An example of the latter fault is when a division by zero occurs, because the error of not guarding against zero was made by the programmer.

*See Table 17.1 for a formal definition of this term.

SOFTWARE RELIABILITY

The above classes of faults are the basis for a software reliability model which allows faults to be present but for which the actual malfunction or symptoms caused by the faults may infrequently occur. This may be taken as a contrary position to the software specialist's traditional view, but is more in keeping with the currently accepted meaning of reliability from system and user viewpoints.

Table 17.1. DEFINITIONS OF SOFTWARE RELIABILITY TERMS

Terms	Definitions
Computer program	A series of instructions or statements in a form acceptable to computer equipment, designed to cause the computer equipment to perform a specified function.
Software	Computer program products in the form of card decks, magnetic tapes, or disks, and all descriptive documentation including specifications, listings, manuals and flow charts.
Software error	A conceptual, syntactic, or clerical discrepancy which causes one or more faults in the software.
Software fault	A specific manifestation of an error. A discrepancy in the software which impairs its ability to function as intended. An error may be the cause of several faults.
Software failure	A software failure occurs when a fault in the computer program is evoked by some input data, resulting in the computer program not correctly computing the required function.
Software reliability	From a system, user, or "macroscopic" viewpoint, the probability that the use of the software does not result in failure of the system to perform as expected by more than a specfied frequency. From a subsystem, developer, or "microscopic" viewpoint, the probability that the software is fault-free.

17.2.3 DIFFERENCES BETWEEN SOFTWARE AND HARDWARE RELIABILITY

Here we highlight some of the differences between software and hardware reliability phenomena, as these indicate new problems the reliability specialist is likely to face when software is a major component of the system. The following list is abstracted for the most part from Ref. 6.

1. Software components do not degrade due to wear or fatigue.
2. No imperfections or variations are introduced in making additional copies of a piece of software (except possibly for a class of easy-to-check copying errors).

3. Software processes are unconstrained by any laws of physics. For example, if a program expresses the force of gravity with the wrong sign, the calculations will produce the effect of anti-gravity.
4. Software interfaces are conceptual rather than physical; there is no easy-to-visualize three-prong plug and its mate.
5. There are many more distinct paths to check in software than in hardware.
6. There are many more distinct entities to check. Any item in a large file may be a source of error.
7. The failure modes are generally different. Software failures generally come with no advance warning, provide no period of graceful degradation, and usually provide no announcement that they have occurred.
8. Hardware failure modes which depend upon material properties have some built-in or inherent safety margins; also redundancy in the form of duplicated components can often be incorporated with relative ease, whereas methods of building "fault tolerance" into software are still difficult and are only beginning to be understood.
9. Repair of a hardware fault generally restores the system to its previous configuration; repair of a software fault does not.

In addition, there are some more qualitative differences which currently make the operational distinction between hardware and software more pronounced.

10. The use of standard parts is much more prevalent in hardware.
11. Management has had a much better intuitive understanding of hardware than of software.
12. It is too easy to make software changes — and too hard to make correct ones, due to the high degree of interdependency within the software.

In spite of these differences, for the most part, the reliability management techniques, mathematical models and statistical methods described in the previous chapters will be found useful for controlling and evaluating software reliability.

17.2.4 SOFTWARE QUALITY CHARACTERISTICS

Another viewpoint is to consider software reliability as a *quality characteristic*, that is, a property specified to be present in the software product to a desired degree. Examples of other quality characteristics are *portability, maintainability, efficiency,* and *modifiability*. Figure 17.4 portrays a *software quality characteristics tree* (Ref. 7) showing the relationships between these characteristics, other more abstract characteristics, and also with more concrete, even quantifiable, quality characteristics. The directions of the relationships shown in Fig. 17.4 by arrows represent logical implication. Thus if the characteristic *reliability* is present to some degree, then *self-containedness, accuracy, completeness, robustness/integrity,* and *consistency* must all necessarily be present to

SOFTWARE RELIABILITY

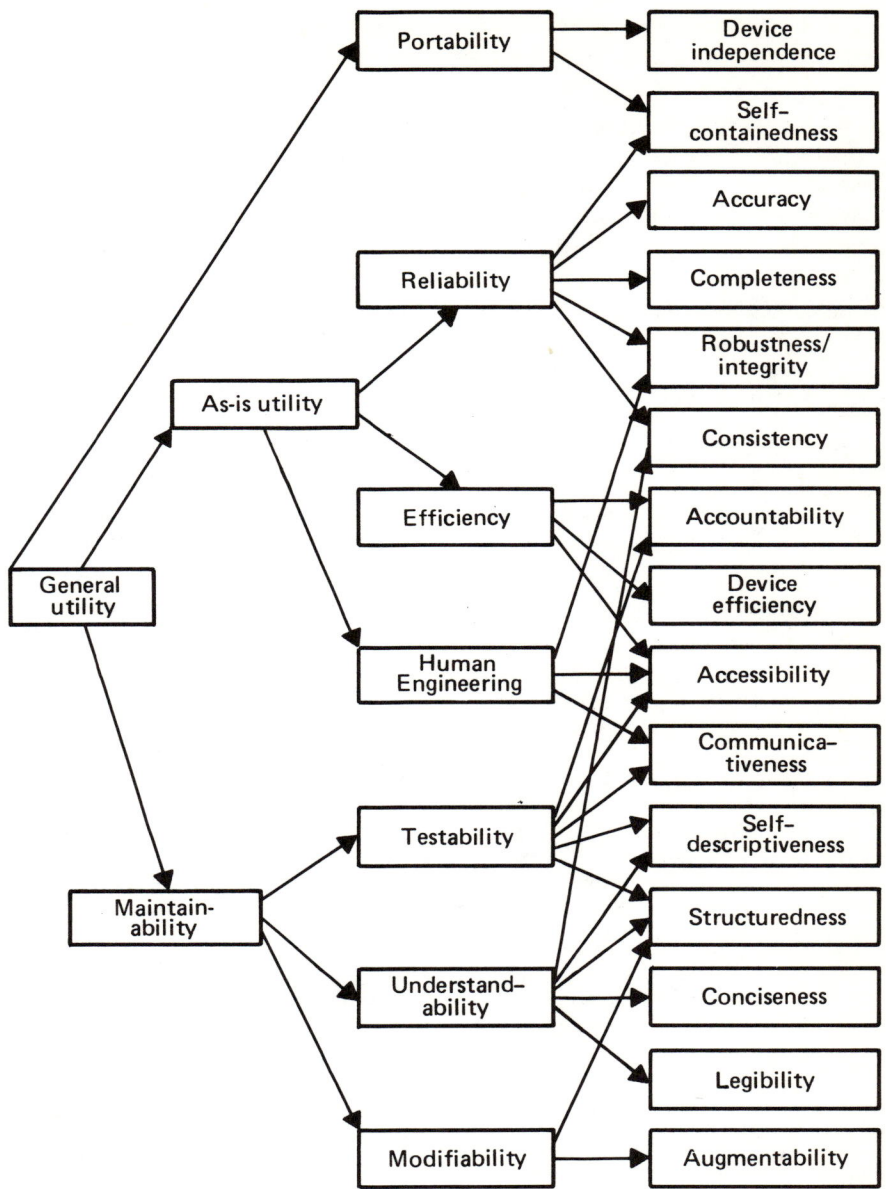

Fig. 17.4 Software quality characteristics tree.

*Reproduced from Ref. 7 with the permission of the Institute of Electrical and Electronics Engineers, Inc.

some degree. Table 17.2 presents definitions, in terms of computer program code, of all of the quality characteristics shown in Fig. 17.4.

It was also shown in Ref. 7 that the application, early in software development, of automated and semi-automated checks of *consistency, robustness* and *self-containedness* would have led to significant improvements in software error detection and correction, as found by detailed analysis of errors in one project. Since the three quality characteristics mentioned are considered necessary for the characteristic *reliability* (Fig. 17.4), the result of this analysis was very consistent with the portrayed association of characteristics. Section 17.3.4.2 further discusses methods of error detection using automated aids or software "tools."

Table 17.2. DEFINITIONS OF SOFTWARE QUALITY CHARACTERISTICS*

Terms	Definitions
Accessibility	Code possesses the characteristic *accessibility* to the extent that it facilitates selective use of its parts. (Examples: variable dimensioned arrays, or not using absolute constants.) *Accessibility* is necessary for *efficiency, testability,* and *human engineering.*
Accountability	Code possesses the characteristic *accountability* to the extent that its usage can be measured.
	This means that critical segments of code can be instrumented with probes to measure timing, whether specified branches are exercised, etc. Code used for probes is preferably invoked by conditional assembly techniques to eliminate the additional instruction words or added execution times when the measurements are not needed.
Accuracy	Code possesses the characteristic *accuracy* to the extent that its outputs are sufficiently precise to satisfy their intended use. Necessary for *reliability.*
Augmentability	Code possesses the characteristic *augmentability* to the extent that it can easily accommodate expansion in component computational functions or data storage requirements. This is a necessary characteristic for *modifiability.*
Communicativeness	Code possesses the characteristic *communicativeness* to the extent it facilitates the specification of inputs and provides outputs whose form and content are easy to assimilate and useful. *Communicativeness* is needed for *testability* and *human engineering.*
Completeness	Code possesses the characteristic *completeness* to the extent that all its parts are present and each part is fully developed.
	This implies that external references are available and required functions are coded and present as designed, etc.

*This table is reproduced from Ref. 7 by permission of the Institute of Electrical and Electronics Engineers, Inc.

SOFTWARE RELIABILITY

Table 17.2. DEFINITIONS OF SOFTWARE QUALITY CHARACTERISTICS (Continued)

Terms	Definitions
Conciseness	Code possesses the characteristic *conciseness* to the extent that excessive information is not present.
	This implies that programs are not excessively fragmented into modules, overlays, functions and subroutines, nor that the same sequence of code is repeated in numerous places, rather than defining a subroutine or macro, etc.
Consistency	Code possesses the characteristic *internal consistency* to the extent that it contains uniform notation, terminology and symbology within itself, and *external consistency* to the extent that the content is traceable to the requirements.
	Internal consistency implies that coding standards are homogeneously adhered to; e.g., comments should not be unnecessarily extensive or wordy at one place, and insufficiently informative at another, that number of arguments in subroutine call matches with subroutine header, etc. External consistency implies that variable names and definitions, including physical units, are consistent with a Glossary; or, there is a one-one relationship between functional flow chart entities and coded routines or modules, etc.
Device Efficiency	Code possesses the characteristic *device-efficiency* for a device to the extent that the operations, functions, or instructions provided by the code are performed without waste of resources with respect to that device (CPU time, I/O channel capacity, core memory, etc.).
	Thus, a program may be device-efficient with respect to one device (say, CPU time) but not another (say, core memory) implying that it is not *efficient* with respect to the overall set of resources it employs.
Device Independence	Code possesses the characteristic *device-independence* to the extent it can be executed on computer hardware configurations other than its current one. Clearly this characteristic is a necessary condition for *portability*.
Efficiency	Code possesses the characteristic *efficiency* to the extent that it fulfills its purpose without waste of resources.
	This implies that choices of source code constructions are made in order to produce the minimum number of words of object code, or that where alternate algorithms are available, those taking the least time are chosen; or that information-packing density in core is high, etc. Of course, many of the ways of coding efficiently are not necessarily efficient in the sense of being cost-effective, since portability, maintainability, etc., may be degraded as a result.
Human Engineering	Code possesses the characteristic *human engineering* to the extent that it fulfills its purpose without wasting the users' time

Table 17.2. DEFINITIONS OF SOFTWARE QUALITY CHARACTERISTICS (Continued)

Terms	Definitions
	and energy, or degrading their morale. This characteristic implies *accessibility, robustness,* and *communicativeness.*
Legibility	Code possesses the characteristic *legibility* to the extent that its function is easily discerned by reading the code. (Example: complex expressions have mnemonic variable names and parentheses even if unnecessary.) *Legibility* is necessary for *understandability.*
Maintainability	Code possesses the characteristic *maintainability* to the extent that it facilitates updating to satisfy new requirements or to correct deficiencies.
	This implies that the code is *understandable, testable* and *modifiable;* e.g., comments are used to locate subroutine calls and entry points, visual search for locations of branching statements and their targets is facilitated by special formats, or the program is designed to fit into available resources with plenty of margins to avoid major redesign, etc.
Modifiability	Code possesses the characteristic *modifiability* to the extent that it facilitates the incorporation of changes, once the nature of the desired change has been determined. Note the higher level of abstractness of this characteristic as compared with *augmentability.*
Portability	Code possesses the characteristic *portability* to the extent that it can be operated easily and well on computer configurations other than its current one.
	This implies that special language features, not easily available at other facilities, are not used; or that standard library functions and subroutines are selected for universal applicability, etc.
Reliability	Code possesses the characteristic *reliability* to the extent that it can be expected to perform its intended functions in a satisfactory manner.
	This implies that the program will compile, load, and execute, producing answers of the requisite accuracy; and that the program will continue to operate correctly, except for a tolerably small number of instances, while in operational use. It also implies that it is complete and externally consistent, etc.
Robustness	Code possesses the characteristic *robustness* to the extent that it can continue to perform despite some violation of the assumptions in its specification.
	This implies, for example, that the program will properly handle inputs out of range, or in different format or type than defined, without degrading its performance of functions not dependent on the non-standard inputs.

SOFTWARE RELIABILITY

Table 17.2. DEFINITIONS OF SOFTWARE QUALITY CHARACTERISTICS (Continued)

Terms	Definitions
Self Containedness	Code possesses the characteristic *self-containedness* to the extent that it performs all its explicit and implicit functions within itself. Examples of implicit functions are initialization, input checking, diagnostics, etc.
Self Descriptiveness	Code possesses the characteristic *self-descriptiveness* to the extent that it contains enough information for a reader to determine or verify its objectives, assumptions, constraints, inputs, outputs, components, and revision status. Commentary and traceability of previous changes by transforming previous versions of code into non-executable but present (or available by macro calls) code are some of the ways of providing the characteristic. *Self-descriptiveness* is necessary for both *testability* and *understandability*.
Structuredness	Code possesses the characteristic *structuredness* to the extent that it possesses a definite pattern of organization of its interdependent parts.
	This implies that evolution of the program design has proceeded in an orderly and systematic manner, and that standard control structures have been followed in coding the program, etc.
Testability	Code possesses the characteristic *testability* to the extent that it facilitates the establishment of verification criteria and supports evaluation of its performance.
	This implies that requirements are matched to specific modules, or diagnostic capabilities are provided, etc.
Understandability	Code possesses the characteristic *understandability* to the extent that its purpose is clear to the inspector.
	This implies that variable names or symbols are used consistently, modules of code are self-descriptive, and the control structure is simple or in accordance with a prescribed standard, etc.
Usability (As-Is Utility)	Code possesses the characteristic *usability* to the extent that it is reliable, efficient and human-engineered.
	This implies that the function performed by the program is useful elsewhere, is robust against human errors (e.g., accepts either integer or real representations for type real variables), or does not require excessive core memory, etc.

17.3 SOFTWARE ERRORS

17.3.1 GENESIS OF SOFTWARE ERRORS

All software errors are in some sense "design" errors. When detected, the faults occurring in programs or documentation caused by the errors can be corrected, and are then no longer present. The relatively high frequency of errors made in software design is due to the fact that most computer programs are more complex than corresponding hardware systems. In an absolute sense, large programs may be considered as the most complex objects ever built by man, some of them having millions of instructions and (correspondingly) hundreds of thousands of decision points. The structural complexity of these large programs is so great that what they can and cannot do has not been well understood.

In recent studies (Ref. 8) evidence has shown that over 60 percent of the errors occur during the requirements formulation, preliminary design, and detailed program design phases of software development, and less than 40 percent arise in the programming or coding phase. On the other hand many errors due to incorrect, inconsistent, incomplete or misunderstood requirements statements are detected, but not necessarily resolved, during subsequent preliminary and detailed design phases. The unresolved deficiencies may, in fact, remain open to the point when the user finds that the software does not perform satisfactorily for his needs during the operations phase, owing to previous inability to ascertain or express these requirements adequately.

The discipline of *software requirements engineering* has shown a belated but rapid growth in the recognition that deficiencies in the specification which describes what the software will do are very difficult and expensive to correct later (Sec. 17.2.1). Table 17.3 expresses some of the problems arising in the requirements formulation phase.

17.3.1.1 Design Errors. What are the causes of errors in design? Answers are elusive because this phase of software development involves an intellectual, creative or intuitive process, often highly dependent upon the personality of the designer. The best answer perhaps is that many errors can be prevented by consideration and avoidance of the following "poor" approaches:

   a. Inadequate simulation. There are very few aids to help the designer to make overall software/hardware tradeoff decisions in order to narrow the number of degrees of freedom available. One published technique is based on the *Extendable Computer System Simulator* (Ref. 12) making it possible to develop a functional simulation of a system in a much shorter time than it takes to develop the complete design itself.

   b. Deficient design representations. Flow charts, the primary method, are too easy to construct in a complicated manner, and hard to understand and maintain. Machine processable structured design languages ("structured pidgin English") are probably more suitable. Program Design Language (Ref. 13) is probably the best example of such a design aid.

   c. Unstructuredness. Structuredness is a general term referring to a design philosophy requiring adherence to a set of rules of enforced standards which embody such techniques as *top-down design, program modularization* or *independence, structured control flow,* and so forth. Each of

SOFTWARE RELIABILITY

Table 17.3. SOME REQUIREMENTS PROBLEM DESCRIPTIONS*

A. Acceptable, but not for current design baseline.
B. Out-of-scope contractually.
C. Probability of satisfying requirement significantly less than unity.
D. Decision criteria, interface characteristics, accuracy criteria, processing rates, error recovery requirements missing, incomplete or inadequately stated.
E. Incorrect requirements:
   1. Timing and/or accuracy requirement not feasible with presently known techniques.
   2. Requirement untestable.
   3. Model does not fit physical situation well enough, or equations incorrect.
   4. Required processing inaccurate, inefficient, illogical, unnecessary or not always possible.
   5. Document references incorrect.
   6. Interpretation not in accordance with updated requirements.
F. Inconsistent or incompatible requirements:
   1. Two locations in specification document give conflicting information.
   2. Referenced paragraph does not exist.
   3. Requirement conventions (e.g., coordinate systems) not consistent with other documentation.
G. New/changed requirements.
H. Requirement unclear or nonsense.
I. Requirement has significant typographical errors or is completely missing in revised specification document.

*Adapted from Ref. 9.

    Refs. 14 and 15 presents excellent detailed discussions on the topic of structuredness in computer program design.

    d. Selection of unstandardized languages. Standardization implies, among other requirements, that rigid configuration controls be kept on compilers, support tools and documentation. An unstandardized language can also be the source of coding errors (see item 2. in Sec. 17.3.1.2). The United States Department of Defense, having recognized language standardization as an important factor in production of reliable software, published Department of Defense Directive (DODD) 5000.29 (Ref. 10) in 1976 which required use of High Order Languages except when an assembly language could be justified. Following this directive, DOD Instruction 5000.31 was issued (Ref. 11), which specified several "approved" languages.

The growing complexity of software/hardware interfaces as a design factor in software also cannot be neglected. Choices of hardware alternatives have been increasing at an enormous rate. As compared to the 1950's the likely number of choices for central processor units or peripherals in the 1970's is about 100 to 1 and will be even higher in the 1980's (Ref. 5).

17.3.1.2 Coding Errors. Errors in coding occur primarily in the following ways:
1. *Typographical Errors.* A programmer incorrectly writes down or copies a statement in the source language; e.g., omitting parentheses, writing down "23" instead of "32", or misspelling a variable name.
2. *Misinterpretation of Language Constructions.* A programmer uses certain language constructions in a way he or she believes is correct, but the compiler* interprets them differently.
3. *Missing or Incorrect Logic.* Assuming that the design specification was correct, a programmer makes an error by omitting a required test for a condition, or (say) incorrectly tests for a condition after a scope is executed whereas if the condition were false, the scope should not have been executed.
4. *Undocumented Assumption.* The programmer makes an error by assuming a design interpretation without telling anyone, when actually the design was ambiguous and allowed two or more interpretations. This is a gray area, since responsibility for the error could be placed entirely on the designer. (For example, the programmer assumes that the first quadrant in the (x,y) plane is defined by $(x>0, y>0)$ whereas the designer should have specified it as $(x \geq 0, y \geq 0)$, according to the conditions of the problem.)
5. *Singularities and Critical Values.* The programmer forgets to test and provide a response to division by zero, singular points or regions of library functions (square roots or logarithms of negative numbers) overflow, round-off errors, etc.
6. *Algorithm Approximation.* A programmer may use an approximation to the correct equations to make their solution tractable, or to a function in order to increase execution speed. The approximations may be insufficiently accurate over the required ranges of the variables. In most cases, responsibility for the error could be placed upon the designer.
7. *Data Structure Defects.* The program is incompatible with the data structure specification; e.g., a table may be specified to contain a maximum number of entries but the program may continue to try to insert entries into the table when it is already full.

---

*A compiler is a program which converts statements in a higher level source language such as FORTRAN to instructions in machine code, which can be directly executed by the computer.

SOFTWARE RELIABILITY

The types of coding errors expressed above are almost never detected during compilation, and perhaps not for many test runs or executions of the program, or may never be detected. There are, fortunately, a growing list of software tools or automated aids which can find many of these errors at reasonable cost. These will be discussed in Sec. 17.3.4.

17.3.2 ERROR CATEGORIZATION

Table 17.4 is a list of error/fault categories from Ref. 8 which has been used successfully on a large state-of-the-art real-time software project. The dual nomenclature "errors/faults" is used since some of the subcategories represent a group of faults which can most easily be expressed by describing the error which caused them. As a further note of explanation, the blank character in the category designator is meant to be used to describe the source of the error. Table 17.5 lists corresponding descriptions of error sources to be used for this purpose.

Table 17.4. SOFTWARE ERROR/FAULT CATEGORIES

A_000		COMPUTATIONAL ERRORS/FAULTS
	A_100	Incorrect operand in equation
	A_200	Incorrect use of parentheses
	A_300	Sign convention error
	A_400	Units or data conversion error
	A_500	Computation produces an over/under flow
	A_600	Incorrect/inaccurate equation used
	A_700	Precision loss due to mixed mode
	A_800	Missing computation
	A_900	Rounding or truncation error
B_000		LOGIC ERRORS/FAULTS
	B_100	Incorrect operand in logical expression
	B_200	Logic activities out of sequence
	B_300	Wrong variable being checked
	B_400	Missing logic or condition tests
	B_500	Too many/few statements in loop
	B_600	Loop iterated incorrect number of times (including endless loop)
	B_700	Duplicate logic
C_000		DATA INPUT ERRORS/FAULTS
	C_100	Invalid input read from correct data file
	C_200	Input read from incorrect data file
	C_300	Incorrect input format
	C_400	Incorrect format statement referenced
	C_500	End Of File encountered prematurely
	C_600	End Of File missing

Table 17.4. SOFTWARE ERROR/FAULT CATEGORIES (Continued)

D_000		DATA HANDLING ERRORS/FAULTS
	D_050	Data file not rewound before reading
	D_100	Data initialization not done
	D_200	Data initialization done improperly
	D_300	Variable used as a flag or index not set properly
	D_400	Variable referred to by the wrong name
	D_500	Bit manipulation done incorrectly
	D_600	Incorrect variable type
	D_700	Data packing/unpacking error
	D_800	Sort error
	D_900	Subscripting error
E_000		DATA OUTPUT ERRORS/FAULTS
	E_100	Data written on wrong file
	E_200	Data written according to the wrong format statement
	E_300	Data written in wrong format
	E_400	Data written with wrong carriage control
	E_500	Incomplete or missing output
	E_600	Output field size too small
	E_700	Line count or page eject problem
	E_800	Output garbled or misleading
F_000		INTERFACE ERRORS/FAULTS
	F_100	Wrong subroutine called
	F_200	Call to subroutine not made or made in wrong place
	F_300	Subroutine arguments not consistent in type, units, order, etc.
	F_400	Subroutine called is nonexistent
	F_500	Software/data base interface error
	F_600	Software user interface error
	F_700	Software/software interface error
G_000		DATA DEFINITION ERRORS/FAULTS
	G_100	Data not properly defined/dimensioned
	G_200	Data referenced out of bounds
	G_300	Data being referenced at incorrect location
	G_400	Data pointers not incremented properly
H_000		DATA BASE ERRORS/FAULTS
	H_100	Data not initialized in data base
	H_200	Data initialized to incorrect value
	H_300	Data units are incorrect
I_000		OPERATION ERRORS/FAULTS
	I_100	Operating system error (vendor supplied)
	I_200	Hardware error
	I_300	Operator error
	I_400	Test execution error
	I_500	User misunderstanding/error
	I_600	Configuration control error

Table 17.4. SOFTWARE ERROR/FAULT CATEGORIES (Continued)

J_000		OTHER
	J_100	Time limit exceeded
	J_200	Core storage limit exceeded
	J_300	Output line limit exceeded
	J_400	Compilation error
	J_500	Code or design inefficient/not necessary
	J_600	User/programmer requested enhancement
	J_700	Design nonresponsive to requirements
	J_800	Code delivery or redelivery
	J_900	Software not compatible with project standards
K_000		DOCUMENTATION ERRORS/FAULTS
	K_100	User manual
	K_200	Interface specification
	K_300	Design specification
	K_400	Requirements specification
	K_500	Test documentation
X0000		PROBLEM REPORT REJECTION
	X0001	No problem
	X0002	Void/withdrawn
	X0003	Out of scope – not part of approved design
	X0004	Duplicates another problem report
	X0005	Deferred

Table 17.5. ERROR SOURCE DESIGNATOR DESCRIPTION

Source ID	Error Source*	Description
0	Requirements	The source of the error was a changing, ambiguous, imprecise, or poorly stated requirement.
1	Design	The source of the error was in preliminary or detailed design.
2	Code	The source of the error was in the implementation of the design as code.
3	Maintenance	The source of the error was an error introduced in the process of trying to correct a previous error.
4	Not known	The source of the error is not known.

*The *source* is not necessarily coincident with the *phase* of the software life cycle, since, for example, the term "Maintenance" includes both the error correction activity of the Operations and Maintenance Phase and the re-design or re-coding necessary to correct an error found during integrated testing, prior to operational use of the software.

17.3.3 FREQUENCIES OF ERROR CATEGORIES

The data given in Table 17.6 are based upon a careful analysis of three major projects as reported in Ref. 8, and show percentages of errors by major categories given in Table 17.4. Categories (C) and (E) are combined (I/O errors) in order to make consistent comparisons of the three projects.

If the percentages are weighted by the numbers of problem reports shown at the bottom of Table 17.6, the resulting average percentages of errors assigned to each category are given in Table 17.7 in order of decreasing frequency.

It cannot of course be claimed that the data given in Table 17.7 are representative. It is to be hoped that current efforts by the U.S. Government* to establish repositories of these types of data will aid in assessments of error frequencies, and consequently provide better estimates of costs attributable to software errors. Section 17.3.5 discusses associated data which should be collected during software development.

17.3.4 METHODS OF ERROR DETECTION AND PREVENTION

In this section some of the techniques and tools for preventing and detecting errors during design and coding phases are discussed. Here we define *tech-*

Table 17.6. PERCENTAGES OF MAJOR ERROR CATEGORIES
FOR THREE LARGE SOFTWARE PROJECTS

Major Error Category	Project I (%)	Project II (%)	Project III Operating System (%)	Project III Applications Software (%)	Project III Simulator Software (%)
Computational (A)†	9.0	1.7	2.5	13.5	19.6
Logic (B)	26.0	34.5	34.6	17.1	20.9
Data I/O (C), (E)	16.4	8.9	8.6	7.3	9.3
Data Handling (D)	18.2	27.2	21.0	10.9	8.4
Interface (F)	17.0	22.5	7.4	9.8	6.7
Data Definition (G)	0.8	3.0	7.4	7.3	13.8
Data Base (H)	4.1	2.2	4.9	24.7	16.4
Other (J)	8.5	0.0	13.6	9.4	4.9
Total Problem Reports Requiring Code Change	2019	405	81	275	225

*Notably Rome Air Development Center, Air Force Systems Command.
†See Table 17.4 for subcategory descriptions.

Table 17.7. MAJOR ERROR CATEGORIES RANKED BY FREQUENCY

	Major Category	%
(B)	Logic	26.2
(D)	Data Handling	18.1
(F)	Interface	16.0
(C), (E)	Data I/O	13.8
(A)	Computational	9.0
(J.)	Other	7.3
(H)	Data Base	6.7
(G)	Data Definition	2.9

niques as the practices and procedures used in development and maintenance of the software system. *Tools* are defined as computer programs which perform tasks which would be tedious or impractical to do manually.

17.3.4.1. Techniques. Table 17.8 presents descriptions or examples of some techniques that have been developed and applied in software developments. Table 17.9 portrays an analysis of the effect of the several preventive and detective techniques listed in Table 17.8 upon Project I by showing the percentages of the total number of errors of a given category which are susceptible to each given preventive or detective technique. For example 26.0 percent of all code errors for Project I were categorized as logic errors, from Table 17.6. Table 17.9 indicates that 13.7 percent of all code errors, or over half the logic errors, could be prevented by use of design standards; 8.2 percent, or about one-third, could be prevented by coding standards, and as much as 23.5 percent, or about nine-tenths, could be prevented by design inspections, and so on. The percentages susceptible to detective techniques are interpreted similarly.

17.3.4.2 Tools. Table 17.10 presents descriptions and examples of some preventive and detective tools that have been used in software developments. In addition, there are *support tools,* which are used to assist some development or test technique. An example is a *dynamic path analyzer,* which determines execution frequency of code segments. Sometimes, perhaps unexpectedly, it is found that most of the execution time is spent in one small section of the code, which may then be optimized for increased execution speed, with a relatively large impact on total program efficiency. This condition may also be indicative of an error, whose location is thereby made visible.

Table 17.11 shows an analysis of the likely effect of these tools had they been applied to Project I.

*Notably Rome Air Development Center, Air Force Systems Command.

Table 17.8. SOME PREVENTIVE AND DETECTIVE TECHNIQUES

Preventive	Description or Examples
Design Standards	1. Use High Order Language (HOL), unless assembly language necessary for storage and timing efficiency.
	2. Use "top-down" hierarchial approach with parallel refinement of requirements and functional descriptions into sub-requirements and sub-functions.
Coding Standards	1. Use specified control structures with no statement labels.
	2. Maximum of 100 executable statements per routine allowed.
Design and Code Inspections	1. Formal or informal presentation of design and code by responsible person.
	2. Review by checklist of standard topics and by sources of most common errors.
Detective	
Algorithm Test	Independently of the primary software, construct strings of critical algorithms in order to test data transfers, sensitivity to input data, timing and accuracy.
Integration Test	Test correctness of interfaces – routine/routine, routine/data base, routine/operating system routines.
Requirements Tests	Termed Validation, Acceptance, System Integration, and Operational Demonstration tests. The latter tests performance in user environment, forecasting problems to be encountered, and necessary modifications.
Path/FCLD/DSET Test	Execution of all "paths," Functional Capabilities List Demonstration, and Data Singularity and Extremes testing at the unit level, where "paths" include at least one execution of a loop's scope.

17.3.5 ERROR DATA COLLECTION

Most of the management procedures and systems for collecting, processing and reporting error data, as discussed in Chapter 3, apply equally to computer software. Some of the problems peculiar to software error data collection arise because of the relative infancy of software engineering techniques utilizing these data. One observation has been that when forms designed primarily for configuration management functions are used to describe the software problem and certify the problem closure, insufficient information is recorded for reliability analysis purposes. This circumstance may have had its origin in the belief that the process of error discovery and correction is of limited usefulness in inferring reliability properties of computer programs (cf. Sec. 17.2.2). The often quoted phrase "testing shows the presence, not the absence of bugs" (Ref. 16), has been

Table 17.9. PERCENTAGE OF CODE ERRORS FOR PROJECT 1 SUSCEPTIBLE TO:

	Preventive Techniques				Detective Techniques				
Major Error Categories	Design Standards	Coding Standards	Design Inspection	Code Inspection	Path/FCLD/ DSET Test	DSET	Algorithm Test	Integration Test	Requirement Test
Computational (A)	4.9	1.4	5.9	4.2	7.9	5.6	6.6	1.5	5.7
Logic (B)	13.7	8.2	23.5	18.9	20.7	14.9	0	6.3	13.6
Data I/O (C), (E)	3.6	3.9	4.5	12.9	16.2	5.8	0	8.2	9.5
Data Handling (D)	0.5	9.3	13.5	13.4	14.0	15.3	0	11.4	10.6
Interface (F)	5.0	3.4	8.6	8.2	7.1	7.4	0	16.9	1.7
Data Definition (G)	0.6	0.1	0.6	0.7	0.7	0.7	0.5	0.8	0
Data Base (H)	0	0	0	2.2	4.1	1.2	1.2	0.6	2.5
Other (J)	0.4	0	1.1	2.2	2.2	0.2	0	0.4	2.1
Totals	28.7%	26.3%	57.7%	62.7%	72.9%	51.1%	8.3%	46.1%	45.7%

Table 17.10. PREVENTIVE AND DETECTIVE TOOLS

Preventive	Description
Simulation	Investigation of logical, computational and environmental (data base, data I/O) aspects using models of software modules. By varying input data and other parameters some errors will be evoked early.
Design Languages	A processable language which facilitates communication of the problem design to other people. It can be written at any level of detail appropriate to the design concept at the time. The chief benefit is that it addresses program logic and can help to prevent missing logic or condition tests, which account for a significant fraction of errors.
Detective	
Code Standards Auditor	Verifies compliance to specified coding standards (see Table 17.8) and locates specific violations. An estimated 26.3 percent of Project I errors were susceptible to prevention with project-specific coding standards (see Table 17.9).
Units Consistency Analyzer	Checks consistency and compatibility of stated parameter units within expressions. Forces the developer to think very carefully about data definitions.
Set/Use Checker	Performs static path analysis to identify variables set but not used and those used but not set. Identifies misspelled variable names.
Compatibility Checker	A general term for a tool which checks for compatibility between two elements; e.g., subroutine calling sequence and header.

unfortunately misleading but influential in this respect. When interpreted as all-or-none logical inference, the quoted phrase is true. However, when well-formulated probabilistic models are used to describe the error discovery and correction process, useful forecasts can be made of, for example, resources needed to correct software errors found in the user environment.

An adequate error data collection system should include the data items listed in Table 17.12.

Clearly, the preceding list, exclusive of references to "software," has general applicability. What are the data items which are most meaningful to reliability analysis of software problems? The items listed in Table 17.12 are not limited to information that would be obtained as a result of an error found in testing, but also refer to requirements or design reviews, as well as code inspections (Ref. 18). In addition, it is convenient to include a list of items which can be associated with each program, subroutine, or other functional module, but which are obtainable from a static analysis of code. These particular data would

SOFTWARE RELIABILITY

Table 17.11. PERCENTAGE OF CODE ERRORS FOR PROJECT I SUSCEPTIBLE TO:

	Preventive Tools		Detective Tools		
Major Error Categories	Simulation	Design Languages	Code Standards Auditor	Units Consistency Analyzer	Set/ Use Checker	Compatibility Checker
Computational (A)	5.8	3.2	1.4	1.4	0?	0.5
Logic (B)	9.5	14.5	2.4	0	2.5	2.5
Data I/O (C), (E)	0.6	4.8	2.8	0	2.7	0
Data Handling (D)	2.6	4.0	8.0	0.1	7.0	0.3
Interface (F)	1.3	5.0	4.2	0	2.0	7.4
Data Definition (G)	0	0.1	0.7	0.5	0.1	0
Data Base (H)	0.8	0	0	0	0	0
Other (J)	0.1	0.1	0	0	0	0
Totals	20.7%	31.7%	19.5%	2.0%	14.3%	10.7%

generally be obtained from a baselined version of the software, and updated periodically or at the next baseline (e.g., test baseline, product baseline, operational baseline). Some of the data descriptions are given in general terms, since the best way of implementing their collection depends upon the user, computer installation, cost, availability of collection tools, and other programmatic

Table 17.12. DATA ITEMS USEFUL FOR RELIABILITY ANALYSIS OF SOFTWARE PROBLEMS

Data Items	Description
A. Requirements/Design Problems	Descriptive information on the type and frequency of problems encountered during the requirements and design period (see Table 17.3 for requirements problem categories and descriptions). Items B – K are also relevant in most cases to Requirements/Design problems.
B. Problem Criticality	An assessment of the importance of the problem to the success of software operation or "mission" completion. This information may be used to decide priorities in correcting each of a group of problems open at a given time, as well as to estimate software availability for defined modes of mission operation. For example, inability to print out a report could result in a lesser than "all-up", but non-critical mode of operation.
C. Test Stress	A quantitative assessment of the amount of code, number of segments, branches (also which branches and which condition caused exercise of each branch), etc., treated by each test case in a test program.
D. CPU Time	CPU Time tied accurately to specific test or re-test runs.
E. Difficulty of Problem Closure	Relative difficulty or description of difficulty encountered in closing a problem. Why was it difficult? Was qualification/experience of individual programmer a factor?
F. Problem Dependency Factor	Identification of a problem introduced as a result of a fix to another problem.
G. Priority Assigned	
H. Calendar Times (and dates):	
1. Delay Time	Time to recognize problem, or start correction activity.
2. Administrative Time	Time to prepare initial report, assess impact, determine priority, assign personnel.
3. Active Correction Time	
a. Time to locate problem.	

SOFTWARE RELIABILITY

Table 17.12. DATA ITEMS USEFUL FOR RELIABILITY ANALYSIS OF SOFTWARE PROBLEMS (Continued)

Data Items	Description
b. Time to prepare design or code change.	
c. Time for retest planning and running tests.	
I. Manhours Expended for Item H3.	Manhours and computer resources expended for documentation and master tape modifications should be recorded also, and if possible the individual contribution of each given software problem to the total expenditure should be assessed.
J. Other data obtainable from baseline documentation or code:	
1. Before and After Difficulty Ratings	Assessment of difficulty to design, code, test, document and implement a given module made at the time first defined and, also later, after module reaches operational status.
K. Static Analysis of Code (Complexity Data) Number of:	
a. Executable Statements in a Module	
b. Branches and nesting level of each	
c. Call statements to application routines	
d. Call statements to operating system or system support routines	
e. Input/Output statements	
f. Computational Statements (assignment statements containing arithmetic operators)	
g. Data handling statements (assignment statements without arithmetic operators)	
h. Loops and nesting level of each	
i. Predicate conditions for exercise of each branch	Example: (N.GT.0..OR. M.GT.0.) has three predicate conditions for TRUE, one for FALSE.
j. Distinct operators, operands	Items j, k, are based on a "software physics" application by M.H. Halstead (Ref. 17). Operators include arithmetic, relational, logical, brackets, etc. Operands are variable names or absolute constants.
k. Uses of operators, operands	

considerations. Also, the relative value of some of the data items is not well-established (e.g., central processor time of operation may be more difficult to obtain and only slightly more informative than wall-clock time).

Applications of data described in Table 17.12 to reliability prediction and estimation models are given in Sec. 17.4.

17.4 SOFTWARE RELIABILITY EVALUATION METHODS

17.4.1 INTRODUCTION

Hecht (Ref. 19) constructively lays down definitions of three relevant numerical indices of software reliability: *reliability prediction, reliability estimation,* and *reliability measurement.*

Reliability *prediction,* as discussed in Sec. 17.4.2, is defined as a statement about the reliability of a computer program that is based upon the evaluation of certain relatively easily measurable properties of code (e.g., numbers of executable statements, numbers of branches, etc.) or from related project data (e.g., number of design problem reports which have been generated before the coding phase begins) which are thought to be correlated with reliability.

The prediction is therefore not directly expressed as a reliability number, but usually in terms of the number of faults per instruction, i.e., an "index" or indicator of reliability. Since the prediction can be made prior to the fault removal process taking place during the test phase, it would be expected to correlate more closely with the number of faults produced rather than with the number remaining after testing is completed.

Because of the empirical nature of the relationships considered useful for reliability prediction, the principal methods of measuring this process are statistical regression techniques. However, some recent attempts have been made to develop a "software physics" approach to remove some of the phenomenology or empiricism from the prediction process (Refs. 17, 20).

The second numerical index, reliability *estimation,* discussed in Sec. 17.4.3, requires measurement of parameters such as duration of test period (during which faults are removed), operational time duration (the period for which the estimate is desired), an assumed relationship between fault discovery rate and fault removal rate, the "operational profile" and other environmental parameters which could influence software reliability. Musa (Ref. 21) develops an extensive model which accounts for these and many other parameters. Here, we will limit ourselves to those parameters which depend upon time. In this case, the reliability estimation process does result in a numerical reliability, or probability of no failure for operational time, t, since rates of error discovery and fault removal are evaluated in terms of the operational usage rate.

Reliability *measurement* (Sec. 17.4.4), deals directly with reliability evaluation of software in its intended operating environment. Consequently, one application of reliability measurement techniques is for the purpose of reliability demonstration, for which some of the methods of Chap. 10 may be used. The newer feature of the model presented in Sec. 17.4.4 is the determination of lower confidence limits on software reliability defined as a weighted average of reliabilities for its components, which are subsets of an input data space.

17.4.2 RELIABILITY PREDICTION

The term *reliability prediction* as used in this chapter represents a class of models based upon a *phenomenological* or *empirical* approach to evaluation of indexes of software reliability. Analogous methods for hardware are widely practiced in prediction of electronic equipment reliability based upon parts counts, individual part stress factors, and equipment application factors, and are often used in preliminary design phases.

The analogy is a tenuous one however, since standardization at the "part" level does not carry over for software. No two lines of code will be alike, and each can contain different faults. Analogous to the equipment level are program modules or standard software packages. It is an accepted fact that well checked-out program modules with wide user experience are more reliable and often are preferred to programs that are designed anew for a given application. However, the contribution to software failures by a given "standard" module is not easily predictable from the apparent trend of previous error discoveries and corrections (Sec. 17.4.3), since, in general, no one facility organizes, evaluates and publishes the error history of the module. Furthermore, and perhaps more important, each application of a "standard" software module will nearly always require extensive "tailoring," such as stripping out unessential code, preparing new interfacing subroutines, recoding for increased speed or fitting into a given size memory, and so forth. As a result, many new faults can be introduced, and the objective of standardized failure rates is substantially compromised even at the module level.

Nevertheless, it is felt that numbers of errors committed, or faults residing in a computer program somehow relate to size, complexity and other measures of varying difficulty of definition and evaluation.

17.4.2.1 Background to the Phenomenological Approach. One of the initial studies of software reliability as a quality characteristic (Sec. 17.2.4) was reported by Rubey and Hartwick (Ref. 22). Their method was to define certain attributes and their metrics, the former being a prose expression of the particular quality characteristic desired of the software, and the latter a mathematical function of parameters thought to relate to or define the attribute.

Another study included the formulation of metrics and their application to a controlled experiment in which two computer programs were independently prepared, both in FORTRAN, to the same specification (Ref. 23). Several metrics were evaluated (one, for example, was expressed as a function of the frequency of GO-TO's (unconditional branch statements) and the number of lines of code skipped over in getting to the target of the GO-TO), and were found to correlate well with the number of faults found in each program.

A detailed presentation of the conceptual framework for the analysis of software reliability in terms of software quality characteristics and their metrics, discussed briefly in Sec. 17.2.4, was given in Ref. 24. Section 17.2.4 also mentions the use of tools and/or techniques for measuring reliability-related quality characteristics, which were found to have a potential for significantly reducing number of errors.

More recently, M.H. Halstead has approached both the problems of code productivity and error generation from a less empirical "software physics" point of view (Ref. 17). Fairly good accuracy (within ± 20 percent) in prediction of errors based upon completed code of several medium to large size projects has been reported* (See also Ref. 20).

17.4.2.2 Early Prediction of Errors. The phenomenological methods used for error prediction have also been applied during requirements and design phases, prior to coding, making it possible to influence some design choices before commitment to code. Figure 17.5 shows that when the total number of errors for a function, as discovered in testing of Project I software, is plotted against the number of design problem reports for the function, a straight line fit through the origin with slope 24 percent and coefficient of determination $r^2 = 0.94$ is obtained.** The term *function* is used to denote a group of routines designed to perform a required task (e.g., telemetry processing, trajectory determination, file management, etc.).

The ratio of approximately 4:1 of design problem reports to errors found in testing, being essentially constant, indicates that the process of detection and correction of design deficiencies was carried out in a fairly consistent manner for Project I. It maybe hypothesized that other software developed by the same facility, using the same application of resources, would show the same ratio of design problem reports to test errors. For other facilities the ratio might not be constant, or if constant might have a different value. One of the uses of this ratio is as a parameter in a model used to perform the cost tradeoff analysis discussed in the following Section.

17.4.2.3 Cost Model. A simple cost model is discussed in the following paragraphs which considers the use of a tool or technique to detect additional errors during the design phase and thereby save some of the greater expense of correcting the errors during the test phase.

---

*Private communication by R. Klobert of Boeing Aerospace Company, Seattle, WA.
**For a discussion of the differences in formulas for computing the slope and coefficient of determination for the usual case and for the case when the constant term of the linear regression function is constrained to be zero, see Ref. 25.

SOFTWARE RELIABILITY

The following cost factors are defined:

C_D = cost per error of correcting a design error* during the design phase.

C_T = cost per error of correcting a design error discovered during the test phase.

C_o = cost of developing and applying a tool or technique which detects an additional fraction P of design errors during the design phase.

Also, let

N_D = number of design errors discovered and corrected in design review during the design phase. (We assume, for simplicity, that each error discovered is permanently corrected.)

N_T = number of design errors discovered and corrected during the test phase.

M_T = total number of errors (design plus coding) discovered during test.

The total cost of discovering and correcting design errors during design and test phases without the new tool or technique is

$$C_1 = C_D N_D + C_T N_T \qquad (17.1)$$

The total number of design errors discovered and corrected during the design phase using the new tool or technique is $N_D(1+P)$. Consequently the number of design errors discovered and corrected during the test phase would then be reduced to $N_T(1-P)$. Therefore, had the tool or technique been used, the cost of discovering and correcting design errors during design and test, plus the cost of the new tool or technique would be

$$C_2 = C_o + C_D N_D (1+P) + C_T N_T (1-P) \qquad (17.2)$$

or

$$C_2 = C_o + C_D N_D + C_T N_T - P(C_T N_T - C_D N_D) \qquad (17.3)$$

The tool or technique will pay for itself if

$$C_1 - C_2 > 0,$$

or if

$$C_o < P(C_T N_T - C_D N_D) \qquad (17.4)$$

EXAMPLE: We use data from Project I, and some assumptions. First, Fig. 17.3 indicates that the average cost of correcting an error during the test phase is about 10 times the cost of correcting it during the design phase. Thus $C_T \simeq 10\, C_D$. For Project I, from Fig. 17.5, the total number of errors discovered during test, M_T, was about 24 percent of the number of errors discovered and corrected in design review

*The term "design error" as used here includes both errors made in the design phase and errors made in specifying requirements.

during the design phase, N_D, i.e., $M_T \simeq 0.24\, N_D$. Furthermore, the number of design errors discovered during test, N_T, was estimated to be about 62 percent of M_T (i.e., about 38 percent of the total number of test errors were errors made when producing code). Thus $N_T \simeq 0.62\, M_T = (0.62)(0.24)\, N_D = 0.15\, N_D$. For an estimate of P, from Tables 17.9 and 17.11 it appears that $P \simeq 0.5$ may be appropriate. Consequently, using the information on Project I derived so far, the inequality (17.4) becomes

$$C_o < 0.5 \left\{ (10\, C_D)(0.15\, N_D) - C_D N_D \right\} \tag{17.5}$$

or
$$C_o < 0.25\, C_D N_D \tag{17.6}$$

Additionally, we know that there were $N_D \simeq 5400$ design problem reports (see also Fig. 17.5, from which this number can be derived). Also a reasonable guess is that $C_D = \$25/\text{error}$. Thus, if

Thus, if
$$C_o < \$33{,}750$$

is satisfied, then the tool or technique would be worthwhile.

17.4.2.4 Prediction of Operational Failures from Preoperational Tests. When a total number of user-operational failures reported for each function of Project I was plotted against the total number of preoperational test errors for the same function, the appearance is as shown in Fig. 17.6. The result, after eliminating certain points indicated as statistical "outliers" was a straight line fit (forced) through the origin with slope 0.084 and coefficient of determination $r^2 = 0.865$. Thus, it appears that for Project I, the numbers of errors for functions as observed during preoperational test tended to persist in the same ratios during operational use.

17.4.3 RELIABILITY ESTIMATION*

One of the uses of *reliability estimation* is to forecast reliability changes which depend upon time. This means that calendar time, central processor usage time, accumulated manhours, etc. are incorporated into the model to allow for correlation of some measure of reliability with these observable parameters.

17.4.3.1 The Basic Model and Assumptions. The model to be considered has been constructed in order to solve the following basic problem: a software package (a related set of computer programs including operating system, application programs and other support software) is tested. Errors are discovered, the faults corrected, and testing continues for a time, or at a certain rate, until it is believed that only a few faults remain in the software. Fundamental questions

---

*Hecht (Ref. 19) defines reliability estimation more generally than is used here, since we are not considering, for example, the effect of changes in hardware configuration or other modifications in the computing environment upon software reliability.

SOFTWARE RELIABILITY

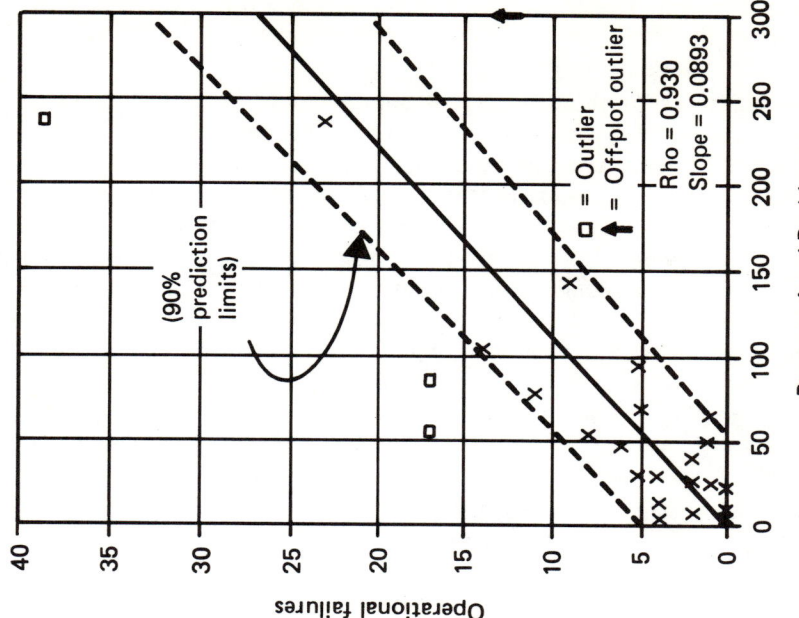

Fig. 17.6 Operational failures vs. preoperational problems.

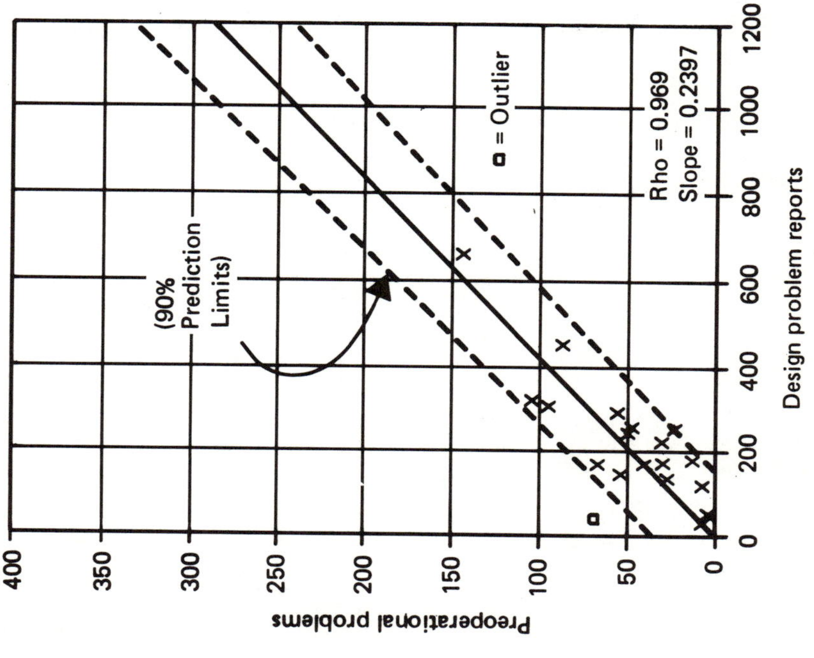

Fig. 17.5 Preoperational problems vs. design problems.

are: How many errors remain to be discovered? If the software is released for operational use, how long will it be before a software failure occurs? How often will the operational system be down because of software failures?

The definition of the time unit is most appropriately believed to be central processor unit usage time, being a measure of instruction execution rate. On the other hand, if CPU time data are unavailable, and an assumption can be made that clock time or calendar time is proportional to CPU time, then the latter time units would also be appropriate. If one were to consider the exposure time in detail, the operational CPU usage time would need to be split into an active and a non-active time in order to properly relate the parameters determined for the development period to the operational usage period. One other way of resolving this problem is to define a unit application in the operational environment, then determine, if possible, the equivalency with development unit applications. This brings out the problem of testing thoroughness in software development (Sec. 17.4.4.4) which would have to be accounted for in defining a "unit application" in development.

Clearly, answers to these questions depend upon the test strategy employed and how it is related to rate or efficiency of error discovery: how soon after discovery errors are corrected and how certain the corrections are, the likelihood of introducing new faults while correcting an error, and the time needed to correct the error.

Referring ahead to Sec. 17.4.4, the degree of testing thoroughness must also be part of the test strategy in order to develop meaningful answers to the previously stated fundamental questions. In other words, testing can be efficient in discovering errors and correcting faults evoked by only part of the Input Data Space. However, if the *operational profile* is such that there is significant probability that program *logic paths* never tested in software development will be used in software operation, then the objective of reducing the software operational failure rate to a satisfactorily low value cannot be met.

Also the *criticality* of the consequences of a fault — at one extreme; e.g., resulting in failure to detect a nuclear reactor out-of-control condition; or at the other extreme; e.g., resulting in a formatting error in a report — must be considered in determining a tolerance in number of faults allowed to remain in the operational software.

These are only a few of the considerations that can determine the usefulness of the type of model discussed here. References 19, 21, and 26-33 represent some of the recent work in evaluation of similar models. As data become better defined and available from newer software projects, significant improvements in selection of good models and analysis techniques can be expected.

17.4.3.2 Estimation of Parameters. The model selected for discussion here is based primarily upon Refs. 26 and 30.

Let t_1, t_2, ..., t_M each represent the time durations of M test periods (ordered in time). Both error discovery and correction take place within each

SOFTWARE RELIABILITY

test period, but all errors found during the ith test period are not necessarily corrected during the same test period. Let η_{i-1} be the cumulative number of errors corrected up through the i-1st test period, with $\eta_0 = 0$.

The number of errors discovered as a result of testing during the ith test period is assumed to have a Poisson distribution (Sec. 6.5) with parameter

$$\lambda \equiv \lambda_i = \omega(N - \eta_{i-1}) \qquad (17.7)$$

where N is the unknown number of errors present before the first test period, and ω is a proportionality constant. In other words the rate of arrival of error discoveries during the ith test period is assumed to be proportional to the number of errors remaining after completion of the i-1st test period. This is just one way of formulating the assumption that errors are more and more difficult to discover as fewer remain.

The techniques for estimating the parameters N, ω of this model are very similar to those of Sec. 11.3. Only the maximum likelihood method will be applied here however.

We define the likelihood function of the observations by

$$L = \text{const.} \prod_{i=1}^{M} (\lambda_i t_i)^{f_i} e^{-\lambda_i t_i} \qquad (17.8)$$

where f_i is the observed number of errors discovered during test period i. Thus, using the assumed relationship (17.7) for λ_i, we have

$$\mathcal{L} \equiv \log L = \log \text{const.} + \sum f_i \log t_i + \log \omega \sum f_i$$

$$+ \sum f_i \log(N - \eta_{i-1}) - \omega \sum t_i (N - \eta_{i-1}) \qquad (17.9)$$

(where all sums are from 1 to M).

Upon differentiating $\mathcal{L}$ with respect to N and ω once and setting the derivatives equal to zero, the two equations obtained to solve for the estimators $\hat{N}$, $\hat{\omega}$, are:

$$\frac{F_M}{\hat{N} + 1 - B/A} = \sum_{i=1}^{M} \frac{f_i}{\hat{N} - \eta_{i-1}} \qquad (17.10)$$

and

$$\hat{\omega} = \frac{F_M/A}{\hat{N} + 1 - B/A} \qquad (17.11)$$

where

$$F_j = \sum_{i=1}^{j} f_i \qquad j = 1, \ldots, M \qquad (17.12)$$

$$A = \sum_{i=1}^{M} t_i \qquad (17.13)$$

$$B = \sum_{i=1}^{M} (\eta_{i-1} + 1) t_i \qquad (17.14)$$

Since observed total number of errors corrected cannot exceed the total number of errors detected, η_i, F_i must satisfy the structural constraint:

$$\eta_i \leq F_i \qquad i = 1, 2, \ldots, M$$

Following the procedures of Sec. 7.5.4, the negative of the inverse of the matrix of second partial derivatives of $\mathcal{L}$ with respect to N, ω yields the asymptotic numerical variance-covariance matrix. The results of this procedure are:

$$\langle \text{Var } \hat{\omega} \rangle = \Sigma_2 / D \qquad (17.15)$$

$$\langle \text{Var } \hat{N} \rangle = F_M / (\omega^2 D) \qquad (17.16)$$

$$\langle \text{Cov}(\hat{\omega}, \hat{N}) \rangle = -A/D \qquad (17.17)$$

where

$$\Sigma_2 = \sum_{i=1}^{M} \frac{f_i}{(\hat{N} - \eta_{i-1})^2} \qquad (17.18)$$

$$D = \frac{F_M}{\hat{\omega}^2} \Sigma_2 - A^2 \qquad (17.19)$$

One complication to any procedure for obtaining estimators for N and ω is that the solution for N should be constrained by:

$$\hat{N} \geq F_M \qquad (17.20)$$

that is, the maximum likelihood estimator for N, the number of errors originally present, should never be less than the number of errors discovered. A rigorous approach to the solution of the maximum likelihood equations under this inequality constraint requires nonlinear programming methods (Ref. 34). However, a practical alternative is to reset $\hat{N} = F_M$ if the maximum likelihood solution for N turns out to be less than F_M, and recalculate $\hat{\omega}$ from (17.11) using $\hat{N} = F_M$. While $\hat{\omega}$ cannot be negative if $\hat{N} \geq F_M$, should ω actually be very small,

SOFTWARE RELIABILITY

the asymptotic formulas (17.15) – (17.17) will fail, which would indicate either that the model is questionable in the particular application, or that an insignificant number of errors (as compared to the original number) are being discovered during each test period for which data are available.

Having obtained estimators for the original number of errors N and the proportionality constant ω, a next step is to estimate the mean time to the next failure. If one can assume that the "unit application" is the same subsequent to the test periods used to estimate N and ω, then the expected time to first failure is given by

$$\hat{\theta} = [\hat{\omega}(\hat{N} - \eta_M)]^{-1} \qquad (17.21)$$

EXERCISE: Derive Equations (17.10) and (17.11) for the maximum likelihood estimators $\hat{N}$, $\hat{\omega}$. Do the same for the variances and covariance, Equations (17.15) – (17.17).

EXERCISE: Determine an approximate formula for the standard deviation of $\hat{\theta}$ (use Eq. (9.91)).

EXAMPLE: The following table lists numbers of errors discovered, f_i, and numbers of errors corrected for each test period, $\eta_i - \eta_{i-1}$, during a preoperational test phase in which the time intervals t_i are all equal and 7 units in duration.

i	f_i	(Cum.) F_i	$\eta_i - \eta_{i-1}$	(Cum.) η_i
1	77	77	11	11
2	72	149	38	49
3	73	222	57	106
4	64	286	57	163
5	58	344	38	201
6	60	404	44	245
7	51	455	53	298
8	47	502	56	354
9	47	549	55	409
10	35	584	61	470
11	41	625	52	522
12	30	655	110	632
13	12	667	5	637
14	14	681	5	642
15	9	690	16	658
16	12	702	10	668
17	10	712	20	688
18	10	722	18	706

From Eqs. (17.13) – (17.17) we have the following results:

$$A = 126.0 \quad B = 47397.0$$

The solutions for N, ω are

$$\hat{N} = 776 \qquad \langle \sigma_{\hat{N}} \rangle = 18.3$$

$$\hat{\omega} = 0.0143 \qquad \langle \sigma_{\hat{\omega}} \rangle = 0.000841$$

$$\langle \hat{\rho}(\hat{N}, \hat{\omega}) \rangle = -0.774$$

and the expected time to the first failure is, from Eq. (16.21)

$$\hat{\theta} = [0.0143 \, (776 - 706)]^{-1}$$
$$= 1.00 \text{ time units}$$

with

$$\langle \sigma_{\hat{\theta}} \rangle \simeq 0.23 \text{ time units}$$

EXERCISE: Let $\lambda_i = \omega T_{i-1} (N - n_{i-1})$

where

$$T_{i-1} = \sum_{j=1}^{i-1} t_j$$

be the total time duration through the i-1st test period. Thus, this error discovery rate function is proportional to the time previously spent in testing (based upon the assumption that test experience results in increasing ability to find errors), but is still also proportional to the number of errors remaining. Work out the maximum likelihood estimators of N, ω analogous to Eqs. (17.10) – (17.14) and the asymptotic variance-covariance matrix, analogous to Eqs. (17.15) – (17.19).

17.4.3.3 Model Validation.
One way of ascertaining how well a particular model for the error discovery rate fits the observed numbers of errors is to calculate

$$\chi^2 = \sum_{i=1}^{M} \frac{(f_i - \hat{\lambda}_i t_i)^2}{\hat{\lambda}_i t_i} \qquad (17.22)$$

where $\hat{\lambda}_i$ represents the numerical value of λ_i after substituting in estimated values for the parameters N and ω. The quantity on the righthand side of Eq. (17.22) has approximately a Chi-square distribution with M-3 degrees of freedom* (*cf*. Sec. 7.4.1).

EXAMPLE: For the previous example $\chi^2 = 7.840$. Since $P(\chi^2_{15} > 7.840) > 90\%$ (using Table A.8), the value of χ^2 is "significantly small," indicating a good fit.

It should be pointed out that Eq. (17.22) suggests that estimators for the parameters involved in λ_i should be found as those which minimize Q, defined by

$$Q = \sum_{i=1}^{M} \frac{(f_i - \lambda_i t_i)^2}{\lambda_i t_i} \qquad (17.23)$$

*Number of degrees of freedom is M-1 minus one degree of freedom for each parameter estimated.

instead of by the maximum likelihood method. Whichever method of estimation is used, the method of this section for testing the goodness-of-fit using the resulting estimators is appropriate.

17.4.4 RELIABILITY MEASUREMENT

The methods discussed in this section deal with reliability evaluation under a relatively fixed set of circumstances, appropriate to an acceptance, or pre-operational demonstration test, sometimes referred to as software *Validation*. If the demonstration is performed using the complete system in the operational environment (real or simulated) for the purpose of official endorsement of the software subsystem as meeting mission requirements, then the software is said to undergo *Certification* (Ref. 35).

17.4.4.1 Partitioning Model. The mathematical development of a simple partitioning model will be carried out here, with examples of its application in the following sections. Many of the ideas expressed here are based on the work by E.C. Nelson (Refs. 8, 28, 29), with contributions by J.R. Brown (Ref. 36). The fundamental assumptions and definitions appropriate to the model to be discussed are as follows:

The basic element of the model is the *Input Data Space E* (also called *input domain* by some authors), consisting of *input data sets*, E_i, $i = 1, 2, ..., N$.

Each E_i can be considered as a vector

$$E_i = (X_{1i}, X_{2i}, \ldots, X_{mi}) \qquad (17.24)$$

of values which together are needed to enable execution of the computer program. The number N of possible input data sets will generally be very large, but finite, since it is essentially the product of the numbers of possible values of each X_{ji}, $j = 1, \ldots, m$.

A partition of the input data space expressed as a union of disjoint subsets (*cf*. Sec. 5.2); i.e.,

$$E = S_1 \cup S_2 \cup \ldots \cup S_k \qquad (17.25)$$

where

$$S_i \cap S_j = \phi \text{ for } i \neq j \qquad (17.26)$$

The probability that input data set E_i is selected, i.e., submitted for execution by the program, is denoted by p_i, $i = 1, 2, ..., N$. The set of probabilities p_i are called the operational profile. The probability that an input data set is chosen from S_j is therefore

$$P(S_j) = \sum_{E_i \epsilon S_j} p_i \qquad (17.27)$$

where

$$\sum_{j=1}^{k} P(S_j) = \sum_{i=1}^{N} p_i = 1 \tag{17.28}$$

Each S_j can be further partitioned into subsets S_j' and S_j'' such that any input from S_j' results in correct execution and any input from S_j'' results in an execution failure. Since $S_j' \cap S_j'' = \phi$,

$$P(S_j) = P(S_j') + P(S_j'') \tag{17.29}$$

For notational simplicity the last three probabilities will hereafter be referred to as P_j, P_j', and P_j'', respectively.

Reliability R is defined as

$$R = \sum_j P_j' = 1 - \sum_j P_j'' \tag{17.30}$$

17.4.4.2 Functional Partition of the Input Data Space. Generally, each functional requirement for a computer program specifies the class of outputs of the program for a defined subset of the input data space E. When programs are designed so that the group of logic paths which produce the desired output can be put into direct correspondence with input data points belonging to the subset G_r, the partition G_r, $r = 1, 2, \ldots, l$ is then termed a *functional* partition.

A *logic path* of a program can be defined as a sequence of adjacent segments, beginning at an entry segment and proceeding by logical transfers to an exit segment. A *segment* is defined as follows:

1. It is a sequence of contiguous executable statements for which all statements in the segment will be executed if and only if the first statement is executed.
2. It begins with a statement to which control can be transferred and ends with a statement which transfers control to an adjacent segment.

An entry segment has no predecessor segments in the program and an exit segment results in termination or return of control to a calling program.

A logic path may be defined also in terms of the sequence of transfers of control that take place. Thus, the occurrence of the first transfer implies that an entry segment is executed followed by execution of an adjacent segment.

Some paths are *structural*, but not logic paths, in that a sequence of potential transfers of control from an entry segment to an exit segment exists, yet a number of such transfers are mutually impossible of execution for *any* input data sets.

We will use L_r to denote the group of logic paths designed to be executed when an input data set is selected from G_r. Using the previous notation, G'_r represents those input data sets of G_r which result in correct execution, and none of the logic paths of the group L_r may be executed when an input data set is selected from G''_r. However, when an input data set from G'_r is selected, only logic paths of L_r will be executed. Therefore, the event that any particular logic path of L_k is executed can be expressed as follows:

$$L_k = (L_k \cap G_1) \cup (L_k \cap G_2) \cup \cdots \cup (L_k \cap G_l)$$

$$= \{L_k \cap (G'_1 \cup G''_1)\} \cup \cdots \cup \{L_k \cap (G'_l \cup G''_l)\}$$

or $$L_k = (L_k \cap G'_k) \cup \{(L_k \cap G''_1) \cup \cdots \cup (L_k \cap G''_l)\} \quad (17.31)$$

since $$L_k \cap G'_r = \phi \text{ for } r \neq k \quad (17.32)$$

In Eq. (17.31) "$(L_k \cap G_1)$" is read: "An input data set is selected from subset G_1 and one of the logic paths of L_k is executed," and so forth. Also, since

$$G'_k \subset L_k$$

then from the result of the first exercise on p. 76

$$L_k \cap G'_k = G'_k \quad (17.33)$$

and furthermore, since

$$G''_i \cap G''_j = \phi, \ i \neq j \quad (17.34)$$

we therefore have

$$P(L_k) = P(G'_k) + \sum_{j=1}^{l} P(L_k \cap G''_j) \quad (17.35)$$

The term $L_k \cap G''_k$ contained within the summation of Eq. (17.35) represents the event that one of the logic paths of L_k is executed, but that the computed correct function differs from required output by more than an allowable tolerance. On the other hand $L_k \cap G''_j$, when $k \neq j$, represents execution of a logic path which computes the wrong function.

The former kind of failure could result for example if the assignment statement $A = B*2$ (B times two) were used in a computer program instead of the correct statement $A = B**2$ (B squared). All input data sets which resulted in $B \neq 2$ would (very likely) result in the function of A (computed by the correct logic path) being incorrect by an amount depending upon the magnitude of B-2.

An exercise involving the latter kind of failure will be given in connection with the triangle problem in Sec. 17.4.4.3.

Now, summing over all groups of logic paths in Eq. (17.35):

$$\sum_{k=1}^{l} P(L_k) = \sum_{k=1}^{l} P(G_k^{'}) + \sum_{k=1}^{l} \sum_{j=1}^{l} P(L_k \cap G_j^{''}) \quad (17.36)$$

The lefthand side of Eq. (17.36) will generally be less than unity, since a program may not terminate at all (be in an endless loop), or may terminate abnormally when some types of faults or "disallowed" states (overflow, etc.) are detected by an operating system. Also, a fault may result in execution of a path that would be logically unexecutable had the fault not been present (the triangle problem example in Sec. 17.4.4.3 has 77 non-logic paths out of a total of 88 structural paths). In these cases then it may happen that *none* of the logic paths corresponding to the $\{G_k\}$ are executed.

From Eqs. (17.30) and (17.36) we may therefore express reliability as

$$R = \sum_{k=1}^{l} P(G_k^{'}) = \sum_{k=1}^{l} P(L_k) - \sum_{k=1}^{l} \sum_{j=1}^{l} P(L_k \cap G_j^{''}) \quad (17.37)$$

In terms of conditional probabilities (Sec. 5.3.1) we can simplify the expression (17.37) to

$$R = \sum_{k=1}^{l} P(L_k) - \sum_{k=1}^{l} \sum_{j=1}^{l} P(L_k | G_j^{''}) P(G_j^{''})$$

$$= \sum_{k=1}^{l} P(L_k) - \sum_{k=1}^{l} \sum_{j=1}^{l} P(G_j^{''} | L_k) P(L_k)$$

or

$$R = \sum_{k=1}^{l} P(L_k) \left[1 - \sum_{j=1}^{l} P(G_j^{''} | L_k) \right] \quad (17.38)$$

but we will not make specific use of the latter representations of R.

17.4.4.3 Interval Partition of the Input Data Space. Like the functional partitions, disjoint intervals of the input variables* are another natural partition of

---

*In general, "intervals" includes discrete sets, primarily integer values, representation of character strings, etc. We are not referring here to the discreteness of real number representation in a computer word.

SOFTWARE RELIABILITY

the input data space. Thus, for example if we define Z_1 as the set of all E_i such that $X_i \leqslant X$ and $Z_2 = \overline{Z}_1$, then $E = Z_1 \cup Z_2, Z_1 \cap Z_2 = \phi$ and therefore Z_1, Z_2 constitute a partition of the input data space E. The usefulness of the Z-partition is that operational usage probabilities can sometimes be more easily expressed for intervals of the input variables than for other partitions.

EXAMPLE: Triangle problem. Given three line segments of integer length, determine whether the triangle formed by the segments does not exist, or is scalene, isosceles or equilateral. For simplicity we assume that the input data space E consists of all triples (I,J,K) such that $1 \leqslant I,J,K \leqslant 4$.

The G-partition can be defined as follows:

Informal specification:

G_1: All (I,J,K) which do not form a triangle
G_2: All (I,J,K) which form a triangle with no two sides equal
G_3: All (I,J,K) which form a triangle with exactly two sides equal
G_4: All (I,J,K) which form a triangle with all three sides equal.

Formal, detailed specification:

G_1: (I,J,K) such that $I + J \leqslant K$ or $I + K \leqslant J$ or $J + K \leqslant I$
G_2: $(I,J,K) \in \overline{G}_1$, and $I \neq J \neq K$
G_3: $(I,J,K) \in \overline{G}_1 \cap \overline{G}_4$ and $I = J$ or $I = K$ or $J = K$
G_4: (I,J,K) such that $I = J = K$

The G-partition of E therefore is as shown in Figure 17.7.

One possible program to solve the triangle problem is shown in Figure 17.8 in the form of a flowchart. This program consists of 88 structural paths but only 11 logic paths. For example any path containing any of segment numbers 2, 4, or 6 and also 8 is not a logic path. Figure 17.9 exhibits the logic paths in the form of strings of segments grouped according to the functional partition G_r.

Suppose segment 14: $I + J \leqslant K$? were mistakenly coded as: $I + J < K$? Upon examination of Fig. 17.8, it will be seen that input data sets (1,1,2) and (2,2,4) belonging to G_1 will now result in executing one of the logic paths of L_3, i.e., an erroneous computation that these sets represent isosceles triangles rather than no triangles. Thus if every input data set has the same probability = 1/64 (the operational profile),

$$P(L_3 \cap G_1'') = P(G_1'') = 2/64$$

but all other $L_k \cap G_j''$ have zero probability.
Hence, from Eq. (17.35)

$$P(L_3) = 24/64 + 2/64 = 26/64$$

$$P(L_1) = 28/64 + 0$$

and otherwise (for $k = 2, 4$)

$$P(L_k) = P(G_k') = P(G_k)$$

$$= \text{(Number of data sets } \epsilon \ G_k)/64$$

Thus, when the program contains the assumed fault,

$$\sum_{k=1}^{4} P(L_k) = 1,$$

since no non-logic paths of the original program will be executed, and

$$R = 62/64 = 0.96875$$

EXERCISE: Suppose that segment 6 of the triangle program were mistakenly coded as "MATCH = NATCH + 3". Assuming that variables used but not set have the value zero, calculate the reliability of the triangle program using the same input data space and operational profile as given in the previous example.

EXERCISE: Rewrite the triangle program so that all structural paths are logic paths. Calculate the reliability of the program resulting from your first attempt.

	1,1,2	1,2,1	2,1,1	1,2,3	1,3,2	3,1,2
	1,1,3	1,3,1	3,1,1	1,2,4	1,4,2	3,2,1
G_1	1,1,4	1,4,1	4,1,1	1,3,4	1,4,3	4,1,2
	2,2,4	2,4,2	4,2,2	2,1,3	2,3,1	4,1,3
				2,1,4	2,4,1	4,2,1
				3,1,4	3,4,1	4,3,1

	2,3,4
	2,4,3
G_2	3,2,4
	3,4,2
	4,2,3
	4,3,2

	1,2,2	2,1,2	2,2,1
	1,3,3	2,3,2	2,2,3
	1,4,4	3,1,3	3,3,1
G_3	2,3,3	3,2,3	3,3,2
	2,4,4	3,4,3	3,3,4
	3,2,2	4,1,4	4,4,1
	3,4,4	4,2,4	4,4,2
	4,3,3	4,3,4	4,4,3

	1,1,1
	2,2,2
G_4	3,3,3
	4,4,4

Fig. 17.7 G-partition of the input data space for the triangle problem.

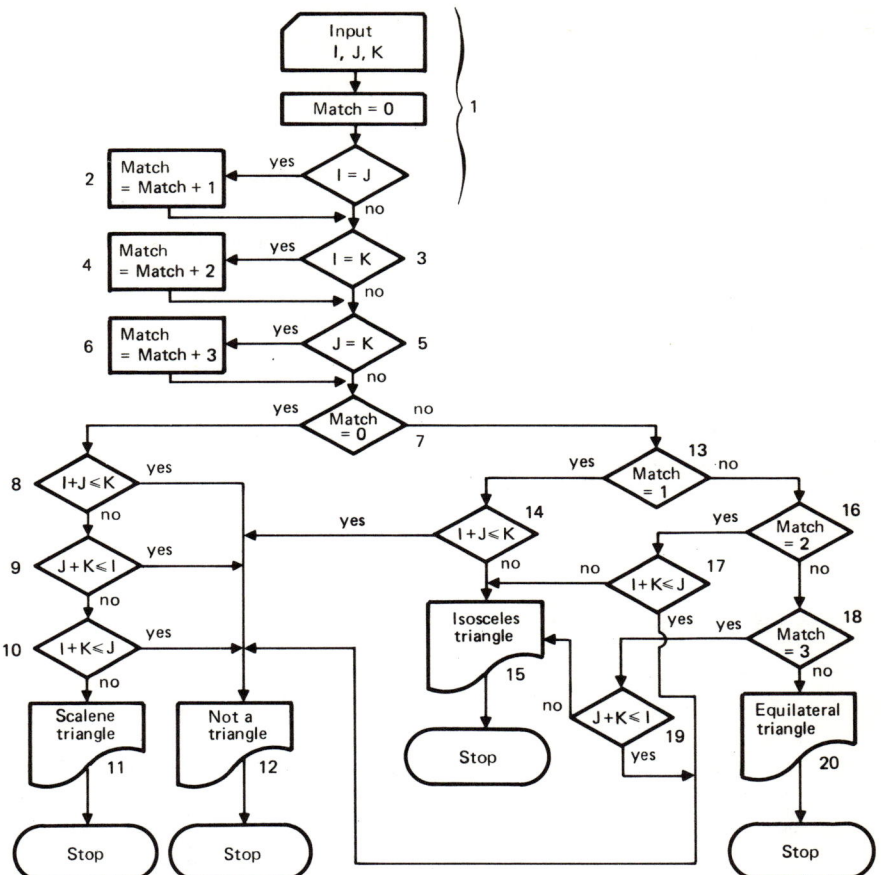

Fig. 17.8 Flow diagram for triangle problem.

$$
L_1 \begin{cases} 1-2-3-5-7-13-14-12 \\ 1-3-4-5-7-13-16-17-12 \\ 1-3-5-6-7-13-16-18-19-12 \\ 1-3-5-7-8-12 \\ 1-3-5-7-8-9-10-12 \\ 1-3-5-7-8-9-12 \end{cases}
$$

$$L_2 \begin{cases} 1-3-5-7-8-9-10-11 \end{cases}$$

$$L_3 \begin{cases} 1-3-5-6-7-13-16-18-19-15 \\ 1-3-4-5-7-13-16-17-15 \\ 1-2-3-5-7-13-14-15 \end{cases}$$

$$L_4 \begin{cases} 1-2-3-4-5-6-7-13-16-18-20 \end{cases}$$

Fig. 17.9 Logic paths for the triangle program

17.4.4.4 Sample Theory for Partition Models.
The discussion on sampling methods in this section is in terms of the general partition S_j, and can therefore be interpreted in terms of the specific G or Z partitions or any other partition which may be usefully defined.

Assume that a preassigned number n_j data points are sampled from S_j, for each j. The sampling is assumed to be simple random; i.e., selection of any point of S_j is unaffected by selection of any other point,* and each has the same chance of being chosen. We merely observe whether a selected point belongs to S_j' or S_j'' so that this type of sampling is simple binomial (Sec. 5.4), and therefore the probability assigned to a sequence of n_j points sampled from S_j is $(P_j'')^{f_j} (P_j')^{n_j - f_j}/(P_j)^{n_j}$, where f_j is the number of points selected from S_j'': those evoking software failure

In general not all S_j may be sampled. If this is the case the estimator R defined below will be biased. Thus if T is the collection of indexes of subsets S_j of the partition which are sampled, and $\overline{T}$ the remaining j's,

$$\hat{R} \equiv 1 - \sum_T (f_j/n_j) P_j \qquad (17.39)$$

and the expected value of $\hat{R}$, denoted by $E(\hat{R})$ is

$$E(\hat{R}) = 1 - \sum_T (P_j''/P_j) P_j = 1 - \sum_T P_j'' \geqslant 1 - \sum_j P_j'' = R \qquad (17.40)$$

where the latter summation is over *all* $j = 1, 2 \ldots$. Therefore, when sampling is incomplete, $\hat{R}$ will be biased on the high side. We now need a measure of the precision of $\hat{R}$.

One such measure is the variance, abbreviated as $V(..)$. Thus, since $V(f_j) = n_j P_j' P_j''/P^2$ (cf. Eq. (6.13)) it can be shown that

$$V_T(\hat{R}) = \sum_T P_j' P_j''/n_j \qquad (17.41)$$

Since, however, $\hat{R}$ is biased for incomplete samples, it is more appropriate to compute the mean-square error, defined by $E[(\hat{R}-R)^2]$. We have

$$E[(\hat{R} - R)^2] = \sum_T \frac{P_j' P_j''}{n_j} + \left(\sum_{\overline{T}} P_j''\right)^2 \qquad (17.42)$$

EXERCISE: Derive formula (17.42). Hint: Similar to the procedure followed on pp. 109–110.

---

*We assume that there are so many points in S_j that the effect of replacement or nonreplacement on the probabilities of selection can be neglected.

SOFTWARE RELIABILITY

Thus the mean-square error can never be less than the number

$$\left(\sum_{\overline{T}} P_j''\right)^2$$

even should the n_j all become large. This makes $E[(\hat{R} - R)^2]$ a measure of sampling thoroughness in a certain sense; however there is no way of calculating it, for although we have information on the sizes of P_j' or P_j'' for $j \in T$, nothing is known about P_j'' for $j \in \overline{T}$, except the crude bounds $0 \leqslant P_j'' \leqslant P_j$. Now it is easy to show that

$$E\left(\sum_T \frac{f_j(n_j - f_j)}{n_j^2(n_j - 1)} P_j^2\right) = V_T(\hat{R}) \qquad (17.43)$$

and therefore, using the crude bounds on P_j'' for $j \in \overline{T}$, an approximate set of numerical* lower and upper bounds on the mean-square error are

$$\sum_T \frac{f_j(n_j - f_j)}{n_j^2(n_j - 1)} P_j^2 \leqslant \left\langle E[(\hat{R} - R)^2] \right\rangle \leqslant \sum_T \frac{f_j(n_j - f_j)}{n_j^2(n_j - 1)} P_j^2 + \left(\sum_{\overline{T}} P_j\right)^2 \qquad (17.44)$$

The existence of the above unbiased estimator for $V_T(\hat{R})$ evidently requires that every $n_j \geqslant 2$.

EXERCISE: Derive formula (17.43).

The variance of

$$\hat{R} = 1 - \sum_j (f_j/n_j) P_j$$

for a complete sample being given by

$$V(\hat{R}) = \sum_j \frac{P_j' P_j''}{n_j} \qquad (17.45)$$

the question arises whether the n_j can be chosen in some manner to make $V(\hat{R})$ as small as possible. Thus for a given

$$n = \sum_k n_k \qquad (17.46)$$

---

*"⟨ ⟩" denotes a numerical estimate of the quantity within.

if we write

$$W = \sum_k \frac{P'_k P''_k}{n_k} + \lambda^2 \left(\sum_k n_k - n\right)$$

then solve the set of equations $\partial W/\partial n_j = 0$ for all j together with the constraint (17.46) we may obtain an extremal solution for the n_j which can then be shown to yield a minimum $V(\hat{R})$. Thus

$$\frac{\partial W}{\partial n_j} = -\frac{P'_j P''_j}{n_j^2} + \lambda^2 = 0$$

Consequently

$$\lambda = \frac{1}{n} \sum_k \sqrt{P'_k P''_k}$$

Therefore

$$n_j \simeq \frac{n\sqrt{P'_j P''_j}}{\sum_k \sqrt{P'_k P''_k}} \qquad (17.47)$$

(The "$\simeq$" is used since n_j is an integer.)

This choice of n_j can be verified to yield a minimum variance. Thus, substituting the solutions for the n_j, the minimum value of the variance becomes

$$V_{min}(R) = \frac{1}{n}\left(\sum_j \sqrt{P'_j P''_j}\right)^2 \qquad (17.48)$$

Of course, unless the values of P''_j are known before sampling takes place, the optimum choice of the n_j cannot be deliberately made. If a preliminary sample were taken, however, and assuming no significant change in the values of P''_j (e.g., as a result of correcting the detected faults), the final sample of n_j from each subset S_j could be chosen to yield a near-minimal variance estimate.

EXAMPLE:

Let $P_1 = 0.5 \ P_2 = 0.3 \ P_3 = 0.2$

Let $P''_1 = 0.1 \ P''_2 = 0.1 \ P''_3 = 0$

(and therefore) $P'_1 = 0.4 \ P'_2 = 0.2 \ P'_3 = 0.2$

SOFTWARE RELIABILITY

Assume that $n = 100$, and that the n_j are chosen proportional to P_j so that $n_1 = 50$, $n_2 = 30, n_3 = 20$. Then

$$V(R) = \frac{(0.1)(0.4)}{50} + \frac{(0.1)(0.2)}{50} + 0$$

$$= 0.00080 + 0.00067 = 0.00147$$

On the other hand, if the n_j had been chosen to yield $V_{min}(\hat{R})$,

$$n_1 = \frac{\sqrt{0.04}\,(100)}{\sqrt{0.04} + \sqrt{0.02}} = \frac{0.2\,(100)}{0.3414} = 53.7 \simeq 59$$

$$n_2 = \frac{0.1414\,(100)}{0.3414} = 41.3 \simeq 41$$

$$n_3 = 0$$

$$V_{min}(R) = \frac{0.04}{59} + \frac{0.02}{41}$$

$$= 0.00068 + 0.00049$$

$$= 0.00117$$

17.4.4.5 Approximate Confidence Limits for R. Several measures of "confidence" for software reliability may be developed from the sampling theory presented in the previous sections. The term "confidence" has also been used to denote degree of representativeness, thoroughness, etc.; of the test strategy or sampling technique as discussed in Ref. 36, but here we will use the term as defined on p. 169 and in App. 8A.

If V denotes an estimator for the variance of $\hat{R}$, then based on asymptotic normality arguments similar to those of Sec. 8.3,

$$\hat{R} \pm z_{\frac{1-\gamma}{2}} \sqrt{V} \qquad (17.49)$$

will be an approximate confidence interval covering R with confidence γ, where z_a is the standard normal deviate exceeded with probability a. The confidence interval (17.49) would be expected to be more correct when the numbers of tests n_j for each subset S_j were large. For smaller numbers of tests $z_{(1-\gamma)/2}$ should be replaced by $t_{n;(1-\gamma)/2}$, the Student t-deviate with n degrees of freedom, exceeded with probability $(1-\gamma)/2$ (p. 169).

We are generally interested in one-sided confidence intervals of the form $\hat{R}_L < R < 1$, however, and for γ confidence, the one-sided intervals for the previous cases would then be

$$\hat{R} - z_{1-\gamma}\sqrt{V} \quad or \quad \hat{R} - t_{n;1-\gamma}\sqrt{V} \qquad (17.50)$$

In any of the above cases, if sampling were incomplete, the intervals cover

$$R = \sum_T P_j' \leq \sum_j P_j'$$

Consequently the true value of R would be covered by the confidence interval with confidence $\gamma' \leq \gamma$.

The magnitude of the difference between γ and γ' is not easily estimated; however, a "safety margin" could be incorporated by using slightly higher values of γ in determining $z_{1-\gamma}$ or $t_{n;1-\gamma}$. These differences are not likely to be important, however, in that the whole procedure already has many elements of approximation.

The approach to finding lower confidence limits on reliability given in the next paragraph avoids the asymptotic normality assumption, and can be termed an "exact" method.

17.4.4.6 Exact Confidence Limit for R. For the software reliability model, since the actual reliability with respect to the jth subset of a partition is $P_j'/P_j \equiv R_j$, the problem becomes one of determining a system of lower confidence limits on the quantity

$$R = \sum_j R_j P_j \qquad (17.51)$$

for all possible outcomes f_j and given sample sizes n_j, $j = 1, 2 \ldots k$.

Since the general formulation of the classical confidence limits problem is cumbersome, the derivation will be given only for the case in which there are two subsets in the partition. In this case

$$R = R_1 P_1 + R_2 P_2 \qquad (17.52)$$

where
$$P_1 + P_2 = 1$$

As mentioned in Sec. 9.2.3, based upon Ref. 37, a lower confidence limit on R for observed numbers of failures f_1, f_2 out of n_1, n_2 tests, respectively, is obtained by minimizing (with respect to X_1, X_2) the function $X_1 P_1 + X_2 P_2$ subject to the equality constraint:

$$\sum_{i_1, i_2 = 0}^{f_1, f_2} \binom{n_1}{i_1} X_1^{n_1 - i_1} (1 - X_1)^{i_1} \binom{n_2}{i_2} X_2^{n_2 - i_2} (1 - X_2)^{i_2} = 1 - \gamma \qquad (17.53)$$

and also the inequalities

$$0 \leq X_1, X_2 \leq 1$$

where γ is the confidence coefficient.

The resulting lower confidence limit is denoted by

$$\hat{R}_L(f_1, f_2; n_1, n_2, \gamma)$$

The quantity summed in Eq. (17.53) is the product of the probabilities of observing exactly i_1 failures out of n_1 tests and i_2 failures out of n_2 tests, where sampling is from the respective subsets S_1, S_2 and are independent observations (both within each subset and between subsets). Not apparent in the summation is a prescribed ordering of the sample outcomes $(f_1, f_2; n_1, n_2)$. The ordering is not unique as shown in Ref. 37, but there are various ways of choosing an ordering which is "more or less" optimum. Here, the meaning of "optimum" is that for every f_1, f_2, each of the confidence limits $\hat{R}_L(f_1, f_2; n_1, n_2, \gamma)$ for the prescribed ordering are at least equal to the corresponding $\hat{R}'_L(f_1, f_2; n_1, n_2, \gamma)$ obtained from some other ordering; i.e., all other things being equal, the larger confidence limits are better.

One way of ordering the sample points is: for any observed (f_1, f_2, n_1, n_2) to include in the summation the probabilities of all points $(i_1, i_2; n_1, n_2)$ such that $0 \leq i_1 \leq f_1$; $0 \leq i_2 \leq f_2$. This ordering will be satisfactory except when n_1 and n_2 are very different. A better ordering is obtained by including all sample points $(i_1, i_2; n_1, n_2)$ for which the estimated reliabilities have the relationship

$$\frac{n_1 - i_1}{n_1} P_1 + \frac{n_2 - i_2}{n_2} P_2 \geq \frac{n_1 - f_1}{n_1} P_1 + \frac{n_2 - f_2}{n_2} P_2$$

which should result in values of $\hat{R}_L$ which are ordered in the same way as the estimated reliabilities, although this assertion could not be verified. In any suitable ordering, however, the sample outcome $(0, 0; n_1, n_2)$ will be the first point, corresponding to the largest $\hat{R}_L$ for the set of possible outcomes.

For this case (17.53) becomes simply

$$X_1^{n_1} X_2^{n_2} = 1 - \gamma \tag{17.54}$$

Using the method of Lagrange multipliers, it is easy to show that the minimum value of $X_1 P_1 + X_2 P_2$ subject to the constraints (17.54) and $0 \leq X_1, X_2 \leq 1$ is

$$\hat{R}_L = n(1-\gamma)^{1/n} \left(\frac{P_1}{n_1}\right)^{\frac{n_1}{n}} \left(\frac{P_2}{n_2}\right)^{\frac{n_2}{n}} \tag{17.55}$$

where $n = n_1 + n_2$ and $P_1 + P_2 = 1$, provided that the righthand side of Eq. (17.55) is no greater than

$$\min (nP_1/n_1, nP_2/n_2) \tag{17.56}$$

If the above condition is not satisfied, then $\hat{R}_L$ is given by

$$\hat{R}_L = \min (P_1 + P_2 (1 - \gamma)^{1/n_2}, P_1(1 - \gamma)^{1/n_1} + P_2) \tag{17.57}$$

which means either $X_1 = 1$ or $X_2 = 1$ in the solution.

The generalization to k subsets in the partition results in

$$\hat{R}_L = n(1 - \gamma)^{1/n} \prod_{j=1}^{k} \left(\frac{P_j}{n_j}\right)^{n_j/n} \tag{17.58}$$

provided that the righthand side of Eq. (17.58) is no greater than

$$\min_i (nP_i/n_i) \tag{17.56'}$$

Otherwise, one or more of the $X_i = 1$, and the solution becomes somewhat more complicated.*

EXAMPLE: $n_1 = 10, n_2 = 4; P_1 = 0.2, P_2 = 0.8; \gamma = 0.90$.

The righthand side of (17.55) is

$$14(0.10)^{1/14}(0.02)^{10/14}(0.20)^{4/14} = 0.45861$$

Since
$$\min (14(0.2)/10, 14(0.8)/4) = 0.28 < 0.45861$$

condition (17.56) is not satisfied, and from (17.57)

$$\hat{R}_L = \min (0.2 + 0.8(0.10)^{1/4}, 0.2(0.10)^{1/10} + 0.8)$$
$$= \min (0.64987, 0.95887) = 0.64987$$

EXERCISE:
a. Work out the details for Eq. (17.58) and the condition (17.56'). Develop a solution procedure for $k = 3$ subsets in the partition.
b. Let $n_1 = 8, n_2 = 4, n_3 = 2; P_1 = 0.2, P_2 = 0.4, P_3 = 0.4; \gamma = 0.90$. Find the solution for $\hat{R}_L$.
Answer: $\hat{R}_L = 0.71502$

When sampling is proportional to the probability of the subset; i.e., $n_j = nP_j$, then

$$\hat{R}_L = n(1 - \gamma)^{1/n} \frac{1}{n} = (1 - \gamma)^{1/n}$$

*This is a change from the 1st printing, which erroneously gave (17.58) as the complete solution, overlooking the constraints $X_i \leq 1$, which are now accounted for by (17.56'). The previous solution gives values of $\hat{R}_L$ no greater than the correct values, so is conservative, with confidence at least γ. The authors were made aware of the correction from an approximate linear programming solution given in an unpublished paper by J. Duran and J. Wiorkowski, Univ. of Texas.

which is the usual lower binomial confidence limit on reliability obtained when n tests are made with zero failures. This is of course equivalent to the case when there is only one subset in the partition.

For the general case when the f_j are not necessarily zero, the method of Sec. 9.2.3 may be used. For this software reliability model, one computes

$$R = \sum_{j=1}^{k} \frac{n_j - f_j}{n_j} P_j \qquad (17.59)$$

and the quantity $F_{eq} = N_m (1-\hat{R})$, representing the "equivalent number of failures," where N_m is the minimum of the n_j. The procedure is then the same as given in the example on p. 227.

EXAMPLE:

$n_1 = 50, f_1 = 5; n_2 = 100, f_2 = 10; P_1 = 0.40, P_2 = 0.60$

$\hat{R} = 0.90 (0.40) + 0.90 (0.60)$
$= 0.90$

Therefore

$N_m = 50$

and

$F_{eq} = 50 (1-0.90) = 5$

and entering Figs. A.1, A.3, and A.4, for confidence $\gamma = 0.50, 0.90, 0.95$, the values of $\hat{R}_L$ are, respectively, 0.89, 0.82 and 0.80.

REFERENCES

1. B.W. Boehm, "Keynote Address: The High Cost of Software," presented at the ONR/OAR/AFOSR Symposium on the High Cost of Software, Monterey, California, September 1973 (also published in the TRW Software Series* TRW-SS-73-08).
2. *Implications of Using Modular Programming,* John Hoskyns and Company, Ltd., Guide No. 1, Hoskyns System Research, Inc., 600 Third Avenue, New York (1973).
3. "That Maintenance Iceberg," *EDP Analyzer,* Vol. 10, No. 10, October 1972.

*TRW Technical Information Center, Mail Station S/1930, One Space Park, Redondo Beach, California 90278, USA.

4. B.W. Boehm, "Software and Its Impact: A Quantitative Assessment," *Datamation,* May 1973 (also published in the TRW Software Series TRW-SS-73-04).
5. B.W. Boehm, "Software Engineering," *IEEE Trans. Comput.,* Vol. C-25, No. 12, pp. 1226-1241, December 1976. Also published in the TRW Software Series TRW-SS-76-08, October 1976.
6. J.H. Manley and M. Lipow, "Findings and Recommendations of the Joint Logistics Commanders' Software Reliability Work Group," Vols. I and II, AFSC TR-75-05, November 1975, Vol. II, pp. 47-48. Available from National Technical Information Service, Department of Commerce, Springfield, Virginia 22151, AD-A018881 and AD-A018882.
7. B.W. Boehm, J.R. Brown, and M. Lipow, "Quantitative Evaluation of Software Quality," *Proceedings, 2nd International Conference on Software Engineering,* October 1976, IEEE Cat. No. 76CH1125-4C, pp. 592-605.
8. T.A. Thayer, M. Lipow, and E.C. Nelson, "Software Reliability Study," TRW Systems Report to Rome Air Development Center, Contract F30602-74-C-0036, March 1976 (also published in the TRW Software Series TRW-SS-76-03).
9. T.E. Bell and T.A. Thayer, "Software Requirements: Are They a Problem?," *Proceedings, 2nd International Conference on Software Engineering,* October 1976, IEEE Cat. No. 76CH1125-4C, pp. 61-68 (previously published in the TRW Software Series TRW-SS-76-04, July 1976).
10. DoD Directive 5000.29, "Management of Computer Resources in Major Defense Systems," 26 April 1976.
11. DoD Instruction 5000.31, "Interim List of DoD Approved High Order Programming Languages (HOL)," 24 November 1976.
12. D.W. Kosy, "The ECSS II Language for Simulating Computer Systems," The Rand Corporation, Report No. R-1895-GSA, December 1975.
13. Caine, Farber & Gordon, Inc., *Program Design Language* (Copyright 1974, 1975).
14. R.C. Tausworthe, *Standardized Development of Computer Software,* Part I, *Methods,* Jet Propulsion Laboratory, California Institute of Technology, Pasadena, California (1976).
15. E. Yourdon, *Techniques of Program Structure and Design,* Prentice-Hall, Inc., Englewood Cliffs, New Jersey (1975).
16. J.N. Buxton and B. Randall (eds.) *Software Engineering Techniques,* NATO Scientific Affairs Division.
17. M.H. Halstead, *Elements of Software Science,* North American Elsevier, New York (1977).
18. M.E. Fagan, "Design and Code Inspections to Reduce Errors in Program Development," *IBM Syst. Journ.,* Vol. 15, No. 3, 1976, pp. 182-211.

19. H. Hecht, "Measurement, Estimation, and Prediction of Software Reliability," *Volume on Software Engineering Techniques*, Infotech International Series (in press; to be published 1977).
20. Y. Funami and M.H. Halstead, "A Software Physics Analysis of Akiyama's Debugging Data," *Proceedings Symposium Computer Software Engineering*, April 20-22 1976, Vol. XXIV, Polytechnic Institute of New York, Microwave Research Institute Symposia Series, pp. 133-138.
21. J.D. Musa, "A Theory of Software Reliability and Its Applications," *IEEE Trans. Software Eng.*, Vol. SE-1, No. 3, pp. 312-327, September 1975.
22. R.J. Rubey and R.D. Hartwick, "Quantitative Measurement of Program Quality," *Proceedings ACM National Conference*, 1968, pp. 671-677.
23. J.R. Brown and M. Lipow, "The Quantitative Measurement of Software Safety and Reliability," TRW Software Series TRW-SS-73-06, in press 1977, originally published as TRW Report No. SD1776, August 1973.
24. B.W. Boehm, J.R. Brown, H. Kaspar, M. Lipow, G.J. MacLeod, and M.J. Merritt, *Characteristics of Software Quality*, North-Holland Publishing Co., New York (in press 1977). Previously published in the TRW Software Series TRW-SS-73-09, 28 December 1973.
25. S.R. Searle, *Linear Models*, John Wiley & Sons, Inc., New York (1971), pp. 95-98.
26. Z. Jelinski and P.B. Moranda, "Software Reliability Research," in *Statistical Computer Performance Evaluation*, W. Freiberger, ed., Academic Press (1972).
27. N.F. Schneidewind, "Analysis of Error Processes in Computer Software," *Proceedings 1975 International Conference on Reliable Software*, April 21-23, 1975, IEEE Cat. No. 75CH0940-7CSR, pp. 337-346.
28. E.C. Nelson, "A Statistical Basis for Software Reliability Assessment," TRW Software Series TRW-SS-73-03, March 1973.
29. E.C. Nelson, "Software Reliability," TRW Software Series TRW-SS-75-05 November 1975.
30. M.L. Shooman, "Operational Testing and Software Reliability Estimation During Program Development," *Record, 1973 IEEE Symposium on Computer Software Reliability*, April 30–May 2, 1973, IEEE Cat. No. 73CH0741-9CSR, pp. 51-57.
31. M.L. Shooman, "Structural Models for Software Reliability Prediction," *Proceedings 2nd International Conference on Software Engineering*, October 13-15, 1976, IEEE Cat. No. 76CH1125-4C, pp. 268-280.
32. A.N. Sukert, "An Investigation of Software Reliability Models," *Proceedings 1977 Annual Reliability and Maintainability Symposium*, January 18-20, 1977, IEEE Cat. No. 77CH1161-9RQC, pp. 478-484.
33. R.W. Wolverton and G.J. Schick, "Assessment of Software Reliability," TRW Software Series TRW-SS-73-04, September 1972.

34. A.V. Fiacco and G.P. McCormick, *Nonlinear Programming: Sequential Unconstrained Minimization Techniques,* John Wiley & Sons, Inc., New York (1968).
35. D.J. Reifer, "A New Assurance Technology for Computer Software," *Proceedings 1976 Annual Reliability and Maintainability Symposium,* January 20-22, 1976, IEEE Cat. No. 76CH01044-7RQC, pp. 446-451.
36. J.R. Brown and M. Lipow, "Testing for Software Reliability," *Proceedings 1975 International Conference on Reliable Software,* April 1975, IEEE Cat. No. 75CH0940-7CSR, pp. 518-527. Also published in the TRW Software Series TRW-SS-75-02, January 1975.
37. R.J. Buehler, "Confidence Intervals for the Product of Two Binomial Parameters," *J. Am. Stat. Assoc.,* Vol. 52, pp. 482-493 (1952).

ADDITIONAL READING

General

Gilb, T., *Software Metrics,* Winthrop Publishing Co, New Jersey (1976).

Myers, G.J., *Software Reliability Principles and Practices,* John Wiley & Sons, Inc., New York (1976).

Symposia on Reliable Software and Software Engineering

Computer Science and Statistics: 8th Annual Symposium on the Interface, Workshop 4 – *Approaches for Programmers to Application Software Validation,* pp. 290–370, Feb. 13–14, 1975, Health Sciences Computing Facility, AV-111, UCLA, Los Angeles, California 90024.

Proceedings Annual Reliability and Maintainability Symposia:
Jan 18–20, 1977 (pp. 478–499), IEEE Cat. No. 77CH1161–9RQC.
Jan 20–22, 1976 (pp. 434–451), IEEE Cat. No. 76CHO–1044–7RQC.
Jan 28–30, 1975 (pp. 476–497), IEEE Cat. No. 75CHO–918–3RQC.

Proceedings 1975 International Conference on Reliable Software, April 21–23, 1975, IEEE Cat. No. 75CH0940–7CSR.

Proceedings of an ACM Conference on Language Design for Reliable Software, March 28–30, 1977, Association for Computing Machinery, P.O. Box 12105, Church St. Station, New York, NY 10249.

Proceedings of the TRW Symposium on Reliable, Cost-Effective, Secure Software, March 20–21, 1974, TRW Software Series, TRW–SS–74-14.

Proceedings 2nd International Conference on Software Engineering, October 13–15, 1976, IEEE Cat. No. 76CH1125–4C.

Proceedings Symposium on Computer Software Engineering, April 20–22, 1976, Vol. XXIV, Polytechnic Institute of New York Microwave Research Institute Symposia Series.

Record 1973 IEEE Symposium on Computer Software Reliability, April 30–May 2, 1973, IEEE Cat. No. 73CH0741–9CSR.

Wortman, D.B., (ed.), "Notes from a Workshop on the Attainment of Reliable Software," University of Toronto, Computer Systems Research Group, Technical Report CSRG–41, September 1974.

Software Requirements Engineering

IEEE Trans. Software Eng., Vol. SE–3, No. 1, January 1977 (special collection on Requirements Analysis, pp. 2–69).

Software Errors

Endres, A., "An Analysis of Errors and their Causes in System Programs," *IEEE Trans. Software Eng.,* Vol. SE–1, No. 2, June 1975, pp. 140–149.

Fosdick, L.D., and Osterweil, L.J., "Data Flow Analysis in Software Reliability," *ACM Computing Surveys,* Vol. 8, No. 3, September 1976, (special issue on Reliable Software, Part I: *Software Validation*), pp. 305–330.

Rudner, B., "Seeding/Tagging Estimates of the Number of Software Errors," New York Polytechnic Institute Report, Electrical Engineering and Electrophysics Department, November 1976.

Shooman, M.L., and Bolsky, M.I., "Types, Distribution, and Test and Correction Times for Programming Errors," *Proceedings 1975 International Conference on Reliable Software,* April 1975, IEEE Cat. No. 75CH0940–7CSR, pp. 347–362.

Program Testing and Test Tools

Gibson, C.G., and Railing, L.R., " Verification Guidelines," TRW Software Series Report No. TRW–SS–71–04, August 1971.

Goodenough, J.B., and Gerhart, S.L., "Toward a Theory of Test Data Selection," *IEEE Trans. Software Eng.,* Vol. SE–1, No. 2, June 1975, pp. 156–173.

Hetzel, W.C., ed., *Program Test Methods,* Prentice-Hall, Inc., Englewood Cliffs, New Jersey (1973).

IEEE Trans. Software Eng., Vol. SE–2, No. 3, September 1976 (special section on testing) pp. 194–231.

Panzl, D.J., "Test Procedures: A New Approach to Software Verification," *Proceedings 2nd International Conference on Software Engineering,* 13–15 October 1976, San Francisco, IEEE Cat. No. 76CH1125–4C.

Reifer, D.J., "Automated Aids for Reliable Software," *Proceedings 1975 International Conference on Reliable Software,* April 1975, IEEE Cat. No. 75CH0940-7CSR, pp. 131–142.

Program Proof of Correctness and Verification Methodologies

Elspas, B., Levitt, K.N., Waldinger, R.J., and Waksman, A., "An Assessment of Techniques for Proving Program Correctness," *ACM Computing Surveys,* Vol. 4, No. 2, June 1972, pp. 97–147.

Hantler, S.L., and King, J.C., "An Introduction to Proving the Correctness of

Programs," *ACM Computing Surveys,* Vol. 8, No. 3, September 1976, (special issue on Reliable Software, Part I: *Software Validation*) pp. 331–353.

Knuth, D.E., *The Art of Computer Programming,* Vol. I, *Fundamental Algorithms,* Second Edition, Addison-Wesley Publishing Co., Reading, Massachusetts, (1973), pp. 14–20.

London, R.L., "A View of Program Verification,"*Proceedings 1975 International Conference on Reliable Software,* April 1975, IEEE Cat. No. 75CH0940–7CSR, pp. 534–545.

Stucki, L.G., and Foshee, G.L., "New Assertion Concepts for Self-Metric Software Validation," *Proceedings 1975 International Conference on Reliable Software,* April 1975, IEEE Cat. No. 75CH0940–7CSR, pp. 59–71.

Program Design, Style and Structured Programming

ACM Computing Surveys, Vol. 6, No. 4, December 1974 (special issue on Programming).

Ashcroft, E., and Manna, F., "The Translation of 'GO–TO' Programs to 'While' Programs," *Proceedings 1971 IFIP Congress,* Ljubljana, Yugoslavia, 23–28 August 1971. American Elsevier Publishing Co., New York (1972).

Beyer, T., "Preprocessors and Programming Language Reform," *Proceedings of Computer Science and Statistics: 8th Annual Symposium on the Interface,* February 13–14, 1975. University of California at Los Angeles, Health Sciences Computing Facility AV–111, UCLA, Los Angeles, California 90024, pp. 306–309.

Brinch Hansen, P., *Operating System Principles,* Prentice-Hall, Inc., Englewood Cliffs, New Jersey (1973).

Dahl, O.J., Dijkstra, E.W., and Hoare, C.A.R., *Structured Programming,* Academic Press, New York (1972).

de Balbine, G., "Using the Fortran Structuring Engine," *Proceedings of Computer Science and Statistics: 8th Annual Symposium on the Interface,* February 13–14, 1975. University of California at Los Angeles, pp. 297–306.

Dijkstra, E.W., "Structured Programming," *Software Engineering Techniques,* NATO Scientific Affairs Division, eds. *Buxton, J.N., and Randall, B.* (1970) pp. 84–88.

Kernighan, B.W., and Plauger, P.J.,*The Elements of Programming Style,* McGraw-Hill Book Co., New York (1974).

Melton, R.A., "Automatically Translating FORTRAN to IFTRAN," *Proceedings of Computer Science and Statistics: 8th Annual Symposium on the Interface,* February 13–14, 1975. University of California at Los Angeles, pp. 291–296.

Mills, H.D., "Top Down Programming in Large Systems," *Debugging Techniques in Large Systems,* (ed.), R. Rustin, Prentice-Hall, Inc., Englewood Cliffs, New Jersey (1971), pp. 43–55.

Mills, H.D., "Mathematical Foundations for Structured Programming," IBM Federal Systems Division Report No. FSC 72–6012, February 1972.

Structured Programming Series: Reports No. TR74–300, under RADC Contract F30602–74–C–0186. (A series of 15 volumes and a final report, dated July 1975, by IBM).

Weinberg, G.M., *The Psychology of Computer Programming,* Van Nostrand Reinhold Company, New York (1971).

APPENDIX

TABLES

Table A.1	Number of tests without failure *vs.* reliability and confidence	545
Table A.2	Minimum size of sample to be tested for a time t to assure a mean life of at least $\hat{\theta}_L$ with confidence $\gamma = 75$ per cent when F is the allowable number of failures	546
Table A.3	Minimum size of sample to be tested for a time t to assure a mean life of at least $\hat{\theta}_L$ with confidence $\gamma = 80$ per cent when F is the allowable number of failures	547
Table A.4	Minimum size of sample to be tested for a time t to assure a mean life of at least $\hat{\theta}_L$ with confidence $\gamma = 85$ per cent when F is the allowable number of failures	548
Table A.5	Minimum size of sample to be tested for a time t to assure a mean life of at least $\hat{\theta}_L$ with confidence $\gamma = 90$ per cent when F is the allowable number of failures	549
Table A.6	Minimum size of sample to be tested for a time t to assure a mean life of at least $\hat{\theta}_L$ with confidence $\gamma = 95$ per cent when F is the allowable number of failures	550
Table A.7	Sample size and criteria to demonstrate reliability at a given confidence	551
Table A.8	Percentage points of the χ_n^2/n distribution	552
Table A.9	Tolerance factors for normal distributions	554

FIGURES

Figure A.1	Upper confidence limit on unreliability (one minus lower confidence limit on reliability) number of trials N, observed failures F, confidence coefficient $\gamma = 0.50$	556
Figure A.2	Upper confidence limit on unreliability (one minus lower confidence limit on reliability) number of trials N, observed failures F, confidence coefficient $\gamma = 0.80$	557
Figure A.3	Upper confidence limit on unreliability (one minus lower confidence limit on reliability) number of trials N, observed failures F, confidence coefficient $\gamma = 0.90$	558
Figure A.4	Upper confidence limit on unreliability (one minus lower confidence limit on reliability) number of trials N, observed failures F, confidence coefficient $\gamma = 0.95$	559
Figure A.5	Upper confidence limit on unreliability (one minus lower confidence limit on reliability) number of trials N, observed failures F, confidence coefficient $\gamma = 0.99$	560
Figure A.6	50 per cent lower confidence limit on system reliability for observed failure combinations of a two-subsystem serial system, N trials per subsystem	561
Figure A.7	90 per cent lower confidence limit on system reliability for observed failure combinations of a two-subsystem serial system, N trials per subsystem	562
Figure A.8	95 per cent lower confidence limit on system reliability for observed failure combinations of a two-subsystem serial system, N trials per subsystem	563

Figure A.9	50 per cent lower confidence limit on system reliability for observed failure combinations of a three-subsystem serial system, N trials per subsystem	564
Figure A.10	90 per cent lower confidence limit on system reliability for observed failure combinations of a three-subsystem serial system, N trials per subsystem	565
Figure A.11	95 per cent lower confidence limit on system reliability for observed failure combinations of a three-subsystem serial system, N trials per subsystem	566

Table A.1. Number of Tests without Failure versus Reliability and Confidence

Reliability	Confidence Level (%)							
	50	60	70	80	90	95	99	99.9
.9999	6,932	9,163	12,040	16,094	23,026	29,957	46,052	69,078
.999	693	916	1,204	1,609	2,303	2,996	4,605	6,908
.998	347	458	602	805	1,152	1,498	2,303	3,454
.997	231	305	401	537	768	999	1,535	2,303
.996	173	229	301	401	575	747	1,149	1,723
.995	138	183	241	321	460	598	920	1,379
.994	115	152	201	267	383	498	765	1,148
.993	99	130	174	229	328	427	657	985
.992	86	114	150	200	287	373	574	860
.991	77	101	134	178	255	332	510	764
.99	69	92	120	160	229	298	459	688
.98	34	45	60	80	114	149	228	342
.97	23	30	40	53	76	99	151	227
.96	17	23	30	39	57	74	113	170
.95	14	18	24	31	45	58	90	135
.94	11	15	20	26	37	49	75	112
.93	10	13	17	22	32	42	64	96
.92	9	11	15	19	28	36	55	83
.91	8	10	13	17	25	32	49	74
.90	7	9	12	15	22	29	44	66
.89	6	8	11	14	20	26	40	60
.88	6	8	10	13	18	24	36	54
.87	5	7	9	12	17	22	33	50
.86	5	7	8	11	16	20	31	46
.85	5	6	8	10	15	19	29	43
.80	3	4	6	7	11	14	21	31
.75	3	4	5	6	8	11	16	24
.70	2	3	4	5	7	9	13	20
.65	2	2	3	4	6	7	11	16
.60	2	2	3	4	5	6	9	14
.50	1	2	2	3	4	5	7	10

Table A.2. Minimum Size of Sample to be Tested for a Time t to Assure a Mean Life of at Least $\hat{\theta}_L$ with Confidence $\gamma = 75\%$ when F is the Allowable Number of Failures †

($\gamma = 75\%$)

F \ $t/\hat{\theta}_L$	1.0	0.5	0.2	0.1	0.05	0.02	0.01	0.005	0.002	0.001	0.0005	0.0002	0.0001
0	2	3	7	14	28	70	139	278	694	1,387	2,773	6,932	13,863
1	4	6	14	28	55	136	270	540	1,347	2,694	5,386	13,464	26,927
2	6	9	21	41	80	198	394	786	1,962	3,922	7,842	19,603	39,205
3	7	12	28	53	104	257	513	1,024	2,557	5,111	10,221	25,549	51,096
4	9	15	34	65	128	316	630	1,257	3,140	6,277	12,551	31,374	62,746
5	11	18	40	77	151	374	745	1,488	3,714	7,426	14,848	37,116	74,230
6	13	21	46	89	175	431	859	1,715	4,283	8,562	17,120	42,795	85,588
7	14	24	53	101	198	488	972	1,941	4,846	9,688	19,373	48,426	96,848
8	16	26	59	113	221	545	1,085	2,165	5,406	10,807	21,609	54,016	108,030
9	18	29	65	124	243	601	1,196	2,388	5,962	11,919	23,832	59,574	119,140
10	19	32	71	136	266	656	1,307	2,609	6,515	13,025	26,045	65,103	130,200
11	21	35	77	147	288	712	1,418	2,830	7,066	14,126	28,247	70,608	141,210
12	23	37	83	159	311	767	1,528	3,050	7,615	15,224	30,441	76,092	152,180
13	24	40	89	170	333	823	1,638	3,269	8,162	16,317	32,627	81,558	163,110
14	26	43	95	182	356	878	1,747	3,487	8,707	17,407	34,807	87,006	174,010

† *Remarks:*
1. It should be noted that only the ratio of t (the test time for each item) and $\hat{\theta}_L$ is needed to enter the table.
2. The quantity $(1 - \gamma)$ is equivalent to the maximum customer's risk at $\hat{\theta}_L$.
3. The last four or five columns above are included to emphasize the magnitude of the sample size required and it is not expected that they will be used without careful consideration of how well the exponential distribution assumption fits the aging process under study.

Tables A.2–A.6 are reprinted with the kind permission of the Editorial Department, Institute of Radio Engineers Inc., from M. Sobel and J. A. Tischendorf, "Acceptance Sampling with New Life Test Objectives," *Proc. 5th National Symposium on Reliability and Quality Control in Electronics,* Philadelphia, January 12–14, 1959, pp. 108–118.

Table A.3. Minimum Size of Sample to be Tested for a Time t to Assure a Mean Life of at Least $\hat{\theta}_L$ with Confidence $\gamma = 80\%$ When F is the Allowable Number of Failures†

($\gamma = 80\%$)

F \ $t/\hat{\theta}_L$	1.0	0.5	0.2	0.1	0.05	0.02	0.01	0.005	0.002	0.001	0.0005	0.0002	0.0001
0	2	4	9	17	33	81	161	322	805	1,610	3,219	8,047	16,094
1	4	7	16	31	61	151	300	600	1,498	2,995	5,989	14,972	29,944
2	6	10	23	44	87	215	429	857	2,141	4,280	8,559	21,396	42,791
3	8	13	30	57	112	278	553	1,105	2,759	5,517	11,032	27,577	55,152
4	10	16	36	70	137	339	675	1,347	3,363	6,723	13,444	33,607	67,212
5	11	19	43	82	161	398	794	1,584	3,956	7,909	15,815	39,532	79,062
6	13	22	49	94	185	457	911	1,819	4,541	9,079	18,154	45,380	90,757
7	15	25	55	106	209	516	1,027	2,050	5,120	10,236	20,469	51,166	102,330
8	16	28	62	118	232	573	1,142	2,280	5,694	11,384	22,764	56,903	113,802
9	18	30	68	130	255	631	1,257	2,509	6,264	12,524	25,042	62,598	125,190
10	20	33	74	142	279	688	1,371	2,736	6,831	13,656	27,307	68,259	136,510
11	22	36	80	154	302	745	1,484	2,961	7,394	14,783	29,559	73,889	147,770
12	23	39	86	166	324	801	1,596	3,186	7,955	15,904	31,801	79,493	158,980
13	25	42	92	177	347	858	1,708	3,410	8,514	17,020	34,033	85,073	170,140
14	27	44	98	189	370	914	1,820	3,632	9,070	18,132	36,257	90,632	181,260

† See remarks on Table A.2.

Table A.4. Minimum Size of Sample to be Tested for a Time t to Assure a Mean Life of at Least $\hat{\theta}_L$ with Confidence $\gamma = 85\%$ When F is the Allowable Number of Failures†

($\gamma = 85\%$)

$t/\hat{\theta}_L$ F	1.0	0.5	0.2	0.1	0.05	0.02	0.01	0.005	0.002	0.001	0.0005	0.0002	0.0001
0	2	4	10	19	38	95	190	380	949	1,898	3,795	9,486	18,971
1	4	8	18	35	68	170	338	675	1,687	3,373	6,746	16,863	33,725
2	6	11	25	49	96	238	474	946	2,363	4,724	9,447	23,616	47,232
3	8	14	32	62	122	303	603	1,205	3,009	6,015	12,029	30,069	60,137
4	10	17	39	75	148	366	729	1,456	3,636	7,269	14,536	36,337	72,672
5	12	20	46	88	173	428	852	1,702	4,250	8,498	16,992	42,476	84,949
6	14	23	52	101	198	489	974	1,944	4,855	9,706	19,409	48,519	97,034
7	15	26	59	113	222	549	1,094	2,183	5,452	10,900	21,797	54,486	108,970
8	17	29	65	125	246	608	1,212	2,420	6,043	12,082	24,160	60,393	120,780
9	19	32	71	138	270	667	1,330	2,655	6,629	13,254	26,502	66,248	132,490
10	21	35	78	150	294	726	1,447	2,888	7,211	14,417	28,828	72,061	144,120
11	22	38	84	162	317	784	1,563	3,119	7,789	15,572	31,138	77,837	155,670
12	24	40	90	174	341	842	1,678	3,349	8,364	16,721	33,436	83,580	167,150
13	26	43	96	186	364	900	1,793	3,578	8,936	17,864	35,722	89,294	178,580
14	28	46	103	198	387	957	1,907	3,806	9,505	19,002	37,997	94,983	189,960

† See remarks on Table A.2.

APPENDIX

Table A.5. Minimum Size of Sample to be Tested for a Time t to Assure a Mean Life of at Least $\hat{\theta}_L$ with Confidence $\gamma = 90\%$ When F is the Allowable Number of Failures†

$(\gamma = 90\%)$

F \ $t/\hat{\theta}_L$	1.0	0.5	0.2	0.1	0.05	0.02	0.01	0.005	0.002	0.001	0.0005	0.0002	0.0001
0	3	5	12	24	47	116	231	461	1,152	2,303	4,605	11,513	23,026
1	5	9	20	40	79	195	390	779	1,946	3,891	7,780	19,449	38,898
2	7	12	28	55	108	268	534	1,066	2,663	5,324	10,646	26,613	53,224
3	9	15	35	69	136	336	670	1,338	3,342	6,683	13,363	33,405	66,609
4	11	19	43	82	162	402	802	1,601	3,999	7,996	15,989	39,970	79,938
5	13	22	49	96	189	467	930	1,858	4,640	9,277	18,552	46,376	92,749
6	14	25	56	109	214	530	1,057	2,110	5,269	10,535	21,067	52,663	105,320
7	16	28	63	122	239	593	1,181	2,358	5,889	11,775	23,545	58,858	117,710
8	18	31	70	135	264	654	1,304	2,603	6,502	12,999	25,993	64,978	129,950
9	20	34	76	147	289	715	1,426	2,846	7,108	14,211	28,417	71,034	142,060
10	22	37	83	160	314	776	1,546	3,087	7,709	15,412	30,818	77,038	154,070
11	23	40	89	172	338	836	1,666	3,326	8,305	16,604	33,202	82,996	165,990
12	25	43	96	184	362	896	1,785	3,563	8,897	17,788	35,569	88,914	177,820
13	27	45	102	197	386	955	1,903	3,799	9,486	18,965	37,922	94,796	189,590
14	29	48	108	209	410	1,014	2,020	4,033	10,071	20,135	40,263	100,650	201,290

† See remarks on Table A.2.

Table A.6. Minimum Size of Sample to be Tested for a Time t to Assure a Mean Life of at Least $\hat{\theta}_L$ with Confidence $\gamma = 95\%$ When F is the Allowable Number of Failures†

($\gamma = 95\%$)

F \ $t/\hat{\theta}_L$	1.0	0.5	0.2	0.1	0.05	0.02	0.01	0.005	0.002	0.001	0.0005	0.0002	0.0001
0	3	6	15	30	60	150	300	600	1,498	2,996	5,992	14,959	29,957
1	6	11	25	48	96	238	475	950	2,373	4,745	9,488	23,720	47,439
2	8	14	33	64	127	316	631	1,261	3,149	6,297	12,593	31,480	62,959
3	10	18	41	80	157	390	777	1,553	3,879	7,755	15,509	38,770	77,538
4	12	21	48	94	186	460	918	1,833	4,579	9,156	18,309	45,770	91,537
5	14	24	56	108	213	529	1,054	2,106	5,259	10,516	21,029	52,568	105,130
6	16	27	63	122	240	596	1,188	2,372	5,925	11,846	23,688	59,215	118,430
7	18	31	70	136	267	661	1,319	2,634	6,578	13,152	26,300	65,744	131,480
8	19	34	77	149	293	726	1,448	2,891	7,222	14,439	28,873	72,177	144,350
9	21	37	84	162	319	790	1,575	3,146	7,857	15,710	31,415	78,531	157,060
10	23	40	90	175	345	854	1,702	3,398	8,486	16,967	33,929	84,816	169,630
11	25	43	97	188	370	916	1,827	3,647	9,110	18,213	36,421	91,043	182,080
12	27	46	104	201	395	979	1,951	3,895	9,728	19,449	38,891	97,219	194,430
13	29	49	111	214	420	1,040	2,074	4,141	10,341	20,675	41,344	103,350	206,690
14	30	52	117	226	445	1,102	2,196	4,385	10,951	21,894	43,780	109,440	218,870

† See remarks on Table A.2.

Table A.7. Sample Size and Criteria to Demonstrate Reliability at a Given Confidence

$\hat{R}_L$ \ N	$\gamma = 0.80$				$\gamma = 0.90$				$\gamma = 0.95$			
	0.90	0.95	0.99	0.995	0.90	0.95	0.99	0.995	0.90	0.95	0.99	0.995
5	12.7	26.1	133	267	15.2	31.2	159	319	17.4	35.7	182	365
10	11.9	24.4	124	249	13.5	27.7	141	283	14.9	30.6	156	313
15	11.5	23.6	120	241	12.7	26.1	133	267	13.9	28.5	145	291
25	11.0	22.6	115	231	12.0	24.6	125	251	12.8	26.3	134	269
50	10.6	21.8	111	223	11.2	23.0	117	235	11.8	24.2	123	247
100	10.3	21.1	107	215	10.7	22.0	112	225	11.1	22.8	116	233
250	10.0	20.5	104	209	10.3	21.1	107	215	10.5	21.6	110	221
500	9.9	20.3	103	207	10.1	20.7	105	211	10.2	20.9	106	213

1. In the life test procedure of Sec. 10.7.1, when

$$\frac{\hat{\theta}}{T} \equiv \frac{\sum_{j=1}^{N} t_j}{NT} \geq \text{the tabled value,}$$

then Reliability $\hat{R}_L$ is demonstrated with Confidence γ. The Sample Size is N; the Required Life is T, and $t_1, \cdots, t_N$ are the observed times-to-failure.

2. Replace $\hat{\theta}$ and N above with $\hat{\theta}_{r,N}$ and r, respectively, for the procedure of Sec. 10.7.2.
3. Replace $\hat{\theta}$ and N above with $\hat{S}/r$ and r, respectively, for the procedure of Sec. 10.8.

Table A.8. Percentage Points of the χ_n^2/n

n	.05	.1	.5	1.0	2.5	5.0	10	20	30	40
1	$.0^639$	$.0^5157$	$.0^439$	$.0^316$	$.0^398$	$.0^239$	.016	.064	.148	.275
2	.001	.001	.005	.010	.025	.052	.106	.223	.356	.511
3	.005	.008	.024	.038	.072	.117	.195	.335	.475	.623
4	.016	.023	.052	.074	.121	.178	.266	.412	.549	.688
5	.032	.042	.082	.111	.166	.229	.322	.469	.600	.731
6	.050	.064	.113	.145	.206	.272	.367	.512	.638	.762
7	.069	.085	.141	.177	.241	.310	.405	.546	.667	.785
8	.089	.107	.168	.206	.272	.342	.436	.574	.691	.803
9	.108	.128	.193	.232	.300	.369	.463	.598	.710	.817
10	.126	.148	.216	.256	.325	.394	.487	.618	.727	.830
11	.144	.167	.237	.278	.347	.416	.507	.635	.741	.840
12	.161	.184	.256	.298	.367	.436	.525	.651	.753	.848
13	.177	.201	.274	.316	.385	.453	.542	.664	.764	.856
14	.193	.217	.291	.333	.402	.469	.556	.676	.773	.863
15	.207	.232	.307	.349	.418	.484	.570	.687	.781	.869
16	.221	.246	.321	.363	.432	.498	.582	.697	.789	.874
17	.234	.260	.335	.377	.445	.510	.593	.706	.796	.879
18	.247	.272	.348	.390	.457	.522	.604	.714	.802	.883
19	.258	.285	.360	.402	.469	.532	.613	.722	.808	.887
20	.270	.296	.372	.413	.480	.543	.622	.729	.813	.890
22	.291	.317	.393	.434	.499	.561	.638	.742	.823	.897
24	.310	.337	.412	.452	.517	.577	.652	.753	.831	.902
26	.328	.355	.429	.469	.532	.592	.665	.762	.838	.907
28	.345	.371	.445	.484	.547	.605	.676	.771	.845	.911
30	.360	.386	.460	.498	.560	.616	.687	.779	.850	.915
35	.394	.420	.491	.529	.588	.642	.708	.795	.862	.922
40	.423	.448	.518	.554	.611	.663	.726	.809	.872	.928
45	.448	.472	.540	.576	.630	.680	.741	.820	.880	.933
50	.469	.494	.560	.594	.647	.695	.754	.829	.886	.937
55	.488	.512	.577	.610	.662	.708	.765	.837	.892	.941
60	.506	.529	.592	.625	.675	.720	.774	.844	.897	.944
70	.535	.558	.618	.649	.697	.739	.790	.856	.905	.949
80	.560	.582	.640	.669	.714	.755	.803	.865	.911	.952
90	.581	.602	.658	.686	.729	.768	.814	.873	.917	.955
100	.599	.619	.673	.701	.742	.779	.824	.879	.921	.958
120	.629	.648	.699	.724	.763	.798	.839	.890	.929	.962
140	.653	.671	.719	.743	.780	.812	.850	.898	.934	.965
160	.673	.690	.736	.758	.793	.824	.860	.905	.939	.968
180	.689	.706	.749	.771	.804	.833	.868	.910	.942	.970
200	.703	.719	.761	.782	.814	.841	.874	.915	.945	.972
250	.732	.746	.785	.804	.832	.858	.887	.924	.951	.975
300	.753	.767	.802	.820	.846	.870	.897	.931	.956	.977
350	.770	.783	.816	.833	.857	.879	.904	.936	.959	.979
400	.784	.796	.827	.843	.866	.887	.911	.940	.962	.981
450	.795	.807	.837	.852	.874	.893	.916	.944	.964	.982
500	.805	.816	.845	.859	.880	.898	.920	.946	.966	.983
750	.839	.848	.872	.884	.901	.917	.934	.956	.972	.986
1000	.859	.868	.889	.899	.914	.928	.943	.962	.976	.988
5000	.936	.939	.949	.954	.961	.967	.974	.983	.989	.995
∞	1	1	1	1	1	1	1	1	1	1

This Table is reprinted by the kind permission of Profs. W. J. Dixon and F. J. Massey, and the McGraw-Hill Book Co., Inc., from *Introduction to Statistical Analysis*, Second Edition (1957).

DISTRIBUTION (n = degrees of freedom)

50	60	70	80	90	95	97.5	99	99.5	99.9	99.95	n
.455	.708	1.07	1.64	2.71	3.84	5.02	6.64	7.88	10.83	12.12	1
.693	.916	1.20	1.61	2.30	3.00	3.69	4.61	5.30	6.91	7.60	2
.789	.982	1.22	1.55	2.08	2.60	3.12	3.78	4.28	5.42	5.91	3
.839	1.011	1.22	1.50	1.94	2.37	2.79	3.32	3.72	4.62	5.00	4
.870	1.03	1.21	1.46	1.85	2.21	2.57	3.02	3.35	4.10	4.42	5
.891	1.04	1.21	1.43	1.77	2.10	2.41	2.80	3.09	3.74	4.02	6
.907	1.04	1.20	1.40	1.72	2.01	2.29	2.64	2.90	3.47	3.72	7
.918	1.04	1.19	1.38	1.67	1.94	2.19	2.51	2.74	3.27	3.48	8
.927	1.05	1.18	1.36	1.63	1.88	2.11	2.41	2.62	3.10	3.30	9
.934	1.05	1.18	1.34	1.60	1.83	2.05	2.32	2.52	2.96	3.14	10
.940	1.05	1.17	1.33	1.57	1.79	1.99	2.25	2.43	2.84	3.01	11
.945	1.05	1.17	1.32	1.55	1.75	1.94	2.18	2.36	2.74	2.90	12
.949	1.05	1.16	1.31	1.52	1.72	1.90	2.13	2.29	2.66	2.81	13
.953	1.05	1.16	1.30	1.50	1.69	1.87	2.08	2.24	2.58	2.72	14
.956	1.05	1.15	1.29	1.49	1.67	1.83	2.04	2.19	2.51	2.65	15
.959	1.05	1.15	1.28	1.47	1.64	1.80	2.00	2.14	2.45	2.58	16
.961	1.05	1.15	1.27	1.46	1.62	1.78	1.97	2.10	2.40	2.52	17
.963	1.05	1.14	1.26	1.44	1.60	1.75	1.93	2.06	2.35	2.47	18
.965	1.05	1.14	1.26	1.43	1.59	1.73	1.90	2.03	2.31	2.42	19
.967	1.05	1.14	1.25	1.42	1.57	1.71	1.88	2.00	2.27	2.37	20
.970	1.05	1.13	1.24	1.40	1.54	1.67	1.83	1.95	2.19	2.30	22
.972	1.05	1.13	1.23	1.38	1.52	1.64	1.79	1.90	2.13	2.23	24
.974	1.05	1.12	1.22	1.37	1.50	1.61	1.76	1.86	2.08	2.17	26
.976	1.04	1.12	1.22	1.35	1.48	1.59	1.72	1.82	2.03	2.12	28
.978	1.04	1.12	1.21	1.34	1.46	1.57	1.70	1.79	1.99	2.07	30
.981	1.04	1.11	1.19	1.32	1.42	1.52	1.64	1.72	1.90	1.98	35
.983	1.04	1.10	1.18	1.30	1.39	1.48	1.59	1.67	1.84	1.90	40
.985	1.04	1.10	1.17	1.28	1.37	1.45	1.55	1.63	1.78	1.84	45
.987	1.04	1.09	1.16	1.26	1.35	1.43	1.52	1.59	1.73	1.79	50
.988	1.04	1.09	1.16	1.25	1.33	1.41	1.50	1.56	1.69	1.75	55
.989	1.04	1.09	1.15	1.24	1.32	1.39	1.47	1.53	1.66	1.71	60
.990	1.03	1.08	1.14	1.22	1.29	1.36	1.43	1.49	1.60	1.65	70
.992	1.03	1.08	1.13	1.21	1.27	1.33	1.40	1.45	1.56	1.60	80
.993	1.03	1.07	1.12	1.20	1.26	1.31	1.38	1.43	1.52	1.56	90
.993	1.03	1.07	1.12	1.18	1.24	1.30	1.36	1.40	1.49	1.53	100
.994	1.03	1.06	1.11	1.17	1.22	1.27	1.32	1.36	1.45	1.48	120
.995	1.03	1.06	1.10	1.16	1.20	1.25	1.30	1.33	1.41	1.44	140
.996	1.02	1.06	1.09	1.15	1.19	1.23	1.28	1.31	1.38	1.41	160
.996	1.02	1.05	1.09	1.14	1.18	1.22	1.26	1.29	1.36	1.38	180
.997	1.02	1.05	1.08	1.13	1.17	1.21	1.25	1.28	1.34	1.36	200
.997	1.02	1.04	1.07	1.12	1.15	1.18	1.22	1.25	1.30	1.32	250
.998	1.02	1.04	1.07	1.11	1.14	1.17	1.20	1.22	1.27	1.29	300
.998	1.02	1.04	1.06	1.10	1.13	1.15	1.18	1.21	1.25	1.27	350
.998	1.02	1.04	1.06	1.09	1.12	1.14	1.17	1.19	1.24	1.25	400
.999	1.02	1.03	1.06	1.09	1.11	1.13	1.16	1.18	1.22	1.23	450
.999	1.01	1.03	1.05	1.08	1.11	1.13	1.15	1.17	1.21	1.22	500
.999	1.01	1.03	1.04	1.07	1.09	1.10	1.12	1.14	1.17	1.18	750
.999	1.01	1.02	1.04	1.06	1.07	1.09	1.11	1.12	1.14	1.15	1000
1.00	1.00	1.01	1.02	1.02	1.03	1.04	1.05	1.05	1.06	1.07	5000
1	1	1	1	1	1	1	1	1	1	1	∞

Note: $\chi^2_{n;\alpha}$ denotes the chi-square deviate (n degrees of freedom) which is *exceeded* with probability α. For example, $\chi^2_{10;0.05}/10 = 1.83$; $\chi^2_{10;0.95}/10 = 0.394$.

Table A.9. Tolerance Factors for Normal Distributions

Factors K such that the probability is γ that at least a proportion $1 - \alpha$ of the distribution will be less than $\bar{x} + Ks'$ (or greater than $\bar{x} - Ks'$), where $\bar{x}$ and s' are estimates of the mean and the standard deviation* computed from a sample of size n

n \ α	$\gamma = 0.75$					$\gamma = 0.90$					$\gamma = 0.95$					$\gamma = 0.99$				
	0.25	0.10	0.05	0.01	0.001	0.25	0.10	0.05	0.01	0.001	0.25	0.10	0.05	0.01	0.001	0.25	0.10	0.05	0.01	0.001
3	1.464	2.501	3.152	4.396	5.805	2.602	4.258	5.310	7.340	9.651	3.804	6.158	7.655	10.552	13.857					
4	1.256	2.134	2.680	3.726	4.910	1.972	3.187	3.957	5.437	7.128	2.619	4.163	5.145	7.042	9.215					
5	1.152	1.961	2.463	3.421	4.507	1.698	2.742	3.400	4.666	6.112	2.149	3.407	4.202	5.741	7.501					
6	1.087	1.860	2.336	3.243	4.273	1.540	2.494	3.091	4.242	5.556	1.895	3.006	3.707	5.062	6.612	2.849	4.408	5.409	7.334	9.540
7	1.043	1.791	2.250	3.126	4.118	1.435	2.333	2.894	3.972	5.201	1.732	2.755	3.399	4.641	6.061	2.490	3.856	4.739	6.411	8.348
8	1.010	1.740	2.190	3.042	4.008	1.360	2.219	2.755	3.783	4.955	1.617	2.582	3.188	4.353	5.686	2.252	3.496	4.287	5.811	7.566
9	0.984	1.702	2.141	2.977	3.924	1.302	2.133	2.649	3.641	4.772	1.532	2.454	3.031	4.143	5.414	2.085	3.242	3.971	5.389	7.014
10	0.964	1.671	2.103	2.927	3.858	1.257	2.065	2.568	3.532	4.629	1.465	2.355	2.911	3.981	5.203	1.954	3.048	3.739	5.075	6.603
11	0.947	1.646	2.073	2.885	3.804	1.219	2.012	2.503	3.444	4.515	1.411	2.275	2.815	3.852	5.036	1.854	2.897	3.557	4.828	6.284
12	0.933	1.624	2.048	2.851	3.760	1.188	1.966	2.448	3.371	4.420	1.366	2.210	2.736	3.747	4.900	1.771	2.773	3.410	4.633	6.032
13	0.919	1.606	2.026	2.822	3.722	1.162	1.928	2.403	3.310	4.341	1.329	2.155	2.670	3.659	4.787	1.702	2.677	3.290	4.472	5.826
14	0.909	1.591	2.007	2.796	3.690	1.139	1.895	2.363	3.257	4.274	1.296	2.108	2.614	3.585	4.690	1.645	2.592	3.189	4.336	5.651
15	0.899	1.577	1.991	2.776	3.661	1.119	1.866	2.329	3.212	4.215	1.268	2.068	2.566	3.520	4.607	1.596	2.521	3.102	4.224	5.507
16	0.891	1.566	1.977	2.756	3.637	1.101	1.842	2.299	3.172	4.164	1.242	2.032	2.523	3.463	4.534	1.553	2.458	3.028	4.124	5.374
17	0.883	1.554	1.964	2.739	3.615	1.085	1.820	2.272	3.136	4.118	1.220	2.001	2.486	3.415	4.471	1.514	2.405	2.962	4.038	5.268
18	0.876	1.544	1.951	2.723	3.595	1.071	1.800	2.249	3.106	4.078	1.200	1.974	2.453	3.370	4.415	1.481	2.357	2.906	3.961	5.167
19	0.870	1.536	1.942	2.710	3.577	1.058	1.781	2.228	3.078	4.041	1.183	1.949	2.423	3.331	4.364	1.450	2.315	2.855	3.893	5.078
20	0.865	1.528	1.933	2.697	3.561	1.046	1.765	2.208	3.052	4.009	1.167	1.926	2.396	3.295	4.319	1.424	2.275	2.807	3.832	5.003

21	0.859	1.520	1.923	2.686	3.545	1.035	1.750	2.190	3.028	3.979	1.152	1.905	2.371	3.262	4.276	1.397	2.241	2.768	3.776	4.932
22	0.854	1.514	1.916	2.675	3.532	1.025	1.736	2.174	3.007	3.952	1.138	1.887	2.350	3.233	4.238	1.376	2.208	2.729	3.727	4.866
23	0.849	1.508	1.907	2.665	3.520	1.016	1.724	2.159	2.987	3.927	1.126	1.869	2.329	3.206	4.204	1.355	2.179	2.693	3.680	4.806
24	0.845	1.502	1.901	2.656	3.509	1.007	1.712	2.145	2.969	3.904	1.114	1.853	2.309	3.181	4.171	1.336	2.154	2.663	3.638	4.755
25	0.842	1.496	1.895	2.647	3.497	0.999	1.702	2.132	2.953	3.882	1.103	1.838	2.292	3.158	4.143	1.319	2.129	2.632	3.601	4.706
30	0.825	1.475	1.869	2.613	3.454	0.966	1.657	2.080	2.884	3.794	1.059	1.778	2.220	3.064	4.022	1.249	2.029	2.516	3.446	4.508
35	0.812	1.458	1.849	2.588	3.421	0.942	1.623	2.041	2.833	3.730	1.025	1.732	2.166	2.994	3.934	1.195	1.957	2.431	3.334	4.364
40	0.803	1.445	1.834	2.568	3.395	0.923	1.598	2.010	2.793	3.679	0.999	1.697	2.126	2.941	3.866	1.154	1.902	2.365	3.250	4.255
45	0.795	1.435	1.821	2.552	3.375	0.908	1.577	1.986	2.762	3.638	0.978	1.669	2.092	2.897	3.811	1.122	1.857	2.313	3.181	4.168
50	0.788	1.426	1.811	2.538	3.358	0.894	1.560	1.965	2.735	3.604	0.961	1.646	2.065	2.863	3.766	1.096	1.821	2.296	3.124	4.096

* s' is the square-root of the *unbiased* estimate of the variance (*cf*. Sec. 8.7).

Reprinted with the kind permission of the publishers, Prentice-Hall, Inc., and the authors, A. H. Bowker and G. J. Lieberman, from *Engineering Statistics* (1959).

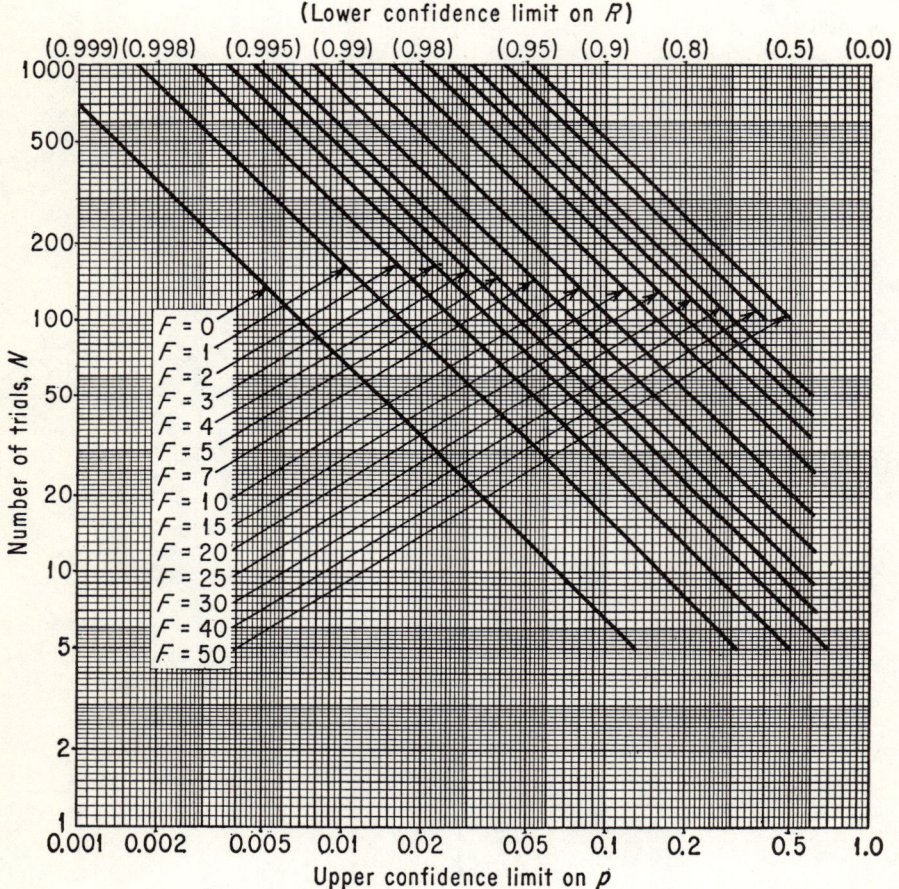

Fig. A.1 Upper confidence limit on unreliability (one minus lower confidence limit on reliability) number of trials N, observed failures F, confidence coefficient $\gamma = 0.50$.

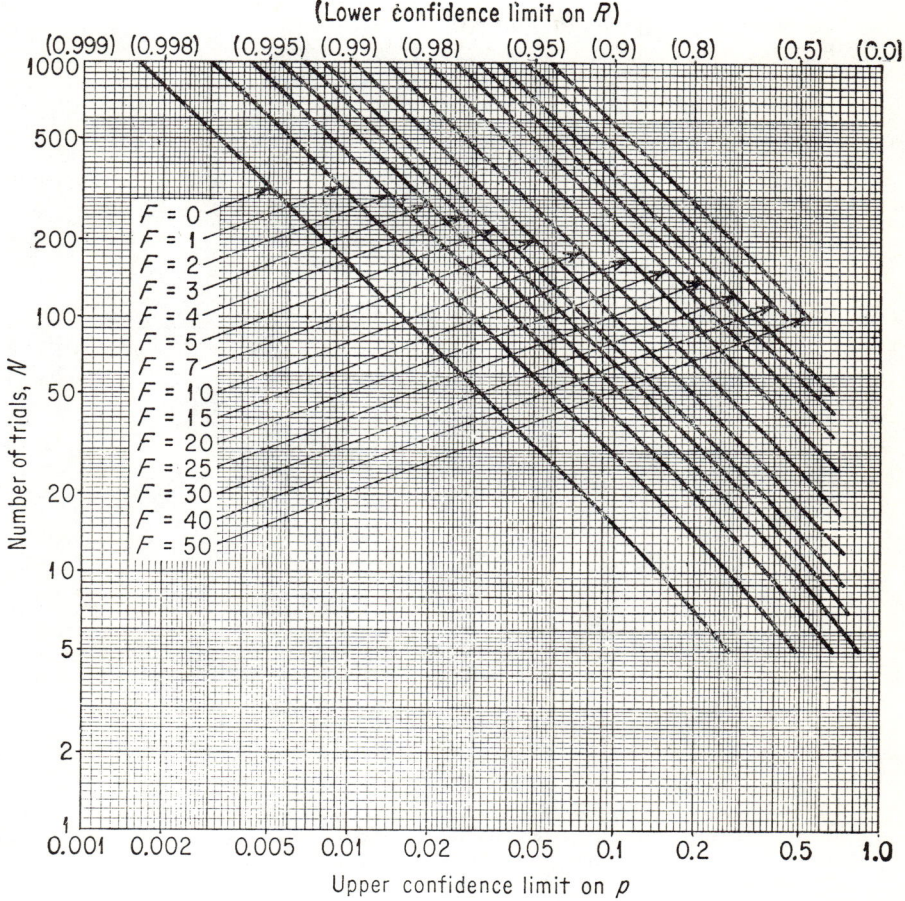

Fig. A.2 Upper confidence limit on unreliability (one minus lower confidence limit on reliability) number of trials N, observed failures F, confidence coefficient $\gamma = 0.80$.

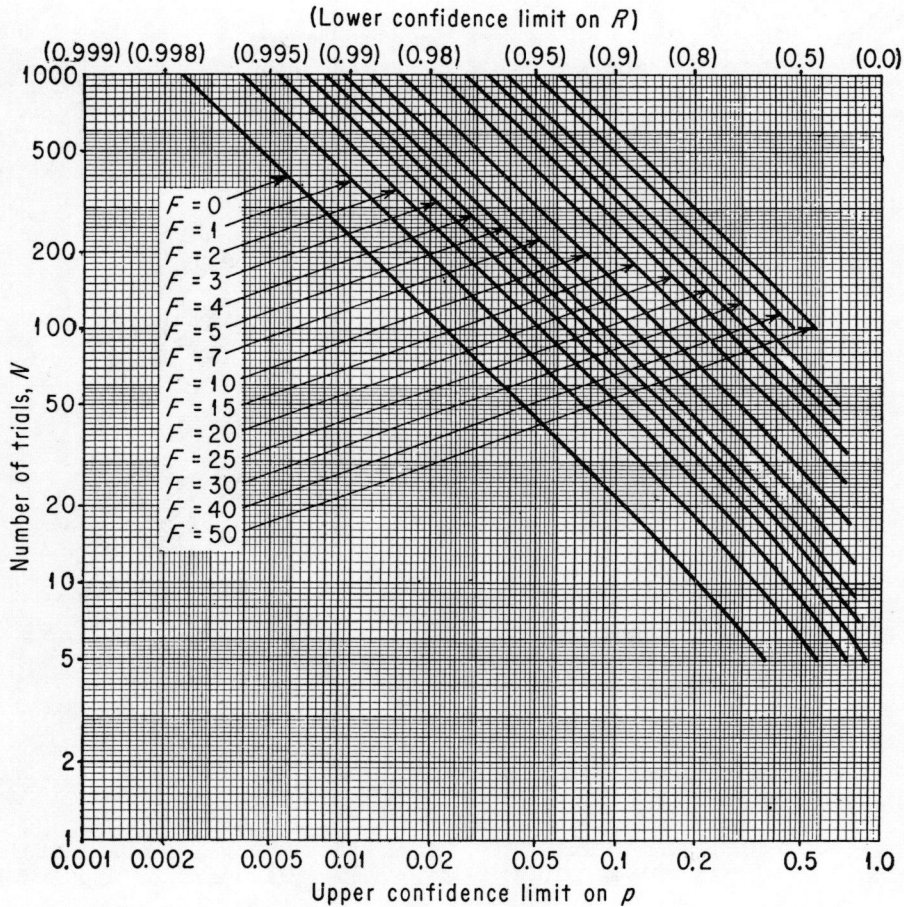

Fig. A.3 Upper confidence limit on unreliability (one minus lower confidence limit on reliability) number of trials N, observed failures F, confidence coefficient $\gamma = 0.90$.

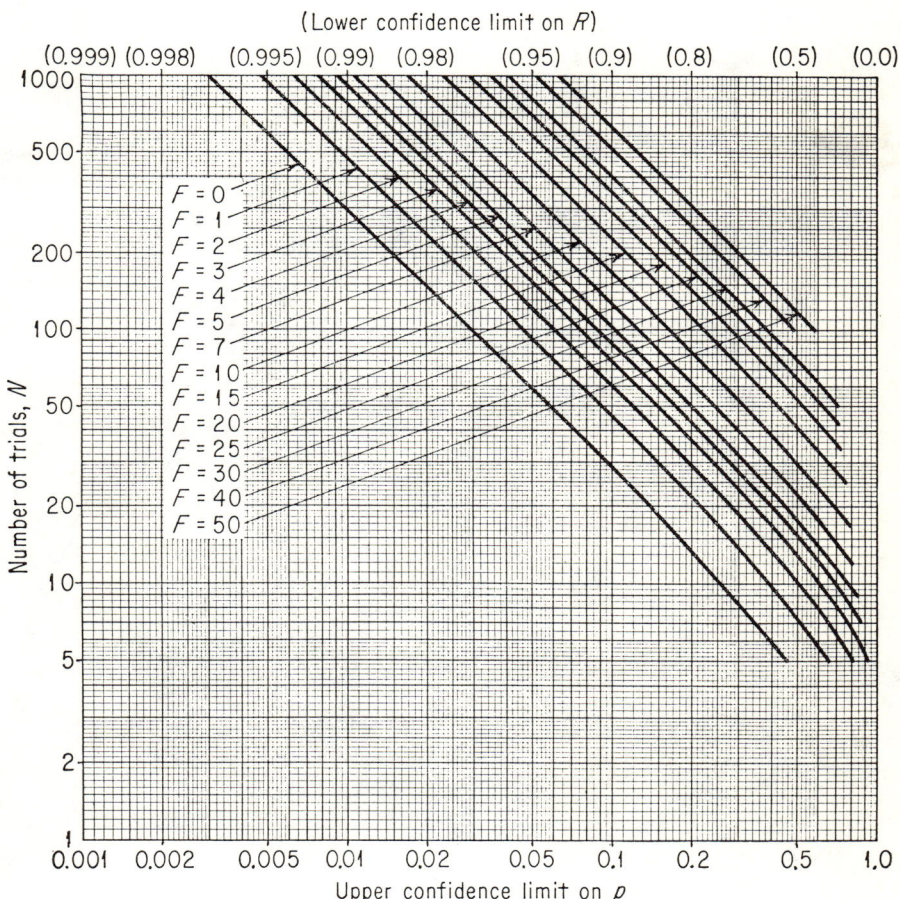

Fig. A.4 Upper confidence limit on unreliability (one minus lower confidence limit on reliability) number of trials N, observed failures F, confidence coefficient $\gamma = 0.95$.

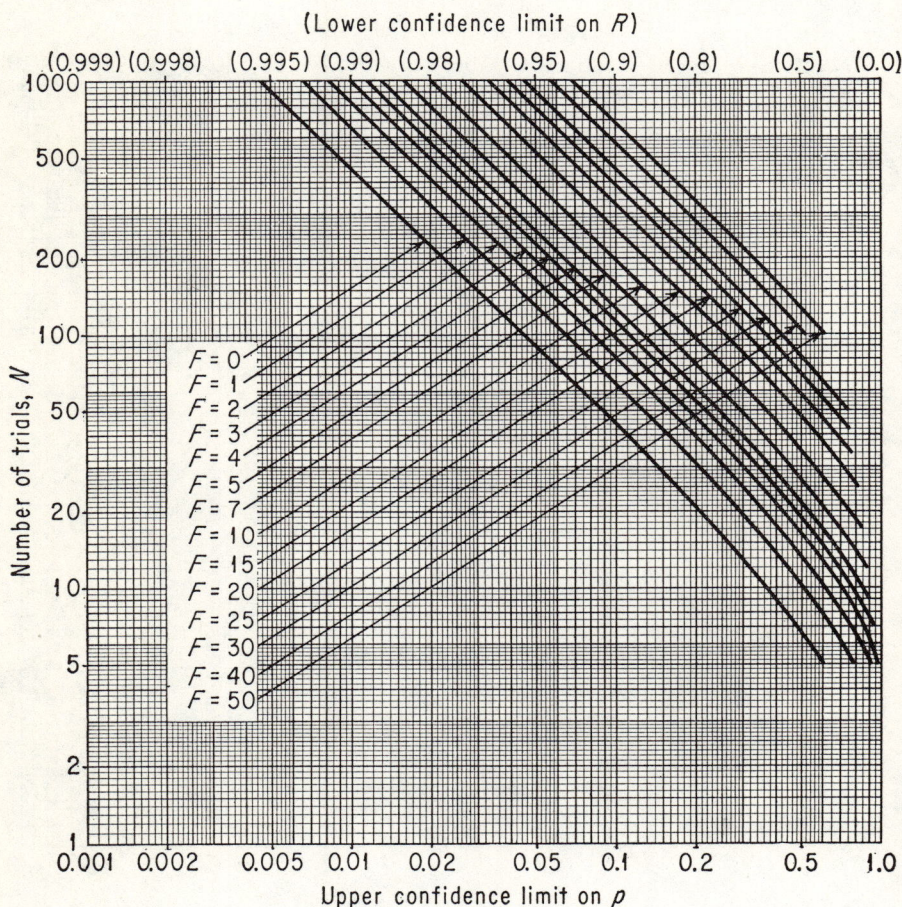

Fig. A.5 Upper confidence limit on unreliability (one minus lower confidence limit on reliability) number of trials N, observed failures F, confidence coefficient $\gamma = 0.99$.

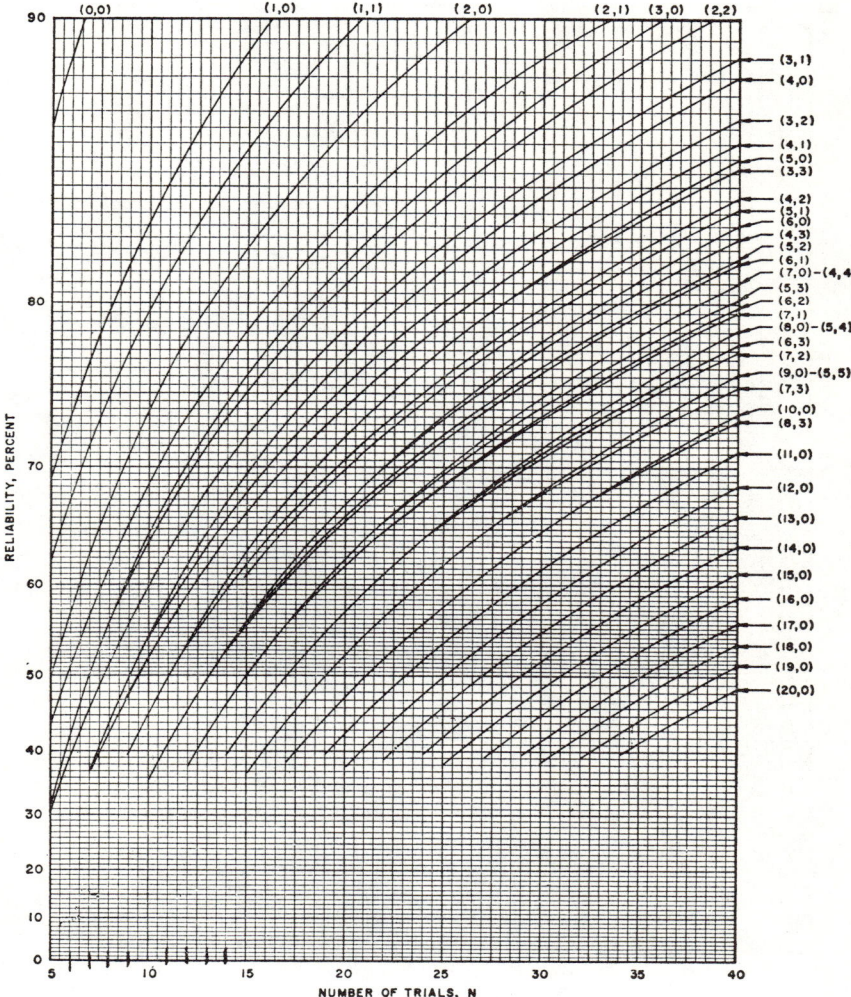

Fig. A.6 50 per cent lower confidence limit on system reliability for observed failure combinations of a two-subsystem serial system, N trials per subsystem. (All permutations of failure combinations are equivalent.)

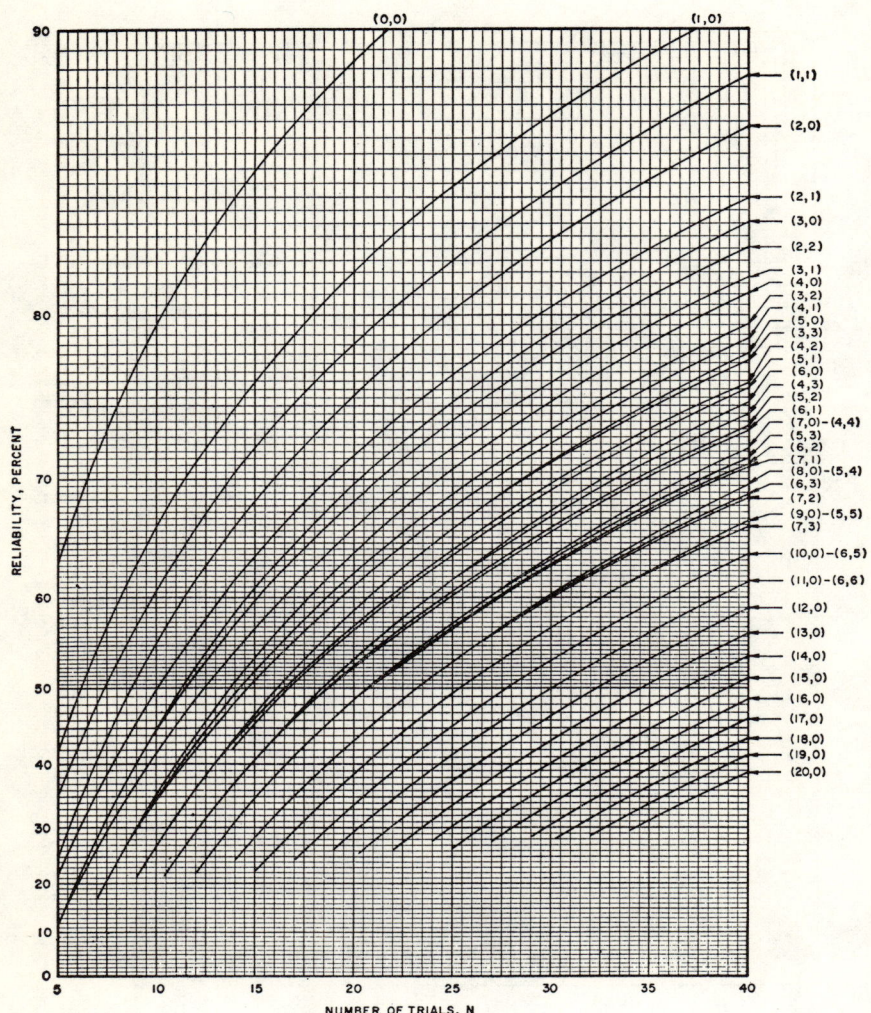

Fig. A.7 90 per cent lower confidence limit on system reliability for observed failure combinations of a two-subsystem serial system, N trials per subsystem. (All permutations of failure combinations are equivalent.)

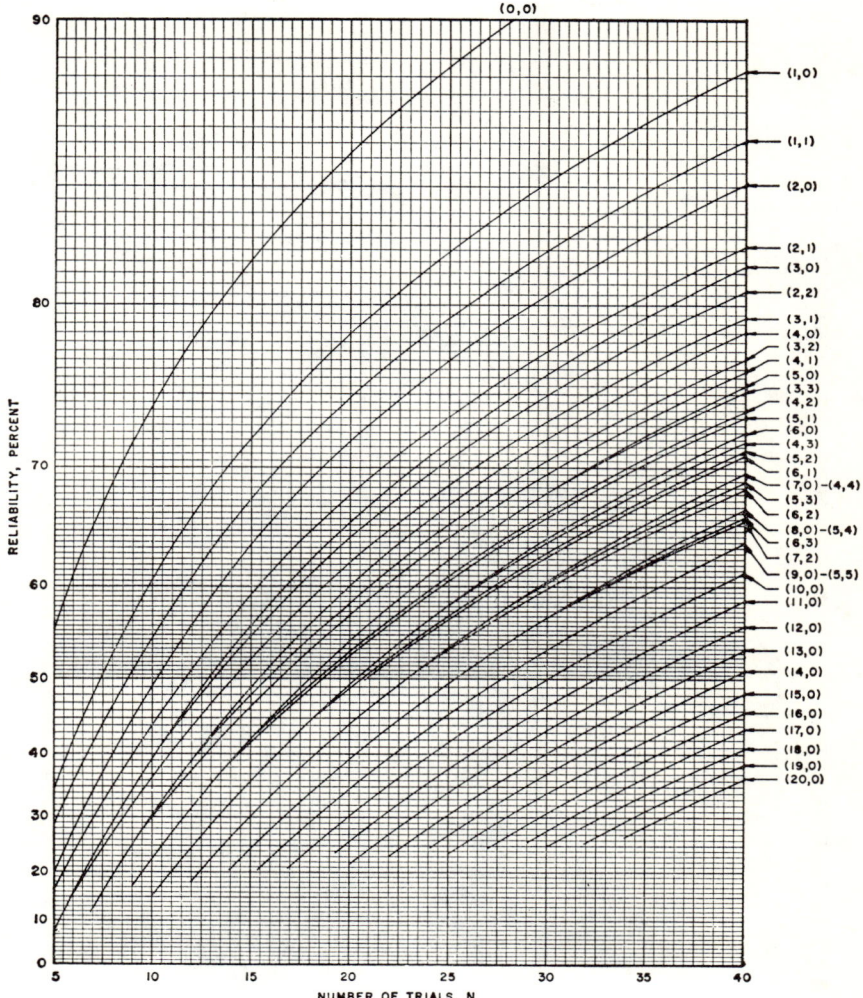

Fig. A.8 95 per cent lower confidence limit on system reliability for observed failure combinations of a two-subsystem serial system, N trials per subsystem. (All permutations of failure combinations are equivalent.)

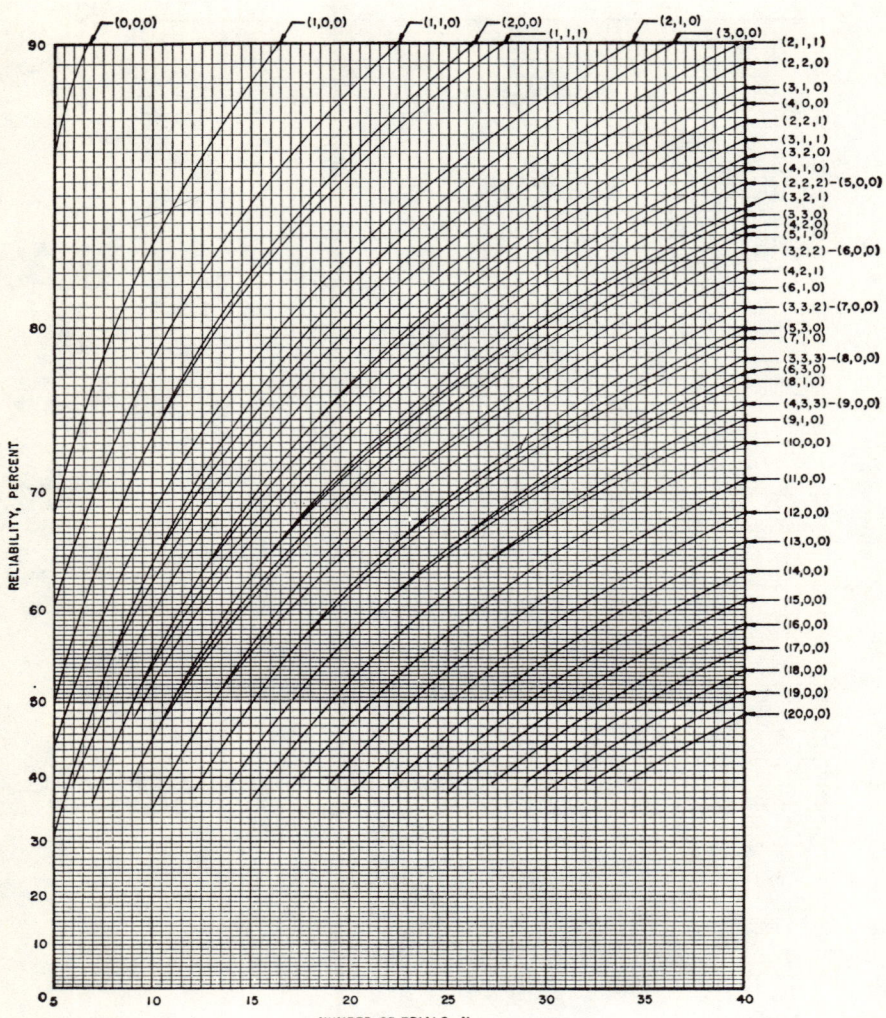

Fig. A.9 50 per cent lower confidence limit on system reliability for observed failure combinations of a three-subsystem serial system, N trials per subsystem. (All permutations of failure combinations are equivalent.)

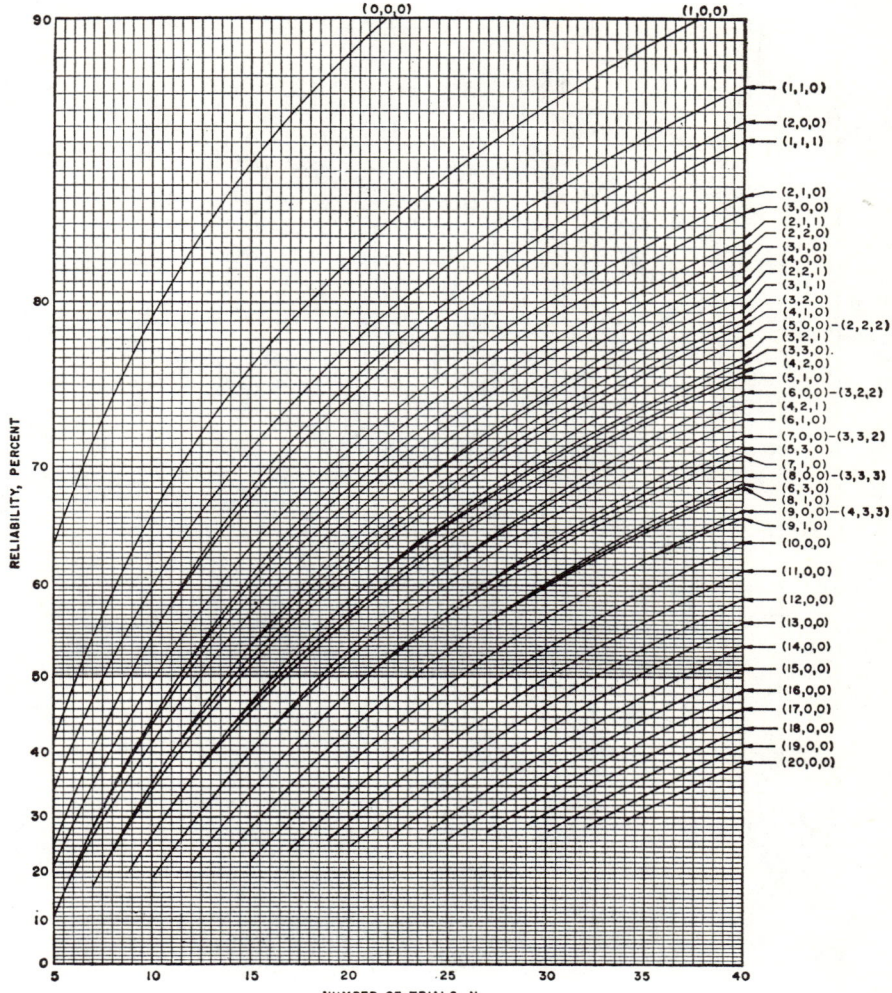

Fig. A.10 90 per cent lower confidence limit on system reliability for observed failure combinations of a three-subsystem serial system, N trials per subsystem. (All permutations of failure combinations are equivalent.)

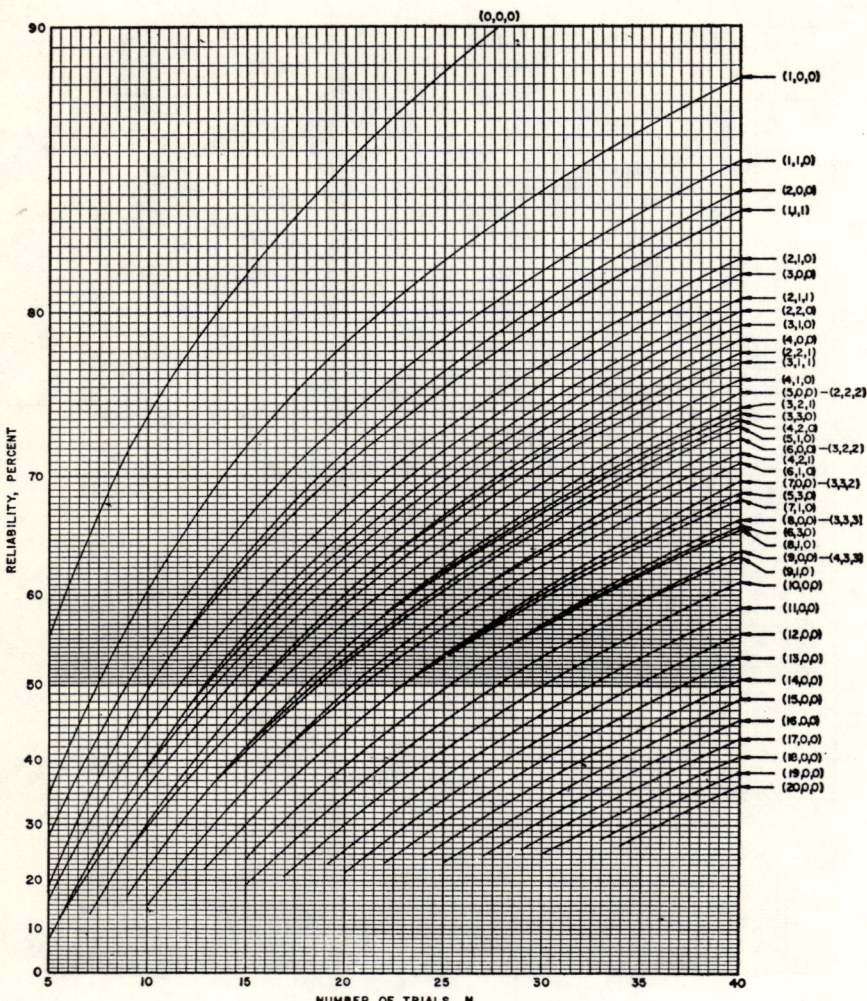

Fig. A.11 95 per cent lower confidence limit on system reliability for observed failure combinations of a three-subsystem serial system, N trials per subsystem. (All permutations of failure combinations are equivalent.)

BIBLIOGRAPHY

Amstadter, B., *Reliability Mathematics: Fundamentals; Practices; Procedures,* McGraw-Hill Book Co., New York (1971).

Barlow, R.E., and Proschan, F., *Mathematical Theory of Reliability,* John Wiley & Sons, Inc., New York (1965).

Barlow, R.E., and Proschan, F., *Statistical Theory of Reliability and Life Testing,* Holt, Rinehart and Winston, New York (1975).

Barlow, R.E., and Scheuer, E.M., *An Introduction to Reliability Theory,* CEIR, Inc. (1969).

Bazovsky, I., *Reliability: Theory and Practice,* Prentice-Hall Inc., Englewood Cliffs, New Jersey (1961).

Billinton, R., *Power System Reliability Evaluation,* Gordon and Breach, New York (1970).

Billinton, R., Ringlee, R.J., and Woods, A.J., *Power System Reliability Calculations,* The M.I.T. Press, Cambridge, Massachusetts (1973).

Blanchard, B.S., Jr., and Lowery, E.E., *Maintainability: Principles and Practices,* McGraw-Hill Book Co., New York (1969).

Buckland, W.R., *Statistical Assessment of the Life Characteristic, A Bibliographic Guide,* Hafner Publishing Co., New York (1964).

Calabro, S.R., *Reliability Principles and Practice,* McGraw-Hill Book Co., New York (1962).

Chorofas, D.N., *Statistical Processes and Reliability Engineering,* D. Van Nostrand Co., New York (1960).

Dummer, G.W., and Griffin, N.B., *Electronic Reliability Calculation and Design,* Pergamon Press, Elmsford, New York (1966).

Dummer, G.W., and Winton, R.C., *An Elementary Guide to Reliability,* Pergamon Press, Elmsford, New York (1968).

Enrick, N.L., *Quality Control and Reliability,* Industrial Press, New York (1966).

Epstein, B., *Mathematical Models for System Reliability,* Publishing House of the Student Association, Technion-Israel Institute of Technology, Haifa, Israel (1964).

Fishman, G.S., *Concepts and Methods in Discrete Event Digital Simulation,* John Wiley & Sons, Inc., New York (1973).

Gedye, G.R., *A Manager's Guide to Quality and Reliability,* John Wiley & Sons, Inc., New York (1968).

Gnedenko, B.V., Belyayev, Y.K., and Solovyev, A.D., *Mathematical Methods of Reliability Theory,* Academic Press, New York (1969).

Goldman, A.S., and Slattery, T.B., *Maintainability,* John Wiley & Sons, Inc., New York (1964).

Greene, A.E., and Bourne, A.J., *Reliability Technology,* John Wiley & Sons, Inc., New York (1972).

Gross, A.J., and Clark, V.A., *Survival Distributions: Reliability Applications in the Biomedical Sciences,* John Wiley & Sons, Inc., New York (1975).

Grouchko, D. (ed.), *Operations Research and Reliability,* Gordon and Breach, New York (1969).

Gryna, F.M., *Reliability Training Text,* 2nd Edition, New York Institute of Radio Engineers (1960).

Hammersley, J.M., and Handscomb, D.C., *Monte Carlo Methods,* John Wiley & Sons, New York (1964).

Haugen, E.B., *Probabilistic Approaches to Design,* John Wiley & Sons, Inc., New York (1968).

Haviland, R.T., *Engineering Reliability and Long Life Design,* D. Van Nostrand Co., New Jersey (1964).

Howard, R.A., *Dynamic Probabilistic Systems,* Vols. I and II, John Wiley & Sons, Inc., New York (1971).

Ireson, W.G., (ed.), *Reliability Handbook,* McGraw-Hill Book Co., New York (1966).

Ireson, W.G., and Grant, E.L. (eds.), *Handbook of Industrial Engineering and Management,* Second Edition, Prentice-Hall, Inc., Englewood Cliffs, New Jersey (1971).

Jardine, A.K.S. (ed.), *Operational Research in Maintenance,* Manchester University Press, Manchester and Barnes & Nobles, Inc., New York (1970).

Kapur, K.C., and Lamberson, L.R., *Reliability in Engineering Design,* John Wiley & Sons, Inc., New York (1977).

Kozlov, B.A., and Ushakov, I.A., *Reliability Handbook,* Holt, Rinehart and Winston, New York (1970).

Landers, R.R., *Reliability and Product Assurance,* Prentice-Hall, Inc., Englewood Cliffs, New Jersey (1963).

Lipson, C., and Sheth, N.J., *Statistical Design and Analysis of Engineering Experiments,* McGraw-Hill Book Co., New York (1973).

Mann, N.R., Shafer, R.E., and Singpurwalla, N.D., *Methods for Statistical Analysis of Reliability and Life Data,* John Wiley & Sons, Inc., New York (1974).

Myers, R.H., Wong, K.L., and Gordy, H.M., *Reliability Engineering for Electronic Systems,* John Wiley & Sons, Inc., New York (1964).

NAVAIR 00–65–502/NAVORD OD–41146, *Reliability Engineering Handbook* (1 June 1964).

Naylor, T.H., Balintfy, J.L., Burdick, D.S., and Chu, K., *Computer Simulation Techniques,* John Wiley & Sons, Inc., New York (1966).

Peters, G.A., *Product Reliability and Safety,* Coiner Publications, Ltd., Washington, D.C. (1971).

Pieruschka, E., *Principles of Reliability,* Prentice-Hall, Inc., Englewood Cliffs, New Jersey (1963).

Polovko, A.M., *Fundamentals of Reliability Theory,* Academic Press, New York (1968).

Roberts, N.H., *Mathematical Methods in Reliability Engineering,* McGraw-Hill Book Co., New York (1964).

Rodgers, W.P., *Introduction to System Safety Engineering,* John Wiley & Sons, Inc., New York (1971).

Rossnagel, W.B., (ed.), *Checklists For Management Engineering, Manufacturing, and Product Assurance,* Vol. I–VIII, Spartan Books, New York (1971).

Sandler, G.H., *System Reliability Engineering,* Prentice-Hall, Inc., Englewood Cliffs, New Jersey (1963).

Shooman, M.L., *Probabilistic Reliability: An Engineering Approach,* McGraw-Hill Book Co., New York (1968).

Smith, C.S., *Quality and Reliability: An Integrated Approach,* Pitman, New York (1969).

Smith, D.J., *Reliability Engineering,* Pitman, New York (1973).

Smith, D.J., and Babb, A.H., *Maintainability Engineering,* Pitman, New York (1973).

Stewart, D.A., *Probability Statistics and Reliability,* Draughtsmen's and Allied Technicians Association, Richmond, Surrey (1966).

Thomason, R., *An Introduction to Quality and Reliability,* Machinery Publishing Company, Brighton (1969).

von Alven, W.H. (ed.), *Reliability Engineering,* Prentice-Hall, Inc., Englewood Cliffs, New Jersey (1969).

Zelen, M. (ed.), *Statistical Theory of Reliability,* University of Wisconsin Press, Madison, Wisconsin (1963).

Proceedings of the Annual Reliability & Maintainability Symposia and their predecessors (available from Annual Reliability & Maintainability Symposium, 6411 Chillum Place, NW, Washington, D.C. 20012, USA, or from the IEEE).

IEEE Transactions on Reliability (monthly).

United States Army Materiel Command Engineering Design Handbook, *Maintainability Engineering Theory and Practice,* AMCP 706-133 (January 1976).

United States Army Materiel Command Engineering Design Handbook, *Quality Assurance – Reliability Handbook,* AMCP 702–3 (October 1968).

INDEX OF AUTHORS
CHAPTERS ONE-SIXTEEN

A

Acheson, M. A., 70, 157, 374
Aitchison, J., 157
Albert, A., 265
Albert, G. E., 328
Allen, W. R., 157
Anscombe, F. J., 432
Aroian, L. A., 265

B

Bailey, N. T. J., 374
Balaban, H. S., 265
Ball, L. W., 374
Bartholomew, D. J., 157, 181
Barton, H. A., 475
Batson, H. C., 374
Beaton, G. N., 19
Bechofer, R., 328
Bellman, R., 265
Birkhoff, G., 111
Birnbaum, Z. W., 157
Bollman, J. H., 265
Bowker, A. H., 111, 169, 204, 207, 327, 497
Box, G. E. P., 374, 420, 424, 427, 431, 432
Breakwell, J. V., 328
Brickley, R. L., 374
Broadbent, S., 328
Brown, H. B., 265
Brown, J. A., 157
Brown, R. W., 35
Budne, T. A., 395, 431
Buehler, R. J., 226, 265
Burr, I. W., 329

C

Carslaw, H. S., 265
Chapman, D. G., 181
Chernoff, H., 329
Cleminshaw, C., 374
Cochran, W. G., 373, 431
Cohen, A. C., 157, 181, 208
Cohen, G. D., 265
Cohen, J., 19
Connor, J. A., 35
Connor, W. S., 208
Cowden, D. J., 111
Cox, D. R., 158, 256
Cox, G. M., 373, 431
Cramér, H., 111, 157, 181, 207, 265, 328
Creveling, C. J., 265
Crow, E. L., 208
Croxton, F. E., 111
Culbertson, J. E., 58

D

Dannemiller, M. C., 71, 303
Davies, O. L., 373, 421, 427, 431
Davis, D. J., 158
Davison, W. R., 35
Day, B. B., 373
Del Priore, F. R., 373
Denning, W. E., 111
Derr, E. H., 374
Dertinger, E. F., 35, 58
DiToro, M. J., 266
Dixon, W. J., 494
Dreste, F. E., 35
Drenick, R. F., 266
Dreyfus, S., 266
Durand, D., 181
Dwight, H. B., 181

E

Eaton, W. R., 58
Eisenhart, C., 328
Eldredge, G. G., 158
Emde, F., 181
Epstein, B., 157, 280, 303, 305, 312, 328

F

Fairfield, J. H., 457
Feller, W., 111, 116, 156, 207, 348
Firstman, Sidney I., 266
Flehinger, B. J., 158

G

Gabriel, K. R., 348
Garbarino, H. L., 35
Gardner, R. S., 208
Garner, N. R., 376, 377, 431
Girshick, M. A., 157
Godfrey, M. L., 158
Goode, Harry H., 19
Goode, H. P., 328
Gordon, R., 266
Gottfried, P., 35
Greenberg, B. G., 208
Greenwood, J. A., 181
Gumbel, E. F., 157, 158
Gunn, W. A., 157

H

Hald, A., 457
Harter, H. L., 329
Hartley, H. O., 181, 424, 432
Hartvigsen, D. E., 374, 376, 377, 431
Hastay, M. W., 328
Herd, G. R., 181
Hicks, C. R., 367
Hildebrand, F. B., 184
Hill, D., 374
Hoel, P. G., 111
Horton, W. H., 374
Howard, W. J., 266
Hunter, J. S., 431, 432
Hutchinson, D. W., 208

J

Jacquemard, F. C., 35
Jaeger, J. C., 265
Jahnke, E., 181

K

Kahn, L. B., 266
Kao, J. H. K., 158, 182, 328
Kemeny, J. G., 111, 480
Kempthorne, O., 373, 431
Koehler, T., 432
Krohn, Charles A., 266
Kuehn, R. E., 19
Kuzmin, W. R., 374

L

Lamb, J. J., 35
Lambert, J. S., 9
Lehman, E. H., Jr., 328
Lemus, F., 367
Leubbert, W. F., 35, 374
Levenbach, G. J., 374
Lewis, P. A., 158
Lieberman, G. J., 111, 169, 204, 207, 327, 328, 329, 497
Lindley, D. V., 329
Lindstrom, D. L., 226
Lipow, M., 71
Lloyd, D. K., 374
Lusser, R., 36, 71

M

MacLane, S., 111
McLean, J. P., 266
Madden, J. H., 226
Mandel, J., 348
Massey, F. J., 356, 494
Matosoff, H. I., 58
Meltzer, Sanford A., 266
Mendenhall, D., 328
Miller, J. C. P., 457
Mirkil, J. L., 111
Moan, O. B., 266
Molina, E. C., 157, 207
Mood, A. M., 111, 157, 181, 348
Moore, P. G., 182
Moranda, P. B., 329
Moriguti, Sigeiti, 329
Morrison, S. James, 266
Mosteller, F., 157
Moskowitz, F., 266
Myers, R. H., 265

N

Nadler, J., 328
Neyman, J., 166
Noether, G. E., 348

O

Och, H. G., 9
Okun, A. M., 19
Ordemann, F. A., 36

P

Page, E. S., 457
Parzen, E., 111
Pertschuk, D. W., 35

R

Raff, M. S., 158
Rao, C. R., 181
Resnikoff, G. J., 207, 328
Riley, J., 313
Rosenblatt, J. R., 70
Ruther, F. J., 36, 457

S

Saito, Kin-ichiro, 158
Sarhan, A. E., 208
Satterthwaite, F. E., 431, 432
Savage, L. J., 157
Scheffé, H., 367
Schlager, Kenneth J., 9
Schneider, L. L., 35
Siegel, S., 111
Smith, L. D., 36
Smith, W. L., 265
Snell, J. L., 111
Sobel, M., 280, 305, 328, 488
Soucy, Chester I., 9
Sparling, Rebecca H., 9
Steck, G. P., 226, 265
Stein, C., 300
Stevens, C. F., 329
Stoller, D. S., 158

T

Thatcher, A. G., 475
Thompson, G. L., 111
Tinus, W. C., 9
Tischendorf, J. A., 328, 488
Titchmarsh, E. C., 157
Tukey, J. W., 397

U

Uspensky, J. V., 111

V

Vaswani, R., 432
Voegtlen, D., 374
Vorhees, H. A., 58
Voule, P. V., 431

W

Wagner, D. H., 266
Wald, A., 157, 299, 328
Wallis, W. A., 328
Walsh, J. E., 208, 329
Warner, W. K., 58
Wehrfritz, Frank W., 266
Weiss, H. K., 348
Welker, E. L., 266
Wheeler, R. E., 348
Whidden, P., 432
Whiteman, I. R., 266
Whitewell, J. C., 432
Wilks, S. S., 157, 402
Willers, F. A., 181
Wilson, B. J., 71
Wilson, K. B., 420, 431
Wohl, J. G., 158

X

Xavier, M. A., 35

Y

Youden, W. J., 432
Yueh, J., 71, 374

Z

Zelen, M., 70, 71, 203, 374

SUBJECT INDEX
CHAPTERS ONE-SIXTEEN

A

Abort sensing system, 243
Absorption law (Table 5.1), 74–75
Acceptance boundary:
   for general binomial sampling plans, 283
   for Wald sequential binomial sampling plans, 297
   for Wald sequential life-test sampling plan, 320
Aitken's δ^2 process, 184
Analysis of variance, 367
   in continuous experimental design, 378
   in determining components of variation, 359
Applicable subsystems:
   for purposes of reliability evaluation, 439
   of solid propellant rocket engines, required characteristics for, 439
Application factor, K_A, 249
Apportionment of reliability (*see* Reliability apportionment)
Arcsine transformation, 156, 192, 193
Associative law (Table 5.1), 74–75
Assurance, statistical (*see* Confidence)
Asymptotic normality, 156, 163 (*see also* Normal approximation)
Asymptotically unbiased estimator, 162
ASN function (*see* Average Sample Number function)
Average Sample Number function:
   for curtailed binomial sampling plan, 287–290
   definition, 286
   maximum value, 300
   method of computation, 287
   for Wald sequential binomial sampling plan, 300

B

Battery:
   analysis of, 81
Battery failure:
   in space probe, 153
Bias (*see also* Unbiased estimator):
   in reliability estimate, 468
   of sample standard deviation, 171
   of sample variance, 171
Binomial coefficient, 85
   identity, use of, 290
Binomial confidence limit, 209
   exact, 212
   (*see also* Confidence limits)
Binomial distribution, 63, 84, 91, 96, 160
   examples, 98, 101
   of failures, 113
   mean, 115
   normal approximation, 116
   relation to negative binomial distribution, 124
   of successes, 113
   variance, 115
Binomial sampling plans:
   application in reliability evaluation of turbo-generator, 450
   boundary points of, 283
   criteria for fixed sample-size or curtailed sampling plan, 295
   optimum plan for given operating characteristic function, 293
   parameters of, 282
   using time-to-failure model, 302
   Wald sequential, 293, 297
Boundary points:
   of binomial sampling plans, 283
Break-in failures, 416 (*see also* Initial failure)
Bruceton analysis (*see* Sensitivity testing)

573

Budget:
  relation to reliability, 3
Burn-in period, 134

C

Canonical equation:
  reduction to 425
Central-limit theorem, 155
  applications:
    to arcsine transformation, 156
    to general function of sample mean, 156
    to Poisson distribution, 156
Central moment (*see* Moments about the mean)
Chain model, 230, 231
Chance failure, 135
Characteristic function, 98
Chebyshev, P. L. (*see* Tchebycheff)
Chi-Square distribution, 198, 201, 315
  normal approximation for, 202
  random variable, 194
  test for goodness of fit, 419
Combining test results:
  validity of, 67
Comma:
  used in place of intersection symbol, 86, 98, 223
Communication, 8
  management's role in, 8
  problems in, 5
  (*see also* Data)
Commutative law (Table 5.1) 74–75
  application, 78
Complementarity relation (Table 5.1), 74–75
  application, 78
Complexity, 7
  of equipment, 5
  organizational, 5
  as reason for unreliability, 3
Component:
  damaged, definition of, 467
  experimental, definition of 467
  improvement by system testing 350
  obsolete definition of, 467
  worn-out definition of, 467
Component interaction:
  environmental, 67
  functional, 66 (*see also* Reliability structure)
Component master list, 31

Component replacement, 271
Component testing:
  advantages, 350
  disadvantages, 350
  in large liquid rocket engines, 463
Conditional distribution (*see* Probability distribution)
Conditional probability, 86, 102
  definition, 79
  in growth model, 334
  for serial system, 222
Confidence coefficient, 194, 279
Confidence levels, 21 (*see also* Confidence coefficient)
  examples of, 21
  need for, 21
Confidence limits:
  on binomial parameter, 169, 191, 206, 209
    in negative binomial sampling, 217
  exact,
    on binomial parameter, 212
    on exponential reliability function, 194
    on gamma reliability function, 201
    on normal reliability function, 203, 205
    on Weibull reliability function, 197
  on mean of normal distribution, 169
  on mean time-to-failure, 191
  based on normal approximation, 168, 169, 191, 192, 194, 196, 197, 198, 200, 205, 206, 229, 343, 482, 483
    limitations, 172
  on the Poisson distribution parameter, 206, 218
  on probability of meeting full duration requirement, 483
  on reliability function:
    exponential distribution 194
    gamma distribution, 200
    of independent parallel system, 207, 239
    of independent serial system, 207, 224, 229
    for multi-parameter distributions, 195
    for single parameter distributions, 190
    two-sided specification, 205
    Weibull distribution, 197
  on reliability growth curve, 343
Confounding effects:
  meaning of, 363
Consistent estimator:
  definition, 162
  example, 162

Continuous probability distributions, 88
Contractual reliability, 15, 26
   demonstration of requirement, 296, 297, 475
     implications, 24
   goals, importance of, 25
Convergence in probability:
   of sample distribution function, 107
   of success ratio, 116
Coordination:
   methods for insuring, 33
Correlation coefficient:
   definition, 104
   of estimators of time-to-failure distribution, 481
   of gamma distribution parameter estimators, 177
   in multinomial distribution, 120
   properties, 104
   of success ratios during reliability growth, 347
   of two subsystem attributes, 223
   of Weibull distribution parameter estimators, 180
Corrosion:
   application of extreme value distribution to, 140
Cost:
   of reliability demonstration:
     to customer, effect on producer, 297
     for nonreplacement sampling procedure, 312
     to producer, 297
     for replacement sampling procedure, 312
   of unreliability, 1
Cost model for reliability demonstration, 312, 327
Covariance, 103–105
   definition:
     continuous distributions, 104
     discrete distributions, 104
     in terms of expected values, 104
   of estimators:
     of gamma distribution parameters, 176
     of Weibull distribution parameters, 179, 180
     of time-to-failure distribution, 480, 481
   of maximum likelihood estimators, 174
   in multinomial distribution, 118
   relation to independence, 104

Criteria (*see also* Ground rules):
   for demonstration of reliability:
     binomial sampling, 295–297
       Figures A.1–A.5, 556-560
       Table A.1, 545
     life testing, 304, 307, 308, 315, 321
       Table 10.4, 326
       Tables A.2–A.7, 564-551
   reliability, use of, 55
   sets of, 22
Critical level of environment, 384
Critical region:
   definition of, for binomial sampling, 283
Cumulative distribution function (*see* Distribution function)
Current data:
   definition of, 444
     with sequential sampling, 453
Curtailed life test, 304
Curtailed sampling, 287
Customers' risk, 296

D

Damaged component:
   definition of, 467
Data:
   analysis and follow-up, 40
   analysis of, for reliability evaluation of liquid rocket engine, 468
   attribute:
     definition, 62
     discussion of, 62
   central file, 40
   collection of, 37
   current:
     definition of, 444
     use of, 444
   deficient handling of, 8
   distribution of, 39
   examples of interpretation of, 21
   homogeneity of, 40
   integrity of, 33, 440
   used in liquid rocket engine reliability evaluation, type of, 467
   from non-representative configurations, 33
   pretest declaration of, 51
   reliability central file, 51
     examples, 52–54
   reliability knowledge, 30
     source of, 32

Data: (Cont.)
  representative:
    method of control, 444
  sequentially generated:
    advantages, 376
    disadvantages, 376
  transformation of, need for, 369
  transmittal of, 8
  variables:
    definition, 62
    discussion of, 62
Death process models, 142
Declaration form, 55
  example:
    for liquid propellant rocket engine, 465
    for solid propellant rocket engine, 441
Declaration policy, 437
  declaration form, 441, 465
    time of submittal, 440
  purpose of, 440
  for reliability evaluation program for a turbo-generator device, 449
Density function (see Probability density function)
Dependence, statistical:
  of trials in growth model, 333
  (see also Independence)
Design and Analysis of Experiments (see Statistical experimentation)
Design change:
  effect on analysis of experiment, 379
  effect on reliability evaluation, 444, 446, 453
Design control, 16
Design feasibility versus reliability, 28
Design reliability:
  importance of, 330
Design review, 16, 27
  check list, example of, 28
  and the design engineer, 28
  general comparisons, 27
Development program:
  mileposts:
    pre-flight rating tests, 435
    qualification tests, 435
Development program complexity:
  effect on statistical experimentation, 375
Dirac delta function, 274
Discrete distributions, 83–87
  (see also Probability distribution and Distribution function)

Distribution, of data, 39
Distribution, probability
  (see Distribution function and Probability distribution)
Distribution function, 90–92, 99
  binomial, 91
  Chi-square, 165
  conditional, 102
  continuous, 90, 136
  discontinuous, 92
  discrete, 91
  extreme-value, 139
  gamma, 148, 165
  for general discrete distribution, 92
  for general time-to-failure distribution, 145
  hazard function, relation to, 135
  inequality relation, 273
  marginal, 99, 100
  normal, 153
  properties, 90, 99, 100
  sample, 106
  unit step function, 92
  (see also Probability distribution)
Distributive law, 78
  of intersection with respect to union (Table 5.1), 74–75
  of union with respect to intersection (Table 5.1), 74–75
Dualization law (Table 5.1), 74–75

E

Edgeworth expansion, 200
Effectivity:
  reliability group's, 19
  strategic, 2
Environmental factors:
  establishing importance of, 369
Environmental testing, 353
  examples of 361, 365, 368, 377
  as part of reliability demonstration requirement, 475
Equivalent events:
  (see Events: equivalence of)
Equivalent full duration test, 468
Error: experimental, measurement, random, sampling:
  risk of, in sampling plan, 279
  two kinds, 295
  variation (see Experimental variation)

INDEX   577

Estimator:
  consistent, 162
  minimum variance, 162
  of reliability function:
    for exponential distribution, 190, 191, 194
    for gamma distribution, 199
    for normal distribution, 203, 205
    for Weibull distribution, 197
  unbiased, 161
  methods of finding, 163
  properties of, 161, 162
Euler's constant, 311, 340, 346
Events, 72
  certain, 73
  combination of, 73
  complement of, 73
  equivalence of, 76
  implication, 76
    probability relation, 78
  impossible, 73, 76
  independence of, 80, 82
  intersection:
    symbol for, 76
      use of comma in place of, 86, 98, 223
  joint, 98
  laws of operation (Table 5.1), 74–75
  mutually exclusive, 76, 79, 82, 85, 118, 242, 244
  probabilities of, 77
  properties of (Table 5.1), 74–75
  success, examples of, 112, 134, 159, 160
  union:
    symbol for, 76
Evolutionary operation, 403
  cycle, definition of, 404
  example, 404
  general method, 404
  to improve "status quo," 403
  information board, 410
  phase, definition of, 404
  as a standard procedure in manufacturing operations, 412
  work sheets, examples of, 406–408
Exclusion of tests from reliability evaluation:
  reasons for, 440, 450, 464
Experimental component:
  definition of, 467
Expectation:
  properties of, 109
  (see also Expected value)

Expected number of trials to reach decision (see Average Sample Number function)
Expected test time:
  in curtailed life test, 305
  in life test:
    nonreplacement procedure, 310
    replacement procedure, 310
Expected value:
  for arcsine transformation, 192
  of constant, 109
  definition, 92
  of a function:
    of one estimator, 192
    of two estimators, 195
  of product of two random variables, 104
  of reliability estimator:
    exponential distribution, 194
    Weibull distribution, 197
  of sample mean, 109
  of sample variance, 110
  of sum, 109
  (see also Mean)
Experiment:
  random, 105
  (see also Statistical experimentation)
Experimental variation, 356, 357, 358, 367, 378, 381, 382, 383, 418
Exponential reliability growth, 333
Exponential distribution, 97, 137, 140, 160 190, 303, 317
  derivation based on death process, 142
  as distribution of sum of squares of two normal variables 318
  hazard function for, 137
  as a limiting case of general time-to-failure distribution 150
  as limiting form of geometric distribution 125
  moment generating function for, 97
  relation to particular form of the hazard function, 140
  role in reliability demonstration, 302, 308, 314, 317, 319
  as a special case of general time-to-failure distribution 150
  truncated, 141
  as the underlying population distribution for statistical experimentation, 368
  Wald sequential sampling plan, application to, 319

Exponential distribution (Cont.)
  Wald sequential sampling plan (Cont.)
    cost model for, 327
    criteria for, 321
    example of, 324
    formulas for, 320–21
    parameters for (Table 10.4), 326
Extreme value distribution:
  application to corrosion problem, 140
  hazard function for, 139
  standard form, 140

F

F-distribution (see F-test)
F-test:
  use in analysis of variance, 367, 378, 380, 382
  use in random balance designs, 392
  use in response surface experimentation, 418
Factorial design, 368
  used in response surface experiment, 420
Failure:
  analysis, follow-up, 46
  analysis, initiation of, 44
  analysis, summary of (Table 3.1), 48–49
  analysis report form, 46
    sample, 46
    use of, 46
  catastrophic:
    in statistical design, 377
  "chance," 63, 135, 147, 152
  consequences of, 1
  cycles to, number of, 352
  definition of, 41
  due to human error, 39
  initial, 63, 134, 151, 152
  mechanisms of, 352
  mode of, 28, 29, 51, 352
  ordinary, 469, 476
  random, 6
  secondary, definition of, 30
  shutdown, 469, 476
  test-to-, 352 (see also Life Testing)
  unassignable, handling of, 446
  "wearout," 63, 135, 147, 151, 152
Failure Criteria:
  for liquid propellant rocket engine, 466
  for solid propellant rocket engine, 442–443
Failure Model, 65, 112
  effect of method of sampling on, 63
  types, determination of, 63

Failure rate:
  change of, example of computation, 460
  example of search for significant difference in, 454
  expression for probability of difference in two sets of data, 454
  test for change in, 451
Failure Report Form:
  distribution of, 43
  example of (Figure 3.1), 42
  use of, 38, 41
Failure Reporting System, 32, 37–45
Frequency function (see Probability density function)

G

Gamma distribution, 97, 148, 171, 199, 201
  as limiting form of negative binomial distribution, 125
  mean, 148
  probability density function, 97, 148, 151
  relation to Chi-square distribution, 165
  as a special case of the generalized time-to-failure distribution, 151
  variance, 148
Gamma Function, 93, 98, 126, 128, 138, 147, 148, 155, 165, 171, 178, 199, 201
  reference to tables of, 178
Generalized time-to-failure distribution, 149
  alternate form, 149
  examples, 152, 153
  initial failure:
    delayed, 152
    in terms of initial condition, 151
  limiting cases, 150–151
  use of Molina's Tables for numerical calculation of, 148
  physical interpretation of parameters, 152
Generating function (see Moment generating function)
Geometric distribution, 87, 122
  limiting form of, 126
Goodness-of-fit-test:
  for fitted response surface, 418, 419, 422, 429
Graeco-Latin Square:
  statistical design for solid propellant rocket engine, 366

INDEX 579

Ground rules, 442, 443, 449, 464, 466
   variety of, for reliability evaluation, 22
   (*see also* Criteria)

H

Hazard function:
   definition, 135
   for exponential distribution, 137
   for extreme value distribution, 139
   for gamma distribution, 138
   a particular form, 140
      reliability demonstration procedure, application to, 307
      properties, 135, 136
   for Weibull distribution, 137
Heterogeneous populations:
   in reliability evaluation, 437
Homogeneity of data, 437, 455
Human engineering, 9
Human error, 39

I

Idempotent law (Table 5.1), 74–75
Independence:
   definition, in terms of:
      conditional probabilities, 80
      distribution functions, 100
      probability density functions, 102
      random variables, 86
   of events, 80
   multiplication of probabilities of general events, 101
   mutual, 80, 106, 222
   pairwise, 80
   of random variables, 86
Independent parallel system, 239, 251, 252
   confidence limit on reliability of, 207, 239
Independent serial system, 221, 251
   confidence limit on reliability of, 207, 224–229
   Figures A.6–A.11, 561-566
Infant mortality, 134
Initial failure, 134, 151
   delayed, 134, 152
Interaction:
   component, 26, 67
   effect on test philosophy, 349
   "enhancing," 68, 242
   environmental effect, 5
   functional effect, 5

Interaction: (Cont.)
   inability of traditional test methods to discover it, 355
   interdependency, 26
   of man and complex systems, 6
   mathematical effect, 4
   in Multiple Balance Designs, 399
   statistical, 356
Interaction effects:
   example of computation, 380
Intersection of events (*see* Events, intersection)
Intervals, types of, and notation for, 89
Inverse binomial (*see* Negative binomial distribution)
Inverse of Matrix, (*see* Matrix)
Involution law (Table 5.1), 74–75

J

Joint distribution function:
   in n random variables, properties of, 99
   in two random variables, 98
Joint event, 98
Joint probability distribution, 336

L

Laplace transform, 93, 97, 98, 147, 155, 275, 276, 277
Law of Large Numbers, 116
Least Square estimators:
   for reliability growth model parameters, 344
Level of development:
   effect on test philosophy, 350
Life testing:
   censored, 314
   based on generalized time-to-failure distribution, 307
   reduction of expected time to completion of, 316
   truncated, 314
   until all items fail, 308
   until r out of N items fail, 310
   use of, 352
Likelihood equation, 164
Likelihood function, 163, 164, 166, 167, 477 (*see also* Log-likelihood function)
Liquid propellant rocket engine:
   criteria for test result classification:
      success, failure, exclusion, 466

Liquid propellant rocket engine: (Cont.)
  reliability evaluation program for, 463
Log of operating time, 47
Log-likelihood function, 164, 167, 170, 171, 477
Logarithmic derivative of gamma function (*see* Psi-function)
Lognormal distribution, 156

M

Main effects, 356
Management:
  its responsibility, 13
Management reports (*see* Reports)
Marginal distribution, 99
  for discrete distribution, 103
Markoff chains, 334
Master List of components:
  use of, 467
Matching moments, method of, 173
Mathematics:
  its role in reliability, 59
Matrix:
  of conditional probabilities, 334
  inverse of, 175
Maximum likelihood estimators:
  minimum variance property, 174
  of reliability growth parameters, 339
  variances and covariances in reliability growth, 342
  (*see also* Maximum likelihood method)
Maximum likelihood method, 163, 170
  applications:
    to binomial distribution, 167
    to exponential distribution, 164
    to gamma distribution, 171
    to normal distribution, 170
    to Poisson distribution, 167
    to Weibull distribution, 177
    special case, 165
  use of in estimation of liquid propellant rocket engine reliability, 476
  use of in estimation of reliability growth, 338
Mean, 92
  of binomial distribution, 115
  of a function of several variables, 261
  of gamma distribution, 148
  of negative binomial distribution, 122
  of reliability estimator:
    exponential distribution, 194
    Weibull distribution, 197

Mean (Cont.)
  of reliability function, 192
  of sample, 108
Missing values, 373
  effect on analysis of random balance experiment, 391
Mode, 94
Moments, 92
  of the binomial distribution, 115
  of a distribution, 95
  about the mean, 92, 93
  for normal distribution, 96
  ordinary, 92
  of a random variable, 95
  of sample distribution:
    k^{th} central moment, 108
    k^{th} ordinary moment, 108
  of sample moments, 108
Moment Generating Function, 95
  for binomial distribution, 115
  applied to binomial random variables, 98
  for discrete distribution, 96
  applied to Poisson random variables, 98
  inversion of, 97
  of square of normal variable, 319
  for sum of independent random variables, 97, 319
Moving average reliability estimate, 474
Multinomial distribution, 118
  applications of, 120, 165, 233, 234
Multiple Balance designs, 394 (*see also* Random Balance designs)
  advantages of, 395
  analysis of, 396
  construction of, 395
  example, 395
Multivariate normal distribution, 156
Mutual independence, 80, 86, 106, 222
  application, 109
  (*see also* Non-mutual independence and Independence)
Mutually exclusive:
  relation to independence, 80
Mutually exclusive events, 244

N

National security, 15
  implications of unreliability to, 2
Negative binomial distribution, 87, 121
  confidence limit on parameter, 217
  limiting form of, 126
  mean, 122

INDEX   581

Negative binomial distribution (Cont.)
  relation to binomial distribution, 124
  variance, 122
Non-mutual independence, 80
  example, 86
Nonreplacement procedure:
  in life testing, 310 ff.
  in Wald sequential life-test, 320
Normal approximation:
  to binomial distribution, 116
  used for calculation of confidence limits,
      168, 169, 191, 192, 194, 196, 197,
      198, 200, 205, 206, 229, 343, 482,
      483
  to gamma distribution, 151
  limitation for small sample size, 172
Normal deviate:
  definition of, 169
Normal distribution, 153–156, 159, 170,
      203, 237, 238, 260, 280, 318
  central limit theorem, 155
  as a limiting form of a generalized time-
      to-failure distribution, 151
  mean of, 90
  properties, 154
  standard deviation of, 90
  standard, 96, 153
  tables, reference to, 155
  tolerance limits, 203
    one-sided (Table A.9), 554, 555
Normal distribution function, 90
  (see also Standard normal distribution
      function)
Normal distribution function (standard),
      91, 117, 149, 154, 169, 195, 203,
      237
Normal probability density function, 95,
      153, 170
  (see also Normal distribution)
  standard, 96

O

Obsolete component:
  definition of, 467
OC function (see Operating Characteristic
      Function)
Operating Mode Factor, 250
Operating Characteristic Function:
  definition, 282
  description of, 282
  example of, 284, 292
  use of, 282–283

Operating Characteristic Function: (Cont.)
  for Wald sequential binomial sampling
      plan, 299
Operational time logging:
  collection of data, 50
  extent of, 47
  purpose of, 47
  sample forms, 47, 50
  use of, 50
Optimum response:
  from evolutionary operation, 403
  from response surface experimentation,
      412
Organization:
  reliability group, placing of, 16
  reliability as an independent group, 18
  reliability personnel qualifications and
      capabilities, 18
  reliability activities:
    performed by engineering, disadvan-
        tages of, 17
    performed by management, disadvan-
        tages of, 17
    performed by quality control, dis-
        advantages of, 17
    performed by a single group, advan-
        tages of, 17

P

Pairwise independence, 80, 261 (see also
      Independence)
Parallel-Serial system, 243, 253
Parallel Systems, 239, 252
  definition in terms of time-to-failure, 242
  partially parallel system, 240–242
  stand-by redundancy, 242
Parameters:
  of binomial sampling plan, 295
  of distribution:
    binomial, 98, 114
    extreme value, 139
    normal, 93
    Poisson, 98
    of time-to-failure, generalized, 150–
        151
    Weibull, 137
Part application, 249
Part Application failure rates, 257
  referenced tabulations of, 249
Pascal distribution (see Negative Binomial
      distribution)
p.d.f. (see Probability density function)

Peripheral testing:
    in demonstration of required reliability, 372
    exclusion of tests in reliability evaluation program, 449
    problems with, 353
    use of, 353
PFRT (*see* Pre-Flight Rating Tests)
Poisson distribution, 87, 126
    confidence limit on parameter of, 218
    examples, 98, 129
    interpretation of, 127
    as limiting form of binomial distribution, 128
    mean, 127
    moment generating function for, 96
    relation to the exponential distribution, 127
    standard deviation, 127
Pre-Flight Rating Tests (PFRT), 435
    (*see also* Proof Tests and Development program)
Prequalification test program:
    purpose of, 361
Probability:
    completely additive property, 77
    conditional, 78, 86
    in rectangle, 99
    of error in making decisions, 295
    of joint event, 99
Probability density element, 89
Probability density function (p.d.f.):
    definition, 88
    for general time-to-failure distribution, 145
    in n variables, 100
    normal, 95, 153
    of age, 271, 274
        examples, 276, 277
        limiting value, 272
    property of, 90
    relation to hazard function, 135
    in two variables, 100
Probability distribution:
    binomial, 63, 84, 96, 113, 160
    chi-square, 165, 315
    conditional, 102
    continuous, 63, 88
    discrete, 63, 83
    exponential, 63, 89, 97, 160, 190, 303, 314, 317
    extreme value, 139, 181

Probability distribution: (Cont.)
    gamma, 64, 165, 171, 199
    geometric, 63, 87, 121
        limiting form, 125
    in n-dimensions, 99
    in two dimensions, 98
    multinomial, 118, 233
    negative binomial, 87, 121
        limiting form, 125
    normal, 63, 153, 159, 170, 195, 203, 237, 238, 280, 318
    of number of trials to k^{th} failure, 87
    of number of trials to reach a decision, 286, 300
    Poisson, 64, 87, 96, 126, 218
    of sample, 106
    of square of normal variable, 318
    underlying, 106
    Weibull, 64, 137, 197, 307
Product rule, 5, 68, 222, 251, 267
    application, 448
Producer's risk:
    relationship to contractual obligations, 296
Proof tests, 377
Psi-function, 180
    approximation to, 171
    derivative of, 180
        approximation to, 176
    reference to tables, 172

Q

Quality Control, 14
    Its relationship to reliability, 17

R

Random Balance design:
    circumstances of use, 390
    construction of, 390
    disadvantages of, 390
    limitations of, 394
    modification to multiple balance designs, 394
    use as "screening" tool, 390
    significance tests for, 392, 397
Random experiment, 105
Random failure, 6
Random sample, 105
Random variables:
    definition, 83

INDEX

Random variables: (Cont.)
  as function of experiment, 105
  independence of, 86, 100
  standardized, 90
Random variation, 355
  in reliability growth, 331
  (*see also* Experimental variation)
Randomness, procedures to achieve, 106
Rate-of-growth model, 338
  generalization, 345–346
Redundancy, 7, 65, 239, 251
  as a concept of design review, 28
  examples, 66, 240, 242, 243, 255
  (*see also* Reliability structure)
Regression theory:
  use of in random balance experimentation, 402
Rejection boundary:
  for general binomial sampling plan, 284–287
  for Wald sequential binomial sampling plan, 297
  for Wald sequential life test sampling plan, 320
Reliability:
  budgeting, 19
  cost of, 14
    difficulty of evaluating, 3
  definition of, 20
    for growth model, 333
  in design, 330
  difficulty of establishing its share in budget, 3
  of independent serial system, 222, 251
    confidence limits, 207, 224–229
      Figs. A.6–A.11, 561–566
  of n-link chain, 234
    maximum value, 236
    minimum value, 236
  need for knowing level of, 2
  as a new methodology, 15–16
  objections to, 14–15
  organizational effectivity, 13
  versus performance, 28
  as a probability concept, 20
  its scope as a subject, 13
  of serial system, 222
    maximum value, 223, 224
  of weakest link serial system, 230
Reliability Action Request Form, 45
  use of, 45

Reliability activities:
  divided responsibility for:
    disadvantages of, 16
  list of, 16
  proving the need for, 14, 15
  reliability estimation, 20
Reliability analysis, 27
  of space vehicle temperature control system, 255–259
Reliability apportionment, 65, 267
  application to solid propellant rocket engine, 438
  its limitations, 26
  reasons for, 25
  use of, 27, 67
Reliability Central Data File:
  examples:
    battery, 53
    nozzle, 52
    valve, 54
Reliability demonstration, 69, 279–327
  application to guided missile, 317
  use of binomial sampling plan in, 281–308
    based on underlying exponential distribution, 303
    Wald sequential binomial sampling plan, 297
  censored life test for, 310, 314
  of a contractual requirement, 15, 296–297, 443, 475
  cost model for, 312, 327
    example of, 312–313, 327
  with electronic components, 27
  based on exponential distribution, 303–327
    applied to more general distributions, 307
  use of binomial sampling methods, 303
  censored life test—r out of N items fail, 310
  test time unequal to required life of equipment, 303
  truncated life test, 314
    and censored life test, 314
  uncensored life test—all items fail, 308
  Wald sequential life test, 319
  importance of, 69
  limitations, 24
  with liquid propellant rocket engines, 27, 463
  producer's risk in, 297

Reliability demonstration (Cont.)
   as related to reliability apportionment, 26
   with solid propellant rocket engines, 443
   test time unequal to required time, 303
   test truncation by acceptance, 316
   time-to-failure sampling:
      attribute data, 302, 303, 468
      variables data:
         all items failing, 308
         r out of N items failing, 310
   true reliability, 24
Reliability Design Change, definition, 444
Reliability design review, 27–30
Reliability education:
   importance of, 9
   methods of, 9
   topics for, 34
      failure report forms, when to use, 38
      human error, 39
      reliability estimate criteria, 40
Reliability Effort function, 269, 270
Reliability estimate:
   used as an absolute measure, 24
   used as a relative measure, 24
Reliability estimation, 63, 65
   of arbitrary time-to-failure distribution, 476
   and confidence levels, 21
   and prediction, 70
   of reliability growth, 70
   technique for solid propellant rocket engine, example of, 446
   (*see also* Reliability evaluation)
Reliability evaluation:
   consistency with reliability demonstration method, 451, 456, 475
   contractual implications, 23
   exclusion of tests from, 440, 464
   in heterogeneous population, 437
   of non-representative configurations, 23
   problems during solid propellant rocket engine development program, 436
   problems for turbo-generator, discussion of, 449–450
   programs:
      for large liquid propellant rocket engines, 463
      for large solid propellant rocket engines, 435
      for turbo-generator, 449

Reliability evaluation: (Cont.)
   scope of data for liquid rocket engines, 467
   by system testing, 351
Reliability Function, 65
   definition, 159
   estimator for:
      exponential distribution, 190
      gamma distribution, 199
      normal distribution, 195, 205
      Weibull distribution, 197
   examples, for:
      binomial distribution, 160
      exponential distribution, 160, 194
      gamma distribution, 201
      normal distribution, 159, 203, 205
Reliability goals, 25
Reliability group:
   activities, 16
   effectivity, 18
   need for formal coordination with remainder of organization, 33
   position in organization, 16
Reliability growth, 24, 330
   difference between, for liquid and solid propellant rocket engines, 70
   measurement of:
      by least-squares estimation, 344
      by maximum likelihood estimation, 338
   non-independence of tests during, 333, 347
   a simple attribute model for, 331–333
      generalization, 337
   a time-to-failure model for, 347
   weighed estimates of, 348
Reliability growth function, 338
   confidence limits on, 343
   generalization, 345–346
Reliability Program Plan:
   advantages of, 34
   as a charter of activities, 34
   contents of, 35
Reliability reporting, 37–58, 444, 474
   from sequential sampling computation, 451, 456, 458
   reliability report form, example of, for solid propellant rocket engine, 445
Reliability reports, 37, 57
   examples:
      component failure summary (Figure 3.10), 56

INDEX 585

Reliability reports (Cont.)
    missile system summary (Figure 3.8), 55
    propulsion system summary (Figure 3.9), 56; (Figure 16.2), 474
    reliability problem status (Figure 3.11), 57
Reliability specifications, 15
Reliability structure diagrams, 245, 251, 252, 253, 256, 259
Reliability structure models, 65, 220–278
    parallel systems, 239, 242
        mixed, 243–249
        partial, 240
        standby, 242
    serial systems, 221, 242
        independent serial systems, 222
        weakest-link systems, 229
            chain model, 231
Renewal equation, 275
Replacement procedure:
    in life testing, 303, 310
    in Wald sequential life test, 320
Response surface:
    the canonical equation, 414
    representation by linear equation, 413
    representation by quadratic equation, 414, 424
Response surface experimentation:
    applications, 416
    comparison with evolutionary operation, 412
    determination of the optimum, 424
        example, 427
    geometrical illustration for two input variables, 413–415
    as a method of finding the optimum response, 412
    method of steepest ascent, 418
        example of, 420
Responsibilities:
    conflict of between reliability and existing groups, 14
    of management, 13–16
    of reliability group, 8
Review:
    of component compatibility, 31
    of design control procedures, 31
    of designs, 27–30
    of materials, 31
    of processes, 31
    of specifications, 31
    of test plans, 31

Risk:
    to customer, 296
    to producer, 296
Risks, α and β, 295
Rocket engine:
    application of extreme-value distribution to corrosion of, 141
    application of statistical experimentation to, 377
    reliability demonstration programs for:
        large liquid propellant, 463
        large solid propellant, 435
    tolerances on performance parameters of, 260, 264

S

Sample, random, 106
Sample distribution, 106
Sample distribution function, 106
    as a random variable, 106
Sample mean, 108
    as consistent estimator for population mean, 162
    definition, 108
    as unbiased estimator for population mean, 161
    (see also Mean)
Sample moment:
    notation, 108
Sample space, 73, 105
    for binomial trials, 129, 209, 283
    continuous, 73, 88
    denumerable, 73
    discrete, 73
    finite, 73
    non-denumerable, 73, 88
    sample point of, 73
    (see also Events)
Sample variance, 108
    relation to unbiased estimator of population variance, 108
    (see also Variance)
Sampling, 69, 105, 129, 279
    binomial, 113, 282
        curtailed, 287
    censored, 106, 310, 314
    partial curtailment of, 304
    purposive, 106
    for reliability demonstration, 279
    quota, 106
    sequential, 106, 131, 297, 319, 450
    simple random, 106

Sampling (Cont.)
  stratified, 106
  truncation of, 293, 301, 314, 320
    reasons for, 281
    (see also Reliability demonstration)
Sampling plans (see Reliability demonstration)
Sampling theory, 105
Satellite:
  application of generalized time-to-failure model to, 153
  SPUTNIK I, 2
  VANGUARD, 2
Scatter diagrams, 391
  (see also Random Balance designs)
Scientific method, 353
  as basis of statistical experimentation, 354
Sensitivity testing, 383–389
  confidence limits, 388
  critical level, 385
  estimator for mean, 386
  estimator for standard deviation, 386
  initial conditions, choice of, 389
  method of, 385
  percentage limits, 388
  reasons for use, 383
  standard deviation of mean, 387
  standard deviation of standard deviation, 387
Sequential sampling, 131, 297, 319
  advantages of, 450
  based on exponential distribution, 319, 457
  based on normal distribution, 457
  in reliability program, 450
  (see also Wald sequential sampling)
Serial-Parallel system, 252
Serial system, 221, 251
  the chain model, 231
  definition of, in terms of time-to-failure, 242
  dependent, 222, 223, 224
  example, 224
  independent, 68, 222, 251
  weakest link, 229, 231
Significant Design Change:
  reasons for and consequences of, in reliability demonstration program, 446
Solar cells, 153, 255

Solid propellant rocket engines:
  applicability of subscale testing to reliability evaluation of, 437
  criteria for test result classifications, 442
  example of poor pre-qualification test program for, 359
  example of statistically designed environmental test program for, 365
    advantages of, 366
  reliability evaluation program for, 435
Space probe, 317
Specifications:
  use of, in reliability program, 436, 450, 463
  review of, for adequacy, 31
SPUTNIK I, 2
Standard deviation, 93
  relation to tolerance, 264
Standard normal deviate, 169, 264
Standard normal distribution, 153–155
Standardized extreme value distribution, 140 (see also Extreme value distribution)
State-of-the-art, 2
State probabilities, 143, 332
Statistical distribution (see Probability distribution)
Statistical experimentation, 349–431
  analysis of simple example, 356
  continuous experimental designs, 376
    application to small rocket engine, 377
  difference from traditional testing methods, 355
  evolutionary operation, 403
  example and consequences of not using statistically designed test plan, 359
  example for electronic equipment, 368
  example for solid propellant rocket engine, 365
  factorial design, 368
    use of fractional factorial design, 377
  importance of, 361
  as means of obtaining correlation between failure rate and environmental level, 370
  need for flexible designs, 376
  need for new statistical designs for industrial use, 375
  need for sequential designs, 376
  a practical sequence of test progression, 371

INDEX 587

Statistical experimentation (Cont.)
    random balance designs, 389
    relationship to the scientific method, 353
    response surface exploration, 412
    its role in reliability, 373
    time-to-failure testing, 368
Stopping rule, 131
Strength distribution, 232
Stress distribution, 232
Stress versus strength analysis:
    examples, 237, 238
    value of, in design analysis, 61
Student t-distribution, 169, 388
Subscale testing:
    of solid propellant rocket engines, 437
Success:
    definition of, 22, 112
        for liquid propellant rocket engine, 466
        for solid propellant rocket engine, 442, 443
        for space vehicle's temperature control subsystem, 255, 256, 258, 259
        for turbo-generator device, 450
    degree of, 22
    interpretation of definition, 22
Survival probability:
    as a function of age, 271
System apportionment:
    use of, in rocket engine reliability evaluation program, 438
    technique for, 267
    (*see also* Reliability apportionment)
Systems engineering, 7
System testing:
    advantages, 350
    disadvantages, 351
    reasons for, in liquid rocket engine development program, 463

T

Tchebycheff's inequality, 93
    applications, 94, 116, 162
Test Planning department, 14
Test result classification:
    ground rules for liquid propellant rocket engines, 466
    ground rules for solid propellant rocket engines, 441
Testing:
    censored, 310, 314

Testing: (Cont.)
    component, 350
    cycle-to-failure, 352
    event, 352
    environmental, 353
    experimental considerations, 354
    overstress, 353
    peripheral, 353, 372
    planning of, 30
    reasons for, 349
    sensitivity, 383
    sequence of, 353, 372
    system, 350
        versus component, 350
    time-to-failure, 352
    (*see also* Life testing)
Time-to-failure:
    definition of reliability in terms of, 134
    derivation of general distribution of, 145
    distribution models, 134–153
        death process, 142
        exponential, 137, 144
        extreme value, 139, 140–142
        gamma, 148
        generalized, 149
        normal, 149
        Weibull, 137
    reliability demonstration, based on, 302
Tolerance factors:
    for normal distribution, one-sided, 203
    (Table A.9), 554
Tolerance limit:
    for normal distribution, one-sided, 203
        application to rocket engine performance evaluation, 378
Tolerances, evaluation of, 263
    relation to standard deviations, 264
Trade-off, of reliability and cost, 14
Transition probability, 143, 334–335
Trials:
    in general binomial sampling, 283
    expected number, to reach decision, 286
    in Wald sequential sampling, 297
    (*see also* Sampling)
Truncated and censored life test, 314
    application to guided missile reliability demonstration, 317–318
Truncation:
    rule for, in Wald sequential life test, 320
    of Wald sequential binomial sampling plan, 301

U

Unassignable failure, 6
   method of handling in reliability evaluation program, 446
Unbiased estimator, 161
   of binomial parameter, 117
      in generalized binomial sampling, 133
      for negative binomial distribution, 124–125
         uniqueness, 125
      for geometric distribution, 124–125
   for parent distribution variance, 108
      relation to sample variance, 108
   of reliability, for arbitrary time-to-failure distribution, 472–473
Underlying distribution, 106
Union of events, 76
   (*see also* Events)
Unit step function:
   as a distribution function, 92
Unreliability:
   consequences of, 1
   cost of, 1
   critical value of, 282, 295
   due to equipment complexity, 5
   general topics, 7
   due to human error, 6
   due to organizational complexity, 5
   prevention of, 7
   reasons for, 3
   (*see also* Failure)
Up-and-down method (*see* Sensitivity testing)

V

VANGUARD satellite, 2
Variability:
   of number of trials to reach decision, 291, 300
   (*see also* Variance)
Variance:
   for arcsine transformation, 193
   of binomial distribution, 115
   components of, 367
   of constant, 109
   definition, 93
   of a function:
      of several random variables, 261
      of a single parameter estimator, 192
      of two parameter estimators, 195
   for gamma distribution, 148

Variance: (Cont.)
   of linear function, 105
   of maximum likelihood estimator:
      for multi-parameter distributions, 174
      for single parameter distributions, 167
   for negative binomial distribution, 123
   of product, 105
   of reliability function, 192
      exponential distribution, 194
      gamma distribution, 200
      normal distribution, 196, 205
      Weibull distribution, 197
   of reliability growth parameter estimators, 342
   of sample distribution, 108
   of sample mean, 110
      application, 162
   of sample standard deviation, 193
   of sample variance, 110, 193
      for normal distribution, 110
   of weighting factors, 482
Variance-Covariance matrix, 174, 175, 180, 480
Variation:
   components of, 358
   random, 355
      consequences of, 364
   (*see also* Experimental variation)
Vendor control, 32

W

Wald sequential binomial sampling plans:
   acceptance and rejection points, formulas for, 298
   Average Sample Number, formula for, 300
   Operating Characteristic function, formulas for, 299
Wald sequential life test:
   example, 324
   expected number of failures, 324
   expected waiting time:
      non-replacement procedure, 324
      replacement procedure, 323
   formulas for, 321
   Operating Characteristic function, 322
Weakest-Link model, 229
   chain model, as particular case of, 230, 231
Wearout failure (*see* Failure, wearout)
Weibull distribution, 137
   application, 307

Weibull distribution (Cont.)
   hazard function for, 137
   mean, 138
   variance, 138
Weighting factor:
   computation of, 469

Weighting factor: (Cont.)
   definition of, 469
   example of use, 472
   need for, 22, 468
   variance of, 482
Worn-out component, definition of, 467